Two Manuals Combined into One Book

Military Marksmanship, all M16 and M4 Series Weapons

And

Technical Manual for 5.56MM, M16A2, M4 and M4A1 Carbine

Department of Defense

Pentagon
Publishing

*Field Manual
No. 3-22.9

Headquarters
Department of the Army
Washington, DC, 12 August 2008

Rifle Marksmanship M16-/M4-Series Weapons

Contents

DISTRIBUTION RESTRICTION: Approved for public release; distribution is unlimited.

***This publication supersedes FM 3-22.9, 24 April 2003.**

Figures

Tables

Preface

This manual provides guidance for planning and executing training on the 5.56-millimeter M16-series rifle (M16A1/A2/A3/A4) and M4 carbine. It is a guide for commanders, leaders, and instructors to develop training programs, plans, and lessons that meet the objectives or intent of the United States Army rifle marksmanship program and FM 7-0.

This manual is organized to lead the trainer through the material needed to conduct training during initial entry training (IET) and unit sustainment training. Preliminary subjects include discussion on the weapon's capabilities, mechanical training, and the fundamentals and principles of rifle marksmanship. Live-fire applications are scheduled after the Soldier has demonstrated preliminary skills.

This manual was revised to include references to new materiel and systems. This revision includes—

- The new Army total marksmanship training strategy, to include specific strategies for the United States Army Reserve (USAR) and the Army National Guard (ARNG).
- Information about the advanced combat optical gunsight (ACOG), the AN/PEQ-15 advanced target pointer/illuminator aiming light (ATPIAL), various thermal sights, and the MK 262 round.
- Information about the alternate qualification record fire courses (known distance [KD] record fire, 25-meter scaled target alternate course, 15-meter scaled target alternate course).
- Information about the rapid magazine change and barricade transition fire for short-range marksmanship (SRM).
- Changes to all of the scorecards.
- Updated terminology.

This publication prescribes DA Form 3595-R (Record Fire Scorecard), DA Form 3601-R (Single Target—Field Fire I Scorecard), DA Form 5239-R (100-, 200-, and 300-Meter Downrange Feedback Scorecard), DA Form 5241-R (Single and Multiple Targets—Field Fire II Scorecard), DA Form 5789-R (Record Fire Scorecard—Known-Distance Course), DA Form 5790-R (Record Fire Scorecard—Scaled Target Alternate Course), DA Form 7489-R (Record Night Fire Scorecard), DA Form 7649-R (Squad Designated Marksman—Record Fire I and II), and DA Form 7650-R (Squad Designated Marksman—Position Evaluation).

This publication applies to the Active Army, the Army National Guard (ARNG)/National Guard of the United States (ARNGUS), and the US Army Reserve (USAR).

Terms that have joint or Army definitions are identified in both the glossary and the text. Terms for which FM 3-22.9 is the proponent FM are indicated with an asterisk in the glossary.

Uniforms depicted in this manual were drawn without camouflage for clarity of the illustration. Unless this publication states otherwise, masculine nouns and pronouns refer to both men and women.

The proponent for this publication is the US Army Training and Doctrine Command. The preparing agency is the United States Army Infantry School (USAIS). You may send comments and recommendations by any means (US mail, e-mail, fax, or telephone) as long as you use DA Form 2028 (Recommended Changes to Publications and Blank Forms) or follow its format. Point of contact information is as follows:

E-mail:	benn.29IN.229-S3-DOC-LIT@conus.army.mil
Phone:	Commercial: 706-545-8623 DSN: 835-8623
Fax:	Commercial: 706-545-8600 DSN: 835-8600
US Mail:	Commandant, USAIS ATTN: ATSH-INB 6650 Wilkin Drive, Bldg 74, Rm 102 Fort Benning, GA 31905-5593

Chapter 1
Marksmanship Training

An effective marksmanship program can be measured by the unit's ability to put effective fire on a target. This chapter provides a proven strategy for establishing and conducting an effective rifle marksmanship training program. The strategy begins with the progressive individual training periods taught during initial entry training (IET) and culminates with advanced rifle marksmanship (ARM) skills. Refresher training is conducted only when necessary.

A Soldier's marksmanship proficiency depends on proper training and application of the basic marksmanship fundamentals. During initial marksmanship training, emphasis is placed on learning the firing fundamentals, which are taught in four phases—preliminary marksmanship instruction (PMI), downrange feedback, field firing, and advanced firing exercises. This prepares Soldiers for advanced optic and laser training for combat-type collective exercises and real-world deployments.

SECTION I. TRAINING STRATEGY

The total Army marksmanship training strategy is the overall concept for integrating resources into a program designed to train, sustain, and improve the individual and collective skills needed to achieve proficiency in individual and collective gunnery tasks. Training strategies for rifle marksmanship are implemented in TRADOC institutions (IET, Noncommissioned Officers Education System [NCOES], Basic Officer Leaders Course [BOLC]) and in units. The overall training strategy is multifaceted and includes supporting strategies that use resources such as publications; ranges; ammunition; and training aids, devices, simulators, and simulations (TADSS). These strategies focus on developing the Soldier and leader skills required for success in combat.

NOTE: See AR 350-1 for specific requirements pertaining to marksmanship training and DA Pam 350-38 for live-fire frequency requirements.

OBJECTIVES

1-1. The procedures and techniques for implementing the total Army rifle marksmanship training strategy are based on the concept that Soldiers must become skilled marksmen. FM 7-0 stresses marksmanship as the paramount Soldier skill. Further, Soldiers should understand common firing principles and be confident in applying their firing skills in combat. Unit leaders accomplish proficiency through practice supervised by qualified instructors/trainers and thorough objective performance assessments.

1-2. The basic firing skills and exercises outlined in this manual must be part of every unit's marksmanship training program. Unit commanders must focus their basic and advanced marksmanship training programs to support their mission-essential task lists (METLs).

MARKSMANSHIP TRAINING STRATEGY

1-3. The following marksmanship training strategy guide contains the tasks that are currently trained in basic rifle marksmanship (BRM) programs, during basic combat training at Army training centers (ATCs), and during infantry one-station unit training (OSUT). It also provides a basis for structuring unit sustainment programs for active Army, Army National Guard, and United States Army Reserve units. Units normally perform diagnostic tests of the tasks and only conduct training on specific periods for Soldiers who must improve their basic firing skills. Unit training is usually conducted in less time than at IET.

1-4. There are two primary components of a marksmanship training strategy: initial training and sustainment training. Both may include individual and collective tasks and skills. Initial training must be taught correctly the first time. A task taught correctly and learned well is retained longer, and skills can be more easily sustained. However, an individual or unit eventually loses skill proficiency if the training is not reinforced. This learning decay depends on many factors, such as the difficulty and complexity of the task. Personnel turnover is a main factor in the decay of collective skills, since the loss of critical team members requires retraining to regain proficiency. If a long period elapses between initial and sustainment sessions or training doctrine is altered, retraining may be required.

INITIAL ENTRY TRAINING

1-5. The training strategy for BRM begins in IET and continues in the unit. Figure 1-1 shows the IET training strategy.

1-6. In IET, Soldiers learn how to maintain a rifle, hit a point target, and apply the four marksmanship fundamentals and other skills needed to engage a target. Once Soldiers understand the weapon and have demonstrated skill in zeroing, additional simulations and live-fire training exercises are conducted to prepare Soldiers for qualification. During these exercises, Soldiers master target types and scenarios of increasing difficulty to develop their proficiency.

1-7. IET culminates in the Soldier's proficiency assessment (which is conducted on the standard record fire range or approved alternates), followed by instruction on advanced firing techniques (a night-fire exercise with iron sights [unassisted] or night vision goggles [assisted]). This evaluation allows leaders to determine the effectiveness of the training.

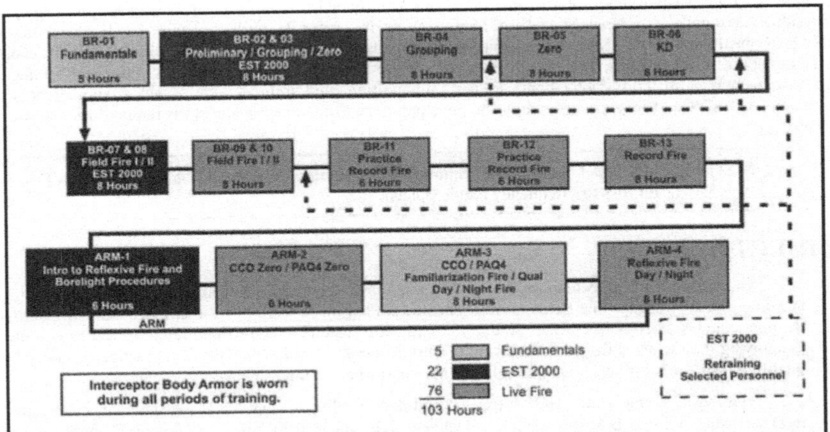

Figure 1-1. Initial entry training marksmanship training strategy.

SUSTAINMENT TRAINING

1-8. Training continues in active Army, National Guard, and Army Reserve units using the same basic skills taught in IET. Units must set up a year-round program to sustain skills and have a plan not only for when they are at their home station, but for when they are deployed as well.

1-9. To sustain the basic marksmanship skills taught in IET, periodic PMI is conducted, followed by instructional and qualification range firing. Key elements include—

- The training of trainers.
- Refresher training of nonfiring skills.
- The use of the Engagement Skills Trainer (EST) 2000, Laser Marksmanship Training System (LMTS), or other devices.
- Sustainment training.
- Remedial training.

NOTE: See Appendix A for more information about TADSS.

1-10. Additional skills trained in the unit include—

- Semiautomatic and automatic fires.
- Night fire.
- Mission-oriented protective posture (MOPP) firing.
- Firing using aiming devices.
- Moving target training techniques.
- Squad designated marksman (SDM) training.
- Short-range marksmanship (SRM).

1-11. These skills are trained and integrated into collective training exercises, such as platoon and squad live-fire situation training exercises (STXs). Figure 1-2 shows a year-round training strategy guide.

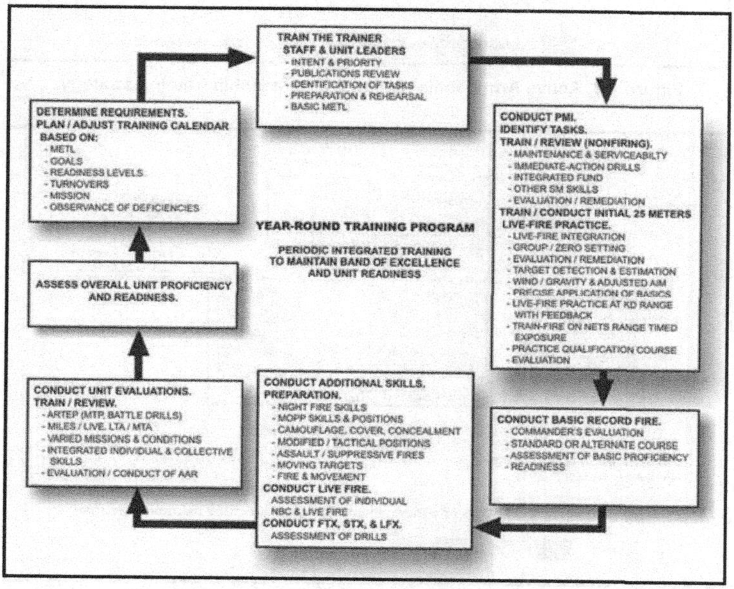

Figure 1-2. Unit marksmanship sustainment strategy.

1-12. General marksmanship knowledge and weapon proficiency are perishable skills. Most units have a readiness requirement that all Soldiers must zero their rifles within a certain time after unit assignment.

Soldiers must confirm the battlesight zeros of their assigned rifles before qualifying. Units should conduct PMI and practice firing throughout the year due to personnel turnover. A year-round marksmanship sustainment program is needed for the unit to maintain the individual and collective firing proficiency requirements to accomplish its mission. The following figures depict marksmanship training programs for active Army home station units (Figure 1-3), National Guard home station units (Figure 1-4), Army Reserve home station units (Figure 1-5), and deployed units (Figure 1-6).

NOTE: While the training strategy depicted in Figures 1-3 through 1-6 is not compulsory, it is strongly recommended for Army-wide application. The strategy has proven success during IET, producing more proficient Soldiers while using less ammunition. Commanders are urged to follow this strategy closely.

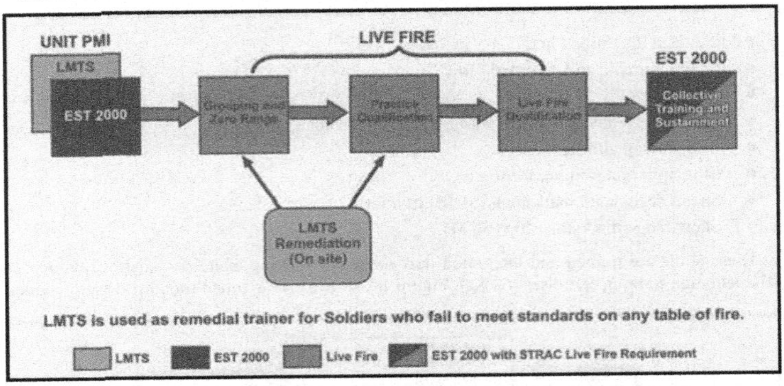

Figure 1-3. Active Army home station marksmanship training strategy.

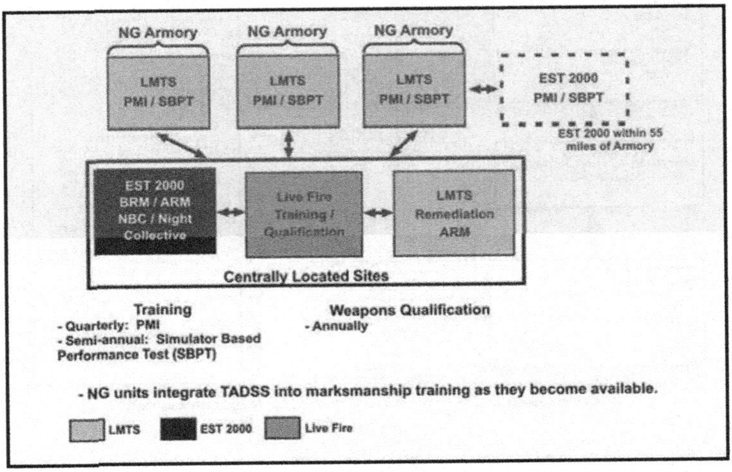

Figure 1-4. National Guard home station marksmanship training strategy.

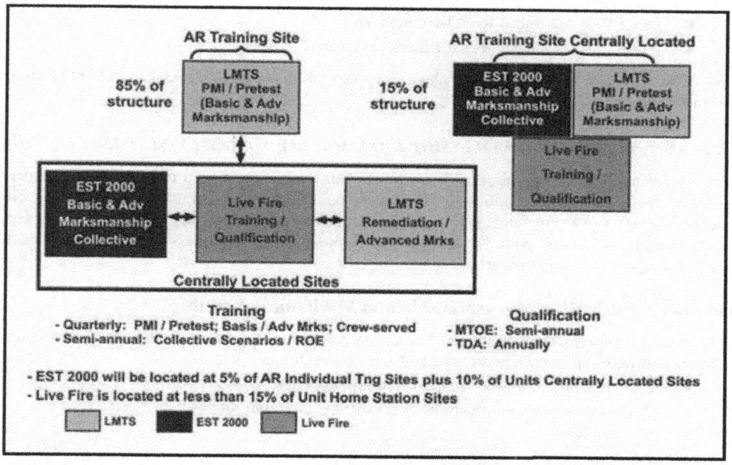

Figure 1-5. Army Reserve home station marksmanship training strategy.

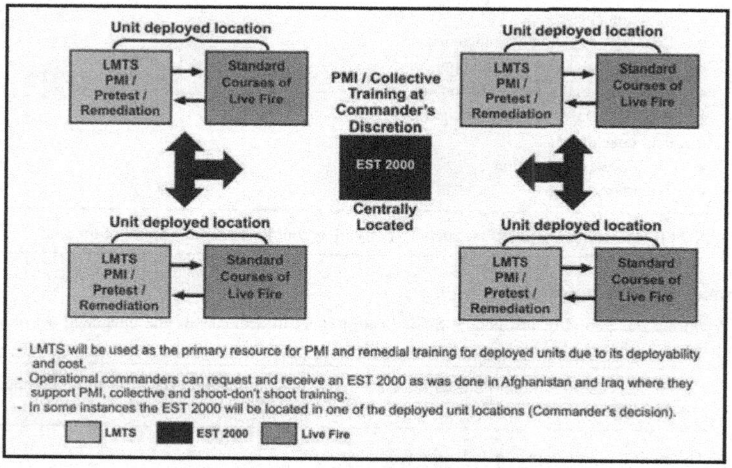

Figure 1-6. Deployed unit marksmanship training strategy.

TRAINING PHASES

1-13. Soldiers progress through five phases of rifle marksmanship training:

- Phase I—Basic Rifle Marksmanship Preliminary Marksmanship Instruction.
- Phase II—Basic Rifle Marksmanship Downrange Feedback Range Firing.
- Phase III—Basic Rifle Marksmanship Field Firing.

- Phase IV—Advanced Rifle Marksmanship.
- Phase V—Advanced Optics, Lasers, and Iron Sights.

1-14. When Soldiers are trained in all phases of rifle marksmanship, a solid sustainment program is key to mission readiness.

PHASE I—BASIC RIFLE MARKSMANSHIP PRELIMINARY MARKSMANSHIP INSTRUCTION

1-15. Understanding the operation and functions of any machine is vital to becoming an expert with that machine. The same theory applies to rifle marksmanship. Soldiers must master weapon maintenance, function checks, and firing fundamentals before progressing to advanced skills and firing exercises under tactical conditions. Armed with this knowledge, a Soldier is able to assess and correct any malfunction to keep the weapon operating properly.

Introduction to Basic Rifle Marksmanship and Mechanical Training

1-16. During this period of instruction, Soldiers are trained to understand the operation and functions of the weapon. This period of instruction includes the following topics:

- General safety rules and weapon clearing procedures.
- Characteristics, components, accessories, and ammunition.
- Disassembly.
- Inspection.
- Lubrication.
- Assembly.
- Function check.
- Loading/unloading the magazine.
- Loading/unloading the weapon.
- Immediate and remedial action.
- Adjusting the front and rear sights.
- Peer coaching.
- Eight cycles of function.
- Troubleshooting.

NOTE: The IET program of instruction (POI) allots four hours for this phase of training.

Marksmanship Fundamentals I

1-17. During this period of instruction, Soldiers are trained to demonstrate the integrated act of shooting using the EST 2000 or LMTS.

NOTE: EST 2000 will not be used in lieu of live-fire qualification except for those outlined in DA Pam 350-38.

1-18. This period of instruction includes the following topics:

- The four fundamentals.
- Basic firing positions.
- Range and safety procedures.
- Dominant eye training.
- Demonstrating the integrated act of shooting during dry-fire exercises.

NOTE: The IET POI allots six hours for this phase of training.

Marksmanship Fundamentals II

1-19. During this period of instruction, Soldiers are trained to demonstrate the integrated act of shooting using the EST 2000 or LMTS.

NOTES: 1. The IET POI allots six hours for this phase of training.

2. See paragraphs 4-76 through 4-80 and Appendix A for more information about EST 2000 and LMTS training.

PHASE II—BASIC RIFLE MARKSMANSHIP DOWNRANGE FEEDBACK RANGE FIRING

1-20. During Phase II, instructors/trainers outline grouping and zeroing procedures and demonstrate how to conduct the three types of known distance (KD) ranges. Downrange feedback provides precise knowledge of what happens to bullets at a given range and allows Soldiers to transition between 25-meter firing and firing on the field fire range. Knowing precisely where bullets hit allows poor firers (with the assistance of instructors/trainers) to improve their performance and good firers to bring their shots to the center of the target. Firers develop the knowledge and skills required to perform with confidence on the field fire range, where only hit-or-miss information is available.

Grouping Procedures

1-21. Grouping is a form of practice firing with two primary objectives: firing tight shot groups and consistently placing those groups in the same location. Frequent use of the EST 2000 or LMTS greatly reduces live-fire grouping time.

Zeroing Procedures

1-22. Zeroing allows firers to use standard issue ammunition to align the sights with the weapon's barrel. When this is accomplished correctly, the aimpoint and the impact point are the same at a given range. For most combat targets, this sight setting provides the greatest probability of a hit with minimum adjustment to the aimpoint. When followed, a properly zeroed rifle for one Soldier is close to zero for another Soldier.

Downrange Feedback

1-23. On KD ranges, Soldiers fire tight shot groups at a known distance and make sight adjustments at a given range while experiencing the effects of wind, gravity, and other environmental factors. The advantage of a KD range is the ability to see precisely where each bullet hits. KD firing is conducted with a single, clearly visible target at a known distance, and the Soldier can establish a position that provides a natural aimpoint on that single target.

NOTES: 1. See Chapter 5 for more information about downrange feedback.

2. See paragraphs 4-76 through 4-80 and Appendix A for more information about EST 2000 and LMTS training.

PHASE III—BASIC RIFLE MARKSMANSHIP FIELD FIRING

1-24. Field fire begins a critical transition from unstressed firing at single targets at known distances to requiring the Soldier to refine techniques for scanning the range for targets, estimating range, and firing quickly and accurately. Phase III includes the following:

- Field Fire I (single timed targets at 75 meters, 175 meters, and 300 meters).
- Field Fire II (single and multiple targets at 75 meters, 175 meters, and 300 meters).
- Practice Record Fire.
- Practice Record Fire II.
- Record Fire.

NOTES: 1. See Chapter 6 for more information about BRM field fire.

2. See paragraphs 4-76 through 4-80 and Appendix A for more information about EST 2000 and LMTS training.

PHASE IV—ADVANCED RIFLE MARKSMANSHIP

1-25. ARM focuses on the techniques and procedures that the Soldier will need to participate in collective training. This phase addresses the following topics:

- Alternate firing positions.
- Burst fire.
- Quick fire.
- Chemical, biological, radiological, and nuclear (CBRN) fire.
- Moving targets.
- SDM.
- Unassisted night record fire.
- M68 close combat optic (CCO).
- Advanced combat optical gunsight (ACOG).
- AN/PAQ-4C infrared (IR) aiming laser.
- Assisted night record fire IR.
- Assisted night record fire (thermal).
- SRM.

NOTE: See Chapter 7 for more information about ARM.

PHASE V—ADVANCED OPTICS, LASERS, AND IRON SIGHTS

1-26. BRM teaches Soldiers to effectively engage the enemy with the basic rifle using the iron sights (primarily during the day). ARM adds additional situations for the firer. To enhance the lethality of night firing, Soldiers participate in training with optics and lasers. This phase addresses the following topics:

- Backup iron sight (BUIS).
- M68 CCO.
- ACOG.
- AN/PAQ-4C IR aiming laser.
- AN/PEQ-2A/B target pointer illuminator/aiming light (TPIAL).
- Thermal weapon sight (TWS).
- AN/PVS-4 night vision sight.

NOTE: See Chapter 8 for more information about advanced optics, lasers, and iron sights.

SECTION II. UNIT MARKSMANSHIP TRAINING PROGRAM

An effective unit marksmanship program reflects the priority, emphasis, and interest of commanders and trainers. This section outlines a rifle marksmanship training program strategy as guidance in establishing and conducting an effective unit training program. The strategy consists of the individual and leader refresher training for maintaining the basic skills learned during IET. It progresses to training advanced and collective skills under near-combat conditions during live-fire STXs.

MISSION-ESSENTIAL TASKS

1-27. Marksmanship proficiency is critical to soldiering and is required for any unit deployed to a wartime theater. All commanders should develop a METL and organize a training program that devotes adequate time to marksmanship. The unit's combat mission must be considered when establishing training priorities. This not only applies to the tasks selected for the unit's METL, but also to the conditions under which the tasks are to be performed. If a unit may be employed in an urban environment, the effects of range, gravity, and wind may not be as important as automatic or burst fire, quick fire, or assault fire. The reverse may be true of a unit that expects to engage the enemy at long ranges.

TRAINING ASSESSMENT

1-28. To conduct an effective marksmanship program, the unit commander must determine the current marksmanship proficiency of all assigned personnel. Constant evaluation provides commanders understanding of where training emphasis is needed. All results are reviewed to determine any areas that need strengthening, along with any individuals that require special attention. Based on this evaluation, marksmanship training programs are developed and executed. Commanders continually assess the program and modify it as required. To develop a training plan and assess the marksmanship program, commanders should use the following tools:

- Direct observation of training.
- Spot checks.
- Review of past training.

1-29. Based on the commander's evaluation, goals, and missions, quarterly, semiannual, or annual training events are identified. Marksmanship programs must be continuous, and to sustain an effective marksmanship program, resources are required. While the unit may only qualify its Soldiers annually or semiannually, test results show that sustainment training is required at least quarterly to maintain marksmanship skills.

DIRECT OBSERVATION OF TRAINING

1-30. Observing and accurately recording performance reveals the status of weapon maintenance, Soldier zero and qualification results, and each Soldier's ability to hit targets. This also allows the commander to identify Soldiers who need special assistance to reach required standards and those who exceed these standards.

SPOT CHECKS

1-31. Spot checks of individual marksmanship performance, such as interviews and evaluations of Soldiers, provide commanders with valuable information about Soldier proficiency and knowledge of the marksmanship tasks.

REVIEW OF PAST TRAINING

1-32. Commanders review past training to gain valuable information for developing a training plan. The assessment should include—

- The frequency and results of training.
- The basic and advanced record fire results.
- The frequency of unit-conducted collective CBRN or night fire training.

COMMANDER'S EVALUATION GUIDE

1-33. The commander's evaluation guide contains three sections:

- Commander's priorities and intent.
- Soldier assessment.
- Trainer assessment.

1-34. The following is an example of a commander's evaluation guide. Commanders can use this guide not only to assess their unit's marksmanship proficiency, but to assess the unit leaders and their ability to effectively implement a marksmanship program. They can also use it to develop NCOs into subject matter experts.

Commander's Priorities and Intent

1-35. When considering their priorities and intent, commanders answer the following questions:

- Have you clearly stated the priority of rifle (small arms) proficiency in your unit? What is it? Do the staff and subordinates support this priority? Is it based on your METL and an understanding of FM 7-0 and FM 7-1?
- Have you clearly stated the intent of record fire? Are leaders evaluating firing performance based on accurately recorded data and results?
- Have you clearly stated that weapon qualification or record fire is one of the commander's opportunities to assess several skills relating to small arms readiness?
- What qualification course will be used to evaluate your unit's marksmanship readiness (small arms)?
  - Is the standard 300-meter, 300-yard KD, or 25-meter scaled target alternate qualification course used?
  - How will it be conducted? Will the prescribed procedures be followed?
  - Who will collect the data?
- Have you clearly stated the purpose and intent of PMI?
  - What skills will PMI address?
  - Will PMI be performance-oriented? Are tasks integrated?

Soldier Assessment

1-36. During Soldier assessment, commanders answer the following questions:

- Do Soldiers maintain their assigned weapons and magazines IAW the TM? Do they have a manual?
- Do Soldiers conduct serviceability checks of weapons and magazines before training? Were maintenance deficiencies corrected?
- Do Soldiers demonstrate an understanding of the weapon's operation, functioning, and capabilities?
- Can Soldiers correctly apply immediate action procedures to reduce weapon stoppages and then continue to fire? Have they demonstrated this during dry-fire exercises?
- Are Soldiers firing their assigned weapons?
  - How often are weapons reassigned between individuals?
  - What is the value of a recorded zero?
- Can Soldiers precisely and consistently apply the four fundamentals of rifle marksmanship? To what standard have they demonstrated their mastery?
  - During a dry-fire exercise?
  - During a live-fire exercise (LFX)?
  - When firing on the 25-meter course?
  - During KD firing?
- Can Soldiers accurately battlesight zero their assigned rifle to standard?
  - Do they understand sight adjustment procedures?
  - Do they record their rifle's zero? How is it done? Why?
  - Do they record the date the Soldier last zeroed his rifle? What is the sight setting? Are these linked? How do you check this?
- Do Soldiers demonstrate their knowledge of the effects of wind and gravity while firing out to 300 meters? What feedback was provided? How?

- Can Soldiers scan a designated area or sector of fire and detect all targets out to 300 meters? If not, why?
- Can Soldiers quickly engage timed single and multiple targets from both supported and unsupported firing positions out to 300 meters? If not, which targets were not engaged? Which were missed? Why?
- During individual and collective training, do Soldiers demonstrate their ability to manage allocated ammunition and to engage all targets? Do they fire several rounds at one target? Which targets? Why?
- Based on an analysis of individual qualification scores, what is the distribution?
  - Are most Soldiers just meeting the minimum acceptable performance (marksman)?
  - Are most Soldiers distributed in the upper half of the performance spectrum (sharpshooter, expert)?
- What is the hit distribution during collective LFXs?
- Do Soldiers demonstrate proficiency during night-fire, target detection and acquisition, and night fire engagement techniques? When using night vision devices (NVDs)?
- Do Soldiers demonstrate individual marksmanship proficiency during MOPP firing conditions? During collective exercises?
- Do Soldiers demonstrate proficiency during moving target engagements? Do they demonstrate proficiency collectively by hitting moving targets at the multipurpose range complex? If not, is moving target training conducted?
- Are marksmanship skills integrated into tactical exercises and unit LFXs? If so, is suppressive fire, rapid semiautomatic fire, and automatic or burst fire conducted? What tasks in the mission training plan are evaluated?
- Based on onsite observations and analysis of training and firing performance, what skills or tasks show a readiness deficiency?
  - What skills need training emphasis? Individual emphasis? Leader emphasis?
  - What are the performance goals?

Trainer Assessment

1-37. During trainer assessment, commanders answer the following questions:

- Who has trained or will train the trainers?
  - What is the subject matter expertise of the cadre?
  - Are they actually training the critical skills?
  - Have they addressed the nonfiring skills first?
  - What aids and devices are being used? Are EST 2000 and LMTS being properly used?
- What administrative constraints or training distracters can you overcome for the junior officer and NCO? At what level are the resources necessary to train marksmanship controlled (time, training aids, weapons, ammunition, ranges)? Do the sergeants do the job they are charged with?

TRAINERS

1-38. Knowledgeable instructors or cadre are the key to marksmanship performance. All commanders must be aware of maintaining expertise in marksmanship instruction/training.

INSTRUCTOR/TRAINER SELECTION

1-39. Institutional and unit instructors/trainers are selected and assigned from the most highly qualified Soldiers. These Soldiers must demonstrate proficiency in all aspects of rifle marksmanship, be proficient in applying these fundamentals, know the importance of marksmanship training, and have a competent and professional attitude. The commander must ensure that selected instructors/trainers can effectively train other Soldiers. Local instructor/trainer training courses and marksmanship certification programs must be established to ensure that instructor/trainer skills are developed.

Cadre/Trainer

1-40. Cadre/trainer refers to a marksmanship instructor/trainer that has more experience and expertise than the firer does. He trains Soldiers in the effective use of the rifle by maintaining strict discipline on the firing line, insisting on compliance with range procedures and program objectives, and enforcing safety regulations. A good cadre/trainer must understand the training phases and techniques for developing marksmanship skills. Each cadre/trainer must have the following attributes:

- Knowledge.
- Patience.
- Understanding.
- Consideration.
- Respect.
- Encouragement.

Knowledge

1-41. Effective cadre/trainers must possess a thorough knowledge of the rifle, proficiency in firing, and a complete understanding of this manual and supporting manuals.

Patience

1-42. Effective cadre/trainers relate to the Soldier calmly, persistently, and patiently.

Understanding

1-43. Cadre/trainers enhance success and understanding by emphasizing close observance of rules and instructions.

Consideration

1-44. Most Soldiers, even those who do not fire well, enjoy firing and begin their training with great enthusiasm. A cadre/trainer who is considerate of Soldiers' feelings and who encourages them throughout their training will find training a pleasant and rewarding duty.

Respect

1-45. An experienced cadre is assigned the duties of instructor/trainer, which classifies him as a technical expert and authority. A good cadre/trainer watches for mistakes and patiently makes needed corrections.

Encouragement

1-46. A cadre/trainer can encourage Soldiers by convincing them to achieve good firing performance through practice. He imparts his knowledge and helps Soldiers gain the practical experience necessary.

Training the Trainer

1-47. Knowledgeable small-unit leaders are key to marksmanship training. This manual and other training publications provide the unit instructor with the required information for developing a good train-the-trainer program.

1-48. The goal of a progressive train-the-trainer program is to achieve a high state of combat readiness. Through the active and aggressive leadership of the chain of command, a perpetual base of expertise is established and maintained.

1-49. The commander should identify unit personnel who have had assignments as marksmanship instructors. These individuals should be used to train other unit cadre by conducting PMI and LFXs for their Soldiers.

1-50. Assistance and expertise from outside the unit may also be available, such as the Army Marksmanship Unit (AMU) at Fort Benning, Georgia. A suggested train-the-trainer program is outlined below:

- Conduct marksmanship diagnostic test.
- Review operation and function of the rifle and ammunition, immediate action, and safety.
- Conduct PMI; review the four fundamentals.
- Review coaching techniques and device usage.
- Establish grouping and zeroing procedures.
- Review effects of wind and gravity when firing as far away as 300 meters (600 meters for ARM).
- Conduct range operations.
- Conduct qualification/record firing.
- Diagnose firing problems.

Duties of the Instructor/Trainer

1-51. The instructor/trainer helps the firer master the fundamentals of rifle marksmanship. He ensures that the firer consistently applies what he has learned. When training the beginner, the instructor/trainer could confront problems, such as fear, nervousness, forgetfulness, failure to understand, and a lack of coordination or determination, which may be compounded by arrogance or carelessness. With all types of firers, the instructor/trainer must ensure that firers are aware of their firing errors, understand the causes, and apply remedies. To perform these duties, he—

- Observes the firer.
- Questions the firer.
- Analyzes the shot group.

Observing the Firer

1-52. To pinpoint errors, the instructor/trainer observes the firer during drills and when firing. If there is no indication of probable error, the firer's position, breath control, shot anticipation, and trigger squeeze are closely observed.

Questioning the Firer

1-53. The firer is asked to detect his errors and to explain his firing procedure (position, aiming, breath control, and trigger squeeze).

Analyzing the Shot Group

1-54. Analyzing the shot group is an important step in detecting and correcting errors. When analyzing a target, the instructor/trainer correlates observations of the firer to probable errors in performance, according to the shape and size of shot groups. A poor shot group is usually caused by more than one observable error.

TRAINER CERTIFICATION PROGRAM

1-55. The certification program sustains the trainers' expertise and develops methods of training. The program standardizes procedures for certifying marksmanship trainers. Trainers' technical expertise must be continuously refreshed, updated, and closely managed.

Training Base

1-56. The training base can expect the same personnel changes as any other organization. Soldiers assigned as marksmanship trainers will have varying degrees of experience and knowledge of training procedures and methods. Therefore, the trainer certification program must be an ongoing process that is tailored to address these variables. At a minimum, formal records should document program progression for each trainer. All marksmanship trainers must complete the four phases of training using the progression steps, and the records should be updated on a quarterly basis.

CERTIFICATION PROGRAM OUTLINE

1-57. Before certification, all trainers must attend all phases of the train-the-trainer program in the following order:

- Phase I—Program Orientation.
- Phase II—Preliminary Marksmanship Training.
- Phase III—Basic Marksmanship Training.
- Phase IV—Advanced Marksmanship Training.

1-58. Then, they conduct all phases to demonstrate their ability to train Soldiers and to diagnose and correct problems. Trainers who fail to attend or do not pass any phase of the diagnostic examination will be assigned to subsequent training.

Phase I—Program Orientation

1-59. During this phase, the trainer must accomplish the following tasks and be certified by the chain of command:

- Be briefed on the concept of the certification program.
- Be briefed on the unit marksmanship training strategy.
- Review the unit marksmanship training outlines.
- Review issued reference material.
- Visit training sites and firing ranges.

Phase II—Preliminary Marksmanship Training

1-60. Phase II should be completed no more than two weeks following the conclusion of Phase I. During Phase II, the trainer demonstrates his ability to master the marksmanship fundamentals, and his performance is reviewed by the chain of command. The results of this review are recorded and maintained on the trainer's progression sheet, which is, in turn, designed IAW the unit SOP.

- Safety, clearing, and unloading procedures.
- Characteristics.
- Capabilities.
- Accessories.
- Ammunition.
- Disassembly.
- Cleaning, inspection, and lubrication procedures.
- Assembly.
- Function check.
- Loading.
- Immediate and remedial actions.
- Four fundamentals.
- Use of EST 2000, LMTS, and other TADSS.
- Sight manipulation.
- Boresighting procedures.
- Zeroing procedures.
- Effect of wind, gravity, and other environmental factors.
- Range determination and estimation.
- Classes of fire.

- Application of fire.
- Range operations, courses of fire, conduct of fire, and fire commands.
- Optics, NVDs, night aiming devices, and thermal devices.
- Scanning techniques.

Phase III—Basic Marksmanship Training

1-61. During this phase, the trainer demonstrates and reinforces what he has learned during Phase II. The trainer sets up and conducts firing on the various ranges. He explains the targets and the zeroing and scoring procedures. The trainer explains the purpose of transition firing, field zeroing procedures, range layout, and the conduct of training on the transition range. This validates that the trainer has gained the knowledge necessary to conduct training. The results of this review are recorded and maintained on the trainer's progression sheet.

Phase IV—Advanced Marksmanship Training

1-62. The final phase of the train-the-trainer program tests the trainer. During this phase, the trainer sets up a range and conducts training for at least one person. If ammunition is available, the trainer conducts a firing exercise. If ammunition is not available, the evaluation is based on the quality of training given.

QUALIFICATION TRAINING

1-63. Although marksmanship is a continuous training requirement, units normally conduct a refresher program before qualification. Soldiers must be well-versed in marksmanship fundamentals and have preparatory marksmanship training before qualification. This applies to qualification for the entire unit or for newly assigned personnel. Trainers must understand that unit rifle marksmanship is not a series of exercises to be trained in a planned sequence as is done during IET, but trainers can use the exercises and POI events covered during IET to identify events that the unit can use for a sustainable and effective unit marksmanship program. The unit must prepare for training by—

- Issuing Soldiers a serviceable weapon.
- Maintaining and replacing bad magazines.
- Issuing and assigning each Soldier his own rifle. Only he will zero and fire the weapon assigned to him.
- Considering available or required resources (targets, ranges, ammunition, training aids, devices, and publications) early in the process.

1-64. Before the Soldier can fire, he must know how to adjust the rifle's sights and should understand ballistics (for example, the effects of wind and gravity on a bullet strike). A refresher training program prevents Soldiers from becoming frustrated and losing confidence, and conserves ammunition and training time. All Soldiers attend this program so they can meet the standards outlined in this manual and supporting manuals.

NOTE: Many individual marksmanship tasks, such as operation and function checks, immediate action, target detection, and dry-fire, do not require live-firing. Live and virtual simulators can be used to reinforce PMI, grouping, zeroing, practice record fire, record fire, CBRN fire, and assisted and unassisted night fire by simulating the LFXs. Building marksmanship confidence by repetition can bring consistency to the unit marksmanship training program.

FEEDBACK

1-65. Feedback must be included in all live-fire training. Soldiers must have precise knowledge of a bullet strike; feedback is not adequate when bullets from previous firings cannot be identified. To provide accurate feedback, trainers ensure that Soldiers triangulate and clearly mark previous shot groups on a zeroing target.

GROUPING AND ZEROING

1-66. The initial LFX should be a grouping exercise, during which Soldiers apply marksmanship fundamentals to obtain tight, consistent shot groups. Following a successful grouping exercise, Soldiers zero their weapons quickly using only a few rounds.

DOWNRANGE FEEDBACK

1-67. After zeroing, downrange feedback should be conducted. If modified field fire or KD ranges are not available, a series of scaled silhouette targets can be used for training on the 25-meter range. A timed-fire scaled silhouette target can add to successful record fire performance, since it represents targets at six different ranges, requires quick response, and allows precise feedback. It also serves as another way to confirm zero and requires the application of the four fundamentals.

FIELD FIRE

1-68. Field fire training is a transitional phase that stresses focusing on a certain area. Soldiers must detect the target as soon as it appears and quickly fire; this is an important combat skill. Soldiers who are exposed to the field fire range before they have refined their basic firing skills cannot benefit from the exercise, and feedback from this exercise is hit-or-miss. If most 175- and 300-meter targets are missed, additional feedback or PMI training should be conducted.

RECORD FIRE

1-69. The intent of record fire is to facilitate the commander's evaluation of several individual tasks and integrated marksmanship skill performances, and to provide unit readiness indicators. The qualification standards are specifically related to a prescribed procedure for the conduct of record fire. Individual performance must be evaluated IAW three components:

- What test was used (standard, known distance, or scaled)?
- How was the test administered?
- How were individual and unit performances distributed (marksman, sharpshooter, and expert) and at what target ranges?

REMEDIAL TRAINING

1-70. For firers who need additional training to meet the requirements, remedial training is conducted using the EST 2000 or LMTS. Soldiers proficient in marksmanship skills can assist in the remedial training effort.

NOTE: Table 1-1 shows the training devices that a commander may use to assist in remedial training and sustain weapon proficiency. See Appendix A for more information about these training devices. The devices replicate, but are not intended to replace, LFXs or qualifications. Active and reserve component units should consult DA Pam 350-38 for regulatory guidance on mandatory live-fire training and qualification events.

Table 1-1. Training simulators, devices, and exercises.

TRAINING DEVICE / EXERCISE	Engagement Skills Trainer (EST 2000)	Laser Marksmanship Training System (LMTS)	Location of Misses and Hits (LOMAH)	Short-Range Training Ammunition (SRTA) and M2 Bolt	Weaponeer
Zero	X	X	X	X	X
Practice Fire	X	X	X	25 m	X
Simulated Record Fire	X	X	X	25 m	X
CBRN Practice	X	X	X	X	X
Simulated CBRN Record	X	X	X	X	X
Unassisted Night Practice	X	X	X	50 m	
Unassisted Simulated Night Record	X	X	X	50 m	
NVD Zero	X	X	X		
NVD Practice	X	X	X	50 m	
NVD Simulated Record	X	X	X	50 m	
Advanced Skills	X	X	X	X	

UNIT LIVE-FIRE EXERCISES

1-71. Unit LFXs are planned, prepared, and performed as outlined in the mission training plan for the infantry platoon and squad. Within the framework of these exercises, Soldiers perform marksmanship tasks under realistic combat conditions.

FUNDAMENTALS

1-72. During training, the fundamentals must apply to combat, as well as to the range. Too often Soldiers disregard the fundamentals while under the pressure of combat, so it is imperative that the Soldier receives feedback regarding his firing results and his use of the fundamentals during collective LFXs. This training should also discuss target acquisition, area fire, quick fire, assuming firing positions, responding to oral fire commands, and safety. To learn SOPs and proper procedures, Soldiers must participate in dry-fire or MILES rehearsals at crawl, walk, and run paces.

EVALUATORS

1-73. During training, enough evaluators must be present to observe each Soldier and provide performance feedback. Evaluators must know the scenario, the location of targets, the friendly plan, and SOPs. They must watch each Soldier to determine if he identifies targets in his sector and successfully engages them. Evaluators must also understand the fundamentals of marksmanship to detect Soldiers' mistakes and review them during the after-action review (AAR).

Chapter 2

Weapon Characteristics, Accessories, and Ammunition

This chapter describes the general components, characteristics, accessories, and ammunition for M16- and M4-series weapons, and includes a brief explanation of how to mount the various accessories.

SECTION I. RIFLES AND CARBINES

All M16-/M4-series weapons are magazine-fed, gas-operated, air-cooled, shoulder-fired 5.56-millimeter weapons. This section describes the general characteristics and components of M16-/M4-series weapons.

CHARACTERISTICS OF M16-/M4-SERIES WEAPONS

2-1. Table 2-1 describes the general characteristics of M16-/M4-series weapons.

Table 2-1. Characteristics of M16-/M4-series weapons.

CHARACTERISTICS	M4-SERIES	M16A2/A3	M16A4	M16A1
WEIGHT (lb)				
Without magazine and sling	6.49	7.78	9.08	6.35
With sling and loaded:				
20-round magazine	7.19	8.48	9.78	6.75
30-round magazine	7.50	8.79	10.09	8.06
Bayonet knife, M9	1.50	1.50	1.50	1.50
Scabbard	0.30	0.30	0.30	0.30
Sling, M1	0.40	0.40	0.40	0.40
LENGTH (in)				
Rifle w/bayonet knife	N/A	44.88	44.88	44.25
Overall rifle length	N/A	39.63	39.63	39.00
Buttstock closed	29.75	N/A	N/A	N/A
Buttstock open	33.0	N/A	N/A	N/A
OPERATIONAL CHARACTERISTICS				
Barrel rifling-righthand 1 twist (in)	7	7	7	12
Muzzle velocity (fps)	2,970	3,100	3,100	3,250
Cyclic rate of fire (rounds per min)	700-900	700-900	800	700-800
MAXIMUM EFFECTIVE RATE OF FIRE (rounds per min)				
Semiautomatic	45	45	45	45-65
3-round burst	90	90 (A2)	90	N/A
Automatic	150-200 A1	150-200 A3	N/A	150-200
Sustained	12-15	12-15	12-15	12-15
RANGE (m)				
Maximum range	3,600	3,600	3,600	2,653
Maximum effective range:				
Point target	500	550	550	460
Area target	600	800	600	N/A

NOTE: For further technical information, refer to TM 9-1005-319-10 and TM 9-1005-249-10.

M4-SERIES CARBINE

2-2. The M4-series carbine consists of the M4, the M4A1, and the M4 modular weapon system (MWS) (Figures 2-1 and 2-2) and features several modifications to equip it for close combat operations:

- The M4-series carbine buttstock has four positions:
  - Closed.
  - ½ open.
  - ¾ open.
  - Fully open.
- The M4 carbine becomes the M4 MWS when the M4 adapter rail system (ARS) is installed (Figure 2-3).
- When operating an M4 or M4 MWS, the firer can move the selector lever (SAFE, SEMI, and BURST) to fire a semiautomatic or a three-round burst.
- The M4A1 is fully automatic.

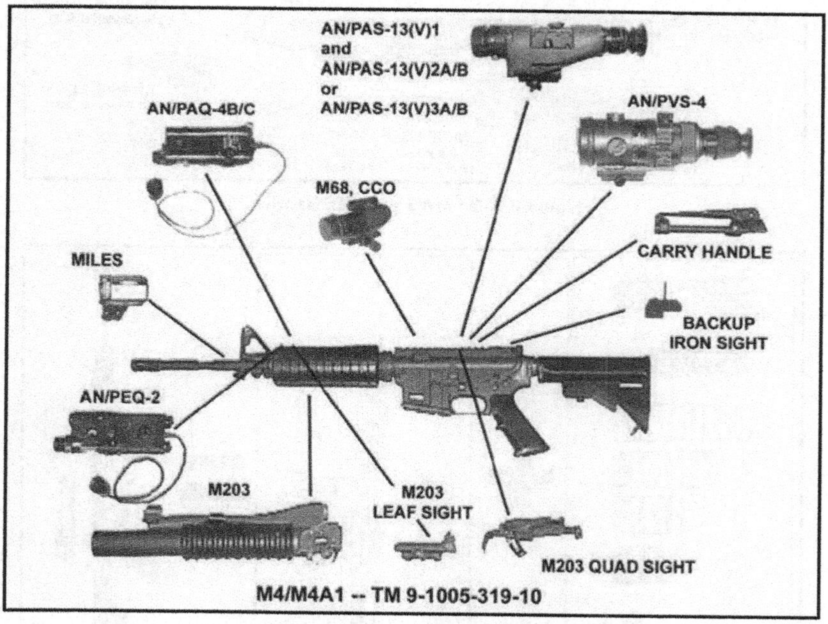

Figure 2-1. M4/M4A1 carbine with accessories.

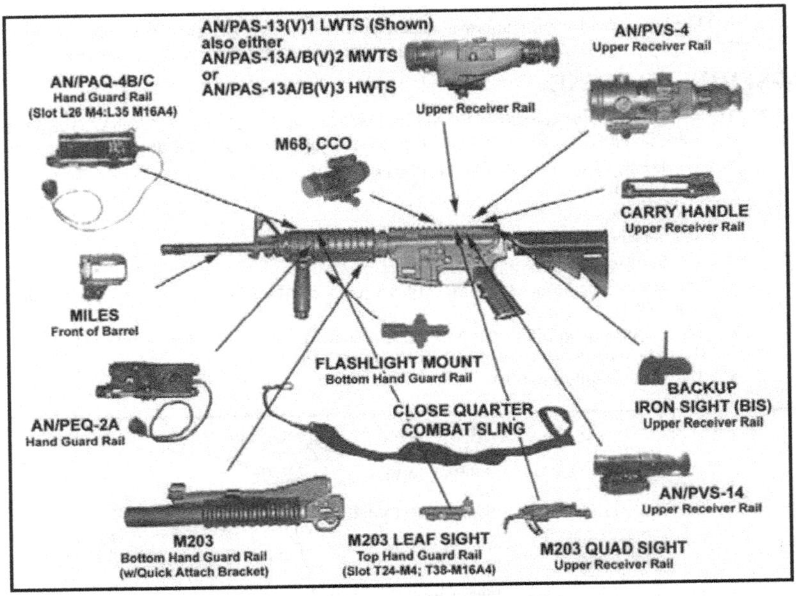

Figure 2-2. M4 MWS with accessories.

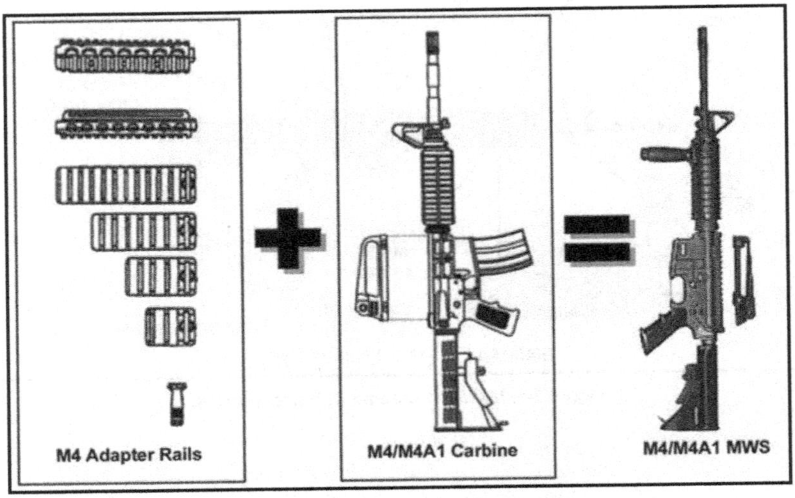

Figure 2-3. M4/M4A1 and M4/M4A1 MWS.

MECHANICALLY ZEROING THE M4/M4A1 CARBINE OR M4 MWS

NOTE: Mechanically zeroing the weapon is only necessary when the weapon zero is questionable, the weapon is newly assigned to the unit, or the weapon sights have been serviced.

2-3. To mechanically zero an M4/M4A1 or M4 MWS—

NOTE: Reference the weapon components using the numbers listed in Figure 2-4.

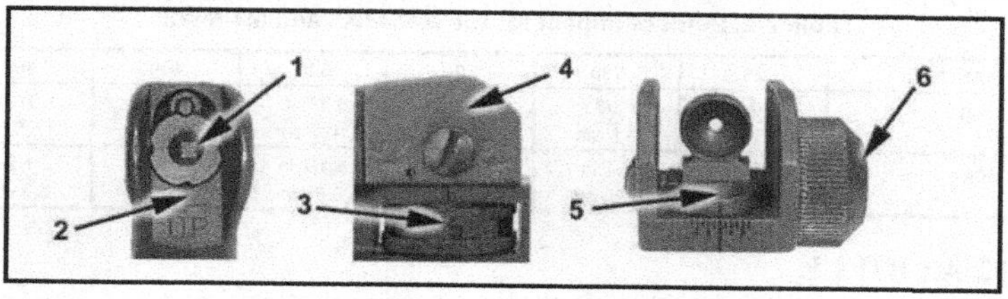

Figure 2-4. M4/M4A1 or M4 MWS mechanical zero.

(1) Adjust the front sightpost (1) until the base of the front sightpost is flush with the front sightpost housing (2).

(2) Turn the elevation knob (3, shown as viewed from above) counterclockwise until the rear sight assembly (4) rests flush with the detachable carrying handle and the 6/3 marking is aligned with the index line (5) on the left side of the carrying handle.

(3) Position the apertures (6) so the unmarked aperture is up and the 0-200 meter aperture is down.

(4) Turn the windage knob (7) to align the index mark (8) on the 0-200 meter aperture with the long center index line on the rear sight assembly.

BATTLESIGHT ZEROING THE M4/M4A1 CARBINE OR M4 MWS

2-4. To battlesight zero an M4/M4A1 or M4 MWS (Figure 2-5)—

NOTE: Reference the weapon components using the numbers listed in Figure 2-5.

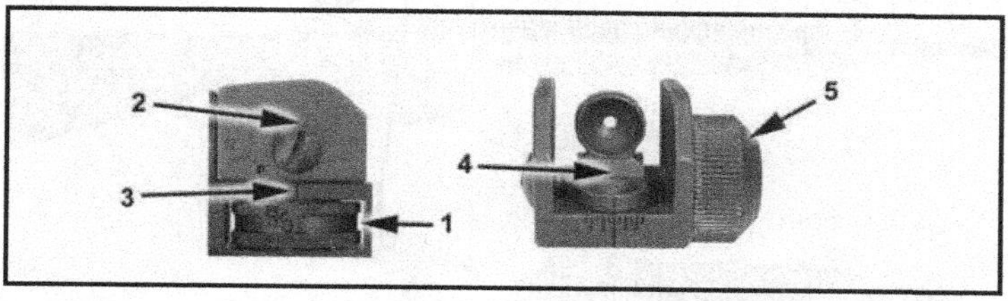

Figure 2-5. M4/M4A1 or M4 MWS battlesight zero.

(1) Turn the elevation knob (1, shown as viewed from above) counterclockwise until the rear sight assembly (2) rests flush with the detachable carrying handle and the 6/3 marking is aligned with the index line (3) on the left side of the detachable carrying handle.

NOTE: The elevation knob remains flush.

(2) Position the apertures (4) so the unmarked aperture is up and the 0-200 meter aperture is down.

(3) Turn the windage knob (5) to align the index mark (6) on the 0-200 meter aperture with the long center index line on the rear sight assembly.

NOTE: The "Z" marking on the elevation knob (employed when using the M4-series weapon's detachable carrying handle) should be ignored. The "Z" marking is only used when the M16A4 is being zeroed.

2-5. Table 2-2 shows how much one click of elevation or windage will move the strike of the round at ranges from 25 to 500 meters.

Table 2-2. Point of impact for the M4/M4A1 and M4 MWS.

RANGE (m)	25	100	200	300	400	500
Elevation	3/4 in 0.9 cm	1 3/8 in 3.5 cm	2 3/4 in 7 cm	4 1/8 in 10.5 cm	5 1/2 in 14 cm	6 7/8 in 17.5 cm
Windage	1/8 in 0.3 cm	1/2 in 1.25 cm	1 in 2.5 cm	1 1/2 in 3.8 cm	2 in 5 cm	2 1/2 in 6.3 cm

M16A2/A3 RIFLE

2-6. Figure 2-6 shows the M16A2/A3 rifle. When operating an M16A2 rifle, the firer can move the selector lever (SAFE, SEMI, and BURST) to fire a semiautomatic or a three-round burst. The M16A3 has the same characteristics as the M16A2, with the exception of the selector lever (SAFE, SEMI, and AUTO) and the addition of the automatic mode.

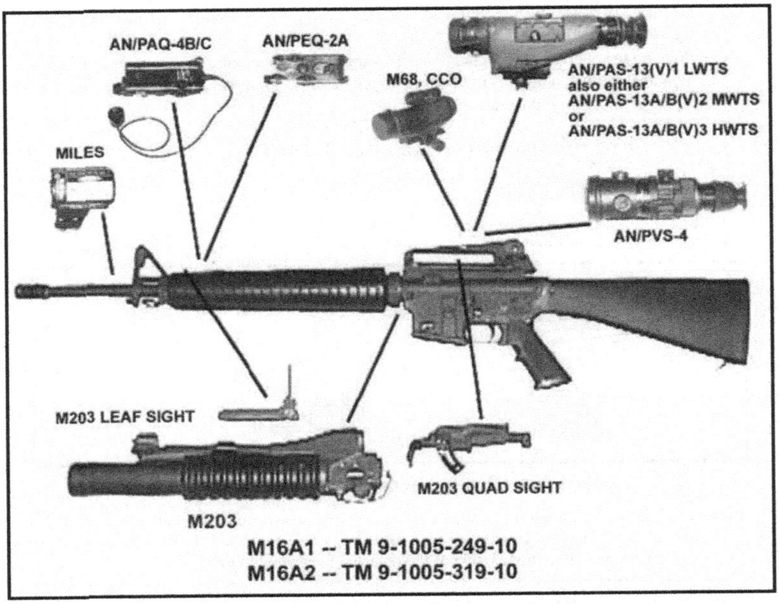

Figure 2-6. M16A2/A3 rifle with accessories.

MECHANICALLY ZEROING THE M16A2/A3 RIFLE

NOTE: Mechanically zeroing the weapon is only necessary when the weapon zero is questionable, the weapon is newly assigned to the unit, or the weapon sights have been serviced.

2-7. To mechanically zero an M16A2/A3 rifle—

NOTE: Reference the weapon components using the numbers listed in Figure 2-7.

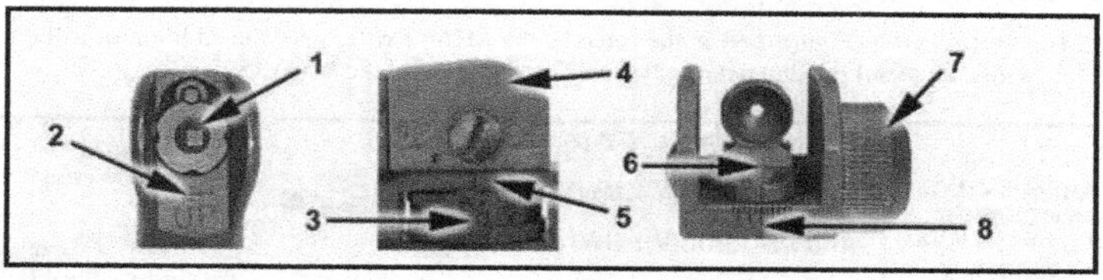

Figure 2-7. M16A2/A3 mechanical zero.

(1) Adjust the front sightpost (1) until the base of the front sightpost is flush with the front sightpost housing (2).
(2) Turn the elevation knob (3, shown as viewed from above) counterclockwise until the rear sight assembly (4) rests flush with the carrying handle and the 8/3 marking is aligned with the index line on the left side of the carrying handle.
(3) Position the apertures (5) so the unmarked aperture is up and the 0-200 meter aperture is down.
(4) Turn the windage knob (6) to align the index mark on the 0-200 meter aperture with the long center index line on the rear sight assembly.

BATTLESIGHT ZEROING THE M16A2/A3 RIFLE

2-8. To battlesight zero an M16A2/A3 rifle—

NOTE: Reference the weapon components using the numbers listed in Figure 2-8.

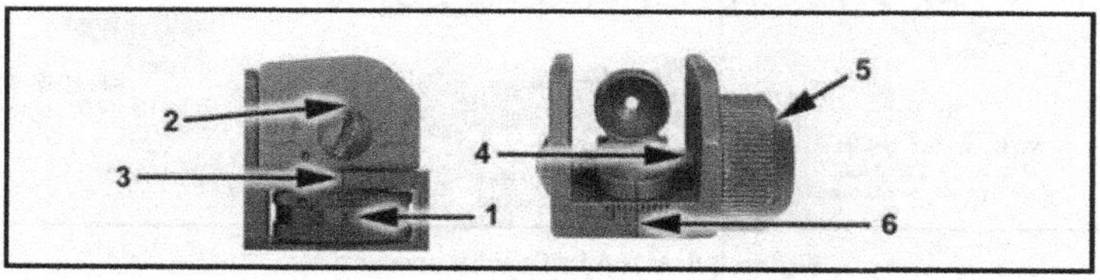

Figure 2-8. M16A2/A3 battlesight zero.

(1) Turn the elevation knob (1, shown as viewed from above) counterclockwise until the rear sight assembly (2) rests flush with the carrying handle and the 8/3 marking is aligned with the index line (3) on the left side of the carrying handle.
(2) Turn the elevation knob one more click clockwise.
(3) Position the apertures (4) so the unmarked aperture is up and the 0-200 meter aperture is down.
(4) Turn the windage knob (5) to align the index mark on the 0-200 meter aperture with the long center index line on the rear sight assembly.

2-9. Table 2-3 shows how much one click of elevation or windage will move the strike of the round at ranges from 25 to 600 meters.

Table 2-3. Point of impact for the M16A2/A3 rifle.

RANGE (m)	25	100	200	300	400	500	600
Elevation	3/4 in 0.9 cm	1 3/8 in 3.5 cm	2 3/4 in 7 cm	4 1/8 in 10.5 cm	5 1/2 in 14 cm	6 7/8 in 17.5 cm	8 1/4 in 20.9 cm
Windage	1/8 in 0.3 cm	1/2 in 1.25 cm	1 in 2.5 cm	1 1/2 in 3.8 cm	2 in 5 cm	2 1/2 in 6.3 cm	3 in 7.6 cm

M16A4 RIFLE

2-10. The M16A4 rifle (Figure 2-9) is the same as the M16A2 rifle, with the addition of a flat top upper receiver (with an integral rail and detachable carrying handle) and the M5 ARS.

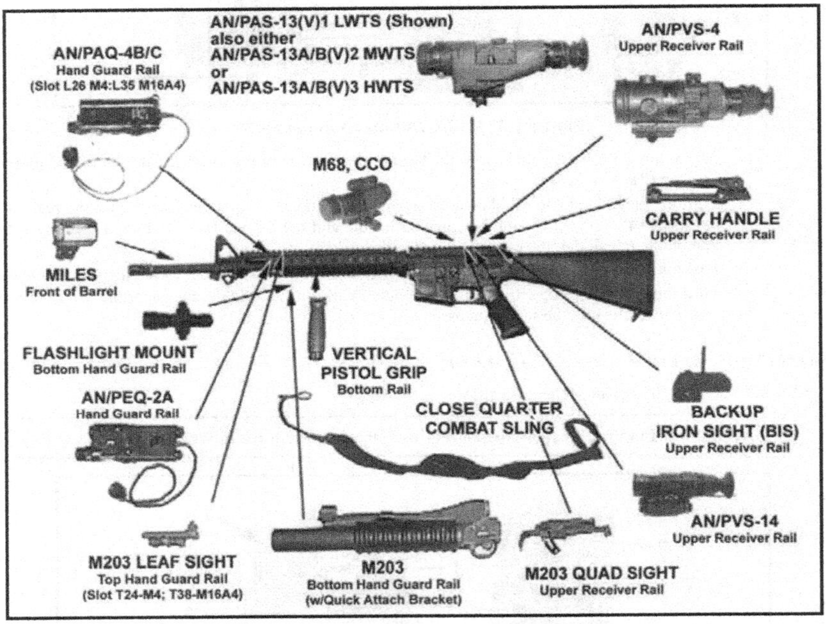

Figure 2-9. M16A4 rifle with accessories.

MECHANICALLY ZEROING THE M16A4 RIFLE

NOTE: Mechanically zeroing the weapon is only necessary when the weapon zero is questionable, the weapon is newly assigned to the unit, or the weapon sights have been serviced.

2-11. To mechanically zero an M16A4 rifle—

NOTE: Reference the weapon components using the numbers listed in Figure 2-10.

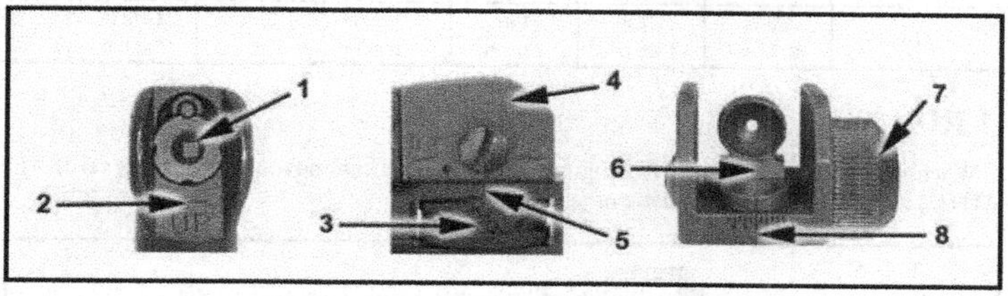

Figure 2-10. M16A4 mechanical zero.

(1) Adjust the front sightpost (1) until the base of the front sightpost is flush with the front sightpost housing (2).

(2) Turn the elevation knob (3, shown as viewed from above) counterclockwise until the rear sight assembly (4) rests flush with the carrying handle and the 6/3 marking is aligned with the index line (5) on the left side of the carrying handle.

(3) Position the apertures (6) so the unmarked aperture is up and the 0-200 meter aperture is down.

(4) Turn the windage knob (7) to align the index mark on the 0-200 meter aperture with the long center index line (8) on the rear sight assembly.

BATTLESIGHT ZEROING THE M16A4 RIFLE

2-12. To battlesight zero an M16A4 rifle—

NOTE: Reference the weapon components using the numbers listed in Figure 2-11.

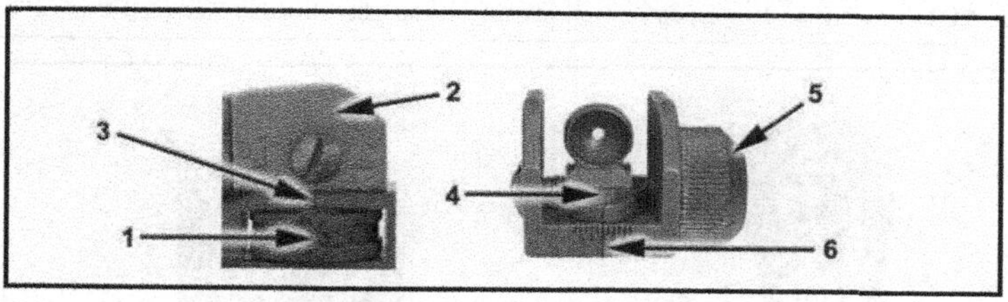

Figure 2-11. M16A4 battlesight zero.

(1) Turn the elevation knob (1, shown as viewed from above) counterclockwise until the rear sight assembly (2) rests flush with the detachable carrying handle and the 6/3 marking is aligned with the index line (3) on the left side of the detachable carrying handle.

(2) Turn the elevation knob two clicks clockwise so the index line on the left side of the detachable carrying handle is aligned with the "Z" on the elevation knob.

(3) Position the apertures (4) so the unmarked aperture is up and the 0-200 meter aperture is down.

(4) Turn the windage knob (5) to align the index mark on the 0-200 meter aperture with the long center index line (6) on the rear sight assembly.

2-13. Table 2-4 shows how much one click of elevation or windage will move the strike of the round at ranges of 25 to 600 meters.

Table 2-4. Point of impact for the M16A4 rifle.

RANGE (m)	25	100	200	300	400	500	600
Elevation	3/4 in 0.9 cm	1 3/8 in 3.5 cm	2 3/4 in 7 cm	4 1/8 in 10.5 cm	5 1/2 in 14 cm	6 7/8 in 17.5 cm	8 1/4 in 20.9 cm
Windage	1/8 in 0.3 cm	1/2 in 1.25 cm	1 in 2.5 cm	1 1/2 in 3.8 cm	2 in 5 cm	2 1/2 in 6.3 cm	3 in 7.6 cm

M16A1 RIFLE

2-14. When operating an M16A1 rifle (Figure 2-12), the firer can move the selector lever (SAFE, SEMI, and AUTO) to fire in the semiautomatic or automatic mode.

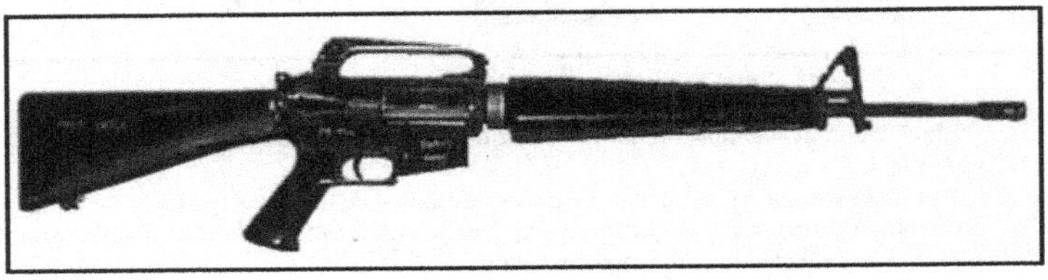

Figure 2-12. M16A1 rifle.

MECHANICALLY ZEROING THE M16A1 RIFLE

NOTE: Mechanically zeroing the weapon is only necessary when the weapon zero is questionable, the weapon is newly assigned to the unit, or the weapon sights have been serviced.

2-15. To mechanically zero an M16A1 rifle—

NOTE: Reference the weapon components using the numbers listed in Figure 2-13.

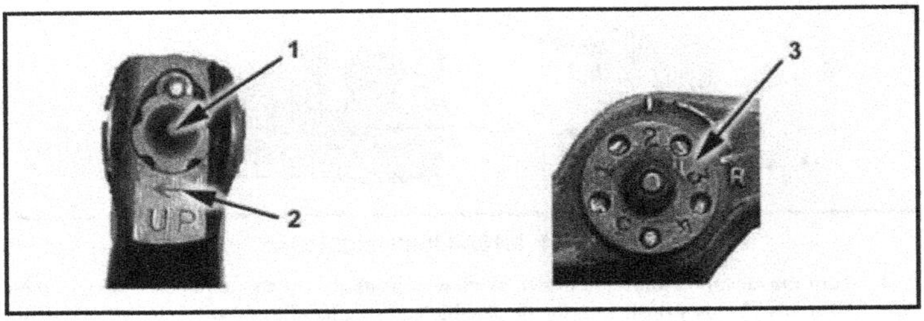

Figure 2-13. M16A1 mechanical zero.

(1) Adjust the front sightpost (1) until the base of the front sightpost is flush with the front sightpost housing (2).
(2) Adjust the front sightpost 11 clicks in the direction of UP.
(3) Turn the rear sight windage drum (3) left until it stops.
(4) Turn the windage drum right 17 clicks so the rear sight is approximately centered.

Battlesight Zeroing the M16A1 Rifle

2-16. Soldiers should use the aperture marked “L” (Figure 2-14) to battlesight zero their weapons.

Figure 2-14. M16A1 battlesight zero.

2-17. Table 2-5 shows how much one click of elevation or windage will move the strike of the round at ranges of 25 to 500 meters using standard sights or the low light level sight system.

Table 2-5. Point of impact for the M16A1 rifle.

STANDARD SIGHTS						
Range (m)	25	100	200	300	400	500
Elevation	17/64 in 0.7 cm	1 3/32 in 2.8 cm	2 13/64 in 5.6 cm	3 9/32 in 8.4 cm	4 3/8 in 11.2 cm	5 15/32 in 14 cm
Windage	17/64 in 0.7 cm	1 3/32 in 2.8 cm	2 13/64 in 5.6 cm	3 9/32 in 8.4 cm	4 3/8 in 11.2 cm	5 15/32 in 14 cm
LOW LIGHT LEVEL SIGHT SYSTEM						
Elevation	23/64 in 0.9 cm	1 3/4 in 3.5 cm	2 3/4 in 7 cm	5 1/4 in 10.5 cm	7 in 17.7 cm	8 3/4 in 22.2 cm
Windage	17/64 in 0.7 cm	1 3/32 in 2.8 cm	2 13/64 in 5.6 cm	3 9/32 in 8.4 cm	4 3/8 in 11.2 cm	5 15/32 in 14 cm

SECTION II. ACCESSORY MOUNTING

The M4/M5 ARS and rail grabbers are designed to mount accessories on M16- and M4-series weapons.

M4/M5 ADAPTER RAIL SYSTEM

2-18. The ARS provides a secure mounting point for various accessories that may be mounted on the weapon's top, left, and right. The M4 ARS (Figure 2-15) consists of lightweight sections that replace the standard handguards on the M4 carbine. The M5 ARS replaces the handguards on the M16A4 rifle.

NOTES:

1. Only the armorer is authorized to remove the ARS from a weapon. However, the user may remove only the lower assembly to perform preventive maintenance checks and services (PMCS). If the top section is removed, the accessories mounted on the rail will lose their zero.
2. To prevent the screw from vibrating loose during firing, apply locking compound to the original screw or use a new screw when reinstalling the ARS.
3. For optics and lasers to retain zero, the ARS must be properly installed. Mounted accessories will not retain zero unless proper procedures are followed, as outlined in TM 9-1005-319-23& P.
4. Accessories may be mounted on the right side of the ARS, but are not currently supported with zeroing procedures. Further, the bottom rail of the ARS will not retain zero. Only accessories that do not require zero retention, such as a flashlight or vertical pistol grip, can be mounted on the bottom rail.

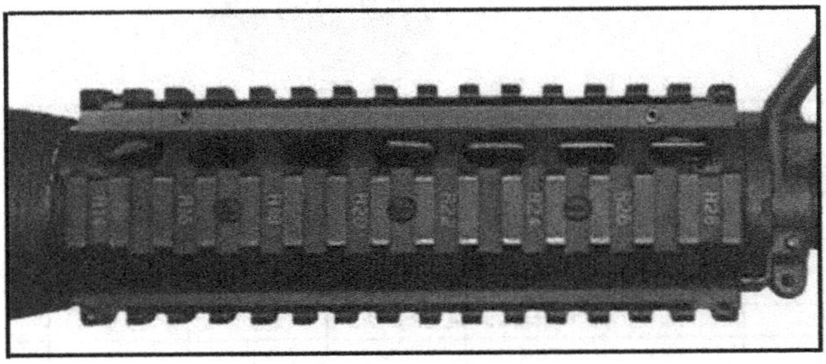

Figure 2-15. Adapter rail system.

Rail Covers

2-19. Rail covers protect the firer's hands from direct contact with the metal parts of the ARS and protect the ARS surfaces from excess wear and damage. The rail covers are available in 11-, 9-, 6-, 5-, and 4-rib sections. For ease of reference, the shorter lengths can be referred to by the number of ribs along their outer surfaces. 11-, 9-, and 5-rib sections are shown in Figure 2-16.

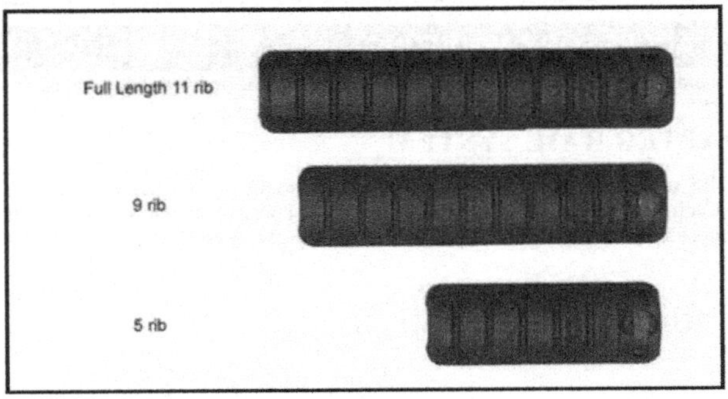

Figure 2-16. M5 rail covers/heat shields.

2-20. ARS rail covers can be quickly attached and detached from the ARS. A spring latch at one end of each rail cover automatically engages cutouts at the end of each rail. To slide the rail cover beyond a cutout or to remove it, apply thumb pressure to the center of the spring latch, and slide it in the desired direction.

NOTES: 1. Keep unused bottom, left, and right rail sections covered with 11-rib rail cover sections.

2. If any accessories are mounted on a rail, cover the remaining rail surface with an appropriately-sized rail cover.

3. If a BUIS is installed, permanently remove the top full length rail cover, and replace it with a shorter rail cover to protect the nonfiring hand when the barrel is hot.

WARNING

When firing at high rates of sustained fire, the barrel and metal components of the ARS can become hot enough to inflict serious burns. Cover exposed metal portions with plastic rail covers. Use the vertical pistol grip (Figure 2-18) during heavy, sustained fire.

Numbered Recoil Grooves

2-21. The even numbered recoil grooves on each rail of the ARS are sequentially numbered within the recoil grooves themselves (Figure 2-17). Each number is preceded by a letter prefix indicating a specific slot on the ARS. The numbers on the top rail have a "T" prefix, while those on the bottom have a "B" prefix. Additionally, the numbers on the rail to the firer's left have an "L" prefix, while those to the firer's right have an "R" prefix. These addresses—

- Assist the user in remounting an accessory in the same position.
- Allow standardization in the location of mounted accessories.
- Identify reference points for discussions on accessory mounting locations.

Figure 2-17. Address markings on the adapter rail system.

NOTE: Within the notches, each ARS also contains holes that are threaded ¼-inch deep, with 20 threads per inch (Figure 2-17). This is the standard thread size for a camera tripod adapter, which is used to attach standard camera or video accessories. For example, an M4 MWS with an NVD mounted may be attached to a standard camera tripod for hands-free support during long periods of surveillance.

VERTICAL PISTOL GRIP

2-22. Each ARS comes with a vertical pistol grip. To install the pistol grip—

NOTE: Reference the weapon components using the numbers listed in Figure 2-18.

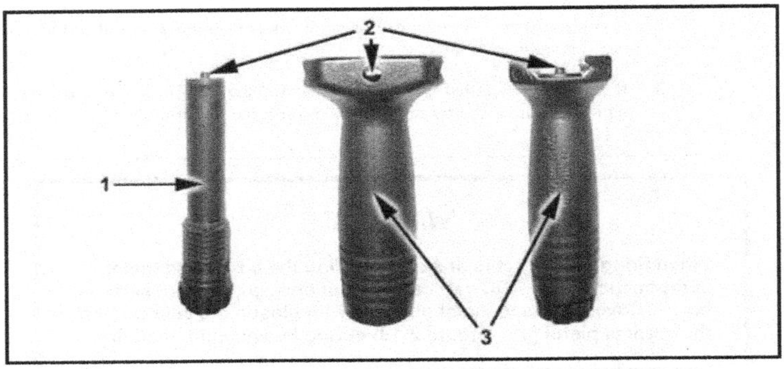

Figure 2-18. Vertical pistol grip.

(1) Remove the rail cover.

(2) Unscrew the pistol grip lock (1) until the tip (2) is no longer visible through the hole in the pistol grip.

(3) Slide the pistol grip (3) onto the ARS.

NOTE: The pistol grip will cover five notches on the ARS.

(4) Align the tip on the top of the pistol grip lock (1) with a notch.

(5) Hand-tighten.

NOTE: For further information about these accessories, refer to TM 9-1005-319-10.

RAIL GRABBERS

2-23. The Insight (Figure 2-19) and Picatinny (Figure 2-21) rail grabbers are designed to mount accessories onto the M16A4 and M4-series weapons. These rail grabbers enable the user to attach accessories on the upper receiver and on all four sides of the ARS, while retaining zero (when installed and tightened properly). Once zeroed, they will retain zero—even when removed from the weapon—as long as the rail grabber is not separated from the accessory and is remounted on the same notch it was zeroed on.

NOTE:
1. Retighten the rail grabber and accessory after firing the first three rounds to fully seat both.
2. The bottom rail of the ARS will not retain zero.

INSIGHT RAIL GRABBER

2-24. The Insight rail grabber (Figure 2-19) can be mounted where the tightening screw (3, Figure 2-19) is on either the left or right side (when top-mounted) or top or bottom (when mounted on the left side) so it does not interfere with weapon operation.

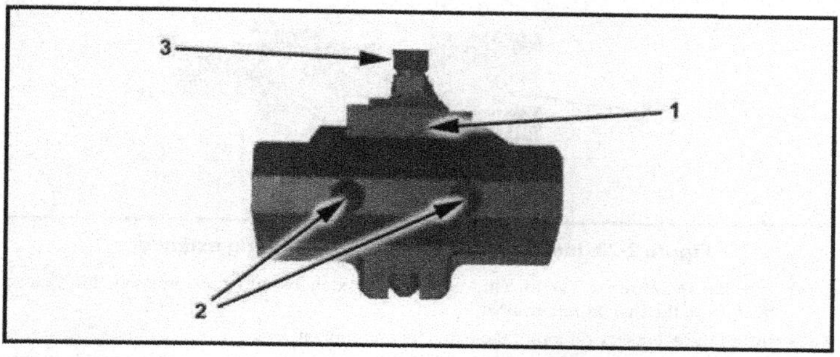

Figure 2-19. Insight rail grabber.

2-25. Unless command-directed, all devices in a unit do not have to be mounted in the same location as long as individual users record or mark (with paint markers or grease pencils) the mounting location on their weapons to avoid unnecessary rezeroing.

CAUTION

Both of the holes (2, Figure 2-19) located in the top of the rail grabber can be used to mount accessories, but the hole closest to the muzzle must be used. This ensures that the majority of the rail grabber is supporting the accessory to prevent damage to the accessory.

NOTES:

1. The Insight rail grabber must fully rest on the ARS in order to retain zero. The locking clamp (1, Figure 2-19) must grasp the ARS, and the screw that tightens the rail grabber must be tightened with a field tool, such as a multipurpose tool.
2. Even if the rail grabber is resting entirely on the ARS, accessories should not make contact with the front sight assembly or the collar of the barrel. The vibrations that occur during firing will interfere with the rail grabber's and accessory's zero retention capabilities.

MILES Training Extender for the Insight Rail Grabber

2-26. The training extender (Figure 2-20) is used to elevate an accessory above the MILES laser during force-on-force training. The training extender is only used when the Insight rail grabber is top-mounted.

2-27. To install the extender and accessories—

NOTE: Reference the weapon components using the numbers listed in Figure 2-20.

Figure 2-20. Insight rail grabber MILES training extender.

(1) Use the thumbscrew (1) to hand-tighten the extender into the mounting hole closest to the muzzle on the Insight rail grabber.

(2) Install the accessory on top of the extender and tighten.

NOTE: The AN/PEQ-2A/B and AN/PAQ-4B/C must be zeroed before and after using the MILES training extender.

Picatinny Rail Grabber

2-28. The Picatinny rail grabber (Figure 2-21) must fully rest on the ARS in order to retain zero. The locking clamp must grasp the ARS, and the torque-limiting knob (2, Figure 2-21) that tightens the rail grabber must be hand-tightened until it clicks twice.

Figure 2-21. Picatinny rail grabber.

CAUTION

Both of the holes located in the top of the rail grabber (1, Figure 2-21) can be used to mount accessories, but the hole closest to the muzzle must be used for the Picatinny rail grabber. This ensures that the majority of the rail grabber is supporting the accessory to prevent damage to the accessory. It also allows the torque-limiting knob to be mounted on either the left or right side (when top-mounted) or top or bottom (when side-mounted) so that it will not interfere with weapon operation.

2-29. Unless command-directed, all devices in a unit do not have to be mounted in the same location as long as individual users record or mark (with paint markers or grease pencils) the mounting location on their weapons to avoid unnecessary rezeroing.

NOTE: Even if the rail grabber is resting entirely on the ARS, accessories should not make contact with the front sight assembly or the collar of the barrel. The vibrations that occur during firing will interfere with the rail grabber's and accessory's zero retention capabilities.

SECTION III. ACCESSORIES

Table 2-6 shows the weapons with which the various accessories are compatible, along with the mounting device and TM number. Table 2-7 shows select characteristics of the various accessories.

Table 2-6. Accessory compatibility and mounting.

ACCESSORY	TECHNICAL MANUAL	M4 MWS , M16A4	M4/M4A1	M16A1/A2/A3
M68, CCO	TM 9-1240-413-13&P	* Upper receiver	* Upper receiver	M16 mount
AN/PVS-14	TM 11-5855-306-10	***Upper receiver	N/A	N/A
AN/PAC-4C	TM 11-5855-301-12&P	** Rail grabber	Bracket assembly	Bracket assembly
AN/PEQ-2A/B	TM 11-5855-308-12&P	** Rail grabber	Bracket assembly	Bracket assembly
AN/PAQ-4	TM 11-5855-261-10	** Rail grabber	Bracket assembly	Bracket assembly
AN/PAS-13B(V1), LWTS	TM 11-5855-312-10	Upper receiver	Upper receiver	M16 mount
AN/PAS-13B(V3), HWTS	TM 11-5855-312-10	Upper receiver	Upper receiver	M16 mount
AN/PAS-13C(V1), LWTS	TM 11-5855-316-10	Upper receiver	Upper receiver	M16 mount
AN/PAS-13C(V3), HWTS	TM 11-5855-316-10	Upper receiver	Upper receiver	M16 mount
AN/PAS-13D(V1), LWTS	TM 11-5855-317-10	Upper receiver	Upper receiver	M16 mount
AN/PAS-13D(V3), HWTS	TM 11-5855-317-10	Upper receiver	Upper receiver	M16 mount
AN/PVS-4(A)	TM 11-5855-213-10	Upper receiver	Upper receiver	M16 mount
BUIS		Upper receiver	Upper receiver	N/A
ACOG		Upper receiver	Upper receiver	M16 mount

* With a half-moon spacer installed.
** Picatinny or Insight rail grabbers may be used.
*** If used in conjunction with the CCO, the CCO will mount on the top rail of the ARS.

Table 2-7. Characteristics of various accessories.

ACCESSORY	WEIGHT	LENGTH	HEIGHT	RANGE
M68, CCO	6.2 oz	4.9 in	2.5 in	300 m
AN/PVS-14	14.8 oz	4.5 in	2.25 in	
AN/PAC-4C	5.78 oz	5.5 in	1.2 in	*600 m +
AN/PEQ-2A/B	7.5 oz	6.4 in	1.2 in	*600 m +
LWTS	4.1 lb	15.5 in	6.25 in	1,600 m +
HWTS	4.5 lb	18 in	6.25 in	1,600 m +
AN/PVS-4	3.8 lb	9.45 in	4.7 in	*600 m
ACOG	15.1 oz	5.8 in	2.17 in	800 m

*Actual range is dependent upon ambient light, NVGs, and background contrast.

M68 CLOSE COMBAT OPTIC

2-30. The M68 CCO (Figure 2-22) is a reflex (nontelescopic) sight that is designed for the "two eyes open" method of sighting, but can be shot with only one eye open. The red dot aiming point follows the horizontal and vertical movement of the firer's eye, while remaining fixed on the target.

NOTES: 1. Retighten the torque-limiting knob after firing the first three rounds to fully seat the M68.

2. No centering or focusing is required beyond 50 meters.

Figure 2-22. M68 close combat optic.

Mounting on the M16A4 Rifle or M4-Series Carbine

2-31. The M68 mounts directly on the integrated rail on top of M16A4 rifles and M4-series carbines (Figure 2-23). The half-moon spacer (1, Figure 23) must be installed to raise the M68 above the front sightpost. The Soldier's preference dictates the notch that the M68 is mounted on. Although any notch is acceptable, testing has shown that the farther away the M68 is from the Soldier's eyes, the better his field of view.

2-32. To mount the M68 on an M16A4 rifle or M4-series carbine (Figure 2-23)—

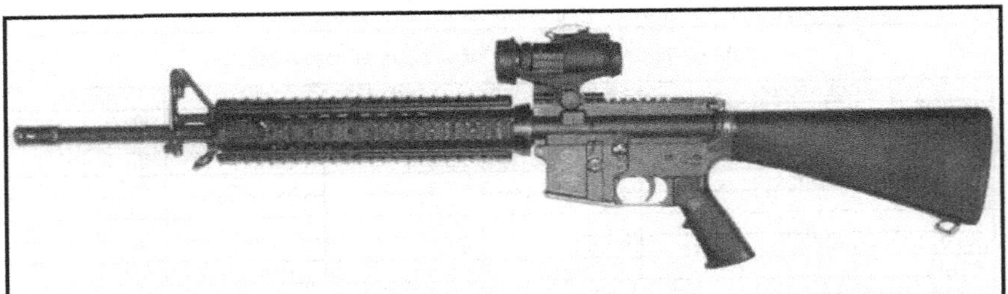

Figure 2-23. Mounting the M68 on an M16A4 rifle or M4-series carbine.

(1) Remove the carrying handle.

(2) Align the locking bar with a notch.

(3) Tighten the torque-limiting knob until it clicks twice.

NOTE: If the M68 is remounted on the same notch, it will retain zero.

Mounting on the M16A1/A2/A3 Rifle

2-33. The M68 mounts on the M16A1 mounting bracket (1, Figure 2-24) that attaches to the carrying handle on the M16A1/A2/A3 rifle. The half-moon spacer should not be installed, but will not hinder firing performance.

2-34. To mount the M68 on an M16A1/A2/A3 rifle—

NOTE: Reference the weapon components using the numbers listed in Figure 2-24.

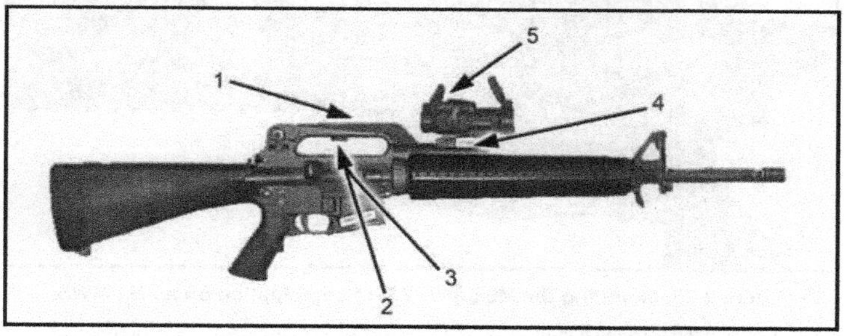

Figure 2-24. Mounting the M68 on an M16A1/A2/A3 rifle.

(1) Firmly hand-tighten the bracket (1), O-ring (2), and machine screw (3).

(2) Align the locking bar (4) under the M68 with the notch in the rail, ensuring that the rotary switch (5) is facing the firer.

(3) Tighten the torque-limiting knob (not shown here) until it clicks twice.

Mounting on the M16A4 Rifle or M4 MWS

2-35. This combination (Figure 2-25) is an effective means of engaging targets during hours of limited visibility. The brightness knob on the M68 should be on the lowest setting (2 or 3) that presents the red dot clearly when viewed through the AN/PVS-14. When employing the AN/PVS-14, Soldiers must consider the following factors:

- The AN/PVS-14 should be mounted where the firer can acquire a good sight picture while performing the integrated act of shooting.
- The M68 can be mounted and zeroed on any slot forward of the AN/PVS-14 as long as the rail grabber fully rests on the ARS and the M68 lens does not rest on the front sightpost.
- The closer the AN/PVS-14 is mounted to the M68, the larger the field of view will be.
- In order to get a clear sight picture with this configuration, fine adjustments must be made to the AN/PVS-14's range focus, gain-control, and diopter.
- The eyecup should be exchanged with the eye guard that is shipped with the AN/PVS-14 to reduce the light signature from the display when not viewing.
- The red dots on settings 2 and 3 project a negligible light signature at night, which can only be seen through an NVD. Settings 4 through 10 project a noticeable signature detectable by opposing forces using NVDs.
- The brighter the dot, the larger the blooming effect becomes in the AN/PVS-14. The blooming effect reduces the Soldier's field of view and prevents him from seeing targets behind the blooming.

2-36. To mount the M68 and AN/PVS-14 on an M16A4 rifle or M4 MWS—

NOTE: Reference the weapon components using the numbers listed in Figure 2-25.

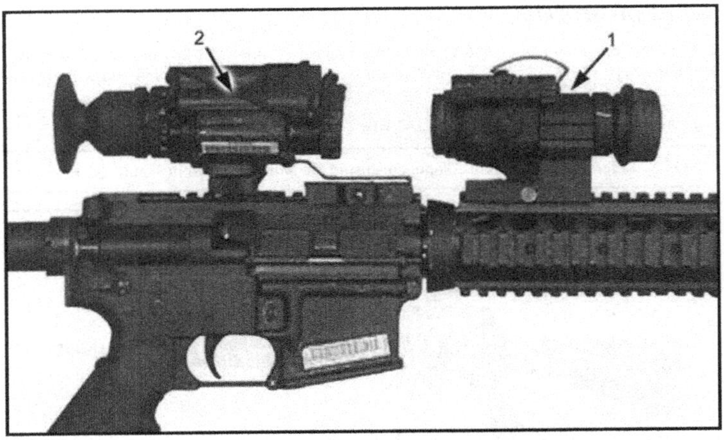

Figure 2-25. Mounting the M68/AN/PVS-14 combination on an M4 MWS.

(1) Remove the carrying handle.
(2) Mount the M68 (1) by tightening the torque-limiting knob.
(3) Mount the AN/PVS-14 (2) where the best field of view is achieved.
(4) Once the preferred location for the M68 is located, zero the M68 to that notch (if different from the notch the M68 was previously zeroed on).

NOTE: For more information about the M68, refer to TM 9-1240-413-13&P. For more information about the AN/PVS-14, refer to TM 11-5855-306-10.

AN/PAQ-4B/C INFRARED AIMING LIGHT

2-37. The AN/PAQ-4B/C IR aiming light (Figure 2-26) projects an IR laser beam that is invisible to the naked eye, but can be seen with NVDs. This aiming light works with the AN/PVS-7-series goggles and the AN/PVS-14. The AN/PAQ-4B/C mounts on various M16-/M4-series weapons with mounting brackets or rail grabbers.

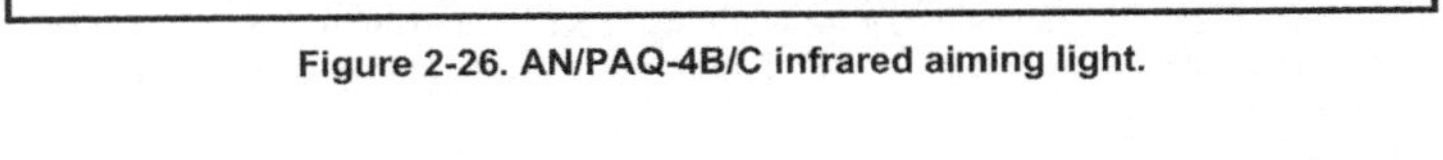

Figure 2-26. AN/PAQ-4B/C infrared aiming light.

MOUNTING ON THE M16A4 RIFLE OR M4 MWS

2-38. The Picatinny (1, Figure 2-27) or Insight (5, Figure 2-27) rail grabber is used to mount the AN/PAQ-4B/C on the ARS.

2-39. To mount the AN/PAQ-4B/C on the ARS—

NOTE: Reference the weapon components using the numbers listed in Figure 2-27.

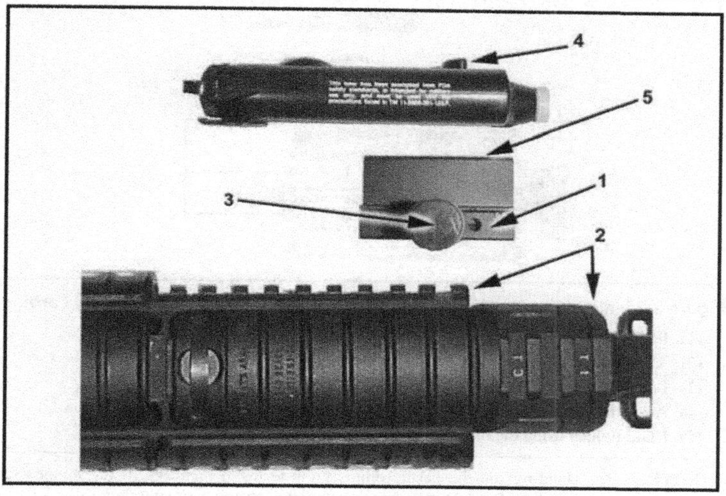

Figure 2-27. Mounting the AN/PAQ-4B/C on the M4 MWS top or left.

(1) Mount the rail grabber all the way forward on the top or either side of the ARS (2), ensuring that it does not extend beyond the end of the ARS.

NOTE: The AN/PAQ-4B/C will not retain zero if the rail grabber extends beyond the end of the integrated rail when mounted.

(2) Tighten the torque-limiting knob (3) until it clicks twice.

(3) Align the thumbscrew (4) on the AN/PAQ-4B/C with the thumbscrew hole in the rail grabber nearest the muzzle.

2-40. The mounting procedures are identical for the M16A4 rifle and M4 MWS. The remote switch should be attached to the weapon where it will not interfere with the functioning of the weapon or hinder the firer's ability to fire the weapon. If the aiming light and rail grabber are removed as a whole unit and mounted onto the same rail, the system will retain zero.

NOTE:
1. If the rail grabber and AN/PAQ-4B/C are separated, the AN/PAQ-4B/C must be rezeroed to the weapon.
2. For further information, refer to TM 11-5855-301-12&P.

MOUNTING ON THE M16A1/A2/A3 RIFLE OR M4 CARBINE

2-41. To mount the AN/PAQ-4B/C on an M16A1/A2/A3 rifle or M4 carbine—

NOTE: Reference the weapon components using the numbers listed in Figure 2-28.

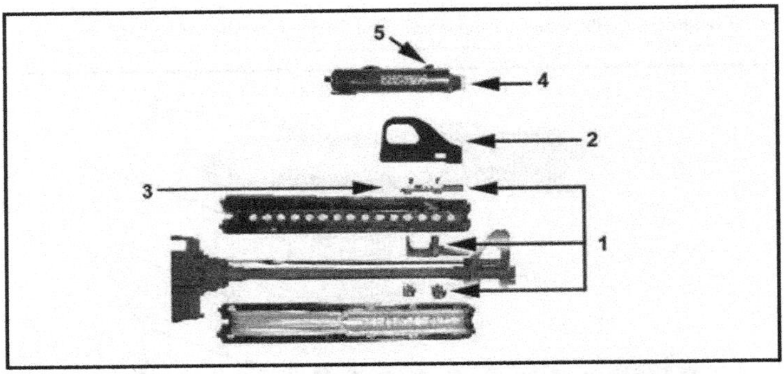

Figure 2-28. Mounting the AN/PAQ-4B/C on the M16A1/A2/A3 rifle and M4 carbine.

(1) Install the bracket assembly (1).

(2) Align the switch lever shroud (2) with the notches on the mounting rail (3).

(3) Lower the ON/OFF switch.

(4) Align the AN/PAQ-4B/C (4) with the notches on the switch lever shroud.

(5) Hand-tighten using the thumbscrew (5).

NOTE:
1. Hand-tighten the plastic thumbscrew to avoid breakage. If the thumbscrew is metal, tool-tightening is recommended to ensure zero retention.
2. To ensure zero retention, retighten the thumbscrew after firing a few rounds.

2-42. The remote switch should be attached to the weapon where it will not interfere with the functioning of the weapon or hinder the firer's ability to fire the weapon.

AN/PEQ-2A/B TARGET POINTER/ILLUMINATOR/AIMING LIGHT

2-43. AN/PEQ-2A and AN/PEQ-2B aiming lights (Figure 2-29) are Class IIIb laser devices that emit a collimated beam of IR light for precise aiming and a separate IR beam for illumination of the target or target area. Both beams can be independently zeroed to the weapon and to each other. The beams can be operated individually or in combination in both high and low power settings.

NOTES:
1. The IR illuminator is equipped with an adjustable bezel to vary the size of the illumination beam based on the size and distance of the target.
2. A safety block is provided for training purposes to limit the operator from selecting high power modes of operation.

2-44. The aiming lights are used with NVDs and can be used as handheld illuminators/ pointers or mounted on the weapon with the included brackets and accessory mounts. In the weapon-mounted mode, the aiming lights can be used to direct fire and to illuminate and designate targets.

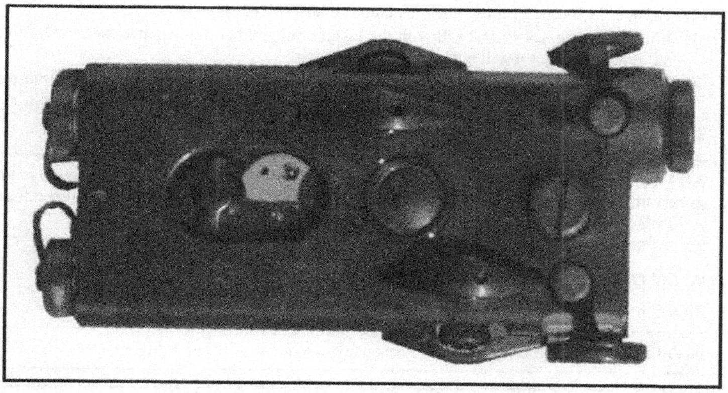

Figure 2-29. AN/PEQ-2A/B target pointer/illuminator/aiming light.

Mounting on the M16A4 Rifle or M4 MWS

2-45. The Picatinny (1, Figure 2-30) or Insight (2, Figure 2-30) rail grabber may be used to mount the AN/PEQ-2A/B on the ARS.

2-46. To mount the AN/PEQ-2A/B on the ARS—

NOTE: Reference the weapon components using the numbers listed in Figure 2-30.

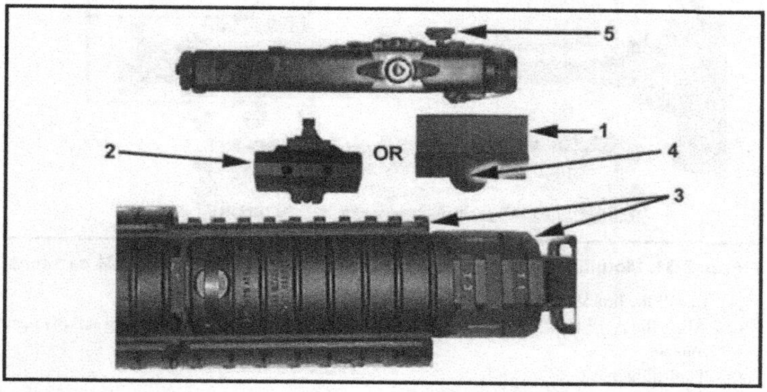

Figure 2-30. Mounting the AN/PEQ-2A/B on the M16A4 rifle and M4 MWS.

(1) Mount the rail grabber all the way forward on the top or either side of the ARS, ensuring that it does not extend beyond the end of the ARS.

NOTE: The AN/PEQ-2A/B will not retain zero if the rail grabber extends beyond the end of the integrated rail when mounted.

(2) Tighten the torque-limiting knob (4) until it clicks twice.

(3) If installing the AN/PEQ-2A/B with the Insight rail grabber, tool-tighten the AN/PEQ-2A/B and rail grabber so that it will not come loose.

(4) Align the thumbscrew (5) on the AN/PEQ-2A/B with the hole that is closest to the front sight assembly located on the top of the rail grabber.

2-47. The mounting procedures are identical for the M16A4 rifle and M4 MWS.

NOTE: If the aiming light and rail grabber are removed as a whole unit and mounted onto the same rail, the system will retain zero. If the rail grabber and AN/PEQ-2A/B are separated, the AN/PEQ-2A/B must be rezeroed to the weapon.

MOUNTING ON THE M16A1/A2/A3 RIFLE OR M4 CARBINE

2-48. To mount the AN/PEQ-2A/B on an M16A1/A2/A3 rifle or M4 carbine—

NOTE: Reference the weapon components using the numbers listed in Figure 2-31.

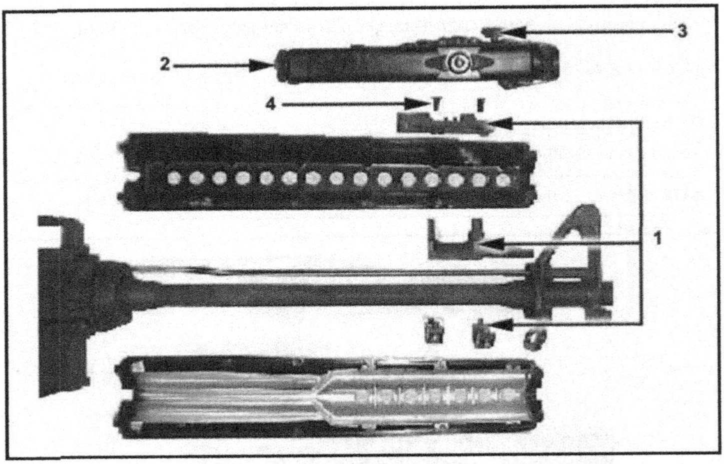

Figure 2-31. Mounting the AN/PEQ-2A/B on M16A1/A2/A3 rifles and M4 carbines.

(1) Install the bracket assembly (1).

(2) Align the AN/PEQ-2A/B (2) thumbscrew (3) with the hole in the mounting rail (4) nearest to the muzzle.

(3) Tool-tighten.

2-49. The remote switch should be attached to the weapon where it will not interfere with the functioning of the weapon or hinder the firer's ability to fire the weapon.

NOTE: To ensure zero retention, retighten the rail grabber and thumbscrew after firing a few rounds.

Use in Conjunction With MILES

2-50. When conducting MILES training with the Insight rail grabber (1, Figure 2-32) or bracket assembly (4, Figure 2-32), the AN/PEQ-2A/B is attached to the M16-/M4-series weapon using the training extender bracket (2, Figure 2-32). The training extender is not required when—

- Mounting the AN/PEQ-2A/B onto the side of the MWS.
- Using the Picatinny rail grabber.

2-51. To attach the training extender, hand-tighten it by turning the thumb wheel (3, Figure 2-32) on the training extender clockwise.

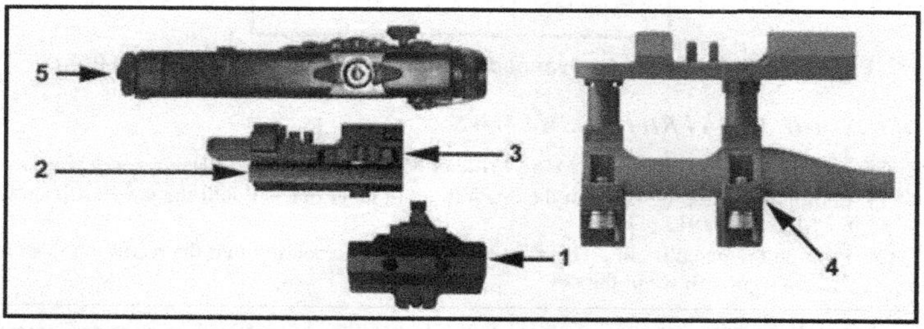

Figure 2-32. MILES training extender bracket installation on M16-/M4-series weapons.

NOTES: 1. The AN/PEQ-2A/B and AN/PAQ-4B/C must be zeroed before and after using the MILES training extender.

2. For further information, refer to TM 11-5855-308-12&P.

AN/PEQ-15 ADVANCED TARGET POINTER/ILLUMINATOR AIMING LIGHT

2-52. The AN/PEQ-15 advanced target pointer/illuminator aiming light (ATPIAL, Figure 2-33) is a multifunctional laser that emits visible or IR light for precise weapon aiming and target/area illumination. This ruggedized system can be used as a handheld illuminator/pointer or can be mounted to weapons equipped with a MIL-STD-1913 rail. Unlike the AN/PAQ-4 and AN/PEQ-2A/2B, the AN/PEQ-15 has an integrated rail grabber molded into the body.

NOTE: The AN/PEQ-15 can be used during force-on-force training in the low power modes only. High power modes can be used on live-fire ranges exceeding 220 meters only. Refer to TM 9-5855-1914-13&P for more information.

2-53. The AN/PEQ-15 ATPIAL's visible aiming laser provides for active target acquisition in low light conditions and close quarters combat situations, and allows users to zero using the borelight without using NVDs. When used in conjunction with NVDs, its IR aiming and illumination lasers provide for active, covert target acquisition in low light or complete darkness.

NOTES: 1. The ATPIAL visible and IR aiming lasers are aligned. A single set of adjusters moves both aiming beams, and the user can boresight/zero using either aiming laser.

2. In addition to momentary and continuous modes of operation, the IR illuminator may be programmed to operate at 1, 2, 4, or 8 pulses per second to eliminate confusion on the battlefield.

Figure 2-33. AN/PEQ-15 advanced target pointer illuminator aiming light.

MOUNTING ON THE M16A4 RIFLE OR M4 MWS

2-54. To mount the AN/PEQ-15 on an M16A4 rifle or M4 MWS (Figure 2-34)—

(1) Loosen the clamping knob on the integral rail grabber bracket until the jaws sufficiently fit over the MIL-STD-1913 rail.

(2) Position the integral rail grabber bracket on the rail, ensuring that the recoil lug is seated in the desired recoil groove of the rail.

NOTE: The ATPIAL may be positioned anywhere on the rail where it is most convenient for the operator; however, ensure that it does not extend beyond the end of the ARS or touch the collar of the barrel. The AN/PEQ-15 will not retain zero if the rail grabber extends beyond the end of the integrated rail when mounted.

---

(3) Turn the clamping knob clockwise to tighten.

(4) Tool-tighten the AN/PEQ-15 onto the MIL-STD-1913 rail to ensure that it does not come loose.

NOTE: Failure to properly secure the AN/PEQ-15 to the rail may lead to zero retention issues.

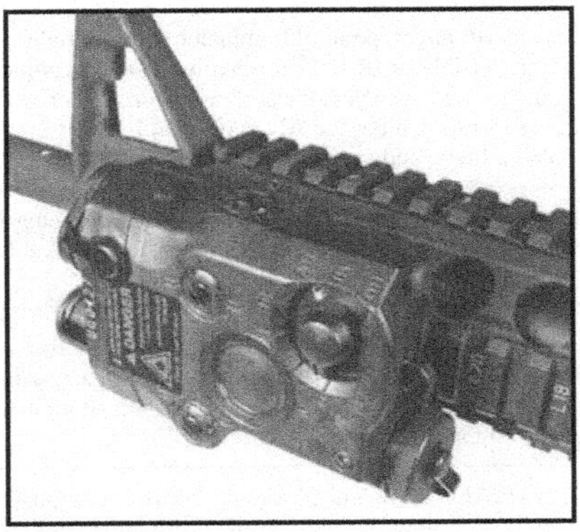

Figure 2-34. AN/PEQ-15 mounted on M4 carbine.

NOTE: If the ATPIAL is removed from the rail, the operator must note the notch that it was zeroed on, and return it to the same position to retain zero.

AN/PAS-13B/C/D (V1) LIGHT WEAPON THERMAL SIGHT AND AN/PAS-13B/C/D (V3) HEAVY WEAPON THERMAL SIGHT

2-55. The AN/PAS-13B/C/D (V1) light weapon thermal sight (LWTS) and the AN/PAS-13B/C/D (V3) heavy weapon thermal sight (HWTS) (Figure 2-35) are silent, lightweight, compact, and durable battery-powered IR imaging sensors that operate with low battery consumption. The AN/PAS-13B/C/D (V1) is used on riflemen's M4s and M16A4s. The AN/PAS-13B/C/D (V3) is used on squad leaders' M4s and M16A4s only.

NOTE: Both the LWTS and the HWTS are referred to henceforth as a singular TWS. For more information, refer to TM 11-5855-312-10, TM 11-5855-316-10, and TM 11-5855-317-10.

2-56. The TWS is capable of target acquisition under conditions of limited visibility, such as darkness, smoke, fog, dust, and haze. The TWS operates effectively both at night and during the day. The TWS is composed of two functional groups: the telescope and the basic sensor.

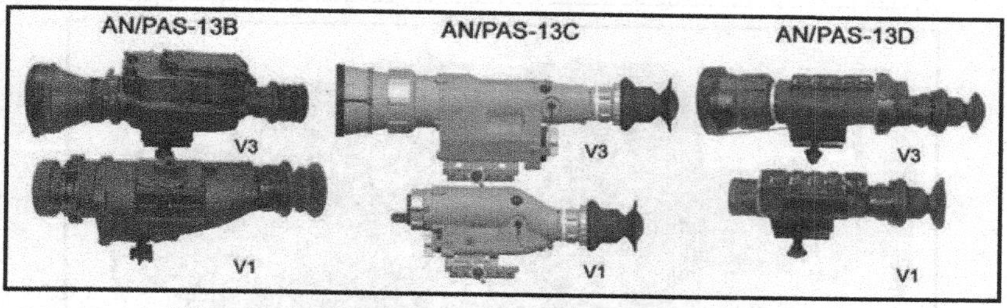

Figure 2-35. AN/PAS-13B/C/D (V1) light weapon thermal sight and AN/PAS-13B/C/D (V3) heavy weapon thermal sight.

MOUNTING THE TWS ON AN M16A4 RIFLE OR M4 CARBINE

2-57. To mount the TWS on an M16A4 rifle or M4 carbine—

NOTE: Reference the weapon components using the numbers listed in Figure 2-36.

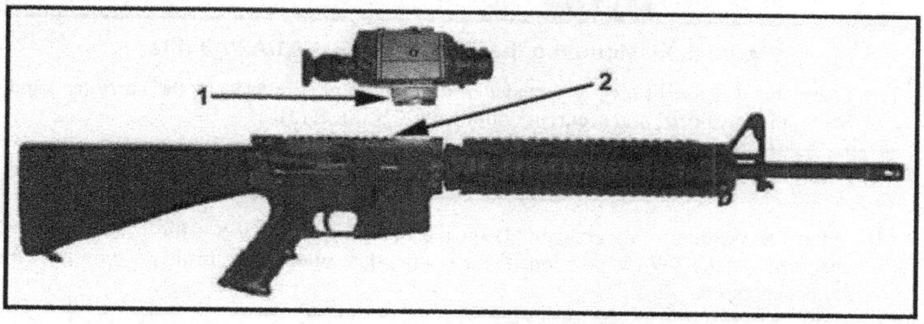

Figure 2-36. Mounting the TWS on M16A4 rifle or M4 carbine.

(1) Align the Picatinny rail grabber with the spacer on the bottom of the TWS with a notch on the integrated rail of the M16A4 rifle or M4 carbine, ensuring that the TWS is positioned to accommodate an effective firing position once the eyecup is depressed.

NOTE: The TWS will not retain zero if the rail grabber extends beyond the end of the integrated rail when mounted.

(2) Tighten the torque-limiting knob clockwise until it clicks twice.

NOTE: Retighten the rail grabber after firing a few rounds to ensure that the sight is fully seated.

2-58. The mounting procedures are identical for the M16A4 rifle and M4 MWS.

MOUNTING THE TWS ON AN M16A1/A2/A3 RIFLE

2-59. To mount the TWS on an M16A1/A2/A3 rifle—

NOTE: Reference the weapon components using the numbers listed in Figure 2-37.

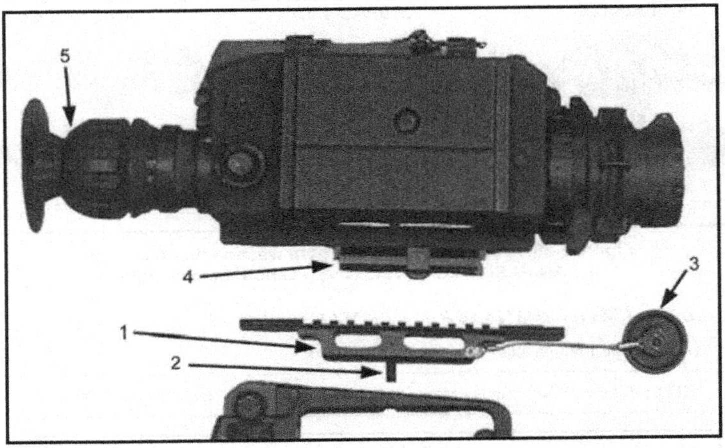

Figure 2-37. Mounting the TWS on an M16A1/A2/A3 rifle.

(1) Insert the weapon bracket's threaded rod (2) through the hole in the carrying handle of the M16A1/A2/A3 rifle, and secure it with the thumb wheel (3).

NOTE: The M16A1/A2/A3 weapon bracket (1) is a standard item in the TWS carrying case.

(2) Align the Picatinny rail grabber (4) on the bottom of the TWS with a notch on the bracket, ensuring that the TWS is positioned to accommodate an effective firing position once the eyecup (5) is depressed.

NOTES:
1. Ensure that the rail grabber fully rests on the bracket when mounting the TWS or the sight will not retain zero.
2. It is not necessary to use a spacer when mounting the HWTS on the M16A1/A2/A3.

AN/PVS-4 NIGHT VISION SIGHT

2-60. The AN/PVS-4 (Figure 2-38) is a passive, battery-operated night sight used for observation and accurate firing of individual-served weapons during nighttime conditions (for example, moonlight, starlight, skyglow) against targets as far away as 400 meters.

NOTE: The sight is less effective when viewing into rain, fog, sleet, snow, smoke, shadows, and other obscurants.

2-61. The sight is portable and mounts to the following weapons:

- M16A1/A2 and M14 rifles.
- M203 and M79 grenade launchers.
- M249 squad automatic weapon (SAW).
- M60 machine gun.
- M67 recoilless rifle.
- M72A1 rocket launcher.

Figure 2-38. AN/PVS-4 night vision sight.

MOUNTING ON THE M16A4 RIFLE OR M4 CARBINE

2-62. To mount the AN/PVS-4 night vision sight on an M16A4 rifle or M4 carbine—

NOTE: Reference the weapon components using the numbers listed in Figure 2-39.

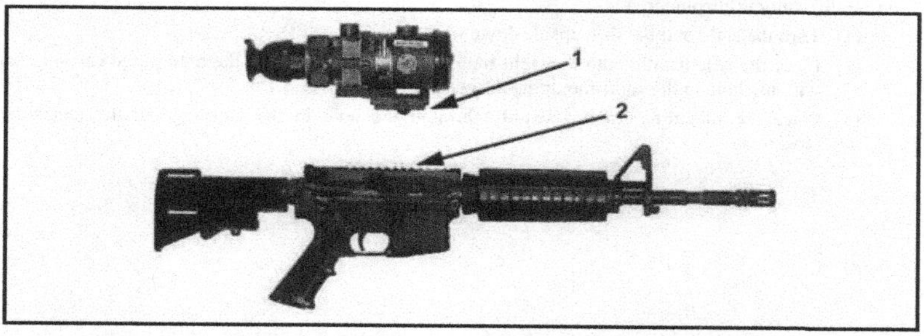

Figure 2-39. Mounting the AN/PVS-4 on an M4 carbine or M4 MWS.

(1) Align the Picatinny rail grabber with a mounting adapter (1) on the bottom of the AN/PVS-4 with a notch on the integrated rail (2) of an M16A4 rifle, M4 carbine, or M4 MWS, ensuring that the AN/PVS-4 is positioned to accommodate an effective firing position once the eyecup is depressed.

NOTE: The AN/PVS-4 will not retain zero if the rail grabber extends beyond the end of the integrated rail when mounted.

---

(2) Tighten the torque-limiting knob clockwise until it clicks twice.

NOTE: Retighten the rail grabber after firing a few rounds to ensure that the sight is fully seated.

2-63. The mounting procedures are identical for the M4 carbine and M4 MWS.

MOUNTING ON THE M16A1/A2/A3 RIFLE

2-64. The AN/PVS-4 is mounted on the carrying handle of the M16A1/A2/A3 rifle (Figure 2-40). To mount the AN/PVS-4 night vision sight on an M16A1/A2/A3 rifle—

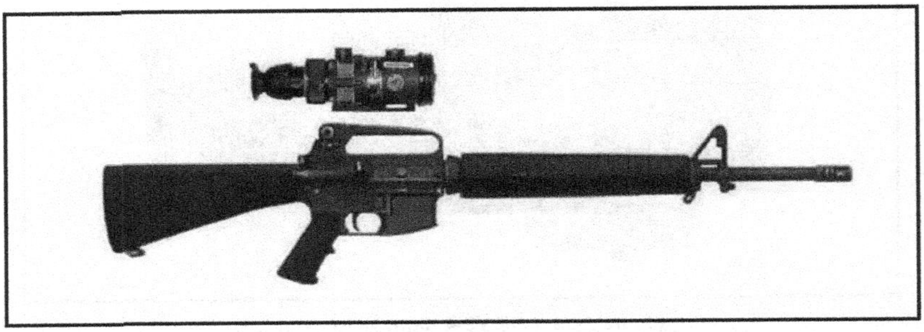

Figure 2-40. Mounting the AN/PVS-4 on an M16A1/A2/A3 rifle.

(1) Position the sight in the groove on the top of the carrying handle.

(2) Align the threaded hole in the base of the sight-mounting adapter over the hole in the handle.

(3) Insert the mounting knob assembly through the hole in the carrying handle, and screw it firmly clockwise into the sight-mounting adapter.

2-65. If difficulty is encountered—

(1) Turn the sight and the rifle upside down.

(2) Place the rifle handle onto the sight-mounting adapter, lining up the hole in the carrying handle with the hole in the sight-mounting adapter.

(3) Place the mounting knob assembly through the hole in the carrying handle, and screw it clockwise.

BACKUP IRON SIGHT

2-66. The BUIS (Figure 2-41) is a semi-permanent flip-up iron sight equipped with a rail-grabbing base. It is intended to remain on the M4 MWS while the M68 CCO is used as the primary means of day fire control. If the M68 fails, the prezeroed BUIS can be flipped up and used to continue the mission. The BUIS provides a backup capability effective when firing 600 meters away and farther and can be installed on the M16A4 rifle and M4 carbine. It provides a sighting capability when all other accessories have been removed, and it can be used to establish approximate zeros for other sighting components without requiring live-fire. Zeros established using this method are only effective to approximately 20 meters and should be refined by a live-fire zero.

2-67. The armorer installs the BUIS on the first notch of the integrated rail, nearest to the charging handle. Before the armorer installs the BUIS, he removes all rail covers/heat shields from the top, except one 4-, 5-, or 6-rib shield. The remaining rail cover/heat shield can be positioned to accommodate accessories and protect the nonfiring hand when the barrel is hot. Once installed and zeroed, the BUIS should be left in the stowed position (collapsing toward the firer, as shown in Figure 2-42) for best durability and minimal interference unless its use is imminent. The BUIS should only be removed by the armorer and remains on the MWS unless the carrying handle/sight is installed.

NOTE: Failure to install the BUIS on the first notch will lead to an improper zero and inaccuracy at longer ranges when the range lever is used.

Figure 2-41. Backup iron sight.

Figure 2-42. Backup iron sight in the stowed position.

ADVANCED COMBAT OPTICAL GUNSIGHT

2-68. The ACOG (Figure 2-43) is designed to provide enhanced target identification and hit probability for the M4A1 or M16-series weapon when firing as far as 800 meters (approximately 870 yards) away. It is designed with dual illuminated technology, using fiber optics for daytime employment and tritium for nighttime and low-light use. The ACOG is a lightweight, rugged, fast, and accurate 3.5 power optic scope. The body is machined from aluminum forgings; both the material and finish are identical to the M4A1. It is internally adjustable to allow the shock of rough handling to be carried by the scope body instead of the adjustment mechanism.

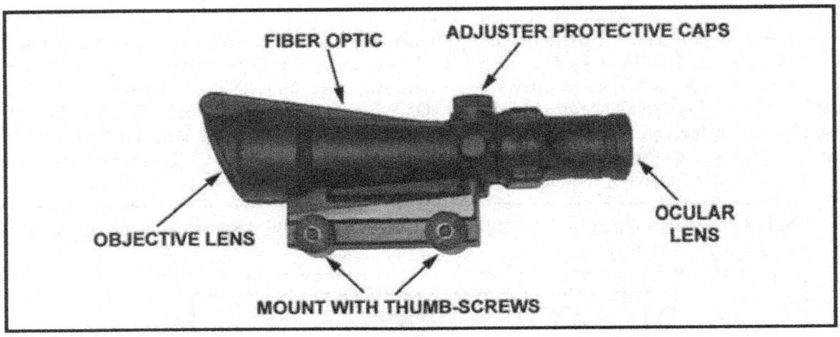

Figure 2-43. Advanced combat optical gunsight.

MOUNTING ON THE M16A4 RIFLE OR M4 CARBINE

2-69. The ACOG can be attached to the flattop easily using the adapter that comes from the factory (Figure 2-44). To attach the ACOG to the flattop with the adapter—

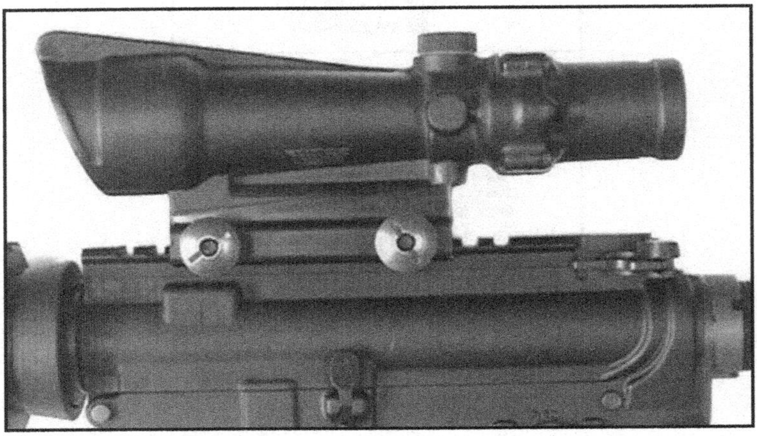

Figure 2-44. Mounting the ACOG on M16A4 rifle and M4 carbine.

(1) Loosen the thumb screws.
(2) Pull the interface clamp bar back against the knobs.
(3) Place the ACOG onto the flattop upper receiver rail surface. Be sure to engage the interface studs on the bottom of the adapter with the grooves on the top mounting surface of the upper receiver.

NOTE: 1. The ACOG can be placed in any of the slots on top of the weapon to allow for eye relief adjustments.

2. To retain zero, place the ACOG into the same slots during each installation.

---

(4) Using a finger, tighten the knobs firmly.
(5) Add another ¼ of a turn using a coin or a blade screwdriver.

NOTE: The slots on the knobs are only for removing the ACOG from the weapon if the knobs have been attached too tightly for hand-removal. If the slots are used to tighten the knobs, the ACOG will be difficult to remove.

MOUNTING ON THE M16A1/A2/A3 RIFLE

2-70. Although the ACOG is designed for use on the flattop rail of M16A4 rifles and M4 carbines, it can be attached to the carrying handle (Figure 2-45). To attach the ACOG to the carrying handle—

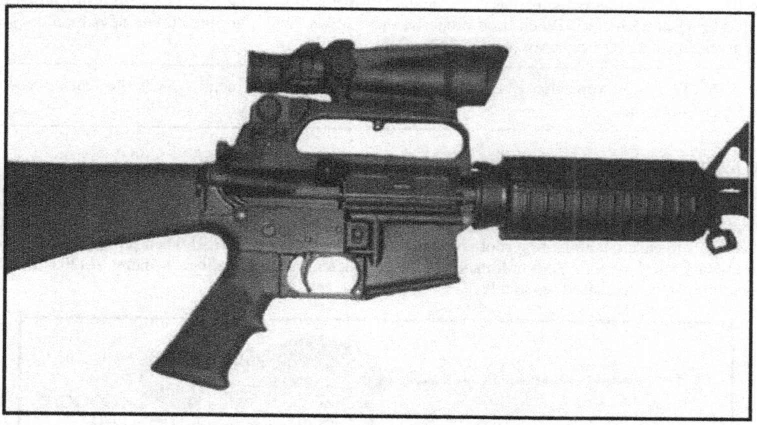

Figure 2-45. Mounting the ACOG on an M16A1/A2/A3 rifle.

(1) Align the forward-most threaded mounting hole on the bottom of the ACOG with the existing hole in the M4/M16 carrying handle.
(2) Press down firmly on the optic until it is seated inside the carrying handle rail.
(3) Insert the locking screw and washer set from the bottom of the carrying handle. Ensure that the locking screw and washer are inserted in the proper order (screw, locking nut, washer), or the screw will eventually shake itself loose. Figure 2-46 shows the proper assembly of the screw set.

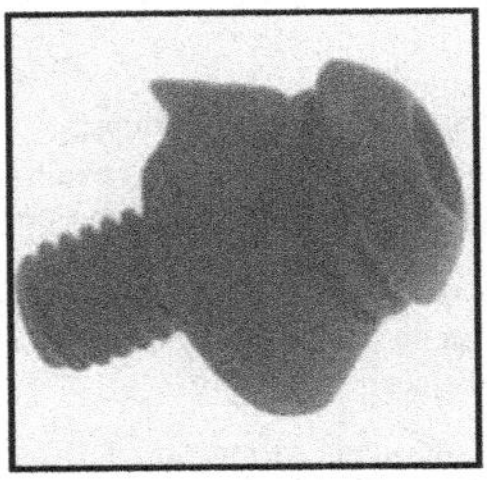

Figure 2-46. ACOG locking screw.

(4) Tighten the locking screw with the factory-supplied ⅛" Allen wrench, and then tighten another ¼ turn.

NOTE: When the ACOG is attached to the carrying handle, the bullet drop compensated (BDC) reticle will be off slightly at extended distances. For example, at 500 meters, the impact will be 5 inches high; at 600 meters, the impact will be 6 inches high. To compensate for this, aim slightly low once the target is ranged beyond 400 meters.

SECTION IV. 10-METER BORESIGHT AND 25-METER ZERO OFFSET

Boresighting is conducted at 10 meters with the borelight, weapon, aiming device, and a 10-meter offset. Each aiming device and weapon combination has a unique 10-meter offset. The 25-meter zero offset target is used when live-firing at 25 meters. When used properly, these offsets will align the aiming device on the selected weapon to engage a target's center of mass at 300 meters.

NOTE: See Appendix F for more information about 10-meter offsets for each weapon combination.

BORELIGHT

2-71. The borelight (Figure 2-47) is an eye-safe laser that is used to zero aiming lasers (such as the AN/PAQ-4 or AN/PEQ-2) without a 25-meter confirmation. The borelight will also boresight optics and iron sights to ensure that the first shot group hits the 25-meter zeroing target when zeroing the weapon. The borelight comes with a 5.56-millimeter, 7.62-millimeter, and .50-caliber mandrel. M203 and MK19 mandrels can be purchased separately.

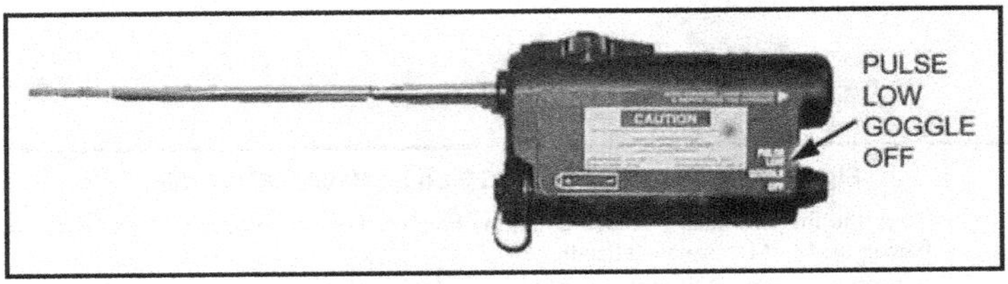

Figure 2-47. Borelight with a 5.56-millimeter mandrel.

2-72. The borelight has four settings:

- OFF.
- GOGGLE.
- LOW.
- PULSE.

OFF SETTING

2-73. This setting is used when the borelight is not in use.

GOGGLE SETTING

2-74. This setting is used when the boresight is being used with NVGs in a tactical environment.

LOW SETTING

2-75. This setting is used during normal operations.

PULSE SETTING

2-76. This setting is used during dry-fire training.

10-METER BORESIGHT

2-77. The 10-meter boresighting target is used in conjunction with the borelight. The 10-meter boresighting target is a 1-centimeter grid system with a crosshair and a circle (Figure 2-48A). The crosshair is the point of aim for the aiming device, and the circle is the point of impact for the borelight.

NOTE: Refer to Chapter 8 for a detailed explanation of boresighting procedures.

25-METER ZERO OFFSET

2-78. The M16A2 25-meter zeroing target is used when live-firing all optics and IR aiming lights at 25 meters, with the appropriate strike zone marked on the target (Figure 2-48B). The M4 zeroing target is only used when zeroing the iron sights on the M4. The M4 25-meter zeroing target will not allow the placement of the correct offset due to the size of the squares, which are 13 millimeters high and wide. The point of aim is always the 300-meter scaled silhouette's center of mass. The designated strike zone is a 4x4 square designating where the rounds should impact when the firer aims at the center of mass.

NOTE: Refer to Chapter 8 for a detailed explanation of the 25-meter offset zeroing procedures.

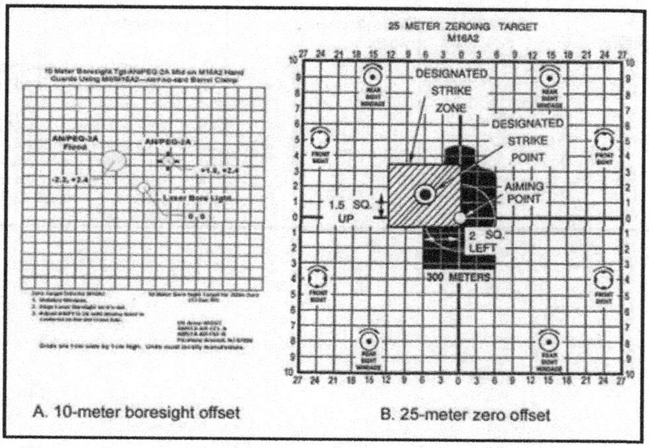

Figure 2-48. 10-meter boresighting target and 25-meter zero offset.

SECTION V. AMMUNITION

This section provides information about different types of standard military ammunition used in M16- and M4-series weapons.

AUTHORIZED AMMUNITION

2-79. Use only authorized ammunition manufactured to U.S. and NATO specifications (Table 2-8).

Table 2-8. Authorized ammunition.

CARTRIDGE/ROUND	IDENTIFICATION	USE	ADDITIONAL INFORMATION
M193 cartridge – 5.56-mm, ball 	Plain tip	The M193 is the standard cartridge for field use with the M16A1 rifle.	The M193 cartridge is a center-fire cartridge with a 55-grain, gilded metal-jacketed, lead alloy core bullet.
M196 cartridge – 5.56-mm, tracer 	Red or orange tip	The M196 cartridge is used only in the M16A1 rifle. Its main uses are for observation of fire, incendiary effect, and signaling.	Soldiers should avoid long-term use of 100 percent tracer rounds, which could cause deposits of incendiary material or chemical compounds that could damage the barrel. When tracer rounds are fired, they are mixed with ball ammunition in a ratio of no greater than one-to-one with a preferred ratio of three or four ball rounds to one tracer round.
M199 cartridge – 5.56-mm, dummy 	Six grooves along the sides of the case beginning about 1/2 inch from its tip	The M199 dummy cartridge is used in all M16-/M4-series weapons during dry-firing and other training.	This cartridge contains no propellant or primer. The primer well is open to prevent damage to the firing pin.
M200 cartridge – 5.56-mm, blank (no projectile) 	Case mouth is closed with a seven-petal rosette crimp, violet tip	The M200 blank cartridge is used in all M16-/M4-series weapons.	N/A
M855 cartridge – 5.56-mm, ball 	Green tip	The M855 cartridge is used in the M16A2/3/4 and in M4-series weapons.	The M855 cartridge has a 62-grain, gilded metal-jacketed, lead alloy core bullet with a steel penetrator. The primer and case are waterproof. This round is also linked and used in the M249. NOTE: This ammunition should not be used in the M16A1 except under emergency conditions, and only at targets less than 90 meters away. The twist of the M16A1 rifling is not sufficient to stabilize the length of the round's projectile.

Table 2-8. Authorized ammunition (continued).

CARTRIDGE/ROUND	IDENTIFICATION	USE	ADDITIONAL INFORMATION
M856 cartridge – 5.56-mm, tracer 	Red tip (orange when linked 4 to 1 for the M249)	The M856 tracer cartridge is used in the M16A2/3/4 and M4-series weapons.	The M856 tracer cartridge has characteristics similar to the M196 tracer, with a slightly longer tracer burnout distance. This cartridge has a 63.7-grain bullet. The M856 does not have a steel penetrator. NOTE: This ammunition should not be used in the M16A1 except under emergency conditions, and only at targets less than 90 meters away. The twist of the M16A1 rifling is not sufficient to stabilize the length of the round's projectile.
M862 cartridge – 5.56-mm, short-range training ammunition (SRTA) 	N/A	The M862 SRTA is used in all rifles and is designed exclusively for training.	The M862 SRTA can be used in lieu of service ammunition on indoor ranges and by units who have a limited range fan that does not allow the firing of service ammunition. If adequate range facilities are not available for sustainment training, SRTA can be used for any firing exercise of 25 meters or less. This includes the 25-meter scaled silhouette, 25-meter alternate qualification course, and quick-fire training. SRTA can also be used for urban operations (UO) training. NOTES: 1. See Appendix A for use of SRTA in training. 2. Although SRTA closely replicates the trajectory and characteristics of service ammunition out to 25 meters, the settings placed on the sights for SRTA could be different for service ammunition. SRTA should not be used to battlesight zero weapons that will fire service ammunition. 3. SRTA ammunition must be used with the M2 training bolt.
M995 cartridge – 5.56-mm, armor-piercing (AP) 	Conventional brass cartridge case Aluminum cup sits at the rear of the projectile (for the purpose of properly locating the penetrator within the projectile)	The M995 cartridge is used by the M249 (SAW), M16/A2/A3/A4, and M4-series weapons. It is intended for use against light armored targets.	The M995 offers the capability to defeat light armored targets at ranges two to three times that of currently available 5.56-mm ammunition. The M995 cartridge consists of a projectile and a propelling charge contained in a brass cartridge case. The projectile is a dense metal penetrator (tungsten carbide) enclosed by a standard gilded metal jacket. The cartridge utilizes a double base propellant. A standard rifle cartridge primer is used in the case to initiate the propelling charge.

TRAJECTORY

2-80. Figures 2-49 through 2-53 show trajectory data for M855 cartridges when fired from M16A2 rifles and M4 carbines.

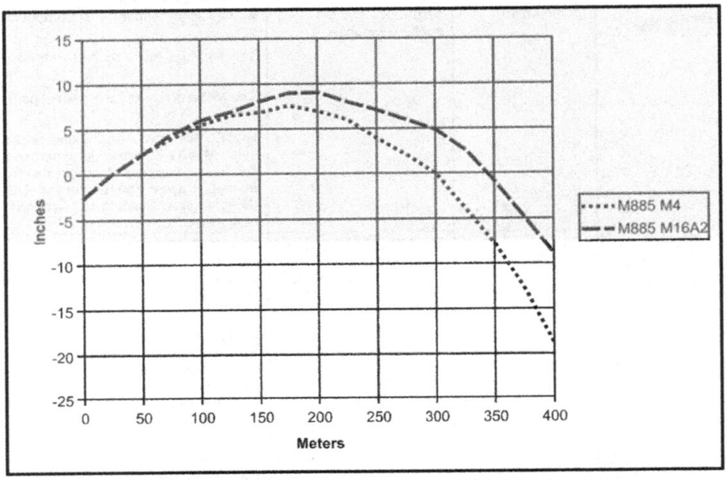

Figure 2-49. M855 drop during 25-meter zeroing (M16A2 at 8/3+1, M4 at 6/3).

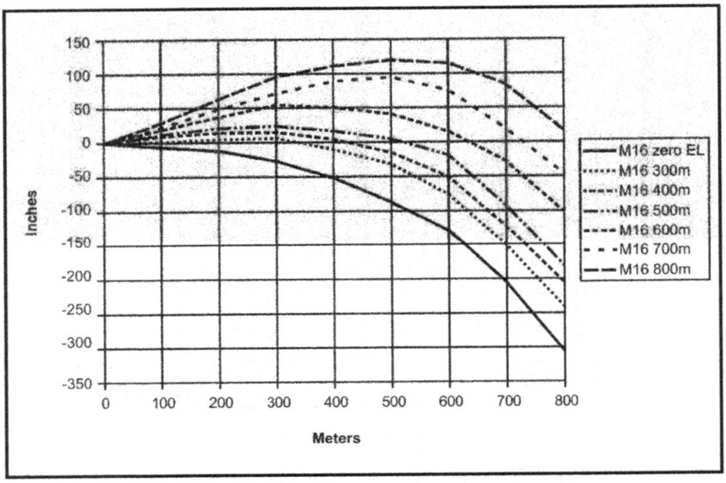

Figure 2-50. Bullet drop of M855 ammunition with M16A2 (8/3).

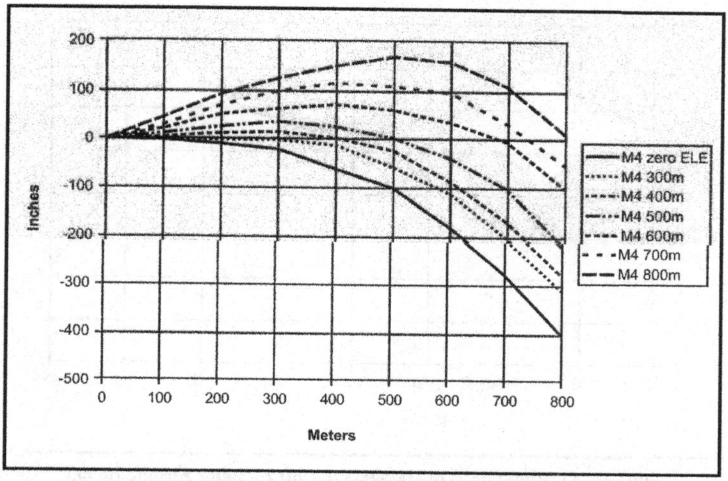

Figure 2-51. Bullet drop of M855 ammunition with M4 (6/3).

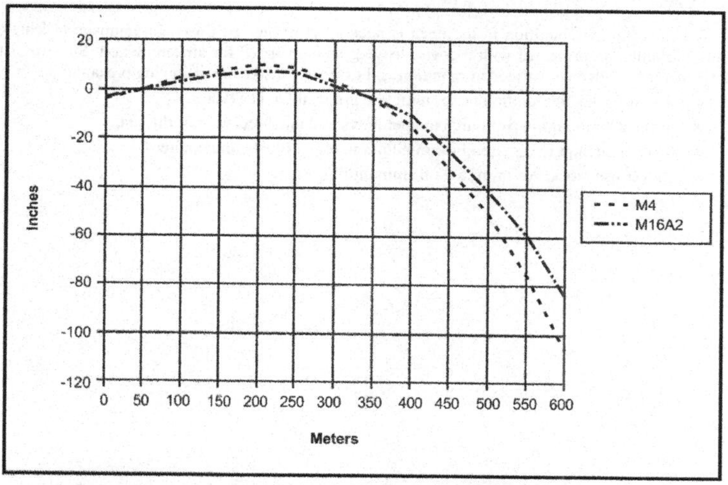

Figure 2-52. M4 carbine and M16A2 rifle bullet trajectory comparison.

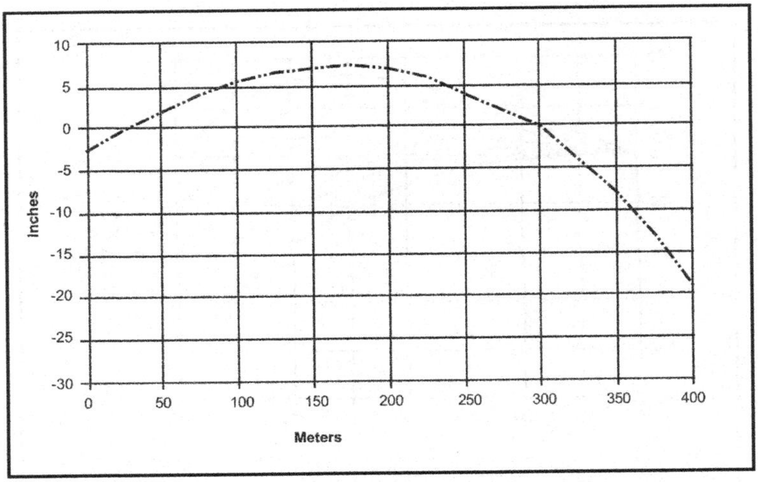

Figure 2-53. Bullet drop of M4/M855 during 25-meter zeroing on 6/3.

STORAGE

2-81. When storing ammunition in the open is necessary, it must be raised on dunnage at least 6 inches from the ground and protected with a cover, leaving enough space for air circulation. Since moisture and high temperatures adversely affect ammunition and explosives, take the following precautions:

- Do not open ammunition boxes until you are ready to use them.
- Protect ammunition from high temperatures and the direct rays of the sun.
- Do not attempt to disassemble ammunition or any of its components.
- Never use lubricants or grease on ammunition.

Chapter 3
Troubleshooting and Destruction

Commanders and unit armorers are responsible for the field level maintenance of weapons and for the destruction of weapons, when necessary. Soldiers are responsible for keeping their weapons clean and operational at all times—in training and in combat—therefore, they should be issued an operator's TM and cleaning equipment for their assigned weapons.

STOPPAGES

3-1. A stoppage is a failure of an automatic or semiautomatic firearm to complete the cycle of operation. The firer can apply immediate or remedial action to clear the stoppage. Some stoppages cannot be cleared by immediate or remedial action and may require weapon repair to correct the problem. A complete understanding of how the weapon functions is an integral part of applying immediate action procedures.

IMMEDIATE ACTION

3-2. Immediate action involves quickly applying a possible correction to reduce a stoppage without performing troubleshooting procedures to determine the actual cause. Apply immediate action only once for a stoppage. If the rifle fails to fire a second time for the same malfunction, inspect the weapon to determine the cause of the stoppage or malfunction, and take the appropriate remedial action. The key word **SPORTS** will help the firer remember the steps for immediate action:

(1) **S**lap gently upward on the magazine to ensure that it is fully seated and that the magazine follower is not jammed.

NOTE: When slapping up on the magazine, be careful not to knock a round out of the magazine into the line of the bolt carrier, causing more problems. Slap only hard enough to ensure that the magazine is fully seated. Ensure that the magazine is locked into place by quickly pulling down on the magazine.

---

(2) **P**ull the charging handle fully to the rear.

(3) **O**bserve the ejection of a live round or expended cartridge.

NOTE: If the weapon fails to eject a cartridge, perform remedial action.

---

(4) **R**elease the charging handle; do not ride it forward.

(5) **T**ap the forward assist assembly to ensure that the bolt is closed.

(6) **S**queeze the trigger and try to fire the rifle.

REMEDIAL ACTION

3-3. Remedial action is the continuing effort to determine the cause of a stoppage or malfunction and attempt to clear the stoppage once it has been identified. To apply the corrective steps for remedial action—

(1) Try to place the weapon on SAFE.

NOTE: A bolt override may not allow the weapon to be placed on SAFE.

---

(2) Remove the magazine.

(3) Lock the bolt to the rear.

(4) Place the weapon on SAFE (if not already done).

MALFUNCTIONS

3-4. Malfunctions are caused by procedural or mechanical failures of the rifle, magazine, or ammunition. Prefiring checks and serviceability inspections identify potential problems before they become malfunctions. This section describes the primary categories of malfunctions.

NOTE: In training, Soldiers must alert other Soldiers and range personnel when experiencing weapon malfunctions.

FAILURE TO FEED, CHAMBER, OR LOCK

3-5. This malfunction can occur when loading the rifle or during the cycle of operation. Once the magazine has been loaded into the rifle, the forward movement of the bolt carrier group could lack enough force (generated by the expansion of the action spring) to feed, chamber, or lock the bolt (Figure 3-1).

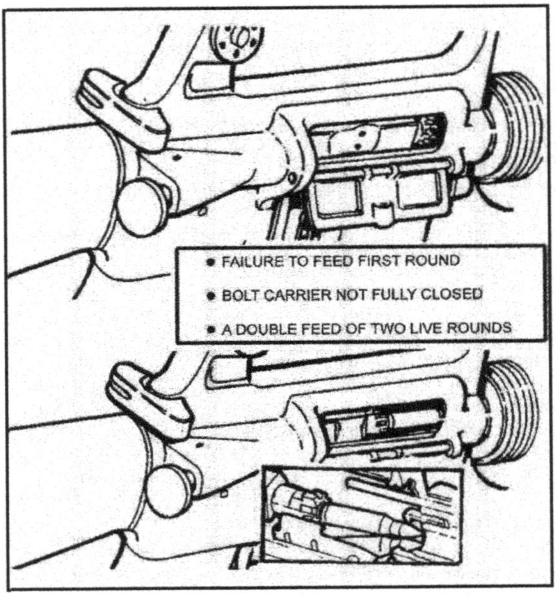

Figure 3-1. Failure to feed, chamber, or lock.

Probable Causes

3-6. The malfunction could be the result of one or more of the following:

- Excess accumulation of dirt or fouling in and around the bolt and bolt carrier.
- Defective magazine (dented, bulged, or a weak magazine spring).
- Improperly loaded magazine.
- Defective round (projectile forced back into the cartridge case, which could result in a stubbed round, or the base of the previous cartridge could be separated, leaving the remainder in the chamber).
- Damaged or broken action spring.
- Exterior accumulation of dirt in the lower receiver extension.
- Fouled gas tube (resulting in short recoil).
- A magazine resting on the ground or pushed forward (causing an improper lock).

Corrective Action

3-7. Applying immediate action usually corrects the malfunction. To avoid the risk of further jamming, watch for ejection of a cartridge and ensure that the upper receiver is free of loose rounds.

3-8. If immediate action fails to clear the malfunction, take remedial action.

NOTE: Do not force the bolt carrier.

3-9. If resistance is encountered (which can occur with an unserviceable round)—

(1) Lock the bolt to the rear.

(2) Remove the magazine.

(3) Clear the malfunction.

3-10. For example, to correct a bolt override (a cartridge has wedged itself between the bolt and charging handle)—

(1) Ensure that the charging handle is pushed forward and locked into place.

(2) Attempt to place the weapon on SAFE.

(3) Secure the rifle, and pull the bolt to the rear until the bolt seats completely into the buffer well.

(4) Turn the rifle upright and allow the overridden cartridge to fall out.

FAILURE TO FIRE CARTRIDGE

3-11. Despite the fact that a round has been chambered, the trigger has been pulled, and the sear has released the hammer, a cartridge may fail to fire. This occurs when the firing pin fails to strike the primer with enough force or when the ammunition is defective. The firer must follow unit safety guidelines until the determining factors of the misfire have been identified and corrected.

Probable Causes

3-12. Excessive carbon buildup on the firing pin (Figure 3-2, A) is often the cause, because the full forward travel of the firing pin is restricted. A defective or worn firing pin can give the same results. Inspection of the ammunition could reveal a shallow indentation or no mark on the primer, indicating a firing pin malfunction (Figure 3-2, B). Cartridges that show a normal indentation on the primer (but did not fire) indicate faulty ammunition or failure of the cartridge to fully seat in the chamber.

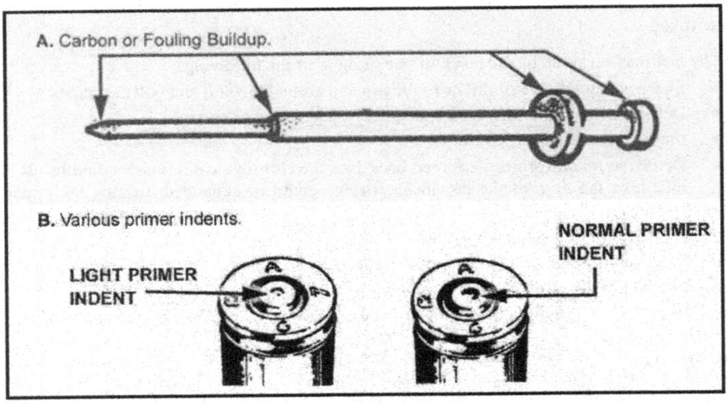

Figure 3-2. Failure to fire.

Corrective Action

3-13. If the malfunction continues—

(1) Inspect the firing pin, bolt, bolt carrier, and locking lug recesses of the barrel extension.

(2) Remove any accumulation of excessive carbon or fouling.

(3) Inspect the firing pin for damage.

3-14. If the round is suspected to be faulty, dispose of it IAW the unit SOP, and ensure that it is reported and returned to the agency responsible for issuing ammunition.

WARNING

If an audible pop or reduced recoil occurs during firing, immediately cease fire. This could be the result of a round being fired without enough force to send the projectile out of the barrel. Do not apply immediate action. Instead, perform the following actions:

1. **Lock the bolt to the rear.**
2. **Place the selector lever in the SAFE position.**
3. **Remove the magazine.**
4. **Visually inspect the bore to ensure that a projectile is not lodged in the barrel.**
5. **If a projectile is lodged in the barrel, do not try to remove it. Turn the rifle in to the unit armorer.**

FAILURE TO EXTRACT

3-15. A failure to extract results when the cartridge case remains in the chamber of the rifle. The bolt and bolt carrier might move rearward only a short distance, but more commonly, the bolt and bolt carrier recoil fully to the rear, leaving the cartridge case in the chamber. A live round is then forced into the base of the cartridge case as the bolt returns in the next feed cycle.

> **WARNING**
>
> **A failure to extract is an extremely serious malfunction, requiring the use of tools to clear. A live round could be left in the chamber and accidentally discharged. If a second live round is fed into the primer of the chambered live round, the rifle could explode and cause personal injury. This malfunction must be properly identified and reported.**
>
> **Ejection failures should not be reported as extraction failures.**

Probable Cause

3-16. Short recoil cycles and fouled or corroded rifle chambers are the most common causes of failures to extract. A damaged extractor or a weak or broken extractor spring can also cause this malfunction.

Corrective Action

3-17. This malfunction is one of the hardest to clear; the severity of the failure determines the corrective action procedures.

NOTE: If the bolt has moved rearward far enough to strip a live round from the magazine in its forward motion, the bolt and bolt carrier must be locked to the rear.

3-18. With the bolt locked to the rear and the weapon on SAFE, perform the following actions:

(1) Remove the magazine and all loose rounds.

(2) Tap the weapon's buttstock on a hard surface to cause the cartridge to fall out of the chamber.

3-19. However, if the cartridge case is ruptured, it can be seized. When this occurs, perform the following actions:

(1) Insert a cleaning rod into the bore from the muzzle end.

(2) Force the cartridge case from the chamber by tapping the cleaning rod against the inside base of the fired cartridge.

3-20. If cleaning and inspecting the mechanism and chamber reveals no defects but failures to extract persist, the extractor and extractor spring should be replaced. If the chamber surface is damaged, the entire barrel must be replaced.

FAILURE TO EJECT

3-21. Ejection of a cartridge is an element in the rifle's cycle of functioning, regardless of the mode of fire. This malfunction occurs when the cartridge is not ejected through the ejection port and either remains partly in the chamber or becomes jammed in the upper receiver as the bolt closes. When the firer initially clears the rifle, the cartridge could strike an inside surface of the receiver and bounce back into the path of the bolt.

Probable Cause

3-22. The cartridge must be extracted before it can eject. Failures to eject can also be caused by a buildup of carbon or fouling on the extractor or by short recoil. Short recoil is usually due to a buildup of fouling in the bolt carrier mechanism or gas tube. Resistance caused by a carbon-coated or corroded chamber can impede the extraction and ejection of a cartridge.

Corrective Action

3-23. While retraction of the charging handle usually frees the cartridge and permits removal, the charging handle must not be released until the position of the next live round is determined.

3-24. If another live round has been sufficiently stripped from the magazine or remains in the chamber, remove the magazine and all live rounds before releasing the charging handle.

3-25. If several malfunctions occur and are not corrected by cleaning and lubricating, replace the ejector spring, extractor spring, and extractor.

OTHER MALFUNCTIONS

3-26. Table 3-1 describes other malfunctions that can occur and the appropriate corrective actions.

Table 3-1. Other malfunctions.

MALFUNCTION	CORRECTIVE ACTION
The bolt fails to lock in the rearward position after the last round in the magazine is fired.	Check for a bad magazine or short recoil.
The bolt fails to lock in the rearward position when the bolt catch has been engaged.	Check the bolt catch; turn in the weapon to the unit armorer.
The weapon fires two or more rounds when the trigger is pulled and the selection lever is in the SEMI position.	This indicates a worn sear, cam, or disconnector. Turn in the weapon to the unit armorer so that he can repair and replace the trigger group parts.
The trigger fails to pull or return after release with the selector set in a firing position.	This indicates that the trigger pin has backed out of the receiver or the hammer spring is broken (A, Figure 3-3). Turn in the weapon to the unit armorer so that he can repair and replace it.
The magazine fails to lock into the magazine well.	Check the magazine and magazine catch for damage (B, Figure 3-3). Turn in to the unit armorer to adjust or replace the catch.
Any part of the bolt carrier group fails to function.	Check for incorrect assembly of components (C, Figure 3-3). Correctly clean and assemble the bolt carrier group, or replace damaged parts.
The ammunition fails to feed from the magazine.	Check for a damaged magazine (D, Figure 3-3). A damaged magazine could cause repeated feeding failures and should be turned in to the unit armorer or exchanged.

NOTE: For more information about troubleshooting malfunctions and replacing components, see organizational and direct support maintenance publications and manuals.

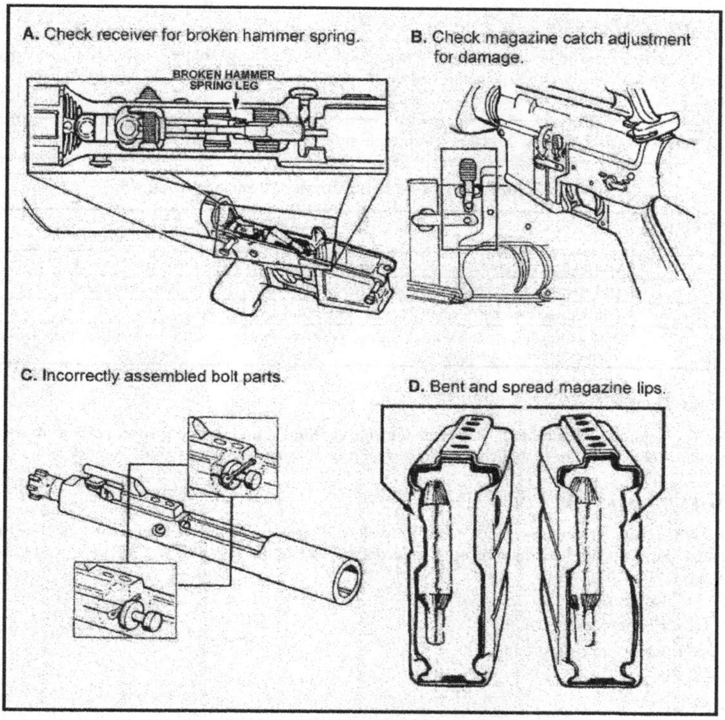

Figure 3-3. Other possible malfunctions.

DESTRUCTION PROCEDURES

3-27. Destruction of any military weapon is authorized only as a last resort to prevent the enemy from capturing or using it. In combat situations, the commander has the authority to destroy weapons, but he must report doing so through the proper channels.

NOTE: Certain procedures outlined require the use of explosives and incendiary grenades. Related principles and the specific conditions under which destruction occurs are command decisions.

Methods of Destruction

3-28. Equipment may be destroyed using several methods. The commander must use his imagination and resourcefulness to select the best method of destruction based on the facilities available. Time is usually critical. Table 3-2 outlines the methods of destruction.

NOTE: If destruction is directed, appropriate safety precautions must be observed.

Table 3-2. Methods of destruction and their applications.

METHOD OF DESTRUCTION	APPLICATION
Mechanical	Use an axe, pick, mattock, sledgehammer, crowbar, or other heavy implement.
Burning	Use gasoline, oil, incendiary grenades, other flammables, or a welding/cutting torch.
Demolition	Use suitable explosives, ammunition or, as a last resort, hand grenades.
Disposal	Bury essential parts, dump them in streams or marshes, or scatter them so widely that recovering them would be impossible.

Degree of Damage

3-29. The method of destruction used must damage equipment and essential spare parts to the extent that they cannot be restored to usable condition (by repair or by cannibalization) in the combat zone.

Priorities of Destruction

3-30. When lack of time prevents completely destroying equipment, Soldiers must destroy the same essential parts on all like equipment. The order in which the parts should be destroyed (priority of destruction) is as follows:

(1) Bolt carrier group.
(2) Upper receiver group.
(3) Lower receiver group.

Chapter 4

Preliminary Marksmanship Instruction

An Infantryman's basic battlefield tool is his weapon. To effectively employ his weapon, he must master marksmanship—from the basics of rifle marksmanship to the advanced stages of target engagement.

Understanding the operation and functions of any machine is vital to becoming an expert with that machine. The same theory applies to rifle marksmanship. Commanders must keep this in mind when setting up a training program.

SECTION I. INTRODUCTION TO BASIC RIFLE MARKSMANSHIP AND MECHANICAL TRAINING

This section covers the mechanical training of M16- and M4-series weapons. With this knowledge, a Soldier is able to assess and correct any malfunction to keep the weapon operating properly. This training program (Table 4-1) introduces Soldiers to BRM and teaches them how to maintain, operate, and correct malfunctions on M16- and M4-series weapons. It also teaches peer coaching responsibilities and sight manipulation, while emphasizing safety.

Table 4-1. Introduction to basic rifle marksmanship and mechanical training.

INTRODUCTION TO BASIC RIFLE MARKSMANSHIP AND MECHANICAL TRAINING
Period 1 (4 hours)
Instructional Intent
• Introduce the Soldiers to BRM and teach them how to maintain, operate, and correct malfunctions on a M16-/M4-series weapon. • Teach peer coaching responsibilities and sight manipulation, while emphasizing safety.
Observables
Soldiers— • Emphasize safety throughout training IAW TM 9-1005-319-10 (refer to this TM for more information). • Clear their weapons IAW this manual and TM 9-1005-319-10. • Identify all components of their weapon IAW TM 9-1005-319-10 (refer to this TM for more information). • Handle and identify 5.56-mm ammunition IAW TM 9-1005-319-10 (refer to this TM for more information). • Understand the eight cycles of function IAW this manual. • Understand the modes of fire IAW this manual. • Disassemble and assemble their weapon IAW TM 9-1005-319-10 (refer to this TM for more information). • Perform a function check on their weapon IAW TM 9-1005-319-10 (refer to this TM for more information). • Maintain, load, and unload their magazines IAW TM 9-1005-319-10 (refer to this TM for more information). • Maintain, load, unload, and clear their weapons IAW TM 9-1005-319-10 (refer to this TM for more information). • Perform SPORTS on their weapon within five seconds IAW TM 9-1005-319-10 (refer to this TM for more information). • Correctly manipulate their sights without assistance IAW TM 9-1005-319-10 (refer to this TM for more information). • Are taught peer-coaching techniques and responsibilities IAW this manual.

NOTE: Soldiers who do not meet the standard will receive remedial training before continuing with subsequent instruction.

CLEARING

WARNING

To maintain safety, the weapon must be cleared before disassembling, cleaning, inspecting, transporting, or storing.

NOTE: Additional mechanical training, including disassembly, maintenance, assembly, loading, and sight manipulation, is available in TM 9-1005-319-10.

4-1. To clear an M16-/M4-series weapon (Figure 4-1), perform the following actions:

(1) Point the muzzle in a designated safe direction or in a clearing barrel.

(2) Attempt to place the selector lever on SAFE. If the selector lever cannot be placed on SAFE because the weapon is not cocked, proceed to Step 3.

(3) Remove the magazine. To perform this procedure—
- Depress the magazine catch button.
- Pull the magazine down.

(4) Lock the bolt open. To perform this procedure—
- Pull the charging handle rearward.
- Press the bottom of the bolt catch.
- Allow the bolt to move forward until it engages the bolt catch.
- Return the charging handle to the full forward position.
- If you have not done so already, place the selector lever on SAFE.

(5) Visually inspect the receiver and chamber to ensure that these areas contain no ammunition.

(6) Allow the bolt to move forward by pressing the upper portion of the bolt catch.

(7) Place the selector lever on SEMI.

(8) Squeeze the trigger.

(9) Pull the charging handle fully rearward and release it, allowing the bolt to return to the full forward position.

(10) Place the selector lever on SAFE.

NOTE: The weapon is now clear.

(11) Close the ejection port cover.

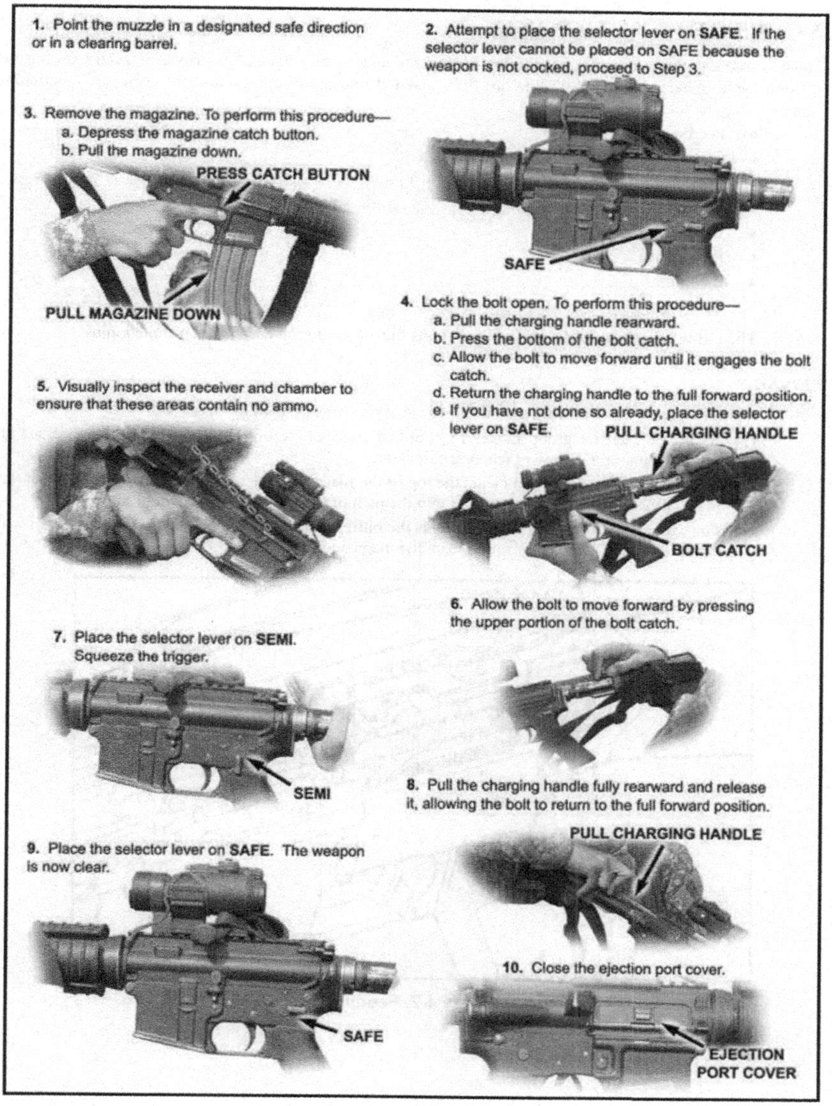

Figure 4-1. Clearing.

CYCLES OF FUNCTIONING

4-2. Soldiers must understand the rifle components and the mechanical sequence of events during the firing cycle. After a loaded magazine has been inserted into the weapon, the eight cycles of functioning begin. They include—

- Feeding.
- Chambering.
- Locking.
- Firing.
- Unlocking.
- Extracting.
- Ejecting.
- Cocking.

4-3. The following paragraphs describe the actions that occur during each cycle of functioning.

FEEDING

4-4. During feeding, the following actions occur. This process is shown in Figure 4-2.

(1) As the bolt carrier group moves rearward, it engages the buffer assembly and compresses the action spring into the lower receiver extension.

(2) When the bolt carrier group clears the top of the magazine, the expansion of the magazine spring forces the follower and a new round into the path of the forward movement of the bolt.

(3) The expansion of the action spring sends the buffer assembly and bolt carrier group forward with enough force to strip a new round from the magazine.

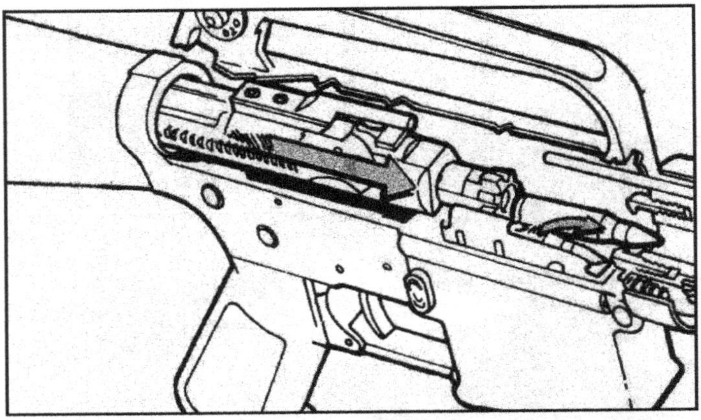

Figure 4-2. Feeding.

CHAMBERING

4-5. During chambering, the following actions occur. This process is shown in Figure 4-3.

(1) As the bolt carrier group continues to move forward, the face of the bolt thrusts the new round into the chamber.

(2) At the same time, the extractor claw grips the rim of the cartridge and the ejector is compressed.

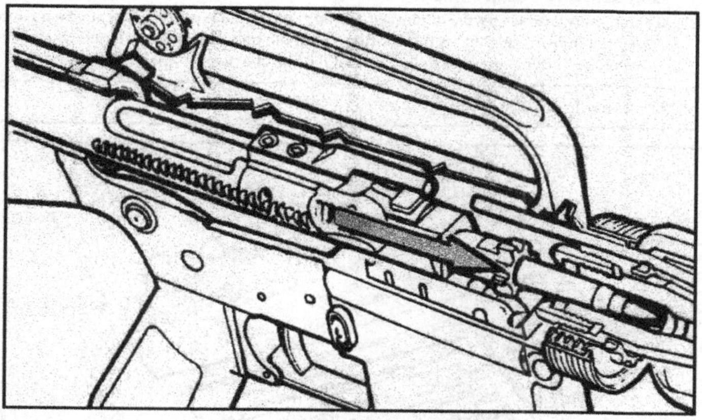

Figure 4-3. Chambering.

LOCKING

4-6. During locking, the following actions occur. This process is shown in Figure 4-4.

(1) As the bolt carrier group moves forward, the bolt cam pin riding in the guide channel in the upper receiver keeps the bolt in its most forward position.

(2) Just before the bolt locking lugs make contact with the barrel extension, the bolt cam pin emerges from the guide channel.

(3) The pressure exerted by the contact of the bolt locking lugs and barrel extension causes the bolt cam pin to move along the cam track (located in the bolt carrier) in a counterclockwise direction, rotating the bolt locking lugs in line behind the barrel extension locking lugs.

NOTE: The rifle is ready to fire.

Figure 4-4. Locking.

FIRING

4-7. During firing, the following actions occur. This process is shown in Figure 4-5.

(1) With a round in the chamber, the hammer cocked, and the selector on SEMI, the firer squeezes the trigger.

(2) The trigger rotates on the trigger pin, depressing the nose of the trigger and disengaging the notch on the bottom of the hammer.

(3) The hammer spring drives the hammer forward.

(4) The hammer strikes the head of the firing pin, driving the firing pin through the bolt and into the primer of the round.

(5) The primer ignites, causing the powder in the cartridge to ignite.

(6) The gas generated by the rapid burning of the powder forces the projectile from the cartridge and propels it through the barrel.

(7) After the projectile has passed the gas port (located on the upper surface of the barrel under the front sight, as shown in Figure 4-5) and before it leaves the barrel, gas enters the gas port and moves into the gas tube.

(8) The gas tube directs the gas into the bolt carrier. It passes down through the key and into a space between the rear of the carrier's bolt cavity and the rear of the bolt itself.

(9) Then, the gas expands.

(10) The bolt is locked into the barrel extension, unable to move forward; the carrier is forced to the rear by the expanding gas.

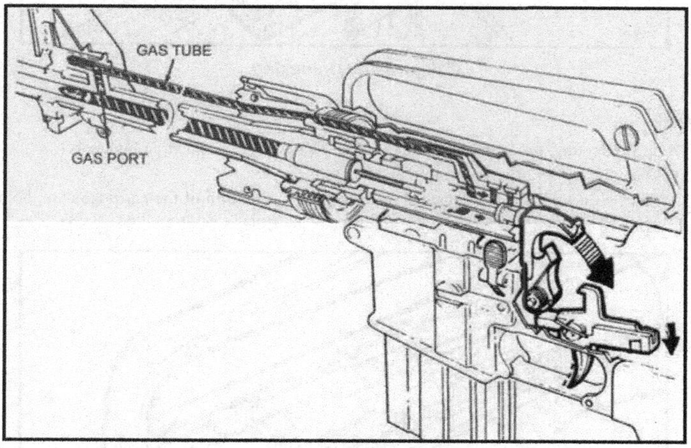

Figure 4-5. Firing.

UNLOCKING

4-8. During unlocking, the following actions occur. This process is shown in Figure 4-6.

(1) As the bolt carrier moves to the rear, the bolt cam pin follows the path of the cam track (located in the bolt carrier).

(2) The cam pin and bolt assembly rotate simultaneously until the locking lugs of the bolt are no longer in line behind the locking lugs of the barrel extension.

Figure 4-6. Unlocking.

EXTRACTING

4-9. During extracting, the following actions occur. This process is shown in Figure 4-7.

(1) The bolt carrier group continues to move to the rear.

(2) The extractor (which is attached to the bolt) grips the rim of the cartridge case, holds it firmly against the face of the bolt, and withdraws the cartridge case from the chamber.

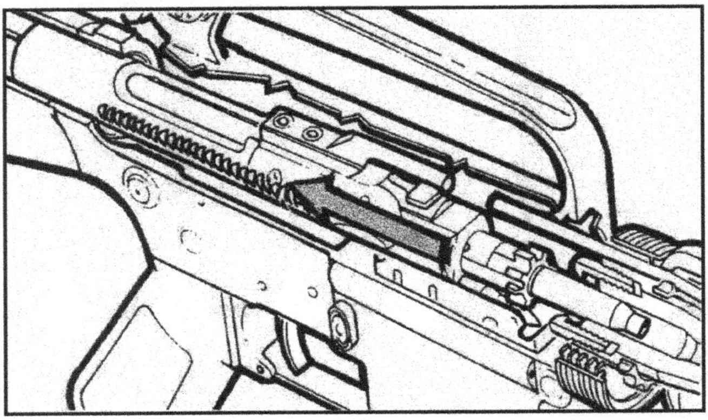

Figure 4-7. Extracting.

EJECTING

4-10. During ejecting, the following actions occur. This process is shown in Figure 4-8.

(1) With the base of a cartridge case firmly against the face of the bolt, the ejector and ejector spring are compressed into the bolt body.

(2) As the rearward movement of the bolt carrier group allows the nose of the cartridge case to clear the front of the ejection port, the cartridge is pushed out by the action of the ejector and ejector spring.

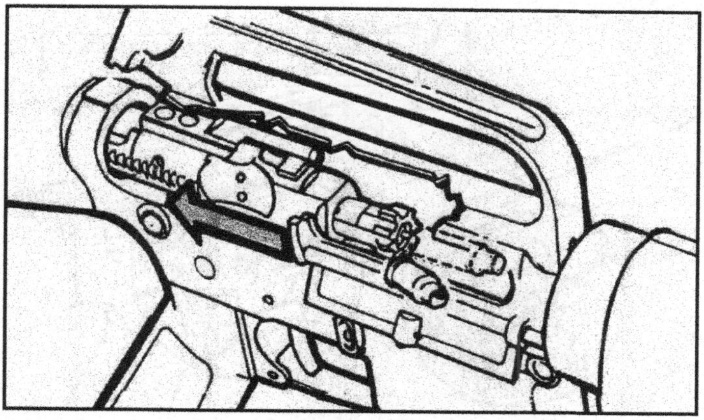

Figure 4-8. Ejecting.

COCKING

4-11. During cocking, the following actions occur. This process is shown in Figure 4-9.

(1) The rearward movement of the bolt carrier overrides the hammer.

(2) The hammer is forced down into the receiver, and the hammer spring is compressed.

NOTE: This action cocks the hammer in the firing position.

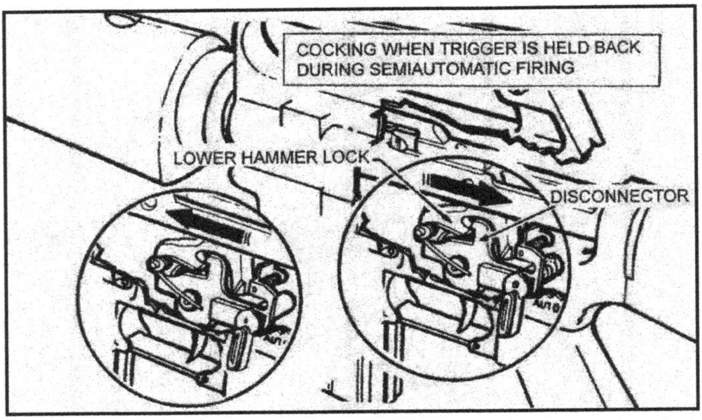

Figure 4-9. Cocking.

MODES OF FIRE

4-12. Weapons may fire using the three modes of fire:

- Semiautomatic (Figure 4-10).
- Automatic (Figure 4-10).
- Burst (Figure 4-11).

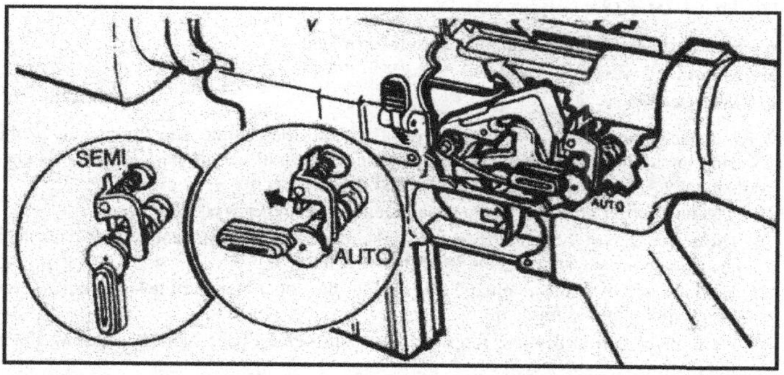

Figure 4-10. Semiautomatic and automatic fire mode selector positions.

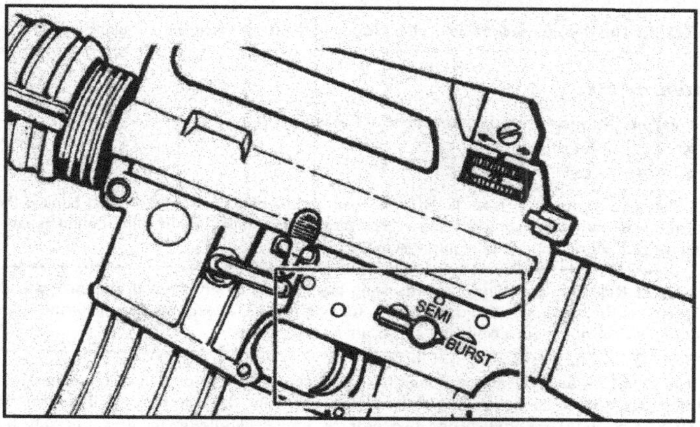

Figure 4-11. Burst fire mode selector position.

SEMIAUTOMATIC MODE OF FIRE

4-13. Weapons that function in the semiautomatic mode of fire are—

- M16A1/A2/A3/A4 rifles.
- M4/M4A1carbines.

4-14. Before a Soldier fires a weapon using the semiautomatic mode of fire, a disconnector (mechanism installed so the firer can fire single rounds) is attached to the trigger and rotated forward by the action of the disconnector spring. When the recoil of the bolt carrier cocks the hammer—

(1) The disconnector engages the lower hook of the hammer and holds it until the trigger is released.
(2) The disconnector rotates to the rear and down, disengaging the hammer and allowing it to rotate forward until caught by the nose of the trigger (this prevents the hammer from following the bolt carrier forward and causing multiple firings).
(3) The trigger must be squeezed again before the next round will fire.

AUTOMATIC MODE OF FIRE

4-15. Weapons that function in the automatic mode of fire are—

- M16A1/A3 rifles.
- M4A1 carbines.

4-16. When the selector lever is in the AUTO position, the rifle continues to fire as long as the trigger is held back and ammunition is in the magazine. The functioning of certain parts of the rifle changes when firing automatically. Once the trigger is squeezed and the round is fired—

(1) The bolt carrier group moves to the rear, and the hammer is cocked.
(2) The center cam of the selector depresses the rear of the disconnector and prevents the nose of the disconnector from engaging the lower hammer hook.
(3) The bottom part of the automatic sear catches the upper hammer hook and holds it until the bolt carrier group moves forward.
(4) The bottom part strikes the top of the sear and releases the hammer, causing the rifle to fire automatically.
(5) If the trigger is released, the hammer moves forward and is caught by the nose of the trigger.

NOTE: This ends the automatic cycle of fire until the trigger is squeezed again.

BURST MODE OF FIRE

4-17. Weapons that function in the burst mode of fire are—

- M16A2/A4 rifles.
- M4 carbines.

4-18. When the selector lever is in the BURST position, the rifle fires a three-round burst if the trigger is held to the rear during the complete cycle. The weapon continues to fire three-round bursts with each separate trigger pull as long as ammunition is in the magazine.

NOTE: Releasing the trigger or exhausting ammunition at any point in the three-round cycle interrupts fire, producing one or two shots. Reapplying the trigger only completes the interrupted cycle; it does not begin a new one. This is not a malfunction.

4-19. The M16A2/4 and M4 disconnectors have a three-cam mechanism that continuously rotates with each firing cycle. Based on the position of the disconnector cam, the first trigger pull (after initial selection of the BURST position) can produce one, two, or three firing cycles before the trigger must be pulled again. The burst cam rotates until it reaches the stop notch.

PEER COACHING

4-20. Peer coaching involves using two Soldiers of equal firing proficiency and experience to assist each other during marksmanship training.

ADVANTAGES AND DISADVANTAGES

4-21. Some problems exist with peer coaching. If the new Soldier does not have adequate guidance, a "blind leading the blind" situation may result, leading to negative training and safety violations. However, when adequate instruction is provided, peer coaching can be helpful even in the IET environment. Since all Soldiers in units have completed BRM, peer coaching within units should yield better results.

BENEFITS

4-22. The pairing of Soldiers can enhance learning for both participants. The coach learns what to look for as he provides guidance to the firer, and the firer has a chance to ask simple questions and to discuss areas that are not understood. Pairing Soldiers who have demonstrated good firing proficiency with those who have firing problems can improve the performance of problem firers.

DUTIES

4-23. The peer coach—

- Constantly checks factors that the firer is unable to observe for himself.
- Prevents the firer from repeating errors.
- Assists the firer in applying marksmanship fundamentals during firing.
- Assists the firer in obtaining a good position and in adjusting sandbags.
- Adds to range safety procedures by helping safety personnel with preliminary rifle checks.
- Watches the firer—not the target—to ensure that the firer maintains proper position, breath control, trigger pressure, and trigger squeeze.

NOTE: There are times when the peer coach may be required to observe the target area and not the firer for example, when field fire targets are being engaged and the firer cannot see where he is hitting or missing targets).

CHECKLIST FOR THE COACH

4-24. Coaches follow these procedures to determine and eliminate rifle and firer deficiencies:

- The coach checks to see that the—
  - Rifle is cleared, and defective parts have been replaced.
  - Ammunition is clean, and the magazine is properly placed in the pouch.
  - Sights are blackened and set correctly for long- or short-range firing.
- The coach observes the firer to see if he—
  - Uses the correct position and properly applies the steady position elements.
  - Properly loads the rifle.
  - Obtains the correct sight alignment (with the aid of an M16 sighting device).
  - Holds his breath correctly (by watching his back).
  - Applies proper trigger squeeze (determines whether he flinches or jerks by watching his head, shoulders, trigger finger, and firing hand and arm).
  - Is tense and nervous. If the firer is nervous, the coach has the firer breathe deeply several times to relax.
- Supervisory personnel and peer coaches correct errors as they are detected. If many common errors are observed, it is appropriate to call the group together for discussion, demonstration of proper procedures, and feedback.

POSITION OF THE COACH

4-25. During an exercise, the coach should be positioned where he can best observe the firer as he assumes position. Then, he moves to various points around the firer (sides and rear) to check the correctness of the firer's position. The coach requires the firer to make adjustments until the firer obtains a correct position.

4-26. When the coach is satisfied with the firing position, he assumes a coaching position alongside the firer. The coach usually assumes a position like that of the firer on the firing side of the firer (Figure 4-12).

Figure 4-12. Prone position of coach (right-handed firer).

SECTION II. MARKSMANSHIP FUNDAMENTALS I

This training program (Table 4-2) reinforces BRM and trains the four fundamentals to standard through dry-firing and simulation circuit training. It also teaches and reinforces range and safety procedures.

Table 4-2. Marksmanship Fundamentals I training program.

MARKSMANSHIP FUNDAMENTALS I
Period 2 (8 hours)
Instructional Intent
• Reinforce BRM I, and train the four fundamentals to standard with hands-on training, simulation, and dry-firing during circuit training with an M16-/M4-series weapon. • Teach and reinforce range and safety procedures.
Observables
Ensure that— • All equipment (helmet, IBA) is fitted properly to maximize training IAW the local SOP. • Live-fire range procedures are replicated and enforced IAW the local SOP. • The four fundamentals are being integrated into all exercises IAW this manual. • All dry-firing is well-aimed using 25-meter zeroing targets, EST, and LMTS. • Peer coaching is being emphasized IAW this manual.
Tasks
• The four fundamentals (IAW with this manual). • Dominant eye training (IAW with this manual). • Basic firing positions (IAW with this manual). • Range and safety procedures (IAW with the local SOP). • Demonstrate the integrated act of shooting during dry-fire exercises utilizing simulators and training devices (IAW this manual).

NOTES: 1. Simulators and training devices are listed in Appendix A of this manual.

2. Soldiers who do not meet the standard will receive remedial training before continuing with subsequent instruction.

INTERCEPTOR BODY ARMOR

4-27. BRM strategy includes the wearing of interceptor body armor (IBA, shown in Figure 4-13), if it is available—minus the throat, collar, and groin attachments—during all BRM periods and concurrent training. Whether or not IBA is worn, marksmanship fundamentals remain the same.

4-28. Prior to BRM training, use an IBA immersion approach so the Soldier can adapt to weight and movement restrictions. For an easier weight transition, incrementally introduce the outer tactical vest (OTV) and front and back small arms protective insert (SAPI) plates.

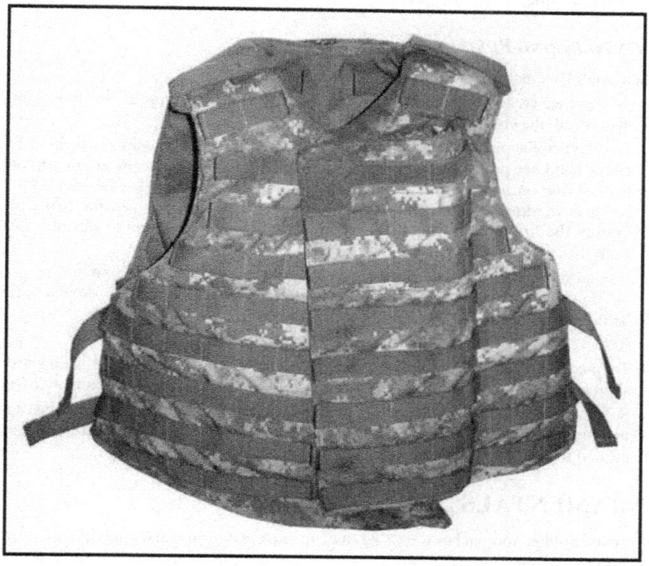

Figure 4-13. Interceptor body armor.

PROPER WEAR AND FIT

4-29. When using IBA, adhere to the following guidelines:

- Have adequate IBA quantities on hand for all Soldiers.
- Properly size IBA to the Soldier by conducting deliberate fit procedures to reduce or eliminate fit and size problems.

NOTE: Improper wear and fit of IBA impedes a Soldier's marksmanship ability.

- Ensure that the SAPI plate size corresponds to the OTV.
- Make sure that nothing else is in the OTV/SAPI compartment.

Wear of Helmets with Interceptor Body Armor

4-30. When using helmets with IBA, adhere to the following guidelines:

- When in the prone position, the IBA's back plate tends to shove the personnel armor system for ground troops (PASGT) helmet over Soldiers' eyes. To minimize the PASGT helmet positioning problem, make sure that the helmet is properly sized and fitted. Female shooters with long hair will find that wearing their hair in a bun adds material between the IBA and helmet, further forcing the helmet down over their eyes. Encourage female Soldiers to wear a short (chin length) haircut or cornrow hairstyle. If the female Soldier chooses not to wear short hair, allow her to wear her hair down when firing. Tightening the suspension harness and sweat band (raising the helmet higher on the head) can lessen interference with the IBA, hair, and helmet.
- The Army combat helmet (ACH) is lighter than the PASGT helmet, has better weight distribution, and contains less material that can impede a Soldier's firing vision when in the prone position. The ACH does not interfere with the IBA or block a Soldier's vision while in the prone position.

Adjustments to Firing Positions

4-31. When using IBA, adhere to the following guidelines:

- To increase comfort and stability while wearing IBA in the prone position, scoop sand or dirt underneath the chest while preparing to fire.
- To alleviate the pain and pressure on elbows and knees that the added weight of IBA causes, use elbow and knee pads. If used in the kneeling position, do not rest the elbow pad on the knee pad; hard plastic on hard plastic is not conducive to a steady position. To help with stability while firing in the kneeling position, squeeze the rifle buttstock between the SAPI plate and bicep. Loosen the firing-side straps and tighten the nonfiring-side straps to shift the SAPI plate away from the firing side.
- Instead of using load-bearing equipment (LBE), attach canteens, ammunition pouches, or first aid pouches directly to the IBA to minimize interference with LBE shoulder straps, IBA, and helmet.
- Reserve IBA firing with throat, collar, or groin protectors for ARM.
- To position themselves more comfortably and be able to reach the handguards, Soldiers of shorter stature may have to increase their body-line-to-rifle axis angle to more of an "L" shape.
- Soldiers should be in a comfortable firing position to leverage the natural point of aim. The more the target and rifle are naturally in line (as when in a relaxed position), the less movement is needed to acquire a proper sight picture.

FOUR FNDAMENTALS

4-32. Before the Soldier approaches the firing line, he must understand and apply the four fundamentals:

- Steady position.
- Aiming.
- Breath control.
- Trigger squeeze.

4-33. Soldiers apply these four fundamentals rapidly and consistently to perform the integrated act of firing. These fundamentals should be practiced while the Soldier is wearing all of his equipment, including his helmet and IBA (if available).

Steady Position

4-34. When the Soldier approaches the firing line, he assumes a comfortable, steady firing position (Figure 4-14). The firer is the best judge of the quality of his position. If he can hold the front sightpost steady through the fall of the hammer, he has a good position. Steady position incorporates the following elements:

- Nonfiring hand grip.
- Rifle's buttstock position.
- Firing hand grip.
- Firing elbow placement.
- Nonfiring-side elbow placement.
- Cheek-to-stock weld.
- Support and muscle relaxation.
- Natural point of aim.

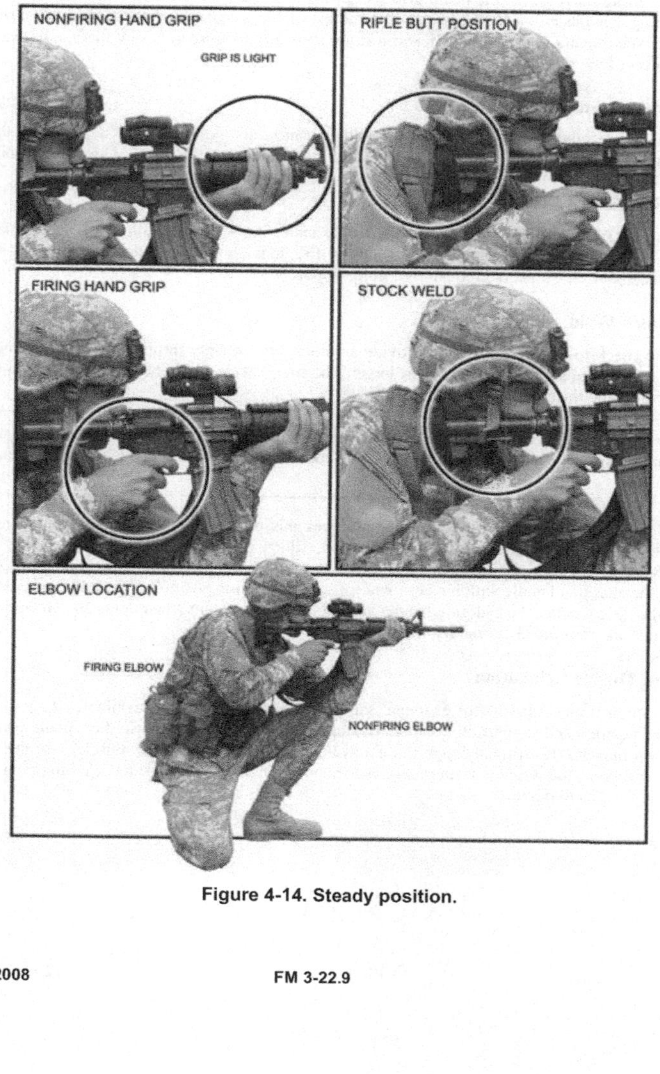

Figure 4-14. Steady position.

Nonfiring Hand Grip

4-35. The weapon's handguard rests lightly on the heel of the nonfiring hand, in the "V" formed by the thumb and fingers.

Rifle's Buttstock Position

4-36. Place the weapon's buttstock into the pocket of the firing shoulder. When wearing IBA, place the weapon's buttstock where the pocket should be; this reduces the effect of recoil and ensures a steady position.

Firing Hand Grip

4-37. The firing hand grasps the pistol grip so that it fits in the "V" formed by the thumb and forefinger. The forefinger is placed on the trigger so that the lay of the weapon is not disturbed when the trigger is squeezed. The remaining three fingers exert a slight rearward pressure to ensure that the buttstock remains in the pocket of the shoulder.

Firing Elbow Placement

4-38. The firing elbow is important in providing balance. Its exact location depends on the firing or fighting position used. Placement of the firing elbow should allow the firer's shoulders to remain level.

Nonfiring-Side Elbow Placement

4-39. The nonfiring-side elbow is positioned firmly under the weapon to allow a comfortable and stable position. When the Soldier engages a wide sector of fire, moving targets, and targets at various elevations, his nonfiring-side elbow should remain free from support.

Cheek-to-Stock Weld

4-40. The cheek-to-stock weld should provide a natural line of sight through the center of the rear sight aperture to the front sightpost and onto the target. The firer's neck should be relaxed, allowing his cheek to fall naturally onto the stock.

NOTE: Proper eye relief is obtained when a Soldier establishes a good cheek-to-stock weld. A small change in eye relief normally occurs each time that the firer assumes a different firing position.

4-41. Through dry-fire training, the Soldier practices this position until he assumes the same cheek-to-stock weld each time he assumes a given position, which provides consistency in aiming. To learn to maintain the same cheek-to-stock weld each time the weapon is aimed, the Soldier should begin by trying to touch the charging handle with his nose when assuming a firing position. The Soldier should be mindful of how the nose touches the charging handle and should be consistent when doing so. This position should be critiqued and reinforced during dry-fire training.

Support and Muscle Relaxation

4-42. When artificial support (for example, sandbags, logs, or stumps) is available, it should be used to steady the position and support the weapon. If support is used properly, the Soldier should be able to relax most of his muscles. If artificial support is not available, the bones—not the muscles—in the firer's upper body must support the weapon. Using muscles to support the rifle can cause muscle fatigue, which in turn, causes the weapon to move.

Natural Point of Aim

4-43. When the Soldier first assumes his firing position, he orients his weapon in the general direction of his target. Then, he adjusts his body to align the weapon and sights with the desired point of aim. When using proper support and consistent cheek-to-stock weld, the Soldier should have his weapon and sights naturally aligned on the target.

4-44. If correct body-rifle-target alignment cannot be achieved, the front sightpost must be held on the target using muscular support and effort. As the weapon fires, muscles tend to relax, causing the front sight to move away from the target, toward the natural point of aim. Adjusting this natural point of aim to the target eliminates this movement. When multiple target exposures are expected or a sector of fire must be covered, the Soldier adjusts his natural point of aim to the center of the expected target exposure area or sector.

AIMING

4-45. Having mastered the task of holding the rifle steady, the Soldier must align the rifle with the target in exactly the same way for each firing. The firer is the final judge as to where his eye is focused. The instructor or trainer emphasizes this point by having the firer focus on the target and then on the front sightpost. He checks the position of the firing eye to ensure that it is in line with the rear sight aperture. The elements of aiming training are as follows:

- Sight alignment.
- Focus of the eye.
- Sight picture.
- Front sightpost.
- Aiming practice.

Sight Alignment

4-46. The weapon must be aligned with the target; to do so, Soldiers place the tip of the front sightpost in the center of the rear sight aperture (Figure 4-15). Any alignment error between the front and rear sights repeats itself for every ½ meter the bullet travels. For example, at the 25-meter line, any error in rifle alignment is multiplied 50 times. If the bullet is misaligned by 1/10 of an inch, it causes a target 300 meters away to be missed by 5 feet.

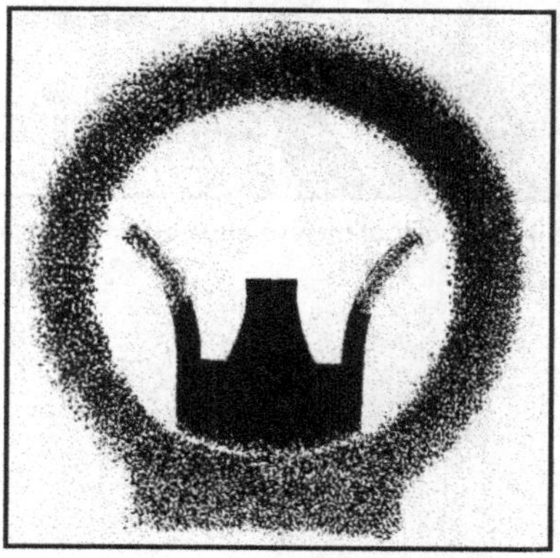

Figure 4-15. Correct sight alignment.

Focus of the Eye

4-47. A proper firing position aligns the eye with the center of the rear sight aperture. When the eye is focused on the front sightpost, the eye's natural ability to center objects in a circle and to seek the point of greatest light (center of the aperture) aid in providing correct sight alignment. For the average Soldier firing at combat-type targets, the eye's natural ability can accurately align the sights. Therefore, the firer can place the tip of the front sightpost on the point of aim, but the eye must be focused on the tip of the front sightpost. This causes the target to appear blurry, while the front sightpost is seen clearly. Two reasons for focusing on the front sightpost are:

(1) Only a minor aiming error should occur, since the error reflects only as much as the Soldier fails to determine the target's center. A greater aiming error can result if the front sightpost is blurry due to focusing on the target or other objects.

(2) Focusing on the tip of the front sightpost aids the firer in maintaining proper sight alignment.

Sight Picture

4-48. Once the Soldier can correctly align his sights, he can obtain a correct sight picture. A correct sight picture has the target, front sightpost, and rear sightpost aligned. The sight picture includes two basic elements: sight alignment and placement of the point of aim. Placement of the point of aim varies, depending on the engagement range. For example, Figure 4-16 shows a silhouette at 300 meters where the point of aim is the center of mass and the sights are aligned for a correct sight picture.

Figure 4-16. Focus of the eye and correct sight picture.

4-49. The side aiming technique can be used to obtain a correct sight picture (Figure 4-17). It involves positioning the front sightpost to the side of the target in line with the vertical center of mass, keeping the sights aligned. The front sightpost is moved horizontally until the target is directly centered on the front sightpost.

Figure 4-17. Side aiming technique.

Front Sightpost

4-50. The front sightpost is vital to proper firing and should be replaced when damaged. The post should be blackened; when it is shiny, the firer cannot focus precisely on the tip of the front sightpost.

Aiming Practice

4-51. Aiming practice is conducted before firing live rounds. During dry-firing, the Soldier should practice sight alignment and placement of the point of aim. Training aids, such as the M15A1 aiming card, can be used to do this.

Breath Control

4-52. While sighted on a target, the firer must be aware of the rifle's movement as a result of breathing. Two breath control techniques are practiced during dry-fire:

- Breath control for engaging single targets.
- Breath control for engaging short-exposure targets.

4-53. As the firer's skills improve and as timed or multiple targets are presented, he must learn to control his breath at any part of the breathing cycle. The coach/trainer ensures that the firer uses both breathing techniques and understands them by instructing him to exaggerate his breathing.

Breath Control for Engaging Single Targets

4-54. When zeroing or when time is available to fire a shot, Soldiers fire when there is a natural respiratory pause, when most of the air has been exhaled from the lungs and before inhaling (Figure 4-18). The shot must be fired before the Soldier feels any discomfort.

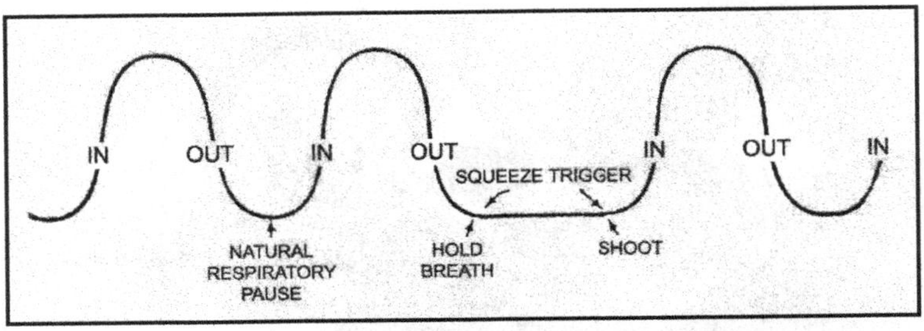

Figure 4-18. Breath control for engaging single targets.

Breath Control for Engaging Short-Exposure Targets

4-55. When employing rapid fire (engaging short-exposure targets), Soldiers stop their breath when they are about to squeeze the trigger (Figure 4-19).

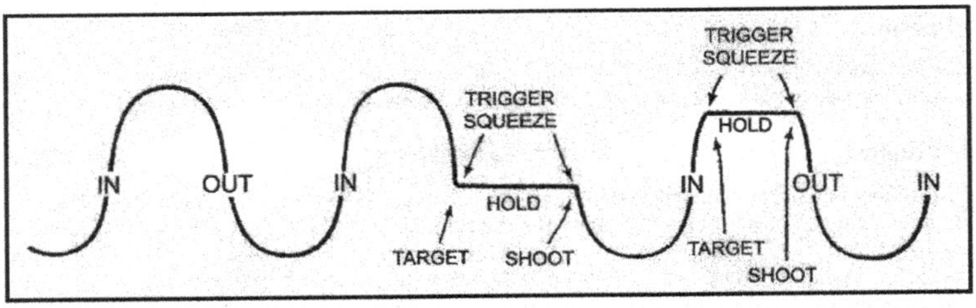

Figure 4-19. Breath control for engaging short-exposure targets.

TRIGGER SQUEEZE

4-56. A steady position reduces disturbance of the rifle during trigger squeeze. If the trigger is not properly squeezed, the rifle will be misaligned with the target at the moment of firing. The elements of trigger squeeze training are as follows:

- Rifle movement.
- Trigger finger.
- Trigger squeeze time.
- Coaching trigger squeeze.
- Wobble area.

Rifle Movement

4-57. Trigger squeeze is important for two reasons:

- Any sudden movement of the finger on the trigger can disturb the lay of the rifle and cause the shot to miss the target.
- The precise instant of firing should be a surprise to the Soldier. If a Soldier knows the exact instant that the rifle will fire, the Soldier will naturally compensate for the weapon's noise and recoil, causing him to miss the target. Soldiers usually tense their shoulders when expecting the rifle to fire; it is difficult to detect since the Soldier does not realize that he is flinching.

Trigger Finger

4-58. The Soldier places his trigger finger (index finger on the firing hand) on the trigger between the first joint and the tip of the finger—not the very end of the finger—and adjusts depending on his hand size and grip. The trigger finger must squeeze the trigger to the rear so the hammer falls without disturbing the lay of the rifle.

4-59. When a live round is fired, it is difficult to see the effect that the trigger pull had on the lay of the rifle. It is important to experiment with many finger positions during dry-fire training to ensure that the hammer is falling with little disturbance to the aiming process.

Trigger Squeeze Time

4-60. The proper trigger squeeze should start with slight pressure on the trigger during the initial aiming process. The firer applies more pressure after the front sightpost is steady on the target and he is holding his breath.

4-61. As the firer's skills increase with practice, he needs less time spent on trigger squeeze. A novice firer can take five seconds to perform an adequate trigger squeeze, but as skills improve, he can squeeze the trigger in a second or less.

Coaching Trigger Squeeze

4-62. The coach/trainer—

- Observes the trigger squeeze, emphasizes the correct procedure, and checks the firer's applied pressure.
- Places his finger on the trigger and has the firer squeeze the trigger by applying pressure to his finger.
- Ensures that the firer squeezes straight to the rear on the trigger, avoiding a left or right twisting movement.
- Observes that the firer follows through and holds the trigger to the rear for approximately one second after the round has been fired.

Wobble Area

4-63. Wobble area is the movement of the front sight around the point of aim when the rifle is in the steadiest position.

4-64. The position must provide for the smallest possible wobble area.

- From a supported position, there should be minimal wobble area and little reason to detect movement. If movement of the rifle causes the front sight to leave the target, more practice is needed.
- From an unsupported position, the firer experiences a greater wobble area than from a supported position. If the front sight strays from the target during the firing process, the firer should hold constant pressure on the trigger and resume as soon as he corrects the sighting.

NOTE: The firer should never try to quickly squeeze the trigger while the sight is on the target. The best firing performance results when the trigger is squeezed continuously and the rifle is fired without disturbing its lay.

DOMINANT EYE TRAINING

4-65. This exercise assists the coach and the firer in determining which eye the firer should use when engaging targets. The firer's dominant eye should be identified early in the training process to prevent unnecessary problems, such as a blurred sight picture or the inability to acquire a tight shot group during the grouping exercise. To perform dominant eye training—

(1) The trainer cuts a 1-inch circular hole in the center of an 8- by 10-inch piece of material (can be anything from paper to plywood).

(2) The trainer positions himself approximately 5 feet in front of the Soldier. The trainer closes his nondominant eye and holds his finger in front of and just below his dominant eye to provide the Soldier with a point of aim.

(3) The Soldier holds the training aid with both hands at waist level and looks with both eyes open at the trainer's open eye. With both eyes focused on the trainer's open eye and arms fully extended, the Soldier raises the training aid between himself and the trainer while continuing to look at the trainer's eye through the hole in the training aid. The Soldier's eye that the trainer sees through the hole in the training aid is the Soldier's dominant eye.

BASIC FIRING POSITIONS

4-66. Basic firing positions are taught during PMI. Other advanced firing positions are added later in training to support tactical conditions. The firing positions used during initial training are—

- Individual foxhole supported.

NOTE: Once the individual foxhole supported fighting position has been mastered, the firer should practice various unsupported positions to obtain the smallest possible wobble area during final aiming and hammer fall. The coach/trainer can check the steadiness of the position by observing movement at the forward part of the rifle, by looking through the Ml6 sighting device, or by checking to see if support is being used. The objective is to establish a steady position under various conditions.

- Prone unsupported firing.
- Prone supported firing.
- Kneeling unsupported.

4-67. These firing positions offer a stable platform for firing the rifle. These positions are used during basic record fire.

Individual Foxhole Supported Firing Position

4-68. This position provides the most stable platform for engaging targets. To assume the individual foxhole supported firing position (Figure 4-20)—

(1) Add or remove dirt, sandbags, or other supports to adjust for your height.
(2) Face the target.
(3) Execute a half-face to the firing side.
(4) Lean forward until the chest is against the firing hand corner of the position.
(5) Place the rifle handguard in the "V" formed by the thumb and fingers of the nonfiring hand.
(6) Rest the nonfiring hand on the material (sandbags or berm) to the front of the position.
(7) Place the weapon's buttstock into the pocket of the firing shoulder.
(8) Rest the firing elbow on the ground outside of the position.

NOTES: 1. When prepared positions are not available, the prone supported position can be substituted.

2. The objective is to establish a steady position under various conditions. The ultimate performance of this task is combat. Although the firer must be positioned high enough to observe all targets, he must remain as low as possible to provide added protection from enemy fire.

Figure 4-20. Individual foxhole supported firing position.

Prone Unsupported Firing Position

4-69. This firing position offers another stable firing platform for engaging targets. To assume the prone unsupported firing position—

(1) Face the target.
(2) Spread the feet a comfortable distance apart.

(3) Drop to the knees, breaking the fall with the weapon's buttstock.

(4) Using the rifle's buttstock as a pivot, roll onto the nonfiring side, placing the nonfiring-side elbow close to the side of the magazine.

(5) For the basic prone unsupported position (Figure 4-21), spread the legs apart, with the inside of the feet flat on the ground. For the alternate prone unsupported firing position (Figure 4-22), bend the firing leg to relieve pressure on the lower back.

(6) Place the weapon's buttstock between the SAPI plate and bicep to stabilize the weapon and absorb recoil.

(7) Grasp the pistol grip with the firing hand.

(8) Lower the firing elbow to the ground.

(9) Place both elbows on the ground to support the upper body.

(10) Rest the rifle in the "V" formed by the thumb and fingers of the nonfiring hand.

(11) Adjust the position of the firing elbow until the shoulders are approximately level.

(12) Pull back firmly on the rifle with both hands.

(13) Obtain a stock weld and relax, keeping the heels close to the ground.

NOTE: To increase comfort and stability while wearing IBA in the prone position, sand or dirt should be scooped underneath the chest while preparing to fire. Elbow and knee pads can be worn to relieve IBA-induced pain and pressure.

Figure 4-21. Basic prone unsupported firing position.

Figure 4-22. Alternate prone unsupported firing position.

PRONE SUPPORTED FIRING POSITION

4-70. To assume the prone supported firing position—

(1) Face the target and drop to the ground, breaking the fall with the weapon's buttstock.

(2) For the basic prone supported firing position (Figure 4-23), spread the legs apart, with the inside of the feet flat on the ground. For the alternate prone supported firing position (Figure 4-24), bend the firing leg to relieve pressure on the lower back.

(3) Use sandbags or any other suitable object to support the handguard. Keep the nonfiring hand free for use on any part of the rifle.

(4) Place both elbows on the ground to support the upper body.

(5) Place the firing hand on the pistol grip.

(6) Place the nonfiring hand on the upper handguard.

(7) Place the weapon's buttstock between the SAPI plate and bicep to stabilize the weapon and absorb recoil.

NOTE: Elbow and knee pads can be worn to relieve IBA-induced pain and pressure.

Figure 4-23. Basic prone supported firing position.

Figure 4-24. Alternate prone supported firing position.

Kneeling Unsupported Firing Position

4-71. To assume the kneeling unsupported firing position (Figure 4-25)—

(1) Keep the left foot in place.

(2) Step back with the right foot.

(3) Drop to the right knee.

(4) Place the left nonfiring hand on the upper handguard with the upper arm (triceps) on the left knee for support.

(5) Place the right firing hand on the pistol grip, with the weapon's buttstock between the SAPI plate and bicep to stabilize the weapon and absorb recoil.

(6) Rest the ball of the right foot firmly on the ground.

(7) Rest the buttock on the heel.

(8) Relax and lean forward into the position to help absorb recoil.

Figure 4-25. Kneeling unsupported firing position.

TRAINING DEVICES AND EXERCISES

4-72. When used alone or in combinations with the appropriate training strategies, training devices and aids can be used to help individuals or squads sustain or practice basic marksmanship skills. They are beneficial when ammunition is limited for training or practice exercises. Some training devices are complex, costly, and in limited supply, while others are relatively simple, cheap, and in large supply.

M15A1 Aiming Card

4-73. This exercise measures the firer's ability to acquire the same sight picture each time he places his iron sights on a target.

NOTE: Refer to Appendix A for a detailed explanation of training aids and devices.

TARGET BOX AND PADDLE EXERCISE

4-74. This exercise checks the consistency of aiming and placement of three-round shot groups in a dry-fire environment. It incorporates the Soldier's position, breathing, and sight picture to simulate a live-fire 25-meter engagement. Further, it reinforces the basic fundamentals, while refining the Soldier's muscle memory during the integrated act of dry-firing.

NOTE: Refer to Appendix A for a detailed explanation of the target box and paddle exercise and training standards.

DIME/WASHER EXERCISE

4-75. The primary purpose of this exercise is to practice trigger control; it is also used to reinforce good body position and breath control. The Soldier must successfully dry-fire his weapon six consecutive times without the dime or washer falling to the ground.

NOTE: Refer to Appendix A for a detailed explanation of the dime/washer exercise and training standards.

SECTION III. MARKSMANSHIP FUNDAMENTALS II

This training program (Table 4-3) reinforces BRM and the four fundamentals, while demonstrating the integrated act of shooting on the EST 2000 or LMTS.

Table 4-3. Marksmanship Fundamentals II training program.

MARKSMANSHIP FUNDAMENTALS II
Period 3 (8 hours)
Instructional Intent
• Reinforce BRM I and II and the four fundamentals, while demonstrating the integrated act of shooting on the EST or LMTS.
Observables
Ensure that— • All fundamentals are emphasized and applied on the EST or LMTS. • Weapon safety is reinforced on the EST or LMTS. • Peer coaching is emphasized during EST or LMTS firing. • All Soldiers who fail to hit six out of nine shots at the 300-meter EST target receive remedial training.
Tasks
• Demonstrate the integrated act of firing while using the EST.

NOTE: Soldiers who do not meet the standard will receive remedial training before continuing with subsequent instruction.

ENGAGEMENT SKILLS TRAINER 2000

4-76. The EST 2000 is a home station, indoor, multipurpose, multilane, small arms, crew-served, shoulder-launched munitions gunnery simulator with superior accuracy and state-of-the-art graphics.

4-77. The EST 2000 is used to provide small arms weapon training on—

- Marksmanship.
- Squad tactical procedures.
- Close-range shoot/don't shoot techniques and skills.

4-78. Commanders should review DA Pam 350-38 for live-fire events that can be executed using the EST 2000.

NOTE: Refer to Appendix A for a detailed explanation of the EST 2000 training simulator.

LASER MARKSMANSHIP TRAINING SYSTEM

4-79. The LMTS helps bridge the existing gap in individual marksmanship training by providing a more widely available, flexible, transportable, and lightweight means of maintaining marksmanship skills. The LMTS large-company suite with the warrior kit is a home station or deployed, multipurpose, modular, scaleable, individual and unit small arms marksmanship sustainment simulator that accommodates the M16-series rifle and M4 carbine.

4-80. The LMTS supports both IET and unit sustainment training for static mounted and dismounted small units. When incorporated into BRM and Marksmanship Fundamentals I and II, it greatly improves Soldier feedback.

NOTES:
1. The LMTS is not intended to replace live-firing, qualification, or the EST 2000.
2. Refer to Appendix A for a detailed explanation of the LMTS.
3. Soldiers failing to maintain a three-centimeter shot group on the LMTS 25-meter target should receive remedial feedback.

Chapter 5

Downrange Feedback

Downrange feedback provides precise knowledge of what happens to bullets at range. It provides for an effective transition between 25-meter firing and firing on the field fire range. Knowing precisely where all bullets are hitting or missing the target, a poor firer (with instructor/trainer assistance) can improve his performance, and a good firer can bring his shots to the target's center.

This chapter contains guidelines for the instructor/trainer to conduct training on ranges that provide detailed feedback from the targets downrange. This chapter contains grouping and zeroing procedures and procedures for conducting the three types of KD ranges.

SECTION I. GROUPING PROCEDURES

This section provides guidelines for conducting a grouping range (Table 5-1). It includes concept, organization, shot group marking, shot group analysis, multiple shot group analysis, and troubleshooting of the fundamentals.

Table 5-1. Grouping procedures.

GROUPING PROCEDURES
Instructional Intent
• Reinforce PMI by performing the integrated act of shooting. • Shoot two consecutive 3-round shot groups within a 4-centimeter circle at 25 meters.
Special Instructions
Ensure that— • The rear sight is on the proper setting (zero; M16A2/3=8/3+1; M16A4=6/3+2; M4=6/3). • The rear sight aperture is set on 300+1, not 800+1. • The M16A1's rear sight is set on the aperture marked L. • The small aperture is being used. • Proper and accurate shot group marking is enforced. • The firer's name is clearly marked on the target.
Observables
• Coaches analyze the firer's fundamentals. • Majority of the round must be inside the circle to be counted. • Two consecutive 3-round groups are shot with 6 of 6 rounds impacting inside of the 4- centimeter circle.

CONCEPT

5-1. Shot grouping is a form of practice firing with two primary objectives: firing tight shot groups and consistently placing those groups in the same location. Grouping exercises can be conducted anywhere that provides precise location of bullet hits and misses, such as a 25-meter live-fire zeroing range, KD range, EST 2000, LMTS, or location of misses and hits (LOMAH) system.

NOTES: 1. Shot grouping should be conducted between dry-fire training and zeroing.

2. The initial live-fire training should be a grouping exercise with the purpose of practicing and refining marksmanship fundamentals.

ORGANIZATION OF A 25-METER GROUPING RANGE

5-2. The organization and conduct of a grouping range are based on the availability of ammunition, number of personnel, and the firing ability of personnel in training.

5-3. To properly conduct a 25-meter grouping range, perform the following actions:

- Divide the unit into firing orders. The first order fires, while the second order coaches.
- Reserve ten firing points to conduct corrective instruction.
- Provide sandbags at each firing point to accommodate supported firing positions.
- Set up the 25-meter grouping range as depicted in Figure 5-1.

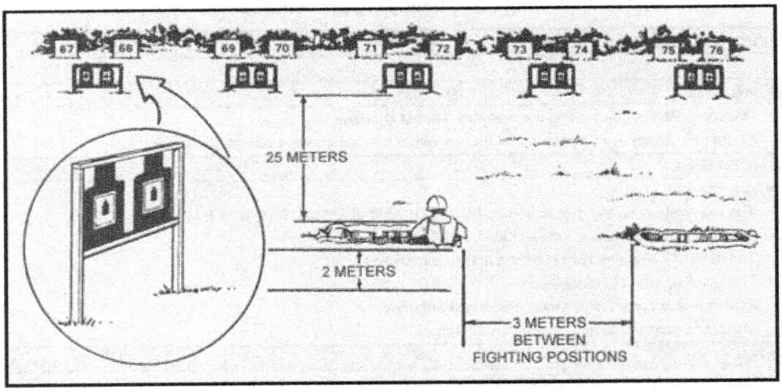

Figure 5-1. 25-meter range.

CONDUCT OF A 25-METER GROUPING FIRING

5-4. Each shot is fired from a supported firing position using the same point of aim (25-meter zeroing target's center of mass). The objective is to fire tight shot groups and to place those shot groups inside a 4-centimeter circle (the actual location of groups on the target is not important).

NOTES: 1. Since this is not a zeroing exercise, few sight adjustments are made unless the shot group is off of or barely on the 25-meter zeroing target.

2. No sight adjustments should be made until the firer can shoot six consecutive shots (two shot groups) inside a 4-centimeter circle. Once this is accomplished, the Soldier is ready to conduct zeroing procedures.

5-5. To conduct a 25-meter grouping firing—

NOTE: Before beginning the 25-meter grouping firing, each Soldier ensures that his sights are set for 25-meter firing.

(1) The Soldier fires a three-round shot group at the 25-meter zeroing target.

(2) The firing line is cleared, and the Soldier and coach move downrange to examine the shot group for fundamental errors, triangulate the shot group, and put the number 1 in the center of the shot group (Figures 5-2 and 5-3).

NOTE: If the shot group is off of the 25-meter zeroing target, the Soldier should mechanically zero the weapon. If the shot group is barely on the target, the Soldier should make a bold adjustment.

(3) The Soldier returns to the firing line and fires a second three-round shot group.

(4) The firing line is cleared, and the Soldier moves downrange to examine the second shot group, triangulate, and mark the center of the shot group with the number 2.

(5) The Soldier groups the two shot groups and marks the center.

5-6. The Soldier repeats Steps 1 through 5 until he places six out of six consecutive rounds inside a 4-centimeter circle. If the Soldier has not grouped in 18 rounds, he should be removed from the firing line and given remedial training before attempting to group again.

NOTE: To be counted, the majority of the round must be inside of the circle.

5-7. Once the Soldier has demonstrated firing proficiency from the supported firing position, grouping exercises can be conducted from the unsupported firing position. For example, 27 rounds are allocated for the grouping exercise (IET only). If the Soldier groups in 18 rounds, he can fire the remaining 9 rounds from the unsupported firing position.

SHOT GROUP MARKING

5-8. If the Soldier is to benefit from this exercise and if the instructor/trainer (or coach) is to provide useful guidance, the Soldier must mark each shot group for a clear record of his firing practice. The instructor/trainer must understand how to analyze shot groups correctly.

5-9. To properly mark the shot groups (Figure 5-2)—

(1) Connect the three bullet holes on the target with a straight line.

(2) Place a number inside of the shot group.

NOTES: 1. The number represents the center of the three shots.

2. When two shots are near one end of the group and the third shot is toward the other end, the number is placed closer to the two near shots (Figure 5-3).

3. This is not a precise marking that requires a measurement, but a procedure to help with shot group analysis.

5-10. The three-round shot group allows the firer's performance to be evaluated.

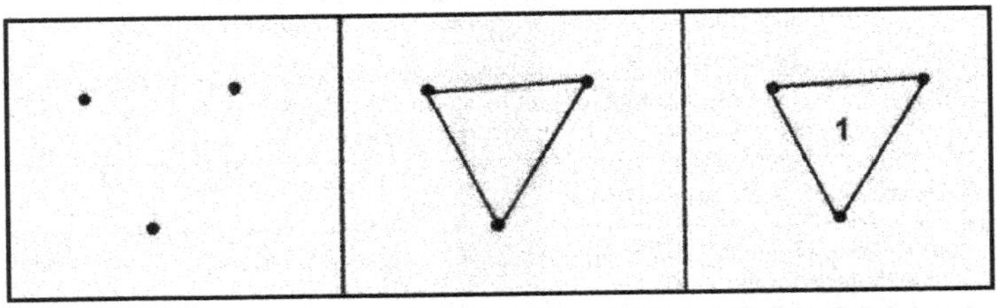

Figure 5-2. Shot group marking.

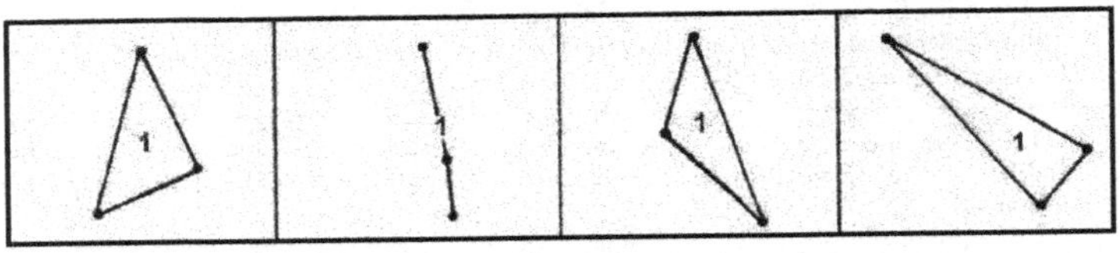

Figure 5-3. Central point of an odd-shaped group.

SINGLE SHOT GROUP ANALYSIS

5-11. The purpose of single shot group analysis is to identify firer errors on the single shots of a shot group so the Soldier can correct these errors while firing the next shot group.

5-12. Shot group analysis begins with the instructor/trainer observing the Soldier while he fires, looking for proper position, aim, trigger squeeze, and breathing. Then, the instructor/trainer analyzes the shot group to confirm problem areas.

NOTE: Coaches should not use shot group analysis without observing the firer.

5-13. The ideal shot group will have all three rounds within a 2-centimeter circle. Three rounds within a 4-centimeter circle is the minimum standard.

NOTE: M16A2 zeroing target squares are .96 centimeter in size, while M4 zeroing target squares are 1.3 centimeters in size.

Match-Grade Performance

5-14. The target shown in Figure 5-4 illustrates a match-grade quality weapon/ammunition combination. This combination places all bullets in almost the same hole and helps detect the firer's slightest errors.

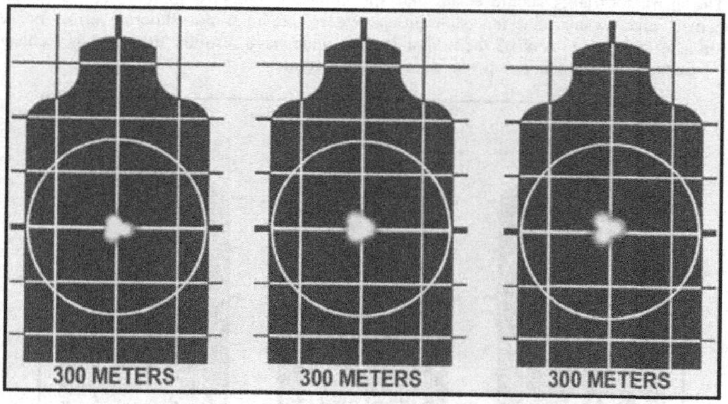

Figure 5-4. 25-meter match grade performance.

2-Centimeter Shot Groups

5-15. The variances of standard weapons and ammunition must be considered during shot group analysis. When firing a standard service weapon/ammunition combination, the dispersion pattern may be up to 2 centimeters apart without human error. The instructor/trainer must ensure the Soldier understands that his weapon or ammunition may not be capable of placing three rounds within a 1-centimeter square.

NOTE: The dispersion pattern is not considered a firer error.

5-16. The targets shown in Figure 5-5 reflect proper 25-meter shot group performances using standard weapon/ammunition combinations.

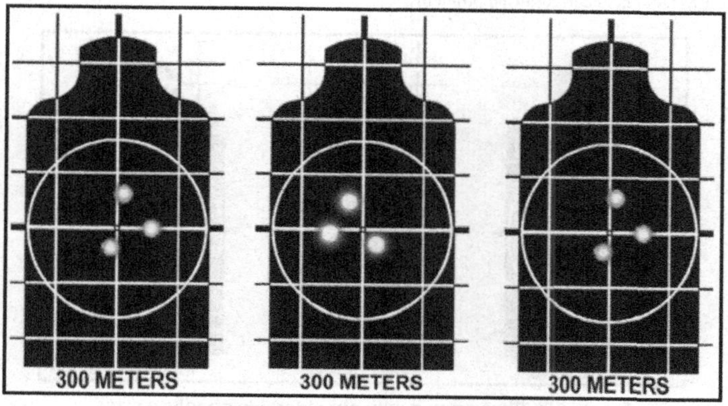

Figure 5-5. Shot groups with no firer error.

3- TO 4-CENTIMETER SHOT GROUPS

5-17. The targets shown in Figure 5-6 represent minimum acceptable firing performances; a better firing performance should be expected.

5-18. The instructor/trainer should ensure that the Soldier is properly applying the four marksmanship fundamentals and explain that this shot group size is due to minor shooting error, not weapon or ammunition performance. Any of these shot groups could have resulted from a minor change in sight picture, breathing, trigger squeeze, position, or an erratic round.

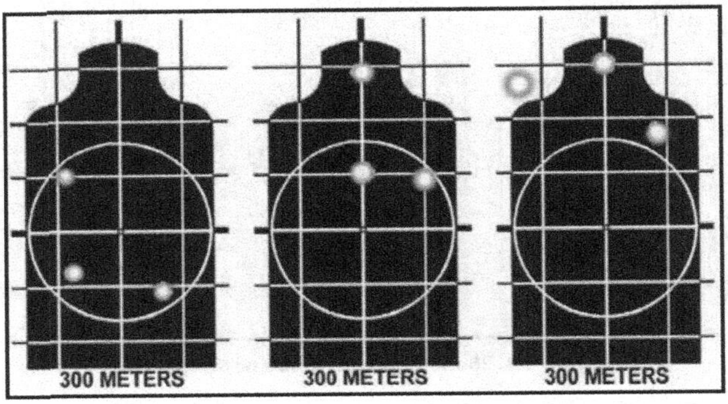

Figure 5-6. Shot groups with minor shooting error.

OVER 4- TO 5-CENTIMETER SHOT GROUPS

5-19. The targets shown in Figure 5-7 represent unacceptable firing performance.

5-20. The instructor/trainer should ensure that the Soldier is properly applying the four marksmanship fundamentals and explain that this shot group size is due to considerable shooting error, not weapon or ammunition performance. Any of these shot groups could have resulted from a change in position, sight picture, breathing, trigger squeeze, position, or an erratic round.

5-21. Soldiers who fire these shot groups should receive dry-fire training or remedial training on the EST 2000 or LMTS to help correct firing problems.

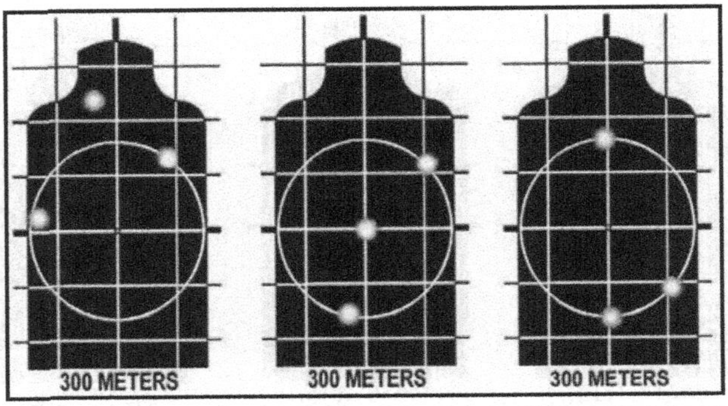

Figure 5-7. Shot groups with considerable shooting error.

Shot Groups Larger than 5 Centimeters

5-22. The targets shown in Figure 5-8 represent unacceptable firing performance; a better firing performance should be expected.

5-23. The instructor/trainer should ensure that the Soldier is properly applying the four marksmanship fundamentals and explain that this shot group size is due to major shooting error, not weapon or ammunition performance. Any of these shot groups could have resulted from a change in position, sight picture, breathing, or trigger squeeze, or the firer may have anticipated the shot.

5-24. Soldiers who fire these shot groups should receive extensive dry-fire training or remedial training on the EST 2000 or LMTS to help correct firing problems.

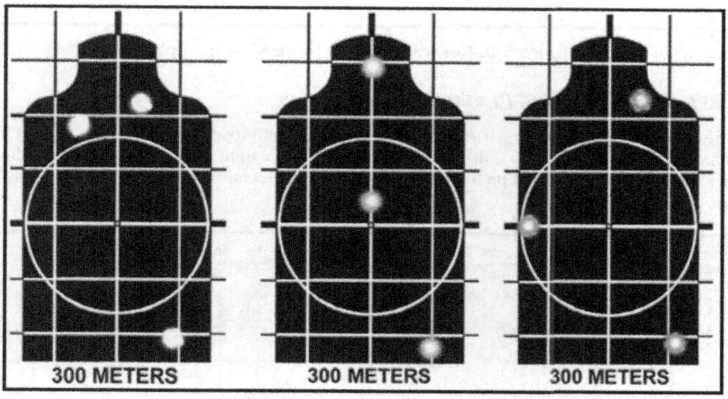

Figure 5-8. Shot groups with major shooting error.

MULTIPLE SHOT GROUP ANALYSIS

5-25. Multiple shot group analysis involves performing ongoing analysis of individual shot groups, while comparing them to each other for consistent aiming.

NOTE: If the Soldier is to benefit from this exercise, and if the instructor/trainer (or coach) is to provide useful guidance, the Soldier must mark each shot group individually and locate the center of more than one shot group.

5-26. To properly mark the shot groups (Figure 5-9)—

(1) Connect each individual shot group on the target with a straight line.

(2) Place a number inside the shot group.

NOTE: The number represents the center of the three shots.

---

(3) Connect the numbers, and place an X in the center.

NOTE: The X represents the center of the shot groups.

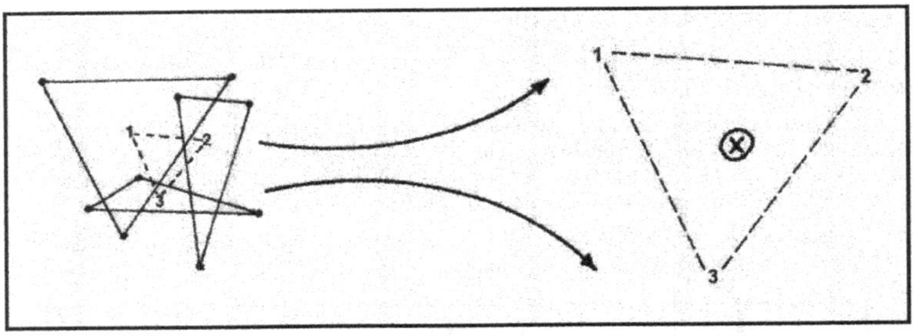

Figure 5-9. Central point of three shot groups.

ACCEPTABLE SHOT GROUPING PERFORMANCE

5-27. The shot groups in Figure 5-10 represent acceptable shot groups (4 centimeters or less) in the same location. The Soldier firing this shot grouping should make a sight change of left 10 and down 4. Any change should be clearly marked on the target and saved for reference. The Soldier is then ready to zero his weapon.

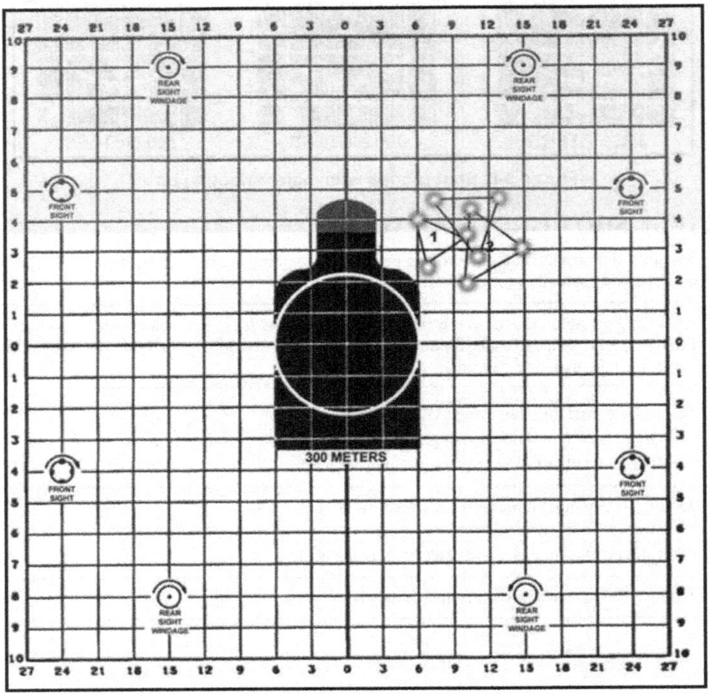

Figure 5-10. Acceptable shot grouping performance.

NOTES: 1. Location of the shot group on the 25-meter target is not important when conducting a grouping exercise. The size and dispersion of the shot groups are the main focus of this exercise.

2. Before the Soldier should be allowed to make any adjustments or start zeroing procedures, two consecutive shot groups must fall within a 4-centimeter circle when fired at 25 meters.

SHOT GROUPS WITH INCONSISTENT AIMING

5-28. The groups in Figure 5-11 indicate that the Soldier firing the shot groups is applying proper firing fundamentals, but is using a different point of aim each time a shot group is fired.

5-29. The instructor/trainer should question the Soldier's understanding of the aiming process and check his position for consistency. The instructor/trainer cannot determine which shot group best represents the firer's zero.

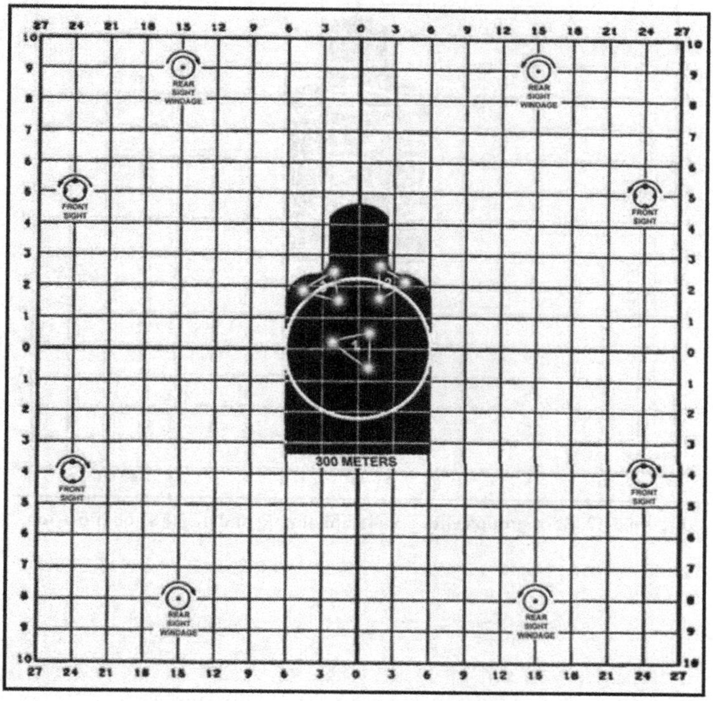

Figure 5-11. Shot groups with inconsistent aiming.

Shot Groups with Consistent Aiming and Major Shooting Error

5-30. The groups in Figure 5-12 indicate consistent aiming, but the Soldier is not applying the four fundamentals properly while firing each shot group.

5-31. The firer should be assigned a coach to troubleshoot his application of the four fundamentals in an attempt to isolate his firing errors.

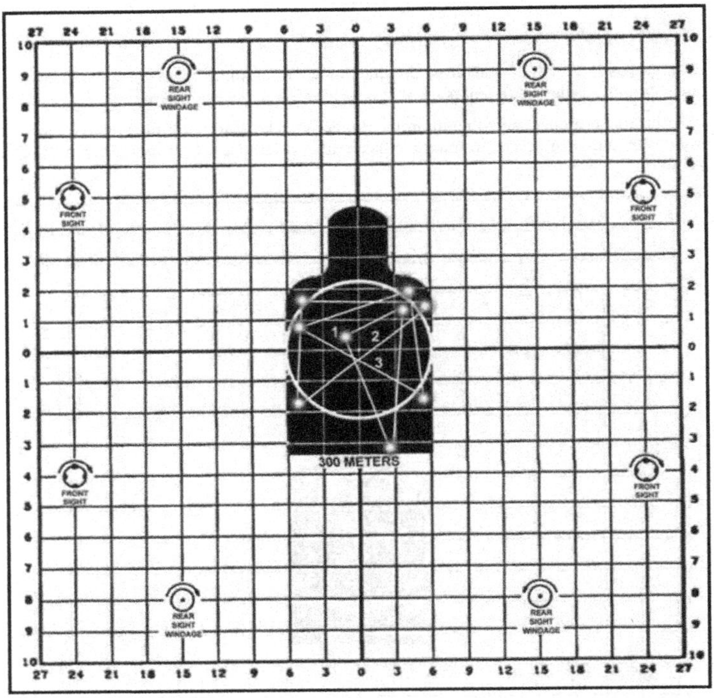

Figure 5-12. Shot groups with consistent aiming and major shooting error.

Shot Groups with Inconsistent Aiming and Major Shooting Error

5-32. The groups shown in Figure 5-13 indicate inconsistent aiming and major shooting error.

5-33. The firer should be assigned a coach to troubleshoot his application of the four fundamentals in an attempt to isolate his firing errors.

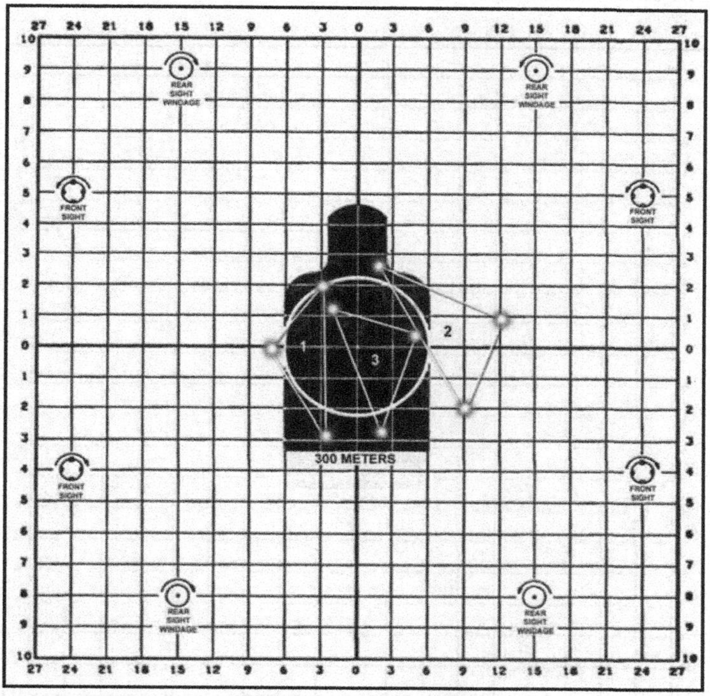

Figure 5-13. Shot groups with inconsistent aiming and major shooting error.

Shot Groups with Improper Vertical Placement

5-34. When viewed as nine shots, the shot groups shown in Figure 5-14 reflect proper horizontal placement of shots, but unsatisfactory vertical dispersion. This indicates a failure to vertically aim at the target's center of mass for each shot.

5-35. The instructor/trainer should check the Soldier's aiming procedure and adherence to marksmanship fundamentals.

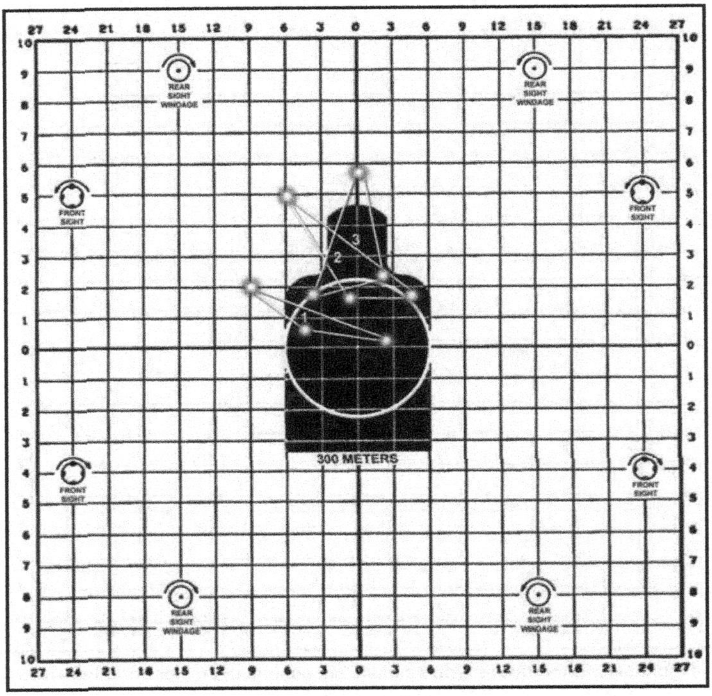

Figure 5-14. Shot groups with improper vertical placement.

Improper Shot Groups on the Edge of the Target

5-36. The shot groups shown in Figure 5-15 are improper groups.

5-37. The Soldier should make a bold sight change to bring the groups closer to the target's center and ensure that the shot groups remain on the 25-meter zeroing target.

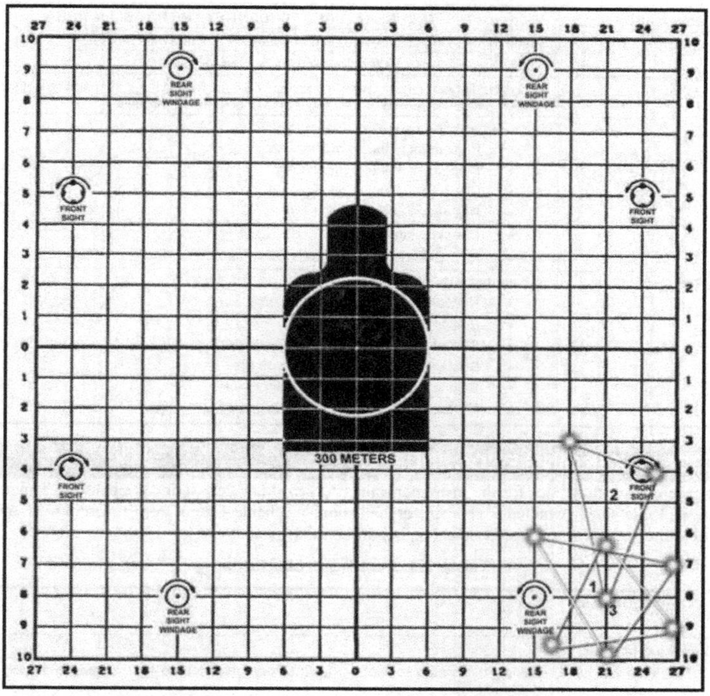

Figure 5-15. Improper shot groups on the edge of the target.

TROUBLESHOOTING THE FUNDAMENTALS

5-38. When troubleshooting the fundamentals, the coach's imagination is the only limiting factor. Table 5-2 outlines the techniques that can be used to identify errors in Soldiers' application of the fundamentals.

Table 5-2. Techniques used to identify errors in Soldiers' application of the fundamentals.

PROBLEM AREA	TECHNIQUE USED TO IDENTIFY THE PROBLEM AREA
Aiming	Attach the M16 sighting device, and observe the Soldier while he fires.
Breathing	Watch the rise and fall of the firer's chest for consistency.
Trigger squeeze	Place your finger over the firer's finger while he fires, feeling for jerking and smooth follow-through. Watch for jerking of the trigger and smooth follow-through.
Position	Observe the following areas for consistency: • Placement of the tip of the nose. • Placement of the trigger finger. • Placement of the nonfiring hand. • Placement of the legs. • Cheek-to-stock weld. • Positioning of equipment.
Other potential problem areas	Ensure that the— • Nonfiring-side eye is not shuttering. • Equipment is fitted properly. • Soldier is not flinching when the trigger is pulled. • Soldier is firing with the dominant eye. • Soldier is wearing glasses (if applicable). • Soldier is maximizing use of the supported position.

SECTION II. ZEROING PROCEDURES

This section provides guidelines for the instructor/trainer to zero M16-/M4-series weapons at 25 meters and at actual range (Table 5-3). It includes concept, organization, mechanical zero, zero recording, 25-meter sight settings, field fire sight settings, and troubleshooting of the fundamentals.

Table 5-3. Zeroing procedures.

ZEROING PROCEDURES
Instructional Intent
• Reinforce PMI while adjusting the confirmed shot groups (center of mass of the 4-centimeter circle with 5 out of 6 consecutive rounds at 25 meters).
Special Instructions
Ensure that— • The rear sight is on the proper setting (zero; M16A2/3=8/3+1; M16A4=6/3+2; M4=6/3). • The rear sight aperture is set on 300+1, not 800+1. • The M16A1's rear sight is set on the aperture marked L. • The small aperture is being used. • Proper and accurate shot group marking is enforced. • The firer's name is clearly marked on the target. • M16A2/A3/A4s zero on M16A2 zeroing targets. • M4s zero on M4 zeroing targets. • M16A1s zero on M16A1 zeroing targets.
Observables
• Coaches analyze the firer's application of the fundamentals. • The majority of the round must be inside of the circle to be counted. • Two consecutive 3-round shot groups are shot with 5 of 6 rounds in the 4- centimeter circle.

PURPOSE

5-39. The purpose of battlesight zeroing is to align the sights with the weapon's barrel given standard issue ammunition. When this is accomplished correctly, the point of aim and point of impact are the same at a given range (250 meters for the M16A1, 300 meters for the M16A2/A3/A4 and M4-series weapons). This sight setting provides the highest hit probability for most combat targets with minimum adjustment to the point of aim.

5-40. When standard zeroing procedures are followed, a properly zeroed weapon for one Soldier is close to the zero for another Soldier. When a straight line is drawn from the target's center to the tip of the front sightpost and through the center of the rear aperture, it makes little difference whose eye is looking along this line. There are many subtle factors that result in differences among individual zeros. Instructors/trainers should emphasize the similarity of individual zeros instead of the differences.

5-41. Most firers can fire with the same zeroed weapon if they properly apply marksmanship fundamentals. This information can be useful in three ways:

(1) If a Soldier has difficulty zeroing and the problem cannot be diagnosed, a good firer could zero the weapon to find the problem and eliminate the weapon as part of the problem.

(2) When a Soldier must fire another Soldier's weapon without opportunity to verify the zero by firing for example, picking up another man's weapon on the battlefield), the weapon will be closer to actual zero if the sights are left unchanged. This information is useful in deciding initial sight settings and recording zeros.

(3) All weapons in the arms room, even those not assigned, should have been previously zeroed by the last Soldier they were assigned to. Zeroing this newly assigned weapon should start with the sights left where they are.

SIGHT VARIANCE

5-42. There is no relationship between the specific sight settings a Soldier uses on his weapon to the sight settings he would use to zero another weapon, which makes it essential that each Soldier zeros the weapon that he is assigned. For example, a Soldier could zero his assigned weapon 10 clicks left of center; when zeroing another weapon, his adjustments could be 10 clicks right of center. This is due to the manufacturing difference between the weapons. Therefore, all newly assigned personnel should be required to zero their weapon as soon as possible after assignment to the unit. The same rule applies anytime a Soldier is assigned a weapon that is returned from field level or sustainment level maintenance, or when the zero is in question.

ORGANIZATION OF A 25-METER ZERO RANGE

NOTES: 1. All Soldiers should successfully group prior to zeroing.

2. If the Soldier is proficient at grouping, two shot groups should be fired to confirm proficiency prior to making any sight adjustments during zeroing procedures.

5-43. To properly conduct a 25-meter zero range, perform the following actions:

- Divide the unit into firing orders. The first order fires, while the second order coaches.
- Reserve firing points to conduct corrective instruction.
- Provide sandbags at each firing point to accommodate supported firing positions.

CONDUCT OF A 25-METER ZERO FIRING

5-44. On the 25-meter zero range, the Soldier applies the fundamentals while consistently aiming at the target's center of mass (A, Figure 5-16). The Soldier fires two separate three-round shot groups (B, Figure 5-16) and groups them. Based on the location of these groups, the Soldier makes the appropriate sight adjustments. Then, the Soldier fires two additional three-round shot groups to confirm that the adjustments have aligned the sights with the center of the target and that the bullets are in the 4-centimeter circle (Figure 5-17).

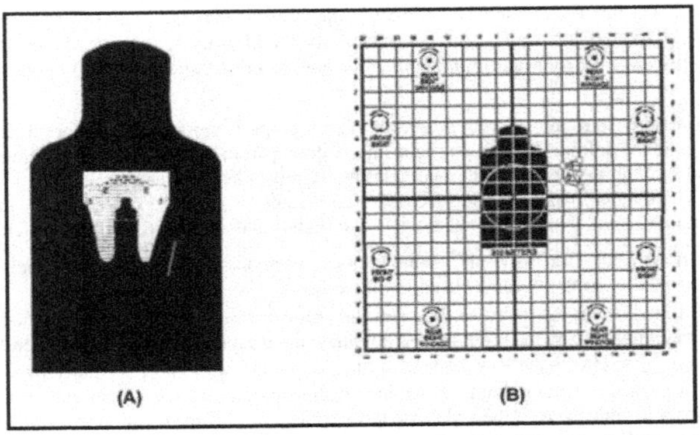

Figure 5-16. Correct aiming (A), initial shot group results (B).

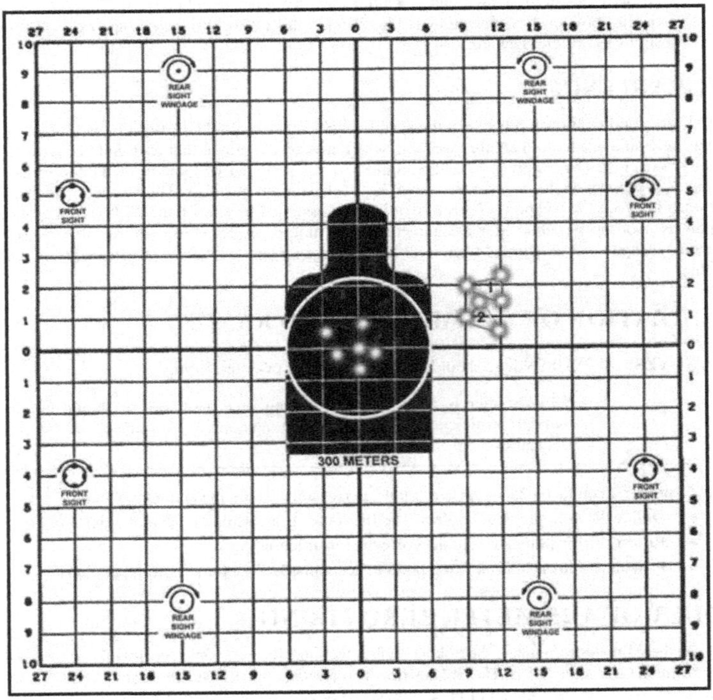

Figure 5-17. Final shot group results.

5-45. To conduct a 25-meter zero range—

NOTES: 1. Each Soldier ensures that his sights are set for 25-meter zeroing.

2. Soldiers fire each shot from a supported firing position using the same point of aim (target's center of mass).

3. Ensure that the correct 25-meter zero target is being used. For M16A1s, use NSN 6920-01-167-1392 (Figure 5-18); for M16A2s, M16A3s, M16A4s, M4s, and M4As, use NSN 6920-01-395-2949 (Figure 5-19; M16A2/A3 is printed on one side, and M16A4/M4/M4A is printed on the other).

(1) The Soldier fires a three-round shot group at the 25-meter zeroing target.

(2) The firing line is cleared, and he moves downrange to examine the shot group for fundamental errors, triangulates the shot group, and puts the number 1 in the center of the shot group.

(NSN 6920-01-167-1392)

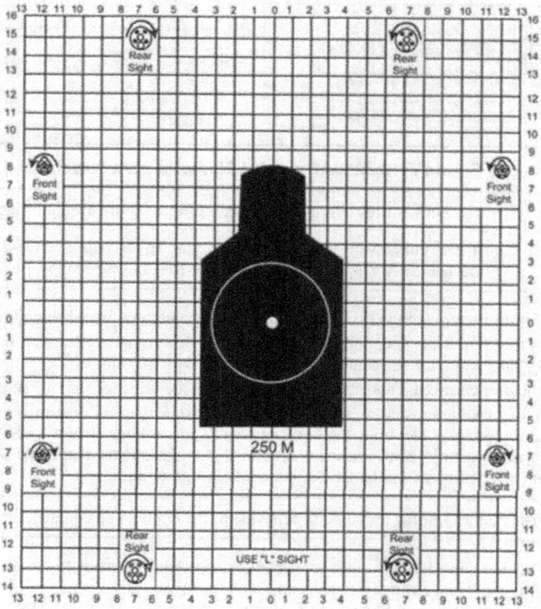

1. Aim at target center. Adjust sights to move shot group center as close as possible to white dot.

2. At completion of zero, rotate rear sight to unmarked aperture and weapon will be battlesight zero for 250 meters.

Figure 5-18. M16A1 25-meter zero target.

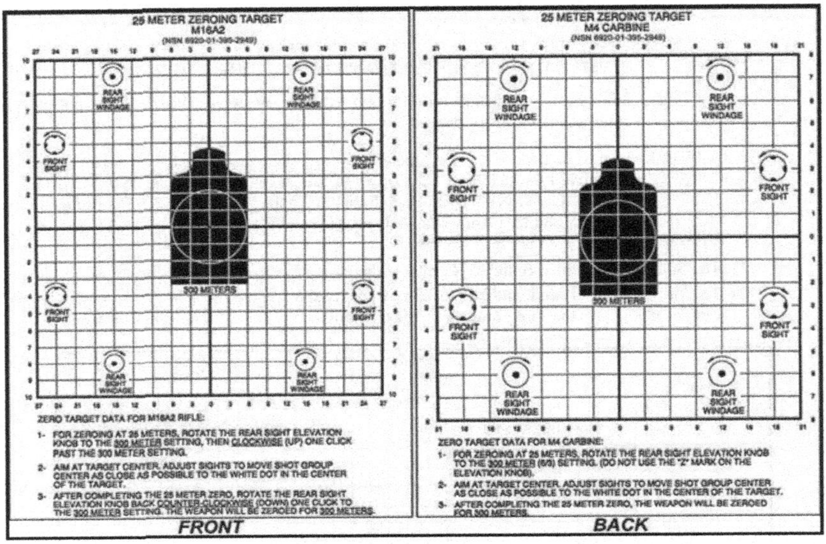

Figure 5-19. M16A2 and M4 25-meter zero target.

NOTES: 1. The Soldier fires two individual shot groups before a sight change is considered.

2. If the initial shot group is not on the target paper, the weapon should be mechanically zeroed before the Soldier fires this weapon again.

---

(3) The Soldier returns to the firing line and fires a second three-round shot group.

(4) The firing line is cleared, and the Soldier moves downrange to examine the second shot group, triangulate, and mark the center of the shot group with the number 2.

(5) The Soldier groups the two shot groups and marks the center of the two shot groups with an X.

(6) If the two shot groups fall within a 4-centimeter circle, the firer determines the sight adjustments he needs to make, identifies the horizontal and vertical lines closest to the X, and reads the 25-meter zeroing target to determine the proper sight adjustments. If the two shot groups do not fall within a 4-centimeter circle, the Soldier continues grouping.

NOTE: The majority of the round must be inside of the circle to be counted.

---

(7) The Soldier annotates any sight adjustments that need to be made to the weapon on the 25-meter zeroing target and ensures that his name is on the target.

(8) If five out of six rounds fell within the 4-centimeter circle, the Soldier is zeroed and can be removed from the firing line. If not, the Soldier returns to the firing line and makes sight adjustments.

5-46. Steps 1 through 8 are repeated until the Soldier places five out of six consecutive rounds inside the 4-centimeter circle. If the Soldier is not zeroed in 18 rounds, he should be removed from the firing line and given remedial training before attempting to zero again.

5-47. Once firing proficiency has been demonstrated from the supported firing position, zeroing exercises can be conducted from the unsupported firing position. For example, 18 rounds are allocated for the zeroing exercise. If the Soldier zeroes in 9 rounds, the Soldier can fire the remaining 9 rounds from the unsupported firing position.

CONDUCT OF A 25-METER ZERO FIRING USING THE LOCATION OF MISSES AND HITS SYSTEM

5-48. When using the LOMAH system on a KD range, zero confirmation is part of the program and will be shot as the first scenario. To conduct a 25-meter zero range using the LOMAH system, the Soldier shoots six rounds at the 175-meter/200-yard target while aiming at the target's center of mass. The outcome is evaluated using the following guidelines:

- If the shot group falls within the 11-inch circle on the LOMAH monitor, the Soldier continues the programmed scenario, which is identical to the downrange feedback scenario without LOMAH.
- If the Soldier shoots a shot group that is 11 inches or smaller but is clearly not zeroed, the instructor/trainer assists the Soldier in making sight adjustments based upon the data provided on the LOMAH monitor.
- If the shot group is not tight (greater than 11 inches), the Soldier should be removed from the firing line and given remedial training on the four fundamentals of marksmanship.

SECTION III. KNOWN DISTANCE RANGE

This section provides guidelines for the instructor/trainer to conduct a KD range and apply the effects of wind and gravity. This section also addresses three types of KD ranges: the standard KD range, the KD record fire range, and the modified field fire range.

NOTE: See Table 5-4 for the current training program.

Table 5-4. Downrange feedback.

DOWNRANGE FEEDBACK
Instructional Intent
• Reinforce PMI while shooting from the prone supported and unsupported firing positions. • Build the Soldier's confidence in his ability to hit where he aims while applying the effects of wind and gravity at range.
Special Instructions
Ensure that— • The effects of wind and gravity are thoroughly explained. • The rear sight is on the proper setting (M16A2/3=8/3; M16A4 and M4=6/3 flush; M16A1=the unmarked aperture, short-range). • The rear sight aperture is set on 300, not 800.
Observables
• Spotters provide correct feedback to firers. • Soldiers hit 8 of 10 targets at 100 meters. • Soldiers hit 14 of 20 targets at 200 meters. • Soldiers hit 5 of 10 targets at 300 meters.

CONCEPT

5-49. A KD range has three primary objectives:

(1) Fire tight shot groups at a known distance.

(2) Make sight adjustments at range (not an objective for IET Soldiers) while experiencing the effects of wind and gravity.

(3) Participate in marksmanship testing.

5-50. KD firing brings the Soldier one step closer to being able to fire during combat. The Soldier is provided information concerning the precise hit-or-miss location of every bullet fired. KD firing is conducted with a single, clearly visible target at a known distance, and the Soldier can establish a position that provides a natural point of aim on that single target. Consider the following:

- On the standard KD range, Soldiers fire at 100-, 200-, and 300-meter targets without any time constraints.
- On the KD record fire range, Soldiers fire at 100-, 200-, and 300-meter targets with time constraints.
- On the modified field fire range, Soldiers fire at 100-, 200-, and 300-meter targets on a standard 50- to 300-meter field fire qualification range.

NOTES: 1. If a qualification range is not available, this exercise may be shot on a standard 75- to 300-meter field fire range. Targets and target frames must be set up to accommodate this training.

2. On ranges that are built in yards instead of meters, the same KD targets will be used. The difference is so small that it does not need to be considered.

- The KD range does not require Soldiers to detect targets, estimate ranges to targets, scan sectors of fire, respond to surprise targets, respond to short-exposure targets, or engage multiple targets.
- An advantage of a KD range is the ability to see precisely where each bullet hits. To benefit from this training, Soldiers must clearly see the results of each firing, whether a group, single shot, or 10-round exercise.

KNOWN DISTANCE TARGET DESCRIPTION

5-51. Downrange feedback training should include detailed explanations of the targets. Consider the following:

- KD targets are large enough to capture all bullets fired. Standard E-type and F-type silhouettes can be used if standard KD targets are not available.
- The 16-centimeter circle on 100-meter targets, the 32-centimeter circle on 200-meter targets, and the 48-centimeter circle on 300-meter targets equate to the 4-centimeter zeroing target at 25 meters. If the Soldier's shot group falls within the 4-centimeter circle at 25 meters, it will fall within the circle on the target being shot. If the round falls outside of the circle, the round will clearly miss the 300-meter target (Figure 5-20).
- An X is located in the bottom portion of the circle to show the firer the target's center of mass, where he must aim when his weapon is zeroed.
- The grid system on the targets in Figure 5-20 equates to that of the 25-meter zeroing target. For example, one click on the front sightpost equals one square on the 25-meter zeroing target and also equals one square on the target being shot. Information similar to that on the zeroing target has been overprinted to help Soldiers apply sight adjustments.

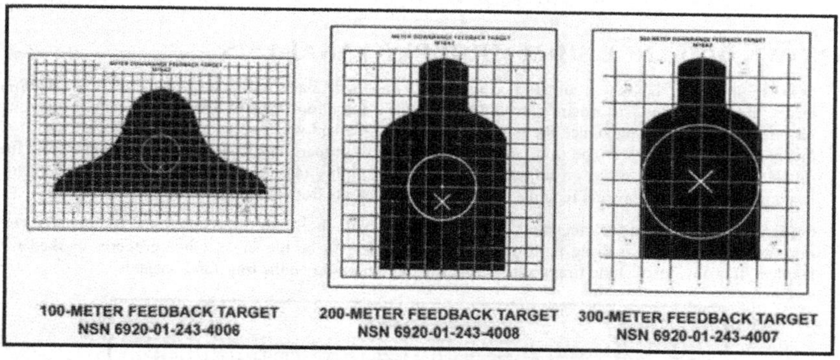

Figure 5-20. Downrange feedback targets.

MARKING KNOWN DISTANCE RANGE TARGETS

5-52. When the initial shot group is fired, target spotters/markers (Figure 5-21) should be placed in each bullet hole, placing the white side on the silhouette and the black side off of the silhouette. This procedure ensures that the firer can see where the rounds impacted and has two benefits:

- Instructors/trainers can observe the firer's performance and focus their attention on the Soldiers having the greatest problems.
- Soldiers are motivated to fire better since their peers can observe their performance.

5-53. On the second and subsequent shot groups, the target spotters/markers should be moved and placed in the holes of the new shot group. The old holes must be pasted using black pastors on black and white pastors on white. Failure to paste all bullet holes makes it difficult to determine one shot group from another.

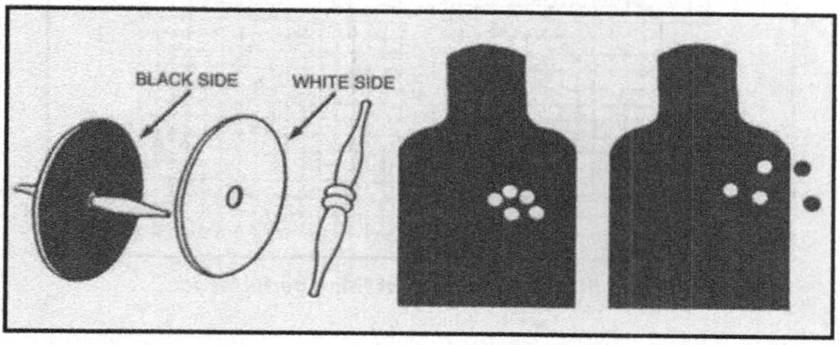

Figure 5-21. Target marking with spotters (markers).

KNOWN DISTANCE SHOT GROUPING ANALYSIS

5-54. Figure 5-22 shows two targets that were both shot with three individual rounds (A). On a pop-up target, these two firing performances would provide the same information to the firing line; each target was hit once and missed twice. Once the targets are properly marked with spotters, it becomes clear why only one round hit either target: The firer on the left is failing to properly apply the four fundamentals; the firer on the right needs to make an adjustment to his iron sights (assuming that wind was not a factor), triangulate the shot group, and read the appropriate adjustments from the target.

5-55. Figure 5-22 shows another two targets that were both shot with three individual rounds (B). On a pop-up target, these two firing performances would appear to be the same. Once properly marked with spotters, it is obvious that the firer on the left needs more training on the four fundamentals.

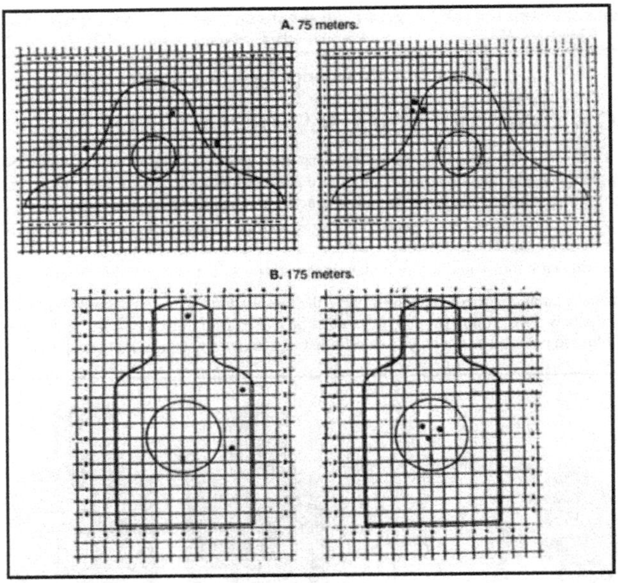

Figure 5-22. Comparison of firing performance.

KNOWN DISTANCE ZEROING

5-56. The 300-meter target can be used to confirm weapon zero or to refine the zero obtained on the 25-meter range. When Soldiers properly compensate for the wind, the zero on this target is more valid than the zero obtained on the 25-meter range. Soldiers should fire two five-round shot groups to confirm zero or three-round shot groups to refine their zero. The pit crews should spot targets after each shot group is fired. If the crosswind exceeds five miles per hour, KD zeroing should not be attempted.

NOTES: 1. IET Soldiers only zero on the 25-meter range.

2. For M16A2/3/4, M4, and M4A1 weapons only: Soldiers should use the unmarked aperture for zeroing and target engagement at all distances on the KD range. When engaging targets beyond 300 meters, the windage knob should be adjusted to the range of the target. 400-meter targets will be set on 4 flush, and 450-meter targets will be set on 4 plus two clicks.

3. For M16A1 rifles only: Soldiers should use the unmarked aperture (short-range) for refining zero at 300 meters. For target engagements beyond the 300-meter line, Soldiers should use the long-range aperture (L).

Elevation and Windage Adjustments at a Distance

5-57. The elevation and windage rule states that one click of elevation or windage moves the strike of the bullet a specific distance at a specific range. At a range of 25 meters, one click of windage moves the strike of the bullet .33 centimeters, and one click of elevation on the front sight moves the strike of the bullet .83 centimeters. To compute the distance (D) one click of elevation (front sight) or windage moves the strike of a bullet at a given range (R)—

(1) Divide the range (expressed in meters) by 25.

(2) Multiply by either .33 centimeters for windage or .83 centimeters for elevation.

Windage: D = R ÷ 25 m x .33 (D = distance in cm)

Elevation: D = R ÷ 25 m x .83 (R = distance in cm)

Windage

5-58. To compute the distance one click of windage moves the strike of the bullet at a range of 300 meters—

(1) Divide 300 meters by 25 meters.

(2) Multiply by .33 centimeters.

D = 300 m ÷ 25 m = 12

12 x .33 = 3.96 cm

Round 3.96 cm up to 4.0 cm.

5-59. One click of windage moves the strike of the bullet 4 centimeters at 300 meters.

NOTE: Tables 5-5 and 5-6 show the amount of change in windage of the strike of the bullet at various ranges.

Elevation—Front Sight

5-60. To compute the distance one click of elevation (front sight) moves the strike of the bullet at a range of 300 meters—

(1) Divide 300 meters by 25 meters.
(2) Multiply by .83 centimeters.

D = 300 m ÷ 25 m = 12

12 x .83 = 9.96 cm

Round 9.96 cm up to 10 cm.

5-61. One click of elevation on the front sight moves the strike of the bullet 10 centimeters at a range of 300 meters.

NOTE: Tables 5-5 and 5-6 show the amount of change in elevation of the strike of the bullet at various ranges.

Elevation—Rear Sight

5-62. The elevation knob adjusts elevation 1.1 inch for each click at 100 meters, with each dot and each number representing one click of elevation.

Table 5-5. M16A2/3 and front sightpost of an M16A4.

DISTANCE (m)	DISTANCE 1 CLICK WILL ADJUST THE POINT OF IMPACT		
	Front Sightpost	Windage Knob	Elevation Wheel
25	.83 cm (3/8 in)	.33 cm (1/8 in)	.5 cm (1/4 in)
50	1.5 cm (5/8 in)	.5 cm (1/4 in)	1.5 cm (1/2 in)
75	2.5 cm (1 in)	1.0 cm (3/8 in)	2.0 cm (3/4 in)
100	3.5 cm (1 3/8 in)	1.5 cm (1/2 in)	2.75 cm (1 in)
150	5.0 cm (2 in)	2.0 cm (3/4 in)	4.0 cm (1 1/2 in)
175	6.0 cm (2 3/8 in)	2.25 cm (7/8 in)	5.0 cm (2 in)
200	6.5 cm (2 5/8 in)	2.5 cm (1 in)	5.5 cm (2 1/4 in)
250	8.5 cm (3 3/8 in)	3.5 cm (1 1/4 in)	7.0 cm (2 3/4 in)
300	10.0 cm (4 in)	4.0 cm (1 1/2 in)	8.5 cm (3 1/4 in)
400	13.5 cm (5 3/8 in)	5.5 cm (2 1/4 in)	11.0 cm (4 1/2 in)
500	17.0 cm (6.69 in)	6.5 cm (2 1/2 in)	14.0 cm (5 1/2 in)
600	20.5 cm (8.05 in)	8.0 cm (3 1/8 in)	16.75 cm (6 1/2 in)
700	24.0 cm (9.45 in)	9.0 cm (3 5/8 in)	19.5 cm (7 1/2 in)
800	27.5 cm (10.83 in)	10.5 cm (4 1/8 in)	22.5 cm (8 3/4 in)
NOTE: All values are rounded off.			

Table 5-6. M4/M4A1 and windage of an M16A4.

DISTANCE (m)	DISTANCE 1 CLICK WILL ADJUST THE POINT OF IMPACT		
	Front Sightpost	Windage Knob	Elevation Wheel
25	1.2 cm (1/2 in)	.5 cm (1/4 in)	.5 cm (1/4 in)
50	2.4 cm (1 in)	1.5 cm (1/2 in)	1.5 cm (1/2 in)
75	3.6 cm (1 1/2 in)	2.0 cm (3/4 in)	2.0 cm (3/4 in)
100	4.8 cm (1 7/8 in)	2.75 cm (1 in)	2.75 cm (1 in)
150	7.2 cm (2 7/8 in)	4.0 cm (1 1/2 in)	4.0 cm (1 1/2 in)
175	8.4 cm (3 3/8 in)	5.0 cm (2 in)	4.0 cm (2 in)
200	9.6 cm (3 3/4 in)	5.5 cm (2 1/4 in)	5.5 cm (2 1/4 in)
250	12.0 cm (4 3/4 in)	7.0 cm (2 3/4 in)	7.0 cm (2 3/4 in)
300	14.4 cm (5 3/4 in)	8.5 cm (3 1/4 in)	8.5 cm (3 1/4 in)
400	19.2 cm (7 1/2 in)	11.0 cm (4 1/2 in)	11.0 cm (4 1/2 in)
500	24.0 cm (9 1/2 in)	14.0 cm (5 1/2 in)	14.0 cm (5 1/2 in)
600	28.8 cm (11 1/4 in)	16.75 cm (6 1/2 in)	16.75 cm (6 1/2 in)
NOTE: All values are rounded off.			

CONDUCT OF A STANDARD KNOWN DISTANCE RANGE

NOTE: If the range is equipped with the LOMAH system, a firing order will be used to operate the LOMAH throughout the period of instruction and will be fired last.

5-63. Standard KD ranges are conducted using the following considerations:

- The standard KD range is conducted with paper targets at 100, 200, and 300 meters to obtain downrange feedback at 75, 175, and 300 meters (Figure 5-23). Soldiers fire shot groups progressively at the targets.
- Half of the bullets are fired from the supported firing position, and the other half are fired from the unsupported firing position.
- The wind speed and direction must be determined before firing, and the firer must know the distance to the target.
- Soldiers mark the targets after firing each shot group. Based on this feedback, Soldiers receive a critique from their instructor/trainer or coach and apply the adjusted point of aim (as necessary).
- The downrange feedback exercise must be conducted within the constraints of time, ammunition, and available ranges.
- If 30 rounds of ammunition are available for training, firing three-round shot groups 10 times is preferable to firing five-round shot groups 6 times.
- Once the Soldier understands the concept for adjusting the point of aim to compensate for the effects of wind and gravity, he is ready to apply his knowledge on the field fire range.

75-Meter Targets

5-64. Instructors/trainers can provide feedback after each round, each three-round shot group, or each five-round shot group on the 100-meter feedback targets. No time limit is placed on the firer. Soldiers fire from the supported firing position and from the unsupported firing position. Then, the targets are marked and evaluated. Feedback consists of a critique of performance, adjustments to the point of aim, effects of wind and gravity, and shot placement. Target spotters mark the bullet holes so hits can be viewed from the firing line.

NOTE: IET Soldiers fire one five-round shot group from the supported firing position and one five-round shot group from the unsupported firing position. They must hit 8 out of 10 targets.

175-METER TARGETS

5-65. Firers engage the 200-meter target using the same downrange procedures as when engaging the 100-meter target.

NOTE: IET Soldiers fire 10 rounds from the supported firing position and 10 rounds from the unsupported firing position. They must hit 14 out of 20 targets.

300-METER TARGETS

5-66. Firers engage the 300-meter target using the same downrange procedures as when engaging the 100-meter target.

NOTE: IET Soldiers fire one five-round shot group from the supported firing position and one five-round shot group from the unsupported firing position. They must hit 5 out of 10 targets.

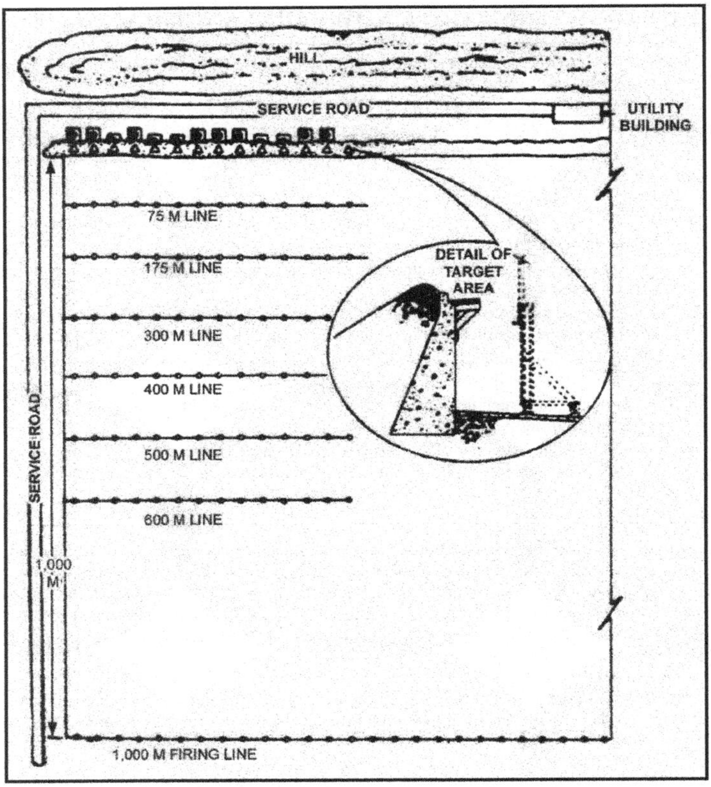

Figure 5-23. Known distance range.

KNOWN DISTANCE RECORD FIRE RANGE

NOTE: See paragraphs 6-79 through 6-82 of Chapter 6 for information about the alternate course KD record fire range.

MODIFIED FIELD FIRE RANGE

5-67. A modified field fire range can be used for downrange feedback. To conduct downrange feedback, minor changes must be made to a standard field fire range. Target frames, like those used on the 25-meter range, are placed on a standard qualification range at 100, 200, and 300 meters. The standard KD range or the KD record fire range can be conducted on the modified field fire range.

NOTE: The firing line must be cleared, moved to the targets for marking, and returned each time a firing order fires.

RECORD OF PERFORMANCE

5-68. During the conduct of downrange feedback, a record of performance should be kept on DA Form 5239-R (100-, 200-, and 300-Meter Downrange Feedback Scorecard) for the AAR.

NOTE: See Appendix B for a sample completed form and the end of this publication for a blank, reproducible copy.

5-69. As Soldiers complete each phase and achieve the performance standard for that range, they should receive a critique. Instructors/trainers must ensure that Soldiers do not progress to a greater range until they become proficient at closer ranges.

25-METER ZERO STANDARD

5-70. A standard E-type silhouette is 48.26 centimeters wide; a circle (angle) that is 48.26 centimeters at 300 meters is 4 centimeters at 25 meters. A Soldier who can fire all bullets in a 4-centimeter circle at 25 meters and adjust the sights for zero will hit the target at ranges as far away as 300 meters (Figure 5-24).

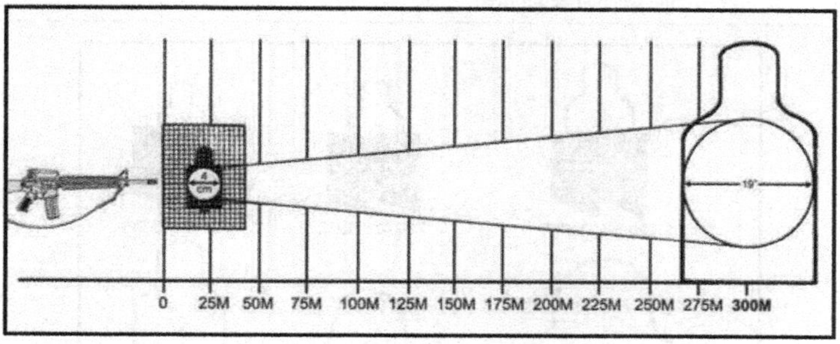

Figure 5-24. 25-meter zero standard.

SECTION IV. EFFECTS OF WIND AND GRAVITY

Marksmanship instructors/trainers should know how the effects of wind and gravity influence the flight of the bullet, and Soldiers should know how to compensate for such bullet displacement. This instruction is appropriate for all marksmanship training and concurrent training.

EFFECTS OF GRAVITY

5-71. Gases created by gunpowder push each round out of the end of the barrel. The barrel must be elevated slightly to allow the round to travel farther, creating an arc. The round will travel straight until it slows down and is gradually pulled to the ground by gravity. Each round fired will be pushed approximately the same distance and will roughly follow the same path.

NOTE: The farther the round travels, the faster it begins to fall.

5-72. When the firer zeroes his weapon, he aligns his line of sight to cross the path of the round at the distance at which he wants to zero his weapon. For example, a 300-meter zero means that the line of sight crosses the path of the round at 300 meters. If the firer engages a target at a distance other than 300 meters (excluding 25 meters), the path of the round hits the target either before or after it crosses the line of sight. If the firer wants his rounds to impact the center of mass, he must adjust his point of aim up or down to account for gravity.

ADJUSTED POINT OF AIM BASED ON GRAVITY

5-73. An adjusted point of aim (Figure 5-25) is intended to increase hit probability when properly presented. However, Soldiers can become confused, which could result in degraded performance. All Soldiers should be taught to aim at the target's center of mass unless they are confident that they know the range to the target. If adjusting the point of aim confuses the Soldier, he should aim at the target's center of mass. These points of aim place the center of each shot group in the target's center of mass (assuming a perfect zero and no firer error).

NOTES:
1. These adjustments are small and should only be applied by competent firers who wish to improve their firing performance.
2. Because the difference between M16- and M4-series weapons is so small and to avoid confusion, the same adjusted points of aim should be used regardless of the weapon being fired.

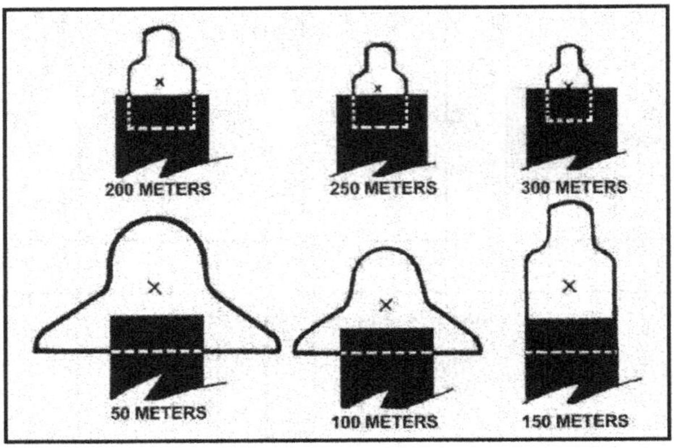

Figure 5-25. M16-/M4-series weapon aiming points.

EFFECTS OF WIND

5-74. Wind affects the bullet similar to the way gravity does: the farther the round travels, the farther the wind will push the round in the direction the wind is blowing. The faster the wind is blowing, the farther the wind will push the bullet.

WIND DIRECTION

5-75. The effects of wind vary depending on changes in wind speed and direction. Wind is classified by the direction it is blowing in relationship to the firer/target line. The clock system is used to indicate wind direction and value (Figure 5-26). This system works as follows:

- Winds that blow from the left (9 o'clock) or right (3 o'clock) are called full-value winds because they have the most effect on the bullet.
- Winds that blow at an angle from the front or rear are called half-value winds because they have about half the effect on the bullet as full-value winds.
- Winds that blow straight into the firer's face or winds that blow straight into the target are termed no-value winds because they have minimal effect on the bullet.

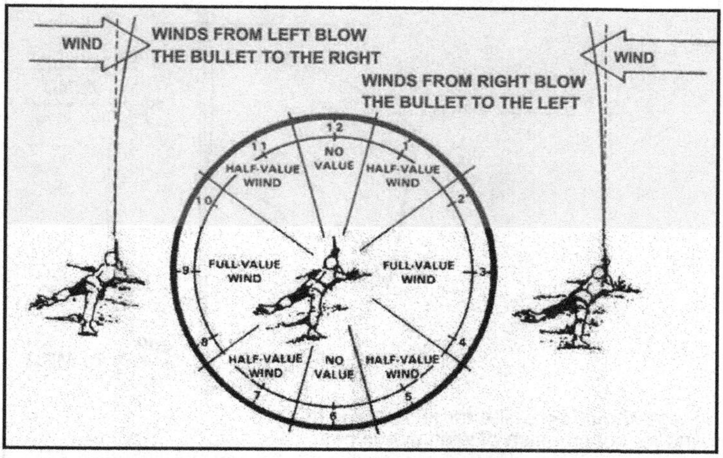

Figure 5-26. Determine wind value using the clock method.

WIND SPEED

5-76. Wind is variable and sometimes quite different at the firing position than at the target's position. Consider the following:

- When wind is blowing hard at the firing line, trees, brush, or terrain could protect the bullet's path.
- Wind can vary by several miles per hour between the time a measurement is taken and when the bullet is fired.

5-77. Therefore, training time should not be wasted trying to teach Soldiers an exact way to measure wind speed. Soldiers should understand that wind can blow a bullet off course, but they should not overcompensate and miss targets by applying too much hold-off.

5-78. A wind gauge can be used for precise measurement of wind velocity. When a gauge is not available, velocity is estimated using one of the following methods:

- Flag method.
- Pointing method.
- Observation method.

Flag Method

5-79. To perform the flag method (Figure 5-27)—

(1) Observe a flag or any cloth-like material hanging from a pole.

(2) Estimate the angle formed at the juncture of the flag and pole.

(3) Divide this angle by the number 4.

NOTE: The answer is the wind velocity expressed in miles per hour.

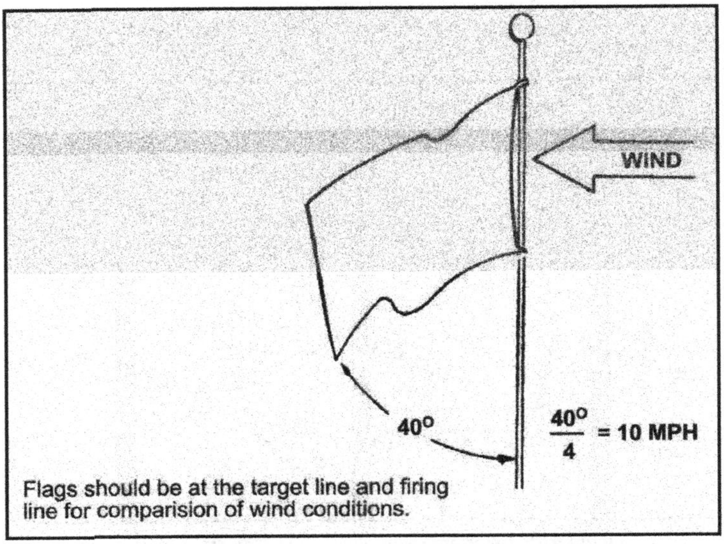

Figure 5-27. Determine wind speed using the flag method.

Pointing Method

5-80. If a flag is not visible, the firer can use the pointing method. To perform the pointing method (Figure 5-28)—

(1) Drop a piece of paper, leaf, or other light material from the shoulder.
(2) Point directly at the place where it lands.
(3) Estimate the angle created by the pointing arm.
(4) Divide this angle by the number 4.

NOTE: The answer is the approximate wind speed at the firing position expressed in miles per hour.

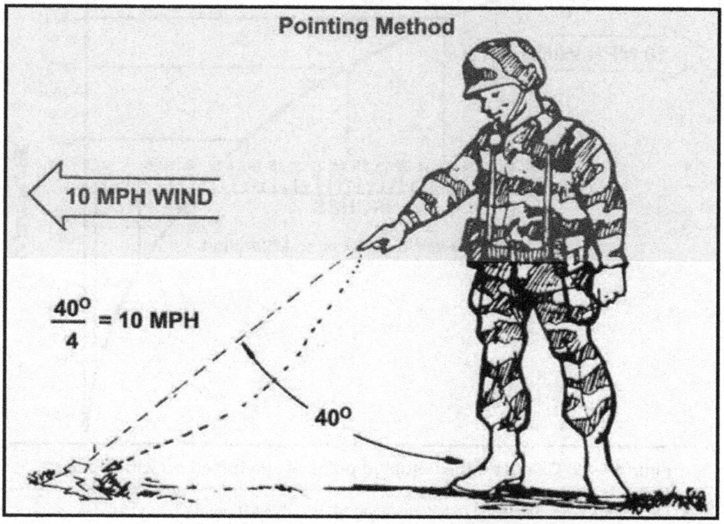

Figure 5-28. Determine wind speed using the pointing method.

Observation Method

5-81. If the flag or pointing methods cannot be used, the following information can assist in determining wind velocities:

- Winds less than 3 miles per hour can barely be felt by the firer, but the presence of slight wind can be determined by drifting smoke.
- Winds of 3 to 5 miles per hour can be felt lightly over the firer's face.
- Winds of 5 to 8 miles per hour constantly move the leaves of trees.
- Winds of 8 to 12 miles per hour raise dust and loose paper.
- Winds of 12 to 15 miles per hour cause small trees to sway.

ADJUSTED POINT OF AIM BASED ON WIND SPEED

5-82. Figure 5-29 illustrates how the effects of wind on the bullet are similar to the effects of gravity—as range increases, the effect of wind increases. For example, a 10 mile-per-hour full-value wind moves an M16A1 (M193) bullet from about ½ of an inch at 25 meters to about 15 inches at 300 meters.

5-83. Table 5-7 displays the wind effects for all conditions—a wind of greater speed increases bullet movement by a uniform amount. For example, a 15 mile-per-hour wind moves the bullet ¾ of an inch at 25 meters and about 22 ½ inches at 300 meters. A half-value wind moves the strike of the round in a 10 mile-per-hour wind ¼ of an inch at 25 meters and 7 ½ inches at 300 meters.

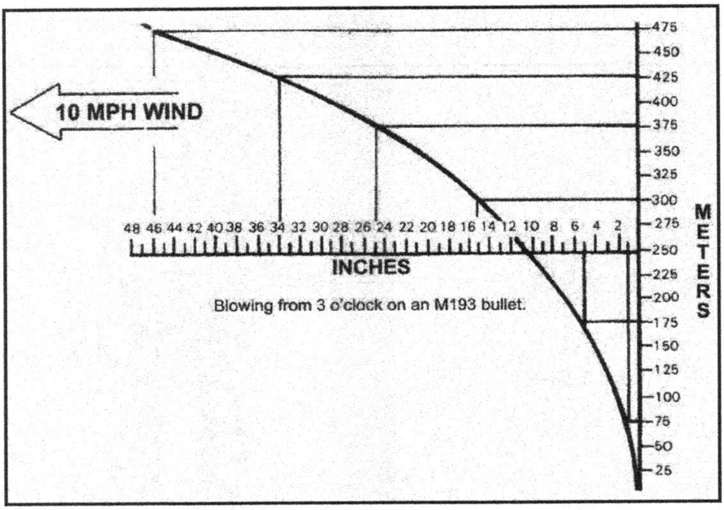

Figure 5-29. Calculate the adjusted point of aim based on wind speed.

NOTE: Table 5-7 can be used to calculate the M193 adjusted point of aim based on wind speed.

Table 5-7. M193 calculated adjusted point of aim based on wind speed (full value).

WIND SPEED (mph)	RANGE (m)								
	25	50	75	100	150	175	200	250	300
	DISTANCE MOVED (in)								
5	1/4	3/8	1/2	1	2	2.5	3.5	5	7.5
10	1/2	3/4	1	2	4	5	7	10	15
15	3/4	1-1/8	1.5	3	6	7.5	10.5	15	22.5

DRIFT FOR A 10 MILE-PER-HOUR WIND USING 5.56-MILLIMETER M855 AMMUNITION

5-84. Table 5-8 illustrates the drift using M855 5.56-millimeter ball ammunition fired in an M16A2 rifle with a 300-meter battlesight zero.

Table 5-8. Drift for 10 mile-per-hour wind using M855 ammunition.

DRIFT FOR 10 mph WIND USING M855 5.56-mm BALL AMMUNITION FIRED IN M16A2 RIFLE WITH A 300-METER BATTLESIGHT ZERO				
Range (m)	**Velocity (fps)**	**Trajectory (in)**	**Drop (in)**	**Drift (in)**
0	3,100	-2.5	0.0	0.0
100	2,751	+4.4	-2.3	1.1
200	2,420	+5.8	-10.2	4.9
300	2,115	0.0	-25.3	11.8
400	1,833	-15.0	-49.5	22.4
500	1,569	-42.9	-86.7	38.0
600	1,323	-88.2	-141.3	59.5
700	1,106	-156.1	-220.9	88.4
800	1,010	-267.7	-339.2	124.9

ADJUSTED POINT OF AIM BASED ON GRAVITY AND WIND SPEED

5-85. Wind has a minor effect on the M16 bullet (relative to the size of the target) at ranges out to 100 meters. When engaging targets in excess of 150 meters in heavy winds, Soldiers adjust the point of aim for the wind to increase the probability of a hit. Wind effects are uniform in relation to speed—that is, a 5 mile-per-hour wind has half the effect of a 10 mile-per-hour wind, and a 20 mile-per-hour wind has twice the effect of a 10 mile-per-hour wind.

5-86. Firers must adjust their points of aim into the wind to compensate for its effects. If they miss a distant target and wind is blowing from the right, they should aim to the right for the next shot. A guide for the initial adjustment is to split the front sightpost on the edge of the target facing the wind (Figure 5-30).

5-87. Newly assigned Soldiers should aim at the target's center of mass for the first shot, and then adjust for wind when they are confident that wind caused the miss. Experienced firers should apply the appropriate hold-off for the first shot, but should follow the basic rule—when in doubt, aim at the center of mass.

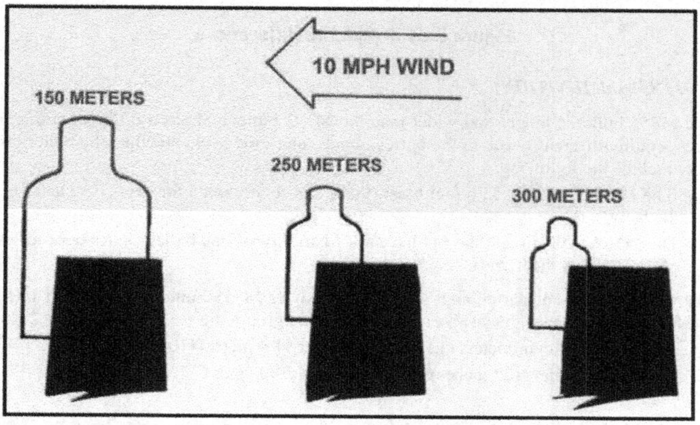

Figure 5-30. M16-/M4-series weapons adjusted point of aim based on wind speed.

SECTION V. BALLISTICS

Commanders and marksmanship trainers must understand some aspects of ballistics to teach the principles of zeroing and engagement of long-range targets. Ballistics is a science dealing with the motion and flight characteristics of projectiles. The study of ballistics in rifles and carbines is divided into three categories:

- Internal ballistics.
- External ballistics.
- Terminal ballistics.

INTERNAL BALLISTICS

5-88. Internal ballistics deals with what happens to the bullet before it leaves the weapon's muzzle.

5-89. The overall dimensions of the combat service 5.56-millimeter cartridges are the same, which allows cartridges to be fired safely in M16-series rifles and M4 carbines, but there are internal differences that affect firing accuracy (Figure 5-31).

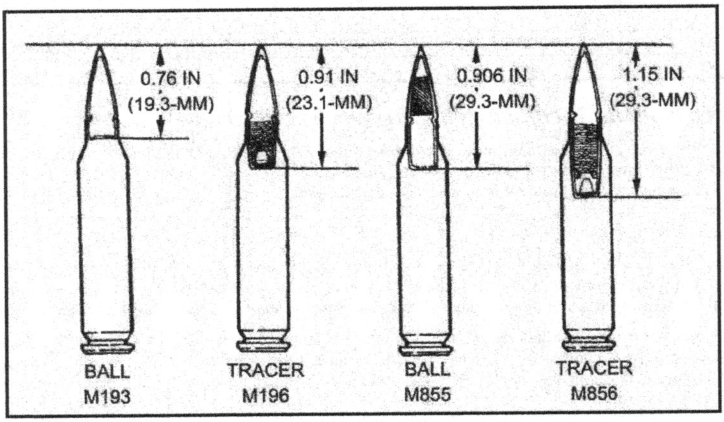

Figure 5-31. Projectile differences.

M855 AND M193 AMMUNITION

5-90. The M855 bullet is longer and wider than the M193 bullet and has a different configuration. These differences require different twists in the barrels, lands, and grooves to stabilize the bullet in flight. These differences include the following:

- The M16A1 has a 1:12 barrel twist (the bullet rotates once for every 12 inches of travel down the barrel).
- The M16A2/A3/A4 and the M4 have a 1:7 barrel twist (the bullet rotates once for every 7 inches of travel down the barrel).

5-91. The M16A1 does not put enough spin on the heavier M855 bullet to stabilize it in flight, causing erratic performance and inaccuracy (Figure 5-32). The shot groups are—

- 30.48 to 35.56 centimeters (12 to 14 inches) at 91.4 meters (100 yards).
- 182.88 centimeters (72 inches) at 274.2 meters (300 yards).

NOTE: Although firing the M855 cartridge in the M16A1 rifle is safe, it should only be used in a combat emergency, and then only for close ranges of 91.4 meters (100 yards) or less.

5-92. The M16A2/A3/A4 rifle and M4/M4A1 carbine fire both M193 and M855 ball ammunition with little difference in accuracy to a range of 500 meters. The M16A2/A3/A4 and M4/M4A1 and their ammunition are more effective than the M16A1 at ranges out to and beyond 500 meters due to better stabilization of the round.

5-93. The three 10-round shot groups in Figure 5-32 (A) were fired by a skilled marksman at a distance of 274.2 meters (300 yards) and 91.4 meters (100 yards) using the same M16A1 rifle.

- At 300 yards, the 25.4-centimeter shot group (shown on the left) was fired (and zeroed) with M193 ammunition.
- The 6-foot shot group (shown on the right) was fired with M855 ammunition.
- At 100 yards, the 35.56-centimeter (14-inch) shot group (shown in the center) was fired with M855 ammunition.

5-94. Figure 5-32 (B) shows two 25.4-centimeter (12-inch) shot groups fired by the same skilled marksman at a distance of 274.2 meters (300 yards) using an M16A2 rifle.

- The shot group on the left was fired and zeroed with M855 ammunition.
- The shot group on the right was fired using M193 ammunition.

NOTE: Both M193 and M855 ball ammunition can be used in training and accurately function in M16A2/3/4 rifles and M4/M4A1 carbines. Due to the different characteristics of each round, zero with the type of ammunition used for training. Do not switch between the types during firing. Do not zero with one type, and then fire the other for any type of training.

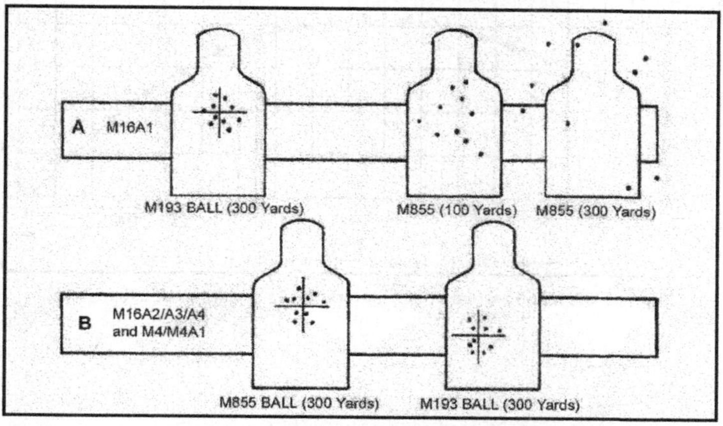

Figure 5-32. Ammunition impact comparison.

5-95. Table 5-9 and Figure 5-33 show the difference between a M16A2/A3/A4 and a M4/M4A1 weapon zeroed with M855 ammunition and then re-zeroed with M193 ammunition at 300 meters. There is practically no difference between the trajectory of the rounds and the impact of the rounds on the target.

Table 5-9. M855 zeroed at 300 meters; M193 re-zeroed at 300 meters using an M16A2/A3/A4 rifle and an M4 carbine.

RANGE (m)	FRONT SIGHT HEIGHT	THETA		BULLET DROP		BULLET LOCATION	
		M193	M855	M193	M855	M193	M855
0	2.568	0.0021987	0.00235575	0	0	-2.57	-2.57
25	2.568	0.0021987	0.00235575	-0.125	-0.14	-0.53	-0.39
50	2.568	0.0021987	0.00235575	-0.52	-0.58	1.24	1.49
75	2.568	0.0021987	0.00235575	-1.21	-1.33	2.71	3.06
100	2.568	0.0021987	0.00235575	-2.1	-2.4	3.99	4.31
125	2.568	0.0021987	0.00235575	-3.18	-3.82	5.07	5.21
150	2.568	0.0021987	0.00235575	-4.91	-5.61	5.51	5.73
175	2.568	0.0021987	0.00235575	-6.89	-7.78	5.69	5.88
200	2.568	0.0021987	0.00235575	-9.16	-10.4	5.58	5.58
225	2.568	0.0021987	0.00235575	-11.8	-13.4	5.11	4.90
250	2.568	0.0021987	0.00235575	-15.4	-16.9	3.67	3.72
275	2.568	0.0021987	0.00235575	-19.2	-20.8	2.04	2.14
300	2.568	0.0021987	0.00235575	-23.44	-25.3	-0.04	-0.04
325	2.568	0.0021987	0.00235575	-28.3	-30.4	-2.74	-2.83
350	2.568	0.0021987	0.00235575	-33.8	-36.0	-6.07	-6.11
375	2.568	0.0021987	0.00235575	-40.13	-42.3	-10.24	-10.09
400	2.568	0.0021987	0.00235575	-46.78	-49.3	-14.72	-14.77

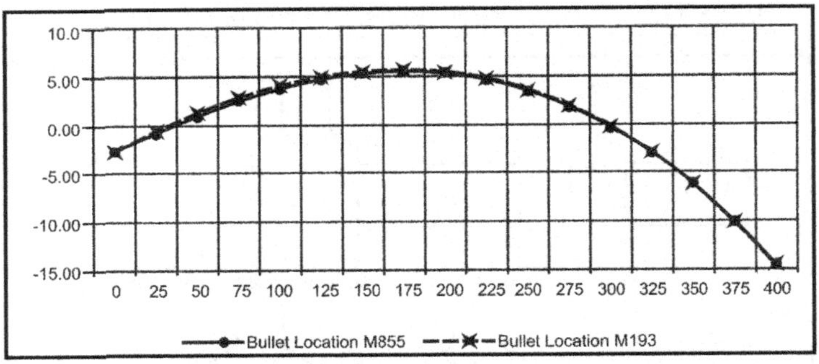

Figure 5-33. M855 zeroed at 300 meters; M193 re-zeroed at 300 meters.

5-96. When zeroing using M855 and M193 ammunition at 25 meters, the differences in the ammunition becomes apparent (Table 5-10 and Figure 5-34). If firing M193 ammunition from an M16A2/A3/A4 rifle or an M4/M4A1 carbine, the weapon should be zeroed with M193 ammunition.

Table 5-10. M855 zeroed at 25 meters; M193 re-zeroed at 25 meters.

RANGE (m)	FRONT SIGHT HEIGHT	THETA		BULLET DROP		BULLET LOCATION	
		M193	M855	M193	M855	M193	M855
0	2.568	0.0023383	0.00235575	0	0	-2.57	-2.57
25	2.568	0.0023383	0.00235575	-0.125	-0.14	-0.39	-0.39
50	2.568	0.0023383	0.00235575	-0.52	-0.58	1.51	1.49
75	2.568	0.0023383	0.00235575	-1.21	-1.33	3.13	3.06
100	2.568	0.0023383	0.00235575	-2.1	-2.4	4.54	4.31
125	2.568	0.0023383	0.00235575	-3.18	-3.82	5.76	5.21
150	2.568	0.0023383	0.00235575	-4.91	-5.61	6.33	5.73
175	2.568	0.0023383	0.00235575	-6.89	-7.78	6.65	5.88
200	2.568	0.0023383	0.00235575	-9.16	-10.4	6.68	5.58
225	2.568	0.0023383	0.00235575	-11.8	-13.4	6.35	4.90
250	2.568	0.0023383	0.00235575	-15.4	-16.9	5.05	3.72
275	2.568	0.0023383	0.00235575	-19.2	-20.8	3.55	2.14
300	2.568	0.0023383	0.00235575	-23.44	-25.3	1.61	-0.04
325	2.568	0.0023383	0.00235575	-28.3	-30.4	-0.95	-2.83
350	2.568	0.0023383	0.00235575	-33.8	-36.0	-4.15	-6.11
375	2.568	0.0023383	0.00235575	-40.13	-42.3	-8.18	-10.09
400	2.568	0.0023383	0.00235575	-46.78	-49.3	-12.52	-14.77

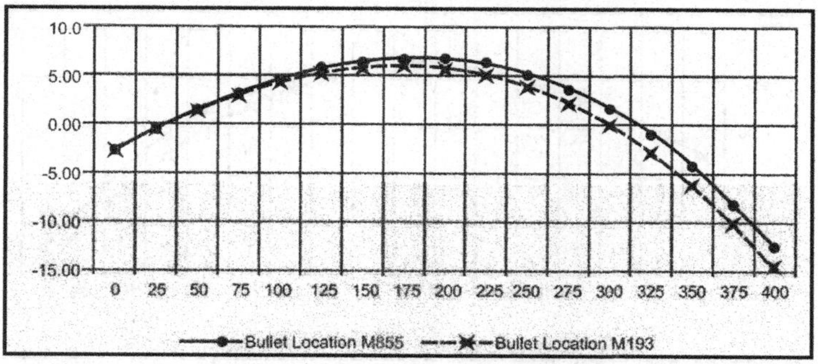

Figure 5-34. M855 zeroed at 25 meters; M193 re-zeroed at 25 meters.

5-97. If an M16A2/A3/A4 rifle or an M4/M4A1 carbine has been zeroed at 25 meters using M855 ammunition and M193 ammunition is fired without re-zeroing, the difference in the impact of the rounds on the target varies between ammunition (Table 5-11 and Figure 5-35).

Table 5-11. M855 zeroed at 25 meters; M193 fired using M855 zero.

RANGE (m)	FRONT SIGHT HEIGHT	THETA		BULLET DROP		BULLET LOCATION	
		M193	M855	M193	M855	M193	M855
0	2.568	0.00235575	0.00235575	0	0	-2.57	-2.57
25	2.568	0.00235575	0.00235575	-0.125	-0.14	-0.37	-0.39
50	2.568	0.00235575	0.00235575	-0.52	-0.58	1.55	1.49
75	2.568	0.00235575	0.00235575	-1.21	-1.33	3.18	3.06
100	2.568	0.00235575	0.00235575	-2.1	-2.4	4.16	4.31
125	2.568	0.00235575	0.00235575	-3.18	-3.82	5.85	5.21
150	2.568	0.00235575	0.00235575	-4.91	-5.61	6.43	5.73
175	2.568	0.00235575	0.00235575	-6.89	-7.78	6.77	5.88
200	2.568	0.00235575	0.00235575	-9.16	-10.4	6.82	5.58
225	2.568	0.00235575	0.00235575	-11.8	-13.4	6.50	4.90
250	2.568	0.00235575	0.00235575	-15.4	-16.9	5.22	3.72
275	2.568	0.00235575	0.00235575	-19.2	-20.8	3.74	2.14
300	2.568	0.00235575	0.00235575	-23.44	-25.3	1.82	-0.04
325	2.568	0.00235575	0.00235575	-28.3	-30.4	-0.73	-2.83
350	2.568	0.00235575	0.00235575	-33.8	-36.0	-3.91	-6.11
375	2.568	0.00235575	0.00235575	-40.13	-42.3	-7.92	-10.09
400	2.568	0.00235575	0.00235575	-46.78	-49.3	-12.25	-14.77

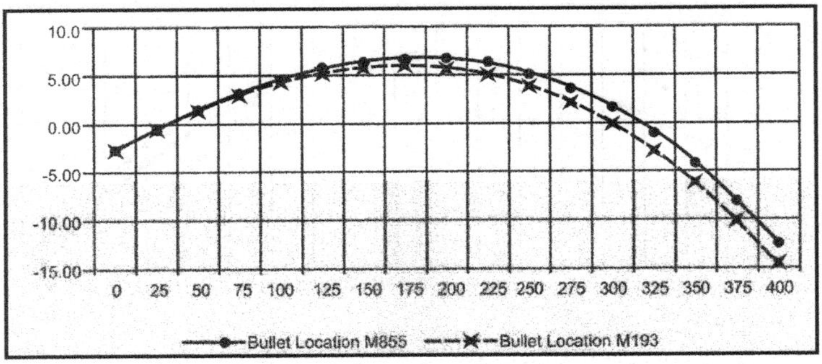

Figure 5-35. M855 zeroed at 25 meters; M193 fired using M855 zero.

5-98. A simple rule of thumb that will preclude any problem is to use only the ammunition specifically designed for each weapon (M193 ball ammunition for M16A1 rifles; M855 ball ammunition for M16A2/3/4 rifles and M4 carbines).

EXTERNAL BALLISTICS

5-99. External ballistics deals with factors affecting the flight path of the bullet between the weapon's muzzle and the target.

5-100. Soldiers must understand the basics of external ballistics so they can make necessary scope adjustments or hold compensations to allow them to hit the target. The external ballistic factors that affect bullet trajectory are:

- Gravity.
- Muzzle velocity.
- Air resistance (drag).
- Altitude or air density.
- Temperature.
- Trajectory.
- Wind.
- Angles.

GRAVITY

5-101. The force of gravity on a bullet is constant regardless of its weight, shape, or velocity. The longer a bullet is in the air or the greater its angle from the vertical, the more effect gravity will have on its trajectory.

NOTE: See paragraphs 5-71 through 5-72 for more information about the effects of gravity.

MUZZLE VELOCITY

5-102. Muzzle velocity is the speed of a bullet as it leaves the barrel, measured in feet per second. Muzzle velocity diminishes as the bullet gets farther away. The bullet reaches its maximum velocity 76 feet from the end of the weapon and slows until it reaches the target.

AIR RESISTANCE (DRAG)

5-103. Air resistance, or drag, immediately produces a slowing effect on a bullet.

ALTITUDE OR AIR DENSITY

5-104. The greater the altitude, the thinner the air and the longer the bullet will travel (with a correspondingly flatter trajectory). Each 5,000-foot elevation will raise the strike of the bullet ½ to 1 minute of angle (MOA).

TEMPERATURE

5-105. Deviation from standard daytime temperature (59 degrees Fahrenheit/15 degrees Celsius) affects bullet trajectory.

Cold Temperatures

5-106. Cold air is denser than warm air; the bullet must travel through more tightly packed air particles. This causes the bullet to lose velocity, causing the bullet to impact lower than intended. Cooler air also causes lower chamber pressure, which reduces the initial velocity.

Hot Temperatures

5-107. Warm or hot temperatures cause the strike of the round to move up.

TRAJECTORY

5-108. When a projectile exits the weapon's muzzle, it drops from the line of departure, otherwise known as the center/bore line. As the projectile travels downrange, air drag decreases the velocity, giving way to the inevitable force of gravity. This effect creates trajectory.

Line of Sight

5-109. The line of sight is an imaginary straight line extending from the firer's eye through the telescopic sight, or rear and front sight, to the target.

Line of Departure

5-110. The line of departure is an imaginary straight line extending from the center of the barrel to infinity.

Zero Range

5-111. Zero range is where the projectile intersects the line of sight. It occurs twice—once on the way up and once on the way down.

Apex

5-112. Otherwise known as midrange trajectory, the apex is the point where the projectile is at its highest in relation to the line of sight.

Bullet Path

5-113. The bullet path is the relationship of a projectile and the line of sight at any given range (normally expressed in inches).

WIND

5-114. External factors influence the trajectory relative to the point of aim, such as wind, altitude, temperature, humidity, and barometric pressure. Wind is by far the most significant. Consider the following effects of wind:

- Because the bullet is moving through the air, the air moves the bullet. Wind deflection is always in the same direction the wind is moving. A wind blowing from the left will move the bullet to the right.
- Deflection decreases as the angle of the wind to the line of flight decreases.

5-115. Effectively reading and correcting for wind effects takes practice, especially at longer ranges where accuracy in correcting is more critical. To shoot accurately in the wind, a firer must know the wind velocity, the wind direction, and the value of deflection at the range at which he is shooting.

NOTE: See paragraph 5-74 for more information on the effects of wind.

ANGLES

5-116. Firing uphill or downhill normally causes the bullet to hit high relative to a horizontal trajectory. If the firer is firing on an angle up or down at a slanted range of 100 meters, the point of impact will be higher than it would be for a level shot of 100 meters. The height depends on the angle.

5-117. Gravity acts on a bullet only during the horizontal component of its flight (the distance from the firer to the target measured as if they were both at the same level). Since the horizontal component will always be less than the slanted range, gravity will not pull the bullet down as far as it would if the range were level.

5-118. Firing uphill or downhill causes the wind to affect the shot over the entire slant range. The firer should aim at the target as if it were 25 meters away and correct for wind as if it were 400 meters away. The correct method for shooting uphill or downhill is to adjust elevation based on the horizontal range and correct for wind deflection based on the slanted range.

TERMINAL BALLISTICS

5-119. Terminal ballistics deals with what happens to the bullet when it comes in contact with the target.

5-120. Bullet penetration depends on the range, velocity, bullet characteristics, and target material. Greater penetration does not always occur at close range with certain materials, since the high velocity of the 5.56-millimeter bullet causes it to disintegrate soon after impact.

BULLET DISPERSION AT RANGE

5-121. Instructors/trainers must have a working knowledge of the effects of bullet dispersion and accuracy at various ranges.

MINUTE OF ANGLE

5-122. An MOA is the standard unit of measurement used in adjusting a weapon's sights and other ballistic-related measurements. It is also used to indicate the accuracy of a weapon.

5-123. A circle is divided into 360 degrees. Each degree is further divided into 60 minutes; therefore, a circle contains 21,600 minutes. An MOA is an angle beginning at the muzzle that covers 2.54 centimeters at a distance of 91.4 meters (Figure 5-36).

5-124. When the range is increased to 182.8 meters, the angle covers twice the distance, or 5.08 centimeters. The rule applies as range increases—7.62 centimeters at 274.2 meters, 10.16 centimeters at 365.6 meters, and so on.

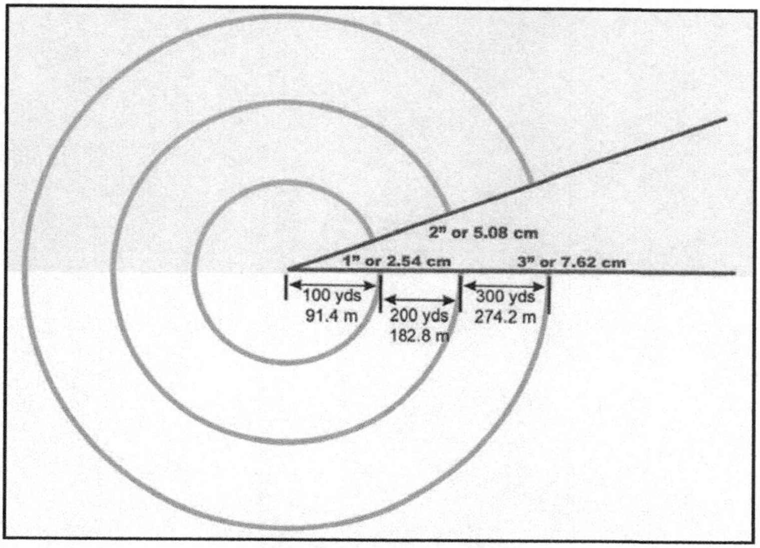

Figure 5-36. Minute of angle.

Increase of Shot Group Size

5-125. Just as the distance covered by an MOA increases each time the range increases, a shot group can be expected to do the same. If there are 2.54 centimeters between bullets on a 25-meter target, there will be an additional 2.54 centimeters of dispersion for each additional 25 meters of range. A 2.54-centimeter shot group at 25 meters (about 3.5 MOA) is equal to a 25.4-centimeter shot group at 250 meters (Figure 5-37).

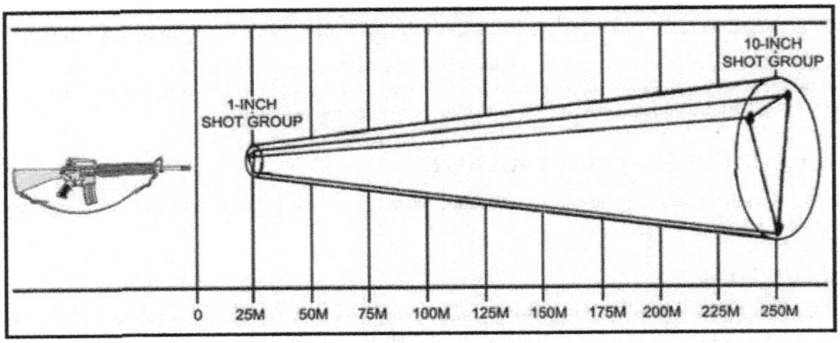

Figure 5-37. Increase in shot group size as range increases.

Chapter 6

Field Fire

Field firing is part of the continued progression in the development of combat shooting skills. This begins the Soldier's critical transition from unstressed firing at single, known distance targets to targets at various ranges for short exposures. It also requires the Soldier to practice and refine previously taught skills.

This chapter introduces the techniques for scanning the range for targets, estimating range, and firing quickly and accurately.

NOTE: See Table 6-1 for the current training program.

Table 6-1. Field Fire I and II training program.

FIELD FIRE I AND II
Instructional Intent
• Reinforce PMI and downrange feedback by detecting and engaging single and multiple timed targets with the M16-/M4-series weapon.
Special Instructions
Ensure that—
• The rear sight is on the proper setting (M16A2/3=8/3; M16A4 and M4=6/3 flush; M16A1=the unmarked aperture, short-range).
• The rear sight aperture is set on 300, not 800.
• The small aperture is being used.
• The range consists of targets at 75, 175, and 300 meters.
Observables
• Coaches are used to analyze the firer's adherence to the fundamentals—not as scorers.
• Soldiers detect and achieve 22 target hits out of 36 timed target exposures (Field Fire I).
• Soldiers detect and achieve 27 target hits out of 44 timed target exposures (Field Fire II).
• Soldiers that don't achieve the standard receive remedial training prior to refiring.

SECTION I. TARGET DETECTION

For most Soldiers, finding the target is a greater problem than hitting it. Target detection is the process of locating, marking, prioritizing, and determining the range to combat targets. Target detection must be conducted as part of individual training and tactical exercises, and must be integrated into day and night LFXs.

NOTE: Refer to TC 25-8 for construction of a target detection range.

LOCATING TARGETS

6-1. The ability to locate a combat target depends on the observer's—

- Position.
- Skill in scanning.
- Ability to observe the area and recognize the type of indicators made by the target.

SELECTION OF A POSITION

6-2. A good position is one that offers maximum visibility of the area, while affording cover and concealment. In this case, the word "position" refers to both the observer's location on the ground and the position of his body at that location.

NOTE: Instructors must continuously refer to and emphasize the importance of the observer's position when conducting practical exercises.

6-3. Depending on the situation, the individual Soldier may or may not select his own position.

- In most defensive situations, the Soldier is told where to prepare his position.
- Some situations (for example, the attack and reorganization on the objective) require the Soldier to select his own position.

SCANNING

6-4. To scan the area, Soldiers use three methods of search:

- Self-preservation method.
- 50-meter overlapping strip method.
- Maintaining observation of the area.

Self-Preservation Method of Search

6-5. When moving into a new area, Soldiers use the self-preservation method of search. To perform this method, use the following techniques:

(1) For approximately 30 seconds, quickly scan the area for enemy activity that may be of immediate danger.

(2) Make quick glances at specific points throughout the area, rather than just sweeping the eyes across the terrain in one continuous panoramic view.

NOTE: The eyes are sensitive to slight movements that occur within the area the eyes are focused on; panoramic views do not allow the eyes to detect the slight movements of a concealed target.

50-Meter Overlapping Strip Method of Search

6-6. If the Soldier fails to locate the enemy during the initial search, he must begin a systematic examination known as the 50-meter overlapping strip method of search. To perform this method—

(1) Begin the search with the area offering the greatest potential danger, the terrain nearest to your position.

(2) Beginning at either flank, systematically search the terrain to your front in a 180-degree arc, 50 meters in depth.

NOTE: Become familiar with the terrain as you search it. Take advantage of peripheral vision by focusing the eyes on specific points as you search from one flank to the other. Make mental notes of prominent terrain features and areas that may offer cover and concealment to the enemy.

---

(3) After reaching the opposite flank, search a second 50-meter strip farther out, but overlapping the first strip by approximately 10 meters.

(4) Continue in this manner until the entire area has been searched.

NOTE: This method should also be used as part of maintaining observation of the area and when the observer has been distracted from his area of responsibility.

Maintaining Observation of the Area

6-7. After completing his detailed search, the Soldier may be required to maintain observation of the area. To perform this method—

- Glance quickly at various points throughout the entire area, focusing the eyes on specific features.
- Always search the area in the same manner to ensure complete coverage of all terrain.

NOTE: Since this quick search may fail to detect the initial movement of an enemy, the observer should periodically repeat the procedures outlined in the 50-meter overlapping strip method of search.

TARGET INDICATORS

6-8. A target indicator is anything that a Soldier (friendly or enemy) does or fails to do that reveals his position. Since these indicators apply equally to both sides of the battlefield, Soldiers must learn to use target indicators to locate the enemy and to prevent the enemy from using the same indicators to locate them. For instructional purposes, these indicators can be grouped into three general areas:

- Sound.
- Movement.
- Improper camouflage.

Sound

6-9. Sounds, such as footsteps, coughing, or equipment noises, provide only a direction and general location, making it difficult to pinpoint a target by sound alone. However, detection of a sound alerts the observer, greatly increasing the possibility that he will eventually locate the target through other target indicators.

Movement

6-10. The degree of difficulty in locating moving targets depends primarily on the speed of movement. Slow, deliberate movements are much more difficult to notice than those that are quick and jerky.

Improper Camouflage

6-11. The lack or improper use of camouflage or concealment reveals the majority of targets detected on the battlefield; alert observers easily notice indicators such as light reflecting from shiny surfaces or a contrast with the background. Three general indicators may reveal a camouflaged or concealed target:

- Shine.
- Regularity of outline.
- Contrast with the background.

Shine

6-12. Metal objects, such as belt buckles, reflect light and act as a beacon to the wearer's position. This is as true at night as it is during the day.

Regularity of Outline

6-13. Humans and most types of military equipment cast outlines that are familiar to all Soldiers. The outlines of rifles, helmets, and vehicles are all easily identified. The reliability of this indicator depends on the visibility and the experience of the observer. On a clear day, most Soldiers can easily identify enemy riflemen or equipment if a distinctive outline is presented. At night or during other periods of poor visibility, seeing outlines is not only more difficult, but inexperienced troops will frequently mistake stumps and rocks for enemy Soldiers. This is an additional reason for Soldiers to become completely familiar with the terrain during periods of good visibility.

Contrast with the Background

6-14. If a Soldier wearing a dark uniform moves into a position in front of a snow bank, the contrast between the white snow and the dark uniform makes him clearly visible. However, if he was wearing a white or light-colored uniform, he would be more difficult to see.

6-15. Contrast with the background is the most difficult target indicator for a Soldier to avoid. During operations in which the Soldier is moving, he is usually exposed to numerous background colors. Since no single type of personal camouflage blends in with all areas, a moving Soldier must be continually aware of the surrounding terrain and vegetation.

MARKING TARGETS

6-16. A Soldier observes two enemy riflemen moving into completely concealed positions, one behind a bush and the other into a depression.

- By selecting a point of aim on the bush, the Soldier should hit the enemy rifleman even though he can't see him. If the target cannot be engaged, the point of aim also allows for quick and accurate engagement once the target is re-exposed.
- The enemy rifleman who moved into the depression provides no distinguishable point of aim, so the Soldier must select a nearby feature as a reference point and determine its distance and general direction from the depression. A reference point provides a general point of aim on a concealed target.

6-17. Of the two, a point of aim is usually the more effective means of delivering accurate fire. The difficulty in using reference points to mark targets moving from one location to another depends on the following factors:

- Number of targets.
- Exposure time to target.
- Spacing of targets.
- Good and poor points of aim.

NUMBER OF TARGETS

6-18. If several targets appear and disappear at approximately the same time, it is very difficult to note each target's point of disappearance.

EXPOSURE TIME OF TARGET

6-19. Usually, moving targets are exposed for only a short period of time, so the observer must be alert to note the point of disappearance for all of the targets. In such situations, the Soldier should mark the location of as many targets as possible before engaging any of them. By doing so, he will know the location of several targets and can engage each of them in rapid succession.

Spacing of Targets

6-20. The greater the interval between targets, the more difficult it is to note each target's movements. When there is considerable distance between targets, the observer should accurately locate and mark the one nearest to his position and note the general area of the others.

Good and Poor Points of Aim

6-21. Good points of aim are easily distinguishable in the surrounding terrain. Targets disappearing behind good points of aim, such as manmade objects and large terrain features, can be easily marked for future reference.

6-22. Poor points of aim are not easily distinguishable within the surrounding terrain. Targets disappearing behind poor points of aim are difficult to mark accurately and are easily lost.

6-23. If two targets offer the same degree of danger to the Soldier, but one disappears behind a good point of aim and the other behind a poor point of aim, the Soldier should mark the location of the target behind the good point of aim and engage the other target first.

RANGE DETERMINATION

6-24. Range determination is the process of finding the distance between two points. In most situations, one of these points will be the Soldier's own position. The other may be a target or prominent feature.

6-25. To accomplish the mission, combat riflemen must be able to accurately determine range. Not only does this affect his combat marksmanship proficiency, it is also required to report information and adjust artillery and mortar fire.

6-26. The methods of range estimation used during this period are—

- 100-meter unit of measure method.
- Appearance of objects method.
- Front sightpost method.

100-Meter Unit of Measure Method

6-27. To perform this method, the Soldier must be able to visualize a distance of 100 meters on the ground.

- For ranges up to 500 meters, he determines the number of 100-meter increments between the two points.
- Beyond 500 meters, the Soldier selects a point halfway to the target, determines the number of 100-meter increments to the halfway point, and then doubles it to find the range to the target.

6-28. During training exercises, the Soldier must become familiar with the effect that sloping ground has on the appearance of a 100-meter increment.

- Ground that slopes upward gives the illusion of greater distance, and observers tend to underestimate a 100-meter increment.
- Ground that slopes downward gives the illusion of shorter distance, and observers tend to overestimate a 100-meter increment.

6-29. Proficiency in the 100-meter unit of measure method requires constant practice. While training this technique, comparisons should be made continually between the range as determined by the Soldier and the actual range as determined by pacing or other more accurate means of measurement.

NOTE: The best training technique is to require the Soldier to pace the range after he has visually determined it. In this way, he discovers the actual range for himself, which makes a much greater impression than if he is simply told the correct range.

6-30. The greatest limitation of the 100-meter unit of measure method is that its accuracy is directly related to the amount of terrain visible to the observer. This is particularly true at longer ranges. If a target appears at a range of 500 meters or more and the observer can see only a portion of the ground between himself and the target, it becomes very difficult to use the 100-meter unit of measure method with any degree of accuracy.

Appearance of Objects Method

6-31. The appearance of objects method is a means of determining range by the size and other details of the object observed. This is a common method of determining distances and is used by most people in their everyday living. For example, a motorist attempting to pass another car must judge the distance of oncoming vehicles based on his knowledge of how vehicles appear at various distances. Suppose the motorist knows that, at a distance of one mile, an oncoming vehicle appears to be 1 inch wide and 2 inches high. Then, any time he sees other oncoming vehicles that fit these dimensions, he knows they are about one mile away.

6-32. The rifleman can use this technique to determine ranges on the battlefield. If he knows the characteristics, size, and detail of personnel and equipment at known ranges, he can compare these characteristics to similar objects at unknown ranges. When the characteristics match, so do the ranges.

6-33. To use the appearance of objects method with any degree of accuracy, the Soldier must be thoroughly familiar with the details of objects as they appear at various ranges. For example, the Soldier should study the appearance of a man standing at a range of 100 meters. He fixes the man's appearance firmly in his mind, carefully noting details of size and the characteristics of uniform and equipment. Next, he studies the same man in a kneeling position and in a prone position. By comparing the appearance of Soldiers in these positions at known ranges from 100 to 500 meters, the Soldier can establish a series of mental images that will help him determine range on unfamiliar terrain.

NOTE: Training should also be conducted in the appearance of other familiar objects, such as weapons or vehicles.

6-34. Because the successful use of this method depends upon visibility, anything that limits visibility (for example, weather, smoke, or darkness) will limit the effectiveness of this method.

Front Sightpost Method

6-35. Using the front sightpost as a scale is another method of estimating range. This method can be used for a quick on-the-spot estimation and engagement.

- Generally, if a man-sized target is ½ of the width of the front sightpost, the target is approximately 300 meters away.
- If the target is ¼ of the width of the front sightpost, the target is approximately 600 meters away.

SECTION II. FIELD FIRE TRAINING

Field fire training provides the transition from unstressed, slow firing at known distances or feedback targets to engaging pop-up silhouettes 50 to 300 meters away. Two basic types of field firing exercises are single and multiple target timed engagements, which use 75-, 175-, and 300-meter targets. Pop-up targets are used to add stress and simulate the short exposure times of combat targets. During field fire training, the firer learns to quickly detect targets, apply SPORTS, and apply the four fundamentals simultaneously.

CONDUCT OF A FIELD FIRE RANGE

6-36. On field fire ranges, Soldiers fire (from the supported and prone unsupported firing positions) at F-type silhouettes 75 meters away and E-type silhouettes 175 and 300 meters away. Initial training begins with single exposed targets and increased time for target exposures (Field Fire I). As Soldiers become proficient, multiple target engagements with shorter exposure times are introduced (Field Fire II).

NOTE: There are two types of electronic pop-up targets used on a field fire range: those that rise to the upright position from the back and those that rise to the upright position from the side. When using targets that rise from the side, Soldiers should be instructed to wait until the target is fully raised before engaging it. If the target is engaged as it is rising, the computer will not register the hit even though the target may fall.

Remedial Training

6-37. Soldiers who miss most targets should be removed from the firing line for remedial training if their problem cannot be corrected. A Soldier who fires at a 300-meter target 10 times and misses it 10 times is not learning, instead he is losing confidence in his ability. The typical Soldier should hit the 300-meter target at least 7 out of 10 times.

Peer Coaching

6-38. Peer coaches should assist Soldiers in observing the strike of rounds and identifying firing problems. If the target is missed and the coach cannot observe the bullet strike, the coach should instruct the Soldier to aim lower for the next shot, expecting to see the strike of the bullet in the ground. With this information, the coach can instruct the Soldier where to aim to hit the target.

Organization

6-39. Live-fire training can be organized in several ways. A unit is divided into two or more firing orders based on the number of personnel to be trained:

- The first order is the firer.
- The second order is the coach.
- The third order is the scorer (if required).

6-40. At the conclusion of each exercise, positions rotate until all orders have fired. Standard field fire scenarios have been developed to provide several target exposures. Although they are recommended for IET, local commanders can develop any variety of more challenging target sequences. Ammunition is allocated based on one round for each target.

Recording

6-41. During live-fire, the Soldier's hit-and-miss performance is recorded to facilitate the instructor/trainer's critiques or to indicate where more training, closer supervision, or remedial training is needed. Two methods are used to record firing performance:

- Manually marked scorecards.
- Automated computer printouts.

Manual Recording

6-42. When manual recording is used, the unit provides Soldiers for recording information on either DA Form 3601-R (Single Target Field Fire I Scorecard) or DA Form 5241-R (Single and Multiple Targets Field Fire II Scorecard).

NOTE: See the end of this publication for blank reproducible copies of these forms.

Automated Recording

6-43. When firing exercises are conducted using automated field fire (AFF) ranges, a computer printout is provided for each firing order. At the conclusion of each firing order, the range noncommissioned officer in charge (NCOIC) completes the printout and ensures that the Soldier identification is matched with each firing point. He adds the Soldier's numbered code to the top of each lane/firing point data column. Based on a one-round allocation for each target exposure, data should be collected on hits, misses, no-fires, and repeated shots to assist the instructor/trainer in assessing firing proficiency.

FIELD FIRE I (SINGLE TIMED TARGET)

6-44. Field Fire I is broken down into four firing tables (Table 6-3).

CONCEPT

6-45. Firing Table 1 helps the firer practice shooting skills and develop the sense of timing and rhythm required to make the transition from KD to field fire. This builds confidence prior to firing the exercises in Firing Tables 2, 3, and 4, and identifies Soldiers who are having difficulty and need reinforcement.

6-46. When firing the exercises in Firing Tables 2, 3, and 4, each Soldier demonstrates his ability to apply the fundamentals of marksmanship during the integrated act of firing by successfully detecting and engaging single timed targets.

CONDUCT

NOTE: When firing record fire, each Soldier must wear the proper uniform: the helmet, LBE, and IBA with all SAPI plates (if available). No other armor is required.

6-47. Each firer receives 54 rounds of 5.56-millimeter ball ammunition, with two 18-round magazines (one magazine each for Firing Tables 1 and 2) and two 9-round magazines (one magazine each for Firing Tables 3 and 4). Table 6-2 depicts the number of rounds that must be fired from each position. Each Soldier must achieve 22 hits out of 36 timed target exposures in Firing Tables 2, 3, and 4. Table 6-3 shows the number of target exposures, target ranges, and exposure times for each firing table.

Table 6-2. Number of rounds that must be fired from each position during Field Fire I.

POSITION	NUMBER OF ROUNDS FIRED
Supported firing position	18
Supported firing position	18
Unsupported firing position	9
Kneeling	9

Table 6-3. Field Fire I firing tables.

FIRING TABLE 1			FIRING TABLE 2			FIRING TABLE 3			FIRING TABLE 4		
ROUND	RANGE (m)	TIME (sec)	ROUND	RANGE (m)	TIME (sec)	ROUND	RANGE (m)	TIME (sec)	ROUND	RANGE (m)	TIME (sec)
1	75	6	1	75	6	1	75	6	1	75	6
2	175	8	2	175	8	2	175	8	2	175	8
3	300	10	3	300	10	3	300	10	3	75	6
4	175	8	4	175	8	4	175	8	4	175	8
5	75	6	5	75	6	5	75	6	5	75	6
6	300	10	6	300	10	6	300	10	6	175	8
7	300	10	7	300	10	7	300	10	7	75	6
8	75	6	8	75	6	8	75	6	8	75	6
9	175	8	9	175	8	9	175	8	9	175	8
10	175	8	10	175	8						
11	300	10	11	300	10						
12	175	8	12	175	8						
13	75	6	13	75	6						
14	300	10	14	300	10						
15	175	8	15	175	8						
16	75	6	16	75	6						
17	300	10	17	300	10						
18	75	6	18	75	6						

FIELD FIRE II (MULTIPLE OR SINGLE TIMED TARGETS)

6-48. Field Fire II consists of three firing tables.

CONCEPT

6-49. Firing Table 1 helps the firer practice shooting skills and develop the sense of timing and rhythm required to make the transition from single timed targets to multiple or single timed fleeting combat targets. This builds confidence prior to firing the exercises in Firing Tables 2 and 3, and identifies Soldiers who are having difficulty and need reinforcement.

6-50. When firing the exercises in Firing Tables 2 and 3, each Soldier demonstrates his ability to apply the fundamentals of marksmanship during the integrated act of firing by successfully detecting and engaging multiple and single timed targets.

CONDUCT

NOTE: When firing record fire, each Soldier must wear the proper uniform: the helmet, LBE, and IBA with all SAPI plates (if available). No other armor is required.

6-51. Every firer receives 54 rounds of 5.56-millimeter ball ammunition, with 10 rounds loaded into one magazine (for Firing Table 1) and 44 rounds loaded into two separate magazines, with 22 rounds each (for Firing Tables 2 and 3). Table 6-4 depicts the number of rounds that must be fired from each position. Each Soldier must achieve 27 hits out of the 44 timed target exposures in Firing Tables 2 and 3. Table 6-5 shows the number of target exposures, target distance, and exposure times for each firing table.

Table 6-4. Number of rounds that must be fired from each position during Field Fire II.

POSITION	NUMBER OF ROUNDS FIRED
Supported firing position	10
Supported firing position	22
Unsupported firing position	22

Table 6-5. Field Fire II firing tables.

FIRING TABLE 1			FIRING TABLE 2			FIRING TABLE 3		
ROUND	RANGE (m)	TIME (sec)	ROUND	RANGE (m)	TIME (sec)	ROUND	RANGE (m)	TIME (sec)
1	75	5	1	175	7	1	75	6
2	175	7	2	75	10	2	175	8
3	75	11	3	300		3	75	13
4	300		4	75	9	4	300	
5	75	9	5	175		5	75	11
6	175		6	300	9	6	175	
7	75	10	7	75	9	7	75	12
8	300		8	175		8	300	
9	175	11	9	175	11	9	175	13
10	300		10	300		10	300	
			11	75	9	11	75	11
			12	175		12	175	
			13	175	11	13	175	8
			14	300		14	75	6
			15	75	5	15	75	11
			16	175	11	16	175	
			17	300		17	75	12
			18	75	9	18	300	
			19	175		19	75	11
			20	75	10	20	175	
			21	300		21	175	13
			22	175	7	22	300	

SECTION III. RECORD QUALIFICATION

The objective of record fire is to access and confirm the individual proficiency of firers and the effectiveness of the training program. Important statistical data, such as qualification ratings and first-time GO rates, are obtained from record fire. These data provide goals for the Soldier and aid the commander in identifying the quality of his training.

PRACTICE RECORD FIRE I AND II

6-52. Although the Soldier receives a practice rating based on the number of target hits, practice record fire should also be considered a valuable training exercise. When practice record fire is correctly conducted, all Soldiers gain valuable experience and become more confident in engaging combat targets.

NOTE: See Table 6-6 for the current training program.

Table 6-6. Practice Record Fire I and II training program.

PRACTICE RECORD FIRE I AND II
Instructional Intent
• Reinforce PMI and KD firing and apply the techniques of target detection by engaging a more difficult course of fire with single and multiple pop-up targets and increased time constraints.
Special Instructions
Ensure that— • The rear sight is on the proper setting (M16A2/3=8/3; M16A4 and M4=6/3 flush; M16A1=the unmarked aperture, short-range). • The rear sight aperture is set on 300, not 800. • The small aperture is being used. • Peer coaching is stressed (Practice Record Fire I only).
Observables
• Soldiers apply all aspects of BRM. • Soldiers hit 23 out of 40 target exposures. • Soldiers that do not meet the standard receive remedial training before refiring. • Practice record fire should be conducted on a different range than record fire.

CONCEPT

6-53. Practice Record Fire I and II should closely resemble all aspects of actual qualification.

CONDUCT

NOTES: 1. When firing record fire, each Soldier must wear the proper uniform: the helmet, LBE, and IBA with all SAPI plates (if available). No other armor is required.

2. Before engaging targets, target detection is accomplished with a dry-fire scenario.

3. If possible, Soldiers should fire Practice Record Fire I and II on different ranges. Soldiers firing Practice Record Fire II on the same range as Practice Record Fire I must fire on a different lane.

6-54. Each firer receives 40 single or multiple target exposures at ranges from 50 to 300 meters and 40 rounds of 5.56-millimeter ball ammunition. Table 6-7 depicts the number of rounds that must be fired from each position. Each Soldier must hit a minimum of 23 out of 40 target exposures.

Table 6-7. Number of rounds that must be fired from each position during Practice Record Fire I and II.

POSITION	NUMBER OF ROUNDS FIRED
Prone supported firing position or (at the unit commander's discretion) the foxhole supported firing position	20
Prone unsupported firing position	10
Kneeling unsupported firing position	10

6-55. Soldiers adhere to the following guidelines:

- Based on the total number of hits achieved in each table, Soldiers are critiqued on the practice record fire score. If a firer consistently misses targets or experiences problems with target detection and range estimation, coaches should point out the shooting error to help correct it.
- Exposure times are three to seven seconds at ranges of 50 to 300 meters. Since it requires one to two seconds for the manually activated target mechanism to raise the target, timing begins when

the target is fully exposed, rather than when the tower operator activates the target switch. When practice record fire is conducted on automated record fire (ARF) ranges, these factors are included in the computer program.

NOTE: Practice Record Fire I allows peer coaching and the use of dummy ammunition. Practice Record Fire II does not allow peer coaching, and dummy ammunition will not be used.

ALIBI FIRING

6-56. Alibi firing should be conducted at the end of each firing table IAW tower operator commands. Alibis are provided during practice record fire for three reasons:

- Malfunction of the weapon.
- Malfunction of the target mechanism.
- Faulty ammunition.

RANGE TRAINING AREAS

6-57. The three range training areas are as follows:

- Orientation area.
- Ready area.
- Retired area.

Orientation Area

6-58. The orientation area is located so that firers cannot see the firing area. Practice record fire orientation includes instructions on the conduct of fire, safety, and range operations, including the procedures used in ready and retired areas.

Ready Area

6-59. The ready area is located near the firing range, but is positioned so that firers cannot see targets on the range. While in this area, the firer blackens the weapon sights, lubricates the weapon, and checks for defects that might cause malfunctions.

Retired Area

6-60. The retired area is about 100 meters behind the ready area. Soldiers completing practice record fire move to the retired area to clean their weapons and be critiqued on their firing performance.

PRACTICE RECORD FIRE STANDARDS

6-61. A firer who fails to qualify on his first try should refire the practice record fire range after his problem has been diagnosed and remedial training has been provided. Practice qualification ratings are shown in Table 6-8.

Table 6-8. Qualification ratings for Practice Record Fire I and II.

QUALIFICATION RATINGS	NUMBER OF TARGETS HIT
Expert	36 to 40
Sharpshooter	30 to 35
Marksman	23 to 29

RECORD OF PERFORMANCE

6-62. Accurate performance data are critical. The firer's score is manually recorded using DA Form 3595-R (Record Fire Scorecard) or automatically documented using a computer printout provided on the automated range. Based on the data recorded, an AAR can be performed by range and firing position to discuss firing performance.

NOTE: See Appendix B for a sample of a completed DA Form 3595-R and the end of this publication for a blank, reproducible copy.

RECORD FIRE

6-63. The intent of record fire is to facilitate the commander's evaluation of several individual tasks and integrated marksmanship skill performances, and to provide unit readiness indicators. The qualification standards are specifically related to a prescribed procedure for conducting record fire. Individual performance must be evaluated IAW two components:

- What record fire test was used (standard, KD, or alternate scaled)?
- How were individual performances distributed (first-time GO rates, percent in each qualification rating)?

NOTE: See Table 6-9 for the current training program.

Table 6-9. Record Fire training program.

RECORD FIRE
Instructional Intent
• Reinforce all phases of BRM. • Allow Soldiers to practice and refine critical marksmanship skills. • Measure the Soldier's complete understanding of BRM.
Special Instructions
Ensure that— • The rear sight is on the proper setting (M16A2/3=8/3; M16A4 and M4=6/3 flush; M16A1=the unmarked aperture, short-range). • The rear sight aperture is set on 300, not 800. • The small aperture is being used. • All targets are operational. • Each Soldier has one 20-round magazine and two 10-round magazines.
Observables
• Soldiers apply all aspects of BRM. • Soldiers hit 23 out of 40 target exposures. • Soldiers that do not meet the standard receive remedial training before refiring.

CONCEPT

6-64. Since all Soldiers must fire the record fire course at least once a year for qualification, the course can provide excellent firing performance evaluations. It also provides excellent diagnostic information for instructor/trainers who are concerned with scheduling training to overcome the most serious firing weaknesses. The standard record fire course should be used for all Soldiers, but there are times when qualification exercises must be conducted on alternate courses.

DEVELOPMENT OF STANDARDS

6-65. Testing and development indicates that the Soldier should hit at least 39 of 40 targets if he applies the marksmanship fundamentals correctly (assuming that target mechanisms have been checked and are functioning). This probability of hit (PH) is provided as a guide to consider the capability of the typical weapon, ammunition, and Soldier firing a standard course (Table 6-10).

Table 6-10. Probability of hits.

RANGE (m)	PH	NUMBER OF TARGETS
50	1.0	6
100	1.0	8
150	1.0	11
200	.99	7
250	.95	5
300	.90	3

6-66. When the IET BRM POI or an adequate unit training program is conducted, the following PH can be expected (Table 6-11).

Table 6-11. Results from an adequate unit training program.

RANGE (m)	TARGETS	LOW PH	AVERAGE PH	HIGH PH
50	6	.80	.95	.98
100	8	.70	.90	.95
150	11	.65	.90	.95
200	7	.45	.70	.90
250	5	.35	.60	.85
300	3	.25	.50	.80
		23 hits	32 hits	37 hits

TARGET FUNCTION

6-67. The first task on a standard record fire course is to ensure that all targets function properly. When in doubt, a lane should be fired to ensure that a bullet strike will activate each target. Sometimes slapping a target with a cleaning rod can cause it to activate, but a bullet impact will not. Hot plastic targets may allow a 5.56-millimeter bullet to pass through without causing sufficient vibration to activate the mechanism, resulting in a requirement to change targets more often, use double targets, or use different silhouettes.

CONDUCT

NOTE: When firing record fire, each Soldier must wear the proper uniform: the helmet, LBE, and IBA with all SAPI plates (if available). No other armor is required.

6-68. Each firer receives 40 single or multiple target exposures at ranges from 50 to 300 meters and 40 rounds of 5.56-millimeter ball ammunition. Table 6-12 depicts the number of rounds that must be fired from each position. Each Soldier must hit a minimum of 23 out of 40 target exposures.

Table 6-12. Number of rounds that must be fired from each position during Record Fire.

POSITION	NUMBER OF ROUNDS FIRED
Prone supported firing position or (at the unit commander's discretion) the foxhole supported firing position	20
Prone unsupported firing position	10
Kneeling unsupported firing position	10

6-69. Soldiers adhere to the following guidelines:

- Credit for targets hit should not be given when rounds are "saved" from difficult targets to be used on easier targets (for example, not firing at the 300-meter target so an additional round can be fired at the 150-meter target).
- When double targets are exposed, the Soldier should fire two rounds. If the first target is missed, he may fire at that same target with the second round.
- Soldiers engage the target that poses the greatest threat first (normally assumed to be the closer target). No scoring distinction is made between near targets and far targets or the sequence in which they are engaged.
- Credit is not given if unused ammunition from one 20-round table is added to a magazine provided for the next table.

Alibi Firing

6-70. Alibi firing is reserved for Soldiers who encounter a malfunctioning target, ammunition, or weapon. A Soldier will not be issued more than 20 rounds for Table 1, 10 rounds for Table 2, or 10 rounds for Table 3. Soldiers who fire 20 rounds, despite a target malfunction, will not be issued additional alibi rounds. There are no alibis for Soldier-induced weapon malfunctions or for targets missed during the application of immediate action procedures.

NOTE: The ammunition allocation and alibi procedures for practice record fire and record fire are conducted the same. The only exception is that coaching is authorized for practice record fire.

6-71. If a weapon or target malfunction occurs—

- The Soldier must apply immediate action procedures and continue to fire the exercise.
- After firing, the Soldier notifies the NCOIC to determine if the ammunition was faulty or if the target malfunctioned.
- The NCOIC verifies the malfunction.
- The Soldier is permitted to fire at that target(s) with the exact number of rounds equal to the target malfunctions. For example, a Soldier had two confirmed target malfunctions at 250 meters. Although he may have had five rounds left from the overall exercise, the Soldier receives only two rounds to engage the two 250-meter target exposures, if repaired, or the nearest target. He is not allowed to fire all five remaining rounds at the two 250-meter target exposures.
- On a computerized range, the tower operator confirms that the target malfunctioned and indicates the number of malfunctions that occurred.

6-72. Inoperable weapons are uncorrectable malfunctions such as the following:

- A broken firing pin.
- Jam caused by a double feed, not by the Soldier.
- Failure to extract due to a broken extractor.
- Round in the bore.

6-73. The Soldier must apply correct immediate action procedures to eliminate stoppages. If a stoppage is determined to be correctable (for example, the Soldier did not apply correct immediate action procedures and, as a result, the Soldier did not engage the required number of targets), the Soldier is at fault.

Troubleshooting Performance

6-74. Onsite observation, detailed analysis and evaluation of individual results, and unit performance identify weaknesses such as the following:

- Unserviceable weapons could cause poor zeroes or failures to fire and, therefore, failures to qualify.
- Some Soldiers may not qualify because of a lack of understanding of immediate action procedures or weapon and magazine maintenance procedures.

- Soldiers who miss targets are not applying the four fundamentals or are not accurately zeroing the weapon.
- Soldiers who do not fire at exposed targets during qualification might be experiencing:
  - Failure to scan the designated area.
  - Lack of ability to detect targets.
  - Lack of ability to shift from one target to another.
  - Failure to manage ammunition.
  - A stoppage.

6-75. Training can then focus on combat tasks, skills, or other factors that address these weaknesses.

Refire

6-76. Qualified weapons personnel or the NCOIC must verify weapon malfunctions before the Soldier can refire the course. Soldiers who erroneously claim a malfunction on the firing line are considered unqualified and refire as a second-time firer. Soldiers who fail to qualify on the first attempt should be given appropriate remedial training and allowed to refire in a few days. When a Soldier refires the course—

- He remains unqualified if he hits 22 targets or less.
- A rating of marksman is awarded for a score of 23 to 40 target hits.
- If automated scoring procedures (if available) allow the Soldier's performance to be stored and retrieved before a weapon malfunction, his performance is added to the score of his first attempt after weapon repair and refire.
- If a Soldier's weapon becomes inoperable and his performance before a malfunction precludes qualification, he is considered unqualified and must refire.

Qualification Ratings

6-77. One point is awarded for each round within or touching some part of the silhouette facing. Qualification ratings are shown in Table 6-13.

Table 6-13. Qualification ratings for Record Fire.

QUALIFICATION RATINGS	NUMBER OF TARGETS HIT
Expert	36 to 40
Sharpshooter	30 to 35
Marksman	23 to 29

Record of Performance

6-78. The record fire range is fired and recorded IAW DA Form 3595-R.

NOTE: See Appendix B for a sample completed scorecard and the end of this publication for a blank, reproducible copy.

SECTION IV. ALTERNATE QUALIFICATION COURSES

Units should conduct weapon qualification on a standard record fire range. Convenience and comfort should not be the prime consideration when choosing a range. Authorized alternate record fire courses are—

- KD record fire range.
- 25-meter scaled target alternate course.
- 15-meter scaled target alternate course.

NOTES: 1. The official records of personnel who use an alternate qualification course are noted to distinguish alternate qualification ratings from standard record fire course ratings. For example, official personnel records are annotated as follows:

JONES, John Q. 000-00-0000 Expert 36 (record fire [RF])

JONES, John Q. 000-00-0000 Expert 38 (known distance alternate course [KDAC])

JONES, John Q. 000-00-0000 Expert 38 (alternate course [AC])

2. The uniform for all alternate qualification courses is a helmet, LBE, and IBA with all SAPI plates (if available). No other armor is required.

3. Firers should engage targets from left to right, from nearest to the farthest (50-meter, 100-meter left, 100-meter center, 100-meter right, 150-meter left, 150-meter right, 200-meter left, 200-meter right, 250-meter, and 300-meter). This ensures that firers do not forget which targets they engaged during qualification. It also alleviates the possibility of shooting each target more than the prescribed number of times.

KNOWN DISTANCE RECORD FIRE RANGE

NOTE: The KD record fire range is used by all components of the U.S. Army, U.S. Army Reserve, and Army National Guard when a record fire range is not available.

6-79. The KD record fire range allows Soldiers to engage targets at range while experiencing time constraints, feedback, and the effects of wind and gravity.

CONDUCT

6-80. To complete this course, Soldiers fire three tables. Table 6-14 depicts these three tables and provides related information, such as time constraints, number of rounds that must be fired, type of target that must be used, and the distance from the firer that the target must be placed.

NOTE: Before firing the course, all Soldiers confirm the zero of their assigned weapons by assuming the prone position and firing six rounds from the 300-meter line. Zero rounds do not count for score.

Table 6-14. Known distance record fire range firing tables and related information.

TABLE	POSITION	TIME CONSTRAINTS	NUMBER OF ROUNDS	TYPE OF TARGET	DISTANCE (m)
Table 1	Prone supported firing position	2 min	20	E-type silhouette target	300
Table 2	Prone unsupported firing position	60 sec	10	E-type silhouette target	200
Table 3	Prone unsupported firing position	60 sec	10	F-type silhouette target	100

QUALIFICATION RATINGS

6-81. Scoring is conducted in the pits, with the results provided after each firing table. One point is awarded for each round within or touching some part of the silhouette facing. Qualification ratings for the KD record fire range are shown in Table 6-15.

Table 6-15. Qualification ratings for the known distance record fire range.

QUALIFICATION RATINGS	NUMBER OF TARGETS HIT
Expert	38 to 40
Sharpshooter	33 to 37
Marksman	26 to 32
Unqualified	25 and below

RECORD OF PERFORMANCE

6-82. The KD record fire range is fired and recorded IAW DA Form 5789-R (Record Fire Scorecard—Known Distance Course).

NOTE: See Appendix B for a sample completed form and the end of this publication for a blank, reproducible copy.

25-METER SCALED TARGET ALTERNATE COURSE

NOTE: The 25-meter scaled target alternate course is used when a standard record fire or KD range is unavailable for weapon qualification.

6-83. The 25-meter scaled target alternate course enables units to test a Soldier's weapon marksmanship proficiency, and firing at scaled silhouettes allows Soldiers to engage targets with time constraints and feedback.

CONDUCT

6-84. To complete this course, Soldiers fire three tables. Table 6-16 depicts these three tables and provides related information, such as time constraints, number of rounds that must be provided, number of silhouettes that must be engaged, and the distance from the firer that the target must be placed.

NOTES:

1. Soldiers should not receive training on target detection or the effects of wind and gravity by engaging targets at 25 meters. These skills are trained by firing at longer distances.
2. If zeroing/grouping exercises are not performed on the day of record fire, all Soldiers confirm the zero of their assigned weapons by firing six rounds of training/sustainment ammunition from the 25-meter line before firing the course. Zero rounds do not count for score.

Table 6-16. 25-meter scaled target alternate course firing tables and related information.

TABLE	POSITION	TIME CONSTRAINTS	NUMBER OF ROUNDS	NUMBER OF SILHOUETTES	ADDITIONAL INFORMATION
Table 1	Prone supported firing position or foxhole supported firing position	120 sec	20-round magazine, two rounds for each silhouette if firing from the prone supported firing position or foxhole supported firing position	10 silhouettes on the same target sheet	No more than two hits for each silhouette will be scored for this table.
Table 2	Prone unsupported firing position	60 sec	10-round magazine, one round for each silhouette	10 silhouettes on the same target sheet	No more than one hit for each target will be scored for this table.
Table 3	Kneeling unsupported firing position	60 sec	10-round magazine, one round for each silhouette	5 silhouettes on the same target sheet (50 to 150 m)	No more than two hits for each target will be scored for this table.

Time Between Firing Positions

6-85. The time between each firing position is not specified, but enough time should be allotted to allow the firer to clear his weapon, quickly change firing positions, and reload before beginning the next firing table.

Duties of Range Personnel

6-86. The following personnel perform range duties:

- Officer in charge (OIC).
- Range safety officer (RSO).
- Firing line safety crew.

Officer In Charge

6-87. The OIC briefs all Soldiers on the proper scoring procedures.

Range Safety Officer

6-88. To facilitate the timely flow of the record fire qualification table, the RSO ensures that enough time is given between firing positions.

Firing Line Safety Crew

6-89. Firing line safety crew personnel—

- Perform as scorers.
- Inform the chief range officer of crossfires.
- Inform the chief range officer of allowable alibis.
- Accurately count hits and misses.
- Count only four hits for each silhouette for score.
- Complete the scorecard.
- Assist the Soldier with target repair.
- Total, sign, and return the completed scorecard to the chief range officer.

SCORING

6-90. One hit is awarded for each round that strikes within or touches some part of the silhouette. If a bullet hole does not touch some part of the scaled silhouette, it is counted as a miss. Ricochets are counted as hits or misses.

6-91. A maximum of four hits for each silhouette on the same target sheet are scored. The same target sheet is used for every 40-round qualification table that a firer completes.

6-92. DA Form 5790-R (Record Fire Scorecard—Scaled Target Alternate Course) is used to score alternate course record fire qualifications.

NOTE: See Appendix B for a sample completed form and the end of this publication for a blank, reproducible copy.

6-93. The NSNs for scaled silhouette targets are—

- 25-meter (NSN 6920-01-1671398).
- 15-meter (NSN 6920-01-1396).

RATINGS

6-94. Qualification ratings for the 25-meter scaled target alternate course are shown in Table 6-17.

Table 6-17. Qualification ratings for the 25-meter scaled target alternate course.

QUALIFICATION RATINGS	NUMBER OF TARGETS HIT
Expert	38 to 40
Sharpshooter	33 to 37
Marksman	26 to 32
Unqualified	25 and below

15-METER SCALED TARGET ALTERNATE COURSE

NOTE: Units are permitted to use the 15-meter scaled alternate course only when standard record fire and KD ranges, and 25-meter scaled target alternate courses are unavailable.

6-95. The 15-meter scaled target alternate course is conducted on a 50-foot indoor range using a .22-caliber rimfire adapter (RFA). Qualification is conducted using the 15-meter alternate course C target (NSN 6920-01-167-1396).

NOTES:
1. See Appendix A for more information about the RFA.
2. Prior to qualification, all Soldiers battlesight zero their weapons using the 15-meter battlesight zeroing target (NSN 6920-01-167-1393).
3. The conduct of fire, scoring, scorecard, and qualification ratings are the same as those used for the 25-meter scaled target alternate course.

Chapter 7

Advanced Rifle Marksmanship

The procedures and techniques for implementing the ARM training program are based on all Soldiers understanding common firing principles, being proficient marksmen, and being confident in applying their firing skills in combat. This chapter concentrates on the advanced techniques and procedures that Soldiers need to participate in collective training during unit live-fire training exercises. This chapter describes advanced firing positions, combat firing techniques, CBRN firing, unassisted night fire, moving target engagements, SRM training, and SDM training.

NOTE: Unit METL and STRAC allocation determine which ARM tasks will be trained.

SECTION I. ADVANCED FIRING POSITIONS

After mastering the four marksmanship fundamentals in the basic firing positions, Soldiers master the four fundamentals while firing from a variety of advanced firing positions. The firer's position may change, but the three remaining fundamentals never change. Ultimately, any firing position that aids the firer in applying the fundamentals is acceptable, as long as the firer applies it consistently to avoid changing his sight picture.

With minor modifications, the dry-fire exercises taught during PMI can effectively train and evaluate a Soldier's ability to apply the fundamentals while in advanced firing positions. Repetitive training (muscle memory) will teach the Soldier the corrections needed to keep the same point of aim while in different firing positions.

NOTE: The act of assuming different firing positions while keeping the same point of aim increases first-time target hits and Soldier survivability.

KNEELING SUPPORTED FIRING POSITION

7-1. This position allows the Soldier to obtain the height necessary to observe many target areas while taking advantage of available cover. This position is the same as the kneeling unsupported firing position, except the Soldier uses some form of support to stabilize his body. Solid cover that can support any part of the body or weapon assists in firing accuracy.

NOTE: See Chapter 4 for more information about the kneeling unsupported firing position.

7-2. To assume the kneeling supported firing position (Figure 7-1)—

(1) Drop to the knee toward the inside of the covered position.

(2) Place the foot toward the outside of cover, with the toes pointing toward the engagement area.

(3) Place the nonfiring hand on the edge of the cover to support the weapon.

(4) Place the firing hand on the pistol grip, with the weapon's buttstock between the SAPI plate and the bicep to stabilize the weapon and absorb recoil.

NOTE: When firing from the strong side, the firer can place the elbow of the firing side on the outside knee to provide stability. When firing from the weak side, the firer should cant the weapon approximately 45 degrees and thrust the hips forward to minimize exposure to the enemy. When placing the weapon against the cover, the firer should take care to prevent the ejection port from becoming obstructed.

Figure 7-1. Kneeling supported firing position.

STANDING UNSUPPORTED FIRING POSITION

7-3. While the standing position provides the least stability, it can be assumed quickly while moving and is a good position for target area observation. Support for any portion of the body or weapon improves stability.

7-4. To assume the standing unsupported firing position (Figure 7-2)—

(1) Face the target.

(2) Step toward the target with the foot closest to the target.

(3) Spread the feet a comfortable distance apart.

(4) Place the firing hand on the pistol grip and the nonfiring hand on the upper handguard.

(5) Place the weapon's buttstock between the SAPI plate of the IBA and the bicep.

NOTE: This action stabilizes the weapon and absorbs recoil.

---

(6) Shift the feet until aiming naturally at the target.

(7) Evenly distribute body weight.

NOTE: More stability can be obtained by adjusting the ammunition pouch to support the nonfiring-side elbow. This allows the weapon's magazine to rest in the nonfiring hand.

Figure 7-2. Standing unsupported firing position.

STANDING SUPPORTED FIRING POSITION AROUND OBSTACLES

7-5. To assume the standing supported firing position when firing around obstacles (Figure 7-3)—

(1) Face the target.

(2) Execute a facing movement to the firing side.

(3) Spread the feet a comfortable distance apart.

(4) Place the firing hand on the pistol grip and the nonfiring hand on the upper handguard.

(5) Place the weapon's buttstock between the SAPI plate and the bicep.

NOTE: This action helps stabilize the weapon and absorbs recoil.

---

(6) Lean into the wall or obstacle, with the nonfiring-side forearm, shoulder, and thigh touching the obstacle for support.

(7) Shift the feet until aiming naturally at the target.

(8) Evenly distribute body weight.

NOTE: When firing from the left side of any obstacle, the firer should take care to prevent the ejection port from becoming obstructed.

Figure 7-3. Standing supported firing position around obstacles.

MODIFIED FIRING POSITIONS

7-6. Soldiers should be encouraged to modify positions by—

- Taking advantage of available cover.
- Using anything that helps to steady the weapon.
- Making changes that allow them to hit more combat targets.

7-7. To provide maximum stability, Soldiers should use prone and supported positions when firing M16- or M4-series weapons in the automatic or burst fire mode. If the weapon is equipped with the ARS, the Soldier should use the vertical pistol grip to further increase control of the weapon. Optional modifications include the following:

- Maximize use of artificial support.
- Grip the weapon firmly, and pull it into the shoulder securely. This helps offset the progressive displacement of weapon/target alignment caused by recoil.
- Use sandbags to support the weapon.
- Position the nonfiring hand on the weapon wherever it provides the most stability and flexibility. The goal is to maintain weapon stability and minimize recoil.
- Form a 5-inch loop with the sling at the upper sling swivel. Grasp this loop with the nonfiring hand, and pull down and to the rear while firing.
- Grasp the small of the stock with the nonfiring hand, and apply pressure down and to the rear while firing.
- Assume the modified supported prone firing position (Figure 7-4). This position uses sandbags to support the handguard and frees the nonfiring hand to hold the magazine steady.

Figure 7-4. Modified supported prone firing position.

URBAN OPERATIONS FIRING POSITIONS

7-8. Although the same principles of rifle marksmanship apply, the selection of firing positions during urban operations (UO) requires some special considerations. During UO, Soldiers may be required to fire—

- Over rooftops.
- Around obstacles.
- From windows.

FIRING OVER ROOFTOPS

7-9. Long-range observation may require Soldiers to occupy positions that are high above the ground. Figure 7-5 shows a Soldier firing over a rooftop, exposing only the parts of his body necessary to engage a target.

Figure 7-5. Firing over a rooftop.

Firing Around Obstacles

7-10. In the urban environment, Soldiers may encounter man-made and natural obstacles. Figure 7-6 shows a Soldier firing around an urban obstacle.

NOTE: Firing around corners could require the Soldier to fire from the opposite shoulder to avoid exposing himself to enemy fire.

Figure 7-6. Firing around an obstacle.

Firing From Windows

7-11. When firing from windows, Soldiers should stay in the shadows and make sure that the weapon's muzzle does not protrude out of the opening (Figure 7-7).

Figure 7-7. Firing from a window.

SECTION II. COMBAT FIRE TECHNIQUES

Combat is the ultimate test of a Soldier's ability to apply the fundamentals of marksmanship and firing skills. Soldiers must apply the marksmanship skills mastered during training, practice, and record fire exercises to many combat situations (for example, attack, assault, ambush, or UO). Although these situations present problems, basic techniques and fundamentals require only two modifications: changes to the rate of fire and alterations in weapon/target alignment.

NOTE: The necessary changes are significant and must be thoroughly taught and practiced before performing LFXs.

RAPID SEMIAUTOMATIC FIRE

7-12. The most important firing technique during fast-moving, modern combat is rapid semiautomatic fire. It is the most accurate technique of placing a large volume of fire on poorly defined targets or target areas, such as short exposure, multiple, or moving targets. To apply rapid semiautomatic fire, the Soldier intentionally fires a quick series of shots into the target area to ensure a high probability of a hit.

NOTE: Increased speed and volume should be sought only after the Soldier has demonstrated expertise and accuracy during slow semiautomatic fire.

Effectiveness and Control of Rapid Semiautomatic Fire

7-13. With proper training, Soldiers can select the appropriate mode of fire: semiautomatic fire, rapid semiautomatic fire, or automatic/burst fire.

NOTE: Leaders must ensure that Soldiers apply proper fire discipline at all times. Even in training, unaimed fire must never be tolerated, especially unaimed automatic fire.

7-14. While Soldiers sacrifice some degree of accuracy to deliver a greater volume of fire, it is surprising how devastatingly accurate rapid semiautomatic fire can be. At ranges beyond 25 meters, rapid semiautomatic fire is superior to automatic fire in all measures: shots per target, trigger pulls per hit, and time to hit. Proper training and repeated practice increases the degree of accuracy.

7-15. Rapid application of the four fundamentals will result in a well-aimed shot every one or two seconds. This technique of fire allows a unit to place the most effective volume of fire in a target area while conserving ammunition. It is the most accurate means of delivering suppressive fire.

Modifications for Rapid Semiautomatic Fire

7-16. Trainers must consider the impact of the increased rate of fire on the Soldier's ability to properly apply the fundamentals of marksmanship and other combat firing skills, such as immediate action procedures.

Marksmanship Fundamentals

7-17. The following paragraphs describe the modifications necessary for Soldiers to apply the four fundamentals when firing in the rapid semiautomatic fire mode.

Steady Position

7-18. Consider the following modifications to achieve a steady position:

- Make sure that the weapon is well-supported to improve accuracy and reduce recovery time between shots.
- Grip the handgrip tightly to reduce recovery time and rapidly shift or distribute fire to subsequent targets.
- When possible, pivot the weapon where the nonfiring hand meets the support.
- Avoid changing the position of the nonfiring hand on the support; it is awkward and time-consuming when rapidly firing a series of shots.

Aiming

7-19. Consider the following recommendations to properly aim the weapon:

- Do not change sighting and stock weld during rapid semiautomatic fire. Keep the cheek on the stock for every shot, align the firing eye with the rear aperture, and focus on the front sightpost.
- When using slow semiautomatic fire, seek a stable sight picture.
- In the fast-moving situations that require rapid semiautomatic fire, accept target movement and unsteady sight picture, and keep firing into the target area until the target is down or there is no chance of a hit.
- Aim every shot.

Breath Control

7-20. Breath control must be modified because the Soldier does not have time to take a complete breath between shots. Consider the following modifications to achieve proper breath control:

- Hold your breath at some point in the firing process.
- Take shallow breaths between shots.

Trigger Squeeze

7-21. To maintain the desired rate of fire, the Soldier has a brief period of time to squeeze the trigger. The firer must cause the weapon to fire in about half of a second or less and still not anticipate the precise moment of firing. Consider the following modifications to achieve proper trigger squeeze:

- Apply initial trigger pressure as soon as a target is identified and while the front sightpost is being brought to the desired point of aim.
- When the front sightpost reaches the point of aim, apply final pressure to cause the weapon to fire almost at once. Apply this additional pressure, also known as final trigger squeeze, without disturbing the lay of the weapon.
- Increase the firing rate by firing, releasing enough trigger pressure to reset the sear, and then immediately firing the next shot. This technique is called rapid trigger squeeze. It eliminates the time used in fully releasing pressure on the trigger and allows the firer to rapidly deliver subsequent rounds.

NOTE: Training and practice sessions are required for Soldiers to become proficient in the technique of rapid trigger squeeze.

7-22. Repeated dry-fire training using simulators, such as the EST 2000 and LMTS, and live-fire practice ensure that the Soldier can squeeze the trigger and maintain a rapid rate of fire consistently and accurately.

Immediate Action Procedures

7-23. To maintain an increased rate of suppressive fire, Soldiers must apply immediate action quickly. Repeated dry-fire practice using blanks or dummy rounds, followed by live-fire training and evaluation, ensures that Soldiers can rapidly apply immediate action procedures while other Soldiers initiate fire.

RAPID SEMIAUTOMATIC FIRE TRAINING

NOTE: Soldiers should be well-trained in all aspects of slow semiautomatic firing before attempting any rapid semiautomatic fire training. Those who display a lack of knowledge of fundamental marksmanship skills should not advance to rapid semiautomatic fire training until these skills are learned and mastered.

7-24. Initial training should focus on the modifications to the fundamentals and other basic combat skills necessary during rapid semiautomatic firing.

NOTE: See Table 7-1 for the current training program.

Table 7-1. Rapid semiautomatic fire training program.

RAPID SEMIAUTOMATIC FIRE TRAINING PROGRAM
Instructional Intent
• Soldiers learn to engage targets using rapid semiautomatic fire and practice rapid magazine changes.
Special Instructions
Ensure that— • The M16A2/A3/A4 rifle's or M4 carbine's rear sight is set on the 0-2 aperture. • The M16A1's rear sight is set on the unmarked aperture. • Soldiers use a 25-meter alternate course C qualification target. • Each Soldier is given four 5-round magazines of 5.56-millimeter ball ammunition. • Soldiers use rapid semiautomatic fire to engage targets. • Each Soldier fires one round at each of the 10 silhouettes on the alternate course C qualification target. • Each Soldier does a rapid magazine change after each magazine is fired. • The first iteration of 10 rounds is fired within a time limit of 40 seconds. • The second iteration of 10 rounds is fired within a time limit of 30 seconds. • Each target is inspected, and the results are posted after each iteration.
Observables
• Coaches continuously analyze the firer's application of the fundamentals. • Each Soldier obtains 14 hits out of 20 silhouette target exposures.

Conduct

7-25. Each Soldier receives four 5-round magazines of 5.56-millimeter ball ammunition. Using rapid semiautomatic fire, the Soldier fires one round at each of the 10 silhouettes on the alternate course C qualification target. Soldiers fire two iterations, performing a rapid magazine change after each magazine is fired. The targets are inspected, and the results are posted after each iteration. Each Soldier must obtain 14 hits out of 20 silhouette target exposures.

7-26. Table 7-2 depicts the two iterations and provides related information, such as time constraints, number of rounds that must be fired, type of target that must be used, and the distance away from the firer that the target must be placed.

Table 7-2. Rapid semiautomatic fire training and related information.

ITERATION	TIME CONSTRAINTS	NUMBER OF ROUNDS	TYPE OF TARGET	DISTANCE (m)
1	40 sec	10	25-m alternate course C qualification target	25
2	30 sec	10	25-m alternate course C qualification target	25

Dry-Fire Exercises

7-27. Repeated dry-fire exercises are the most efficient means to ensure that Soldiers can apply modifications to the fundamentals. Multiple dry-fire exercises are needed, emphasizing a rapid shift in position and point of aim, followed by breath control and fast trigger squeeze.

NOTES: 1. Blanks or dummy rounds may be used to train rapid magazine changes and immediate action procedures.

2. The Soldier should display knowledge and skill during dry-fire exercises before attempting LFXs.

Live-Fire Exercises

7-28. There are two types of LFXs:

- Individual.
- Collective.

Individual

7-29. To conduct an individual LFX—

- Ensure that the emphasis is on each Soldier maintaining a heavy volume of accurate fire.
- Keep weapon downtime (during immediate action and rapid magazine changes) to a minimum.
- Begin by firing at shorter ranges, progressing to longer ranges as Soldiers display increased proficiency.
- Shorten exposure or engagement times and increase the number of rounds to simulate the need for a heavy volume of fire.
- Provide downrange feedback to determine the accuracy of fire.

Collective

7-30. Rapid semiautomatic fire should be the primary means of delivering fire during a collective LFX. To conduct a collective LFX, ensure that the emphasis is on performing staggered rapid magazine changes, maintaining a continuous volume of fire, and conserving ammunition.

AUTOMATIC OR BURST FIRE

NOTE: Automatic or burst fire should be trained only after the Soldier has demonstrated expertise during slow and rapid semiautomatic fire.

7-31. When applying automatic or burst fire, Soldiers deliver the maximum number of rounds (one to three rounds per second) into a designated target area while rapidly applying the four fundamentals. This specialized technique of delivering suppressive fire may not apply to most combat engagements.

NOTE: The M16A1/A3 rifle and M4A1 carbine have fully automatic settings. The M16A2/A4 rifle and M4 carbine use a three-round burst capability.

Effectiveness and Control of Automatic or Burst Fire

7-32. Automatic or burst fire is inherently less accurate than semiautomatic fire. The first fully automatic shot fired may be on target, but recoil and a high cyclic rate of fire often combine to place subsequent rounds far from the desired point of impact. Even controlled (three-round burst) automatic or burst fire may place only one round on the target. Because of these inaccuracies, it is difficult to evaluate the effectiveness of automatic or burst fire, and even more difficult to establish absolute guidelines for its use.

Factors for Use of Semiautomatic Versus Automatic or Burst Fire

7-33. Trainers must ensure that Soldiers understand the capabilities and limitations of automatic or burst fire. They must know when it should and should not be used.

Semiautomatic Fire

7-34. M16 rifles and M4 carbines should normally be employed in the semiautomatic fire mode.

7-35. Depending on the tactical situation, Soldiers should employ the semiautomatic fire mode in the following conditions:

- Ammunition is in short supply, or resupply may be difficult.
- Single targets are being engaged.
- Widely spaced multiple targets are being engaged.
- The target is located more than 50 meters away.
- The effect of bullets on the target cannot be observed.
- Artificial support is not available.
- Targets may be effectively engaged using semiautomatic fire.

Automatic or Burst Fire

7-36. In some combat situations, the use of automatic or burst fire can improve survivability and enhance mission accomplishment. Clearing buildings, final assaults, FPF, and ambushes may require limited use of automatic or burst fire.

7-37. Depending on the tactical situation, Soldiers should employ automatic or burst fire in the following conditions:

- Ammunition is readily available, and there are no problems with resupply.
- Closely spaced multiple targets are located 50 meters away or less.
- Maximum fire is immediately required at an area target.
- Tracers or some other means can be used to observe the effect of bullets on the target.
- Leaders can maintain adequate control over weapons firing in the automatic fire mode.
- Good artificial support is available.
- The initial sound of gunfire disperses closely spaced enemy targets.

MODIFICATIONS FOR AUTOMATIC OR BURST FIRE

7-38. Automatic or burst fire is inherently less accurate than semiautomatic fire. Trainers must consider the impact of recoil and the high cyclic rate of fire on the Soldier's ability to properly apply the fundamentals of marksmanship and other combat firing skills, such as immediate action procedures and rapid magazine changes.

Marksmanship Fundamentals

7-39. The following paragraphs describe the modifications necessary for Soldiers to apply the four fundamentals when firing in the automatic fire mode.

Steady Position

7-40. Consider the following modifications to achieve a steady position:

- Make sure that the weapon is well-supported.
- Grip the weapon a little more firmly and pull it into the shoulder a little tighter than when in the semiautomatic fire mode.

NOTE: This support and increased grip help offset the progressive displacement of weapon/target alignment caused by recoil.

- To provide maximum stability, assume the modified supported prone firing position (Figure 7-4).

NOTE: If the weapon is equipped with the ARS, use the vertical pistol grip to further increase control of the weapon.

Aiming

7-41. Consider the following recommendations to properly aim the weapon:

- Do not change sighting and stock weld during automatic or burst fire. Keep the cheek on the stock for every shot, align the firing eye with the rear aperture, and focus on the front sightpost.
- Although recoil may disrupt this process, try to apply the aiming techniques throughout recoil.

Breath Control

7-42. Breath control must be modified because the Soldier does not have time to take a complete breath between shots. Consider the following modifications to achieve proper breath control:

- Hold your breath at some point in the firing process.
- Take shallow breaths between shots.

Trigger Squeeze

7-43. Training and repeated dry-fire practice aid the Soldier in applying proper trigger squeeze during automatic firing. LFXs enable him to improve this skill.

M16A2/3/4 Rifles and M4 Carbines

7-44. Until the weapon fires, trigger squeeze is applied in the normal manner. To use the burst fire mode—

(1) Hold the trigger to the rear until three rounds are fired.
(2) Release pressure on the trigger until it resets.
(3) Reapply pressure for the next three-round burst.

NOTES:
1. Do not slap or jerk the trigger. Squeeze it, and then quickly release pressure.
2. Depending on the position of the burst can when the selector is moved to the burst fire mode, the weapon may fire one, two, or three rounds when the trigger is held to the rear for the first time. If the weapon fires only one or two rounds, quickly release pressure on the trigger and squeeze again, holding it to the rear until a three-round burst is completed.

M16A1 Rifles

7-45. Until the weapon fires, trigger squeeze is applied in the normal manner. Because three-round bursts are the most effective rate of fire, pressure on the trigger should be released as quickly as possible. To use the burst fire mode, keep the index finger on the trigger, but quickly release pressure to prevent an excessive number of rounds from being fired in one burst. With much dry-fire practice, the Soldier can become proficient at delivering three-round bursts with the squeeze/release technique.

Immediate Action

7-46. To maintain an increased rate of suppressive fire, Soldiers must apply immediate action quickly. Repeated dry-fire practice using blanks or dummy rounds, followed by live-fire training and evaluation, ensures that Soldiers can rapidly apply immediate action procedures.

Rapid Magazine Changes

7-47. Rapid magazine changes are vital in maintaining automatic or burst fire. Rapid magazine changes must be correctly taught and practiced during dry-fire and live-fire exercises until the Soldier becomes proficient.

AUTOMATIC OR BURST FIRE TRAINING

NOTE: Soldiers should be well-trained in all aspects of slow semiautomatic firing before attempting any automatic training. Those who display a lack of knowledge of fundamental skills should not advance to automatic or burst fire training until these skills are learned.

7-48. Initial training should focus on the modifications to the fundamentals and other basic combat skills necessary during automatic firing.

7-49. Unit training is vital to properly applying this technique. Soldiers must be taught the advantages and disadvantages of automatic and burst firing so they know when it should be used. Without this knowledge, Soldiers tend to switch to the automatic or burst fire mode in life-threatening situations.

NOTE: See Table 7-3 for the current training program.

Table 7-3. Automatic or burst fire training program.

AUTOMATIC OR BURST FIRE TRAINING PROGRAM
Instructional Intent
• Soldiers learn the advantages and disadvantages of automatic or burst fire.
Special Instructions
Ensure that— • The M16A2/A3/A4 rifle's or M4 carbine's rear sight is set on the 0-2 aperture. • The M16A1's rear sight is set on the unmarked aperture. • Soldiers use a 25-meter alternate course C qualification target. • Each Soldier is in a proper modified automatic/burst firing position. • Each Soldier is given two 15-round magazines of 5.56-millimeter ball ammunition. • Each Soldier fires one 3-round burst at each of the 10 silhouettes on the alternate course C qualification target. • Each Soldier does a rapid magazine change after each magazine is emptied.
Observables
• Each Soldier obtains five target hits. • Soldiers demonstrate control of the weapon in the automatic/burst fire mode.

Conduct

7-50. Each Soldier receives two 15-round magazines of 5.56-millimeter ball ammunition. Each Soldier fires one 3-round burst at each of the 10 silhouettes on the alternate course C qualification target, performing a rapid magazine change after each magazine is emptied. Each Soldier must obtain five target hits.

7-51. Table 7-4 depicts automatic or burst fire training and provides related information, such as number of rounds that must be fired, type of target that must be used, and the distance away from the firer that the target must be placed.

Table 7-4. Automatic or burst fire training and related information.

FIRING POSITION	NUMBER OF ROUNDS	TYPE OF TARGET	DISTANCE (m)
Modified automatic/burst firing position	30, one 3-round burst at each of the 10 silhouettes	Alternate course C qualification target	25

Dry-Fire and Live-Fire Exercises

7-52. Repeated dry-fire exercises are the most efficient means to ensure that Soldiers can apply modifications to the fundamentals. Multiple dry-fire exercises are needed, emphasizing a stable position and point of aim, followed by breath control and appropriate trigger squeeze.

NOTES: 1. Blanks or dummy rounds may be used to train trigger squeeze, rapid magazine changes, and immediate action procedures.

2. The Soldier should display knowledge and skill during dry-fire exercises before attempting LFXs.

SUPPRESIVE FIRE

7-53. Suppressive fire is precisely aimed at a definite point or area target. Some situations may require a Soldier to place suppressive fire into a wide area (for example, wood line, hedgerow, or small building) while, at other times, the target may be a smaller area (for example, a bunker or window). Suppressive fire is used to control the enemy and the area he occupies. It is employed to kill the enemy or to prevent him from observing the battlefield, effectively using his weapons, or moving.

EFFECTIVENESS AND CONTROL OF SUPPRESSIVE FIRE

7-54. Many Soldiers have difficulty delivering effective suppressive fire when they cannot see a definite target, only likely locations or general areas where the enemy is known to exist. Even though definite targets cannot be seen, most suppressive fire should be well-aimed.

7-55. When controlling suppressive fires, two factors must be considered:

- Point of aim.
- Rate of fire.

Point of Aim

7-56. Suppressive fire should be well-aimed, sustained, semiautomatic fire. Although lacking a definite target, the Soldier must be taught to control and accurately deliver fire within the limits of the suppressed area. As when engaging a point target, the weapon sights are used, with the front sightpost placed so each shot impacts within the desired area.

Rate of Fire

7-57. During most phases of live-fire training (for example, grouping, zeroing, qualifying), shots are delivered using slow semiautomatic fire (one round every 3 to 10 seconds). During training, this allows a slow and precise application of the fundamentals. Successful suppressive fire requires a faster, but sustained, rate of fire. Soldiers may need to fire full automatic or bursts (13 rounds per second) for a few seconds to gain initial fire superiority. Rapid semiautomatic fire (one round every one or two seconds) allows the firer to sustain a large volume of accurate fire while conserving ammunition.

MODIFICATIONS FOR SUPPRESSIVE FIRE

7-58. The tactical situation dictates the most useful rate of fire, but the following must be considered:

- Marksmanship fundamentals.
- Rapid magazine changes.
- Ammunition conservation.

Marksmanship Fundamentals

7-59. As the stress of combat increases, some Soldiers may fail to apply the fundamentals of marksmanship. This factor contributes to reduced accuracy and effectiveness. While some modifications are appropriate, the basic fundamentals should be applied and emphasized—regardless of the rate of fire or combat stress. Strategies to enhance marksmanship skills under combat stress include shooting in the prone position, as opposed to standing.

7-60. Factors that contribute to combat stress are:

- Environmental.
- Operational.

Environmental

7-61. Environmental stressors have been shown to degrade marksmanship accuracy up to 20 percent. Such stressors include—

- Heat.
- Altitude.

Operational

7-62. Operational stressors have been shown to degrade marksmanship accuracy from 17 percent to 136 percent. Such stressors include—

- MOPP gear.
- Tasks that require carrying rucksacks, litter patients, and other equipment on the body.
- Sleep deprivation.

Rapid Magazine Changes

7-63. One of the keys to sustained suppressive fire is reloading the weapon rapidly. Rapid magazine changes must be correctly taught and practiced during dry-fire and live-fire exercises until the Soldier becomes proficient. Small-unit training exercises must be conducted so Soldiers who provide suppressive fire practice staggered magazine changes.

Ammunition Conservation

7-64. Automatic or burst fire should be used sparingly and only to gain initial fire superiority. Depending on the tactical situation, the rate of fire should be adjusted so that a minimum number of rounds are expended. Accurate fire conserves ammunition, while preventing the enemy from placing effective fire on friendly positions.

SUPPRESSIVE FIRE TRAINING

NOTE: See Table 7-5 for the current training program.

Table 7-5. Suppressive fire training program.

SUPPRESSIVE FIRE TRAINING PROGRAM
Instructional Intent
• Soldier learns to suppress targets using suppressive fire.
Special Instructions
Ensure that—
• The M16A2/A3/A4 rifle's or M4 carbine's rear sight is set on the 0-2 aperture.
• The M16A1's rear sight is set on the unmarked aperture.
• Soldiers use a 25-meter scaled landscape target.
• Each Soldier is given two 9-round magazines and one 12-round magazine of 5.56-millimeter ball ammunition.
• Each Soldier is in a proper supported firing position.
• Each Soldier fires 9 rounds at the open window area of the target using rapid semiautomatic fire with the first 9-round magazine.
• Each Soldier fires 12 rounds at the fence or hedgerow area of the target using rapid semiautomatic fire with the 12-round magazine.
• Each Soldier fires three 3-round bursts at the tank turret area of the target using the automatic/burst fire mode with the second 9-round magazine.
Observables
• Each Soldier achieves 5 hits inside the open window area within 18 seconds.
• Each Soldier achieves 10 hits inside the dotted lines surrounding the fence or hedgerow area within 24 seconds.
• Each Soldier achieves 3 hits inside the tank turret area within 24 seconds.

7-65. Figure 7-8 shows a landscape target suitable for suppressive fire training. When this type of target is used, trainers must develop a firing program to include areas of engagement and designated target areas. At 25 meters, this target provides the firer with an area to suppress without definite targets to engage.

Figure 7-8. Landscape target.

Conduct

7-66. Each Soldier receives two 9-round magazines and one 12-round magazine of 5.56-millimeter ball ammunition. The Soldier engages three areas of a 25-meter scaled landscaped target: the open window area, the fence or hedgerow area, and the tank turret area. Each Soldier achieves 5 hits inside of the open window area, 10 hits inside of the dotted lines surrounding the fence or hedgerow area, and 3 hits inside of the tank turret area.

7-67. Table 7-6 depicts suppressive fire training and provides related information, such as number of rounds that must be fired, type of target that must be used, and the distance away from the firer that the target must be placed.

Table 7-6. Suppressive fire training and related information.

FIRING POSITION	TYPE OF TARGET	AREA OF TARGET ENGAGED	NUMBER OF ROUNDS	TYPE OF FIRE	TIME CONSTRAINTS
Supported firing position	25-m scaled landscape target	Open window	9	Rapid semiautomatic	18 sec
		Fence or hedgerow	12	Rapid semiautomatic	24 sec
		Tank turret	9, in three 3-round bursts	Automatic/burst	24 sec

QUICK FIRE

7-68. The two main techniques of directing fire with a rifle or carbine are—

- Aim using the sights.
- Use weapon alignment, instinct, bullet strike, or tracers to direct the fire.

7-69. The preferred technique is to use the sights, but sometimes quick reflex action is required. Quick fire, also known as instinctive firing or quick kill, is a technique used to deliver fast, effective fire on surprise personnel targets 25 meters away or less.

EFFECTIVENESS AND CONTROL OF QUICK FIRE

7-70. Quick fire techniques are appropriate when Soldiers are presented with close, suddenly appearing, surprise enemy targets; or when close engagement is imminent.

NOTE: Fire may be delivered in the SEMIAUTO or AUTOMATIC/BURST fire mode. For example, a point man in a patrol may carry the weapon on AUTOMATIC/BURST. This may also be required when clearing a room or bunker. Initial training should be in the SAFE mode.

7-71. Two techniques of delivering quick fire are:

- Aimed.
- Pointed.

7-72. The difference in the speed of delivery of these two techniques is small. Pointed quick fire can be used to fire a shot about one-tenth of a second faster than aimed quick fire. The difference in accuracy, however, is more pronounced:

- A Soldier well-trained in pointed quick fire can hit an E-type silhouette target at 15 meters, although the shot may strike anywhere on the target.
- A Soldier well-trained in aimed quick fire can hit an E-type silhouette target at 25 meters, with the shot or burst striking 5 inches from the center of mass.

7-73. This variance of target hit for this type of engagement reinforces the need for well-aimed shots.

7-74. Pointed and aimed quick fire should be used only when a target cannot be engaged fast enough using the sights in a normal manner. These techniques should be limited to targets appearing at 25 meters or less. Modern short-range combat (SRC) techniques emphasize carrying the weapon with the buttstock high so that the weapon sights can be brought into display as quickly as firing a hasty unaimed shot. In extremely dangerous moments, special reaction teams (SRTs) commonly advance with weapons shouldered, aiming as they advance.

Aimed

7-75. When using this technique, a Soldier can accurately engage a target at 25 meters or less in one second or less.

7-76. To use aimed quick fire (Figure 7-9)—

(1) Bring the weapon to the shoulder.

(2) With the firing eye, look through or just over the rear sight aperture.

(3) Use the front sightpost to aim at the target.

(4) Quickly fire a single shot.

Figure 7-9. Aimed quick fire.

Pointed

7-77. When using this technique, a Soldier can engage a target at 15 meters or less in less than one second.

7-78. To use pointed quick fire (Figure 7-10)—

(1) Keep the weapon at your side.

(2) Keeps both eyes open, and use instinct and peripheral vision to line up the weapon with the target.

(3) Quickly fire a single shot or burst.

Figure 7-10. Pointed quick fire.

Factors for Use of Controlled Pairs Versus Burst Fire

7-79. Tactical considerations dictate whether controlled pairs or burst fire is most effective in a given situation.

Modifications for Quick Fire

7-80. Trainers must consider the impact of the increased rate of fire on the Soldier's ability to properly apply the fundamentals of marksmanship and other combat firing skills.

Marksmanship Fundamentals

7-81. Quick fire techniques require major modifications to the four fundamentals of marksmanship. Initial training in these differences, followed by repeated dry-fire exercises, will be necessary to prepare the Soldier for live-fire.

Steady Position

7-82. The quickness of shot delivery prevents the Soldier from assuming a stable firing position. Consider the following modifications:

- Fire from the present position when the target appears.
- If moving, stop.
- Do not make adjustments for stability and support before the round is fired.

Aimed

7-83. Consider the following modifications:

(1) Pull the weapon's buttstock into the pocket of the shoulder as the cheek comes in contact with the stock.
(2) Firmly grip the weapon with both hands, applying rearward pressure.
(3) Place the firing eye so that it looks through or just over the rear sight aperture.
(4) Place the sight on the target.

Pointed

7-84. Consider the following modifications:

- Pull the weapon into the side.
- Firmly grip the weapon with both hands, applying rearward pressure.

Aiming

7-85. This fundamental must be highly modified because the Soldier may not have time to look through the rear sight, find the front sight, and align it with the target.

NOTE: When using either aiming technique, bullets may tend to impact above the desired location. Repeated live-fire practice is necessary to determine the best point of aim or the best focus. Such practice should begin with the Soldier using a center of mass aim.

Aimed

7-86. Consider the following modified procedure:

(1) Initially focus on the target.

(2) Place the firing eye so that it looks at the target through or just over the rear sight aperture.

(3) Using peripheral vision, locate the front sightpost and bring it to the center of the target.

NOTE: Focus remains on the front sightpost throughout the aiming process.

---

(4) When the front sightpost is in focus, fire a controlled pair.

Pointed

7-87. Consider the following modifications:

- Place the focus on or slightly below the center of the target as you align the weapon with it, and fire the weapon.
- Use your instinctive pointing ability and peripheral vision to aid in proper alignment.

Breath Control

7-88. This fundamental has little application to the first shot of quick fire. The round must be fired before a conscious decision can be made about breathing. If subsequent shots are necessary, breathing must not interfere with the necessity of firing quickly. When possible, use short, shallow breaths.

Trigger Squeeze

7-89. Consider the following modifications:

(1) Apply initial pressure as weapon alignment is moved toward the target.

(2) Exert trigger squeeze so when weapon/target alignment is achieved, the rounds are fired at once.

7-90. Perfecting rapid trigger squeeze requires much training and practice.

QUICK FIRE TRAINING

NOTE: Only Soldiers in basic training will conduct quick fire training. SRM will be conducted at the unit level. See Section VI of this chapter for more information about SRM training.

7-91. The key to the successful employment of both quick fire techniques is practice. Both pointed and aimed quick fire must be repeatedly practiced during dry-fire training. LFXs provide further skill enhancement and illustrate the difference in accuracy between the two techniques.

NOTE: See Table 7-7 for the current training program.

Table 7-7. Quick fire training program.

QUICK FIRE TRAINING PROGRAM
Instructional Intent
• Soldiers learn how to engage targets using the quick fire techniques.
Special Instructions
Ensure that— • The M16A2/A3/A4 rifle's or M4 carbine's rear sight is set on the 0-2 aperture. • The M16A1's rear sight is set on the unmarked aperture. • Each Soldier is given two 10-round magazines. • Each Soldier engages 10 target exposures of 2 seconds each at 15 meters using the first 10-round magazine. • Each Soldier moves to the 25-meter line and engages 10 target exposures of 2 seconds each at 25 meters using the second 10-round magazine.
Observables
• Each Soldier achieves 7 hits out of 10 target exposures at 15 meters. • Each Soldier achieves 5 hits out of 10 target exposures at 25 meters.

Conduct

7-92. Each Soldier receives two 10-round magazines. Each Soldier must achieve 7 target hits out of 10 target exposures at 15 meters and 5 target hits out of 10 target exposures at 25 meters.

7-93. Table 7-8 depicts quick fire training and provides related information, such as the number of target exposures, distance from the firer, number of rounds that must be fired, and time constraints.

Table 7-8. Quick fire training and related information.

NUMBER OF TARGET EXPOSURES	DISTANCE	NUMBER OF ROUNDS	TIME CONSTRAINTS
10	15	10	2 sec per target exposure
10	15	10	2 sec per target exposure

SECTION III. CHEMICAL, BIOLOGICAL, RADIOLOGICAL, AND NUCLEAR FIRING

All Soldiers must effectively fire their weapons to accomplish combat missions in a CBRN environment. With proper training and practice, Soldiers gain confidence in their ability to effectively hit targets in full MOPP equipment. MOPP firing proficiency must be part of every unit's training program.

MISSION-ORIENTED PROTECTIVE POSTURE EQUIPMENT FIRE

7-94. Firing weapons is only part of overall CBRN training. Soldiers must be familiar with CBRN equipment, its use, and proper wear before they progress to learning the techniques of MOPP firing.

Modifications for Mission-Oriented Protective Posture Fire Training

7-95. Trainers must consider the impact of MOPP equipment (for example, hood or mask, gloves, overgarments) on the Soldier's ability to properly apply the fundamentals of marksmanship and combat firing skills.

Operation and Function

7-96. Many actions are affected by MOPP equipment: handling the weapon, performing operation and function checks, loading and unloading, and cleaning. Consider the following modifications:

- Movements are slowed; tasks take longer to complete and often require more effort.
- Vision is impaired.
- Care is needed to avoid damaging MOPP equipment and possible exposure to lethal agents.

7-97. Because of the great differences between MOPP Level 1 and MOPP Level 4, Soldiers must be trained in all aspects of weapon operation and maintenance while practicing at the highest MOPP level. Only through repeated training and practice can the Soldier be expected to perform tasks efficiently.

Immediate Action

7-98. Consider the following modifications:

- Under normal conditions, a Soldier should be able to clear a stoppage in 3 to 5 seconds. Under increased MOPP levels, however, this may take as long as 10 seconds to successfully complete.
- Mask (with or without hood) and gloves must be worn. Care must be taken not to snag or damage the gloves or dislodge the hood or mask during movements.
- Vision is limited to what can be seen through the mask lenses or faceplate. Peripheral vision is severely restricted. The lenses or faceplate may be scratched or partly fogged, further restricting vision.

NOTE: Soldiers requiring corrective lenses must be issued insert lenses before training.

- Scanning movements may be restricted by the hood or mask.

7-99. These factors could adversely affect the Soldier's ability to quickly and accurately detect targets. Dry-fire practice under these conditions is necessary to reduce time and streamline actions. Until Soldiers can instinctively apply immediate action to stoppages, they should practice using dummy or blank ammunition.

Marksmanship Fundamentals

7-100. Although the four marksmanship fundamentals remain valid during MOPP firing, some modifications may be needed to accommodate the equipment.

Steady Position

7-101. Consider these modifications:

- Due to the added bulk of the overgarment, adjust firing positions for stability and comfort.
- Stand, crouch, or squat during dry- and live-fire to reduce bodily contact with contaminated ground or foliage.
- Understand that a consistent spot or stock weld is difficult to maintain due to the shape of the protective mask. This requires the firer to hold his head in an awkward position to place the eye behind the sight.

Aiming

7-102. Wearing a protective mask may force firers to rotate (cant) the weapon to see through the rear aperture. This ideal aiming procedure (Figure 7-11) should be the initial procedure taught and practiced. If this cannot be achieved, a canted sight picture may be practiced.

Ideal Aiming Procedure

7-103. To perform the ideal aiming procedure—

(1) Rotate the weapon as little possible to see through and line up the sights.

(2) Place the center tip of the front sightpost on the ideal point of aim.

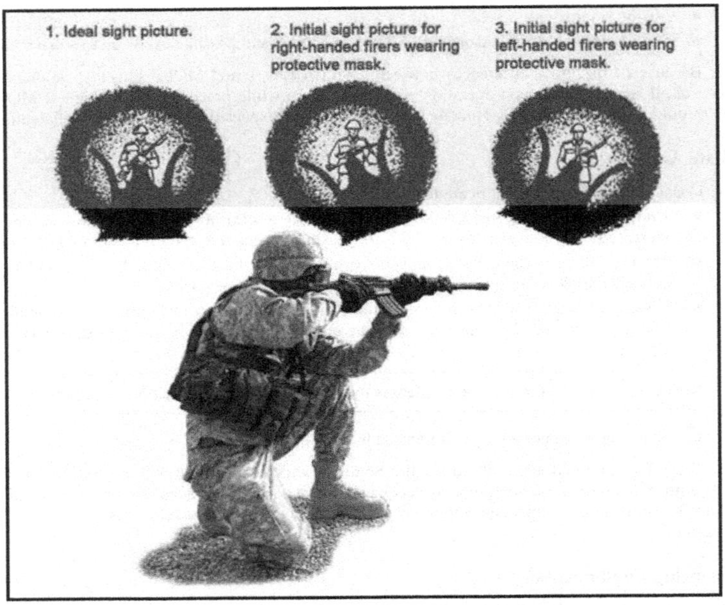

Figure 7-11. Sight picture when canting the rifle while wearing a protective mask (75-meter target).

Canted Sight Picture

NOTE: The normal amount of cant has a limited influence on rounds fired at ranges of 75 meters or less. Rifle ballistics cause the strike of the bullet to impact low in the direction of the cant at longer ranges.

7-104. Due to a shift in bullet strike and the many individual differences in sight alignment when wearing a protective mask, it is important to conduct downrange feedback training at ranges beyond 75 meters on KD ranges. This allows Soldiers to determine the aiming adjustments needed to hit the center of the target.

7-105. Figure 7-12 shows what might be expected for a right-handed firer engaging a target at 175 meters with a certain amount of cant; it depicts the adjustment in point of aim needed to move the bullet strike to the center of the target. Figure 7-13 shows what might be expected for a right-handed firer engaging a 300-meter target.

NOTE: The adjustments in point of aim for left-handed firers are the opposite of those shown in Figures 7-12 and 7-13.

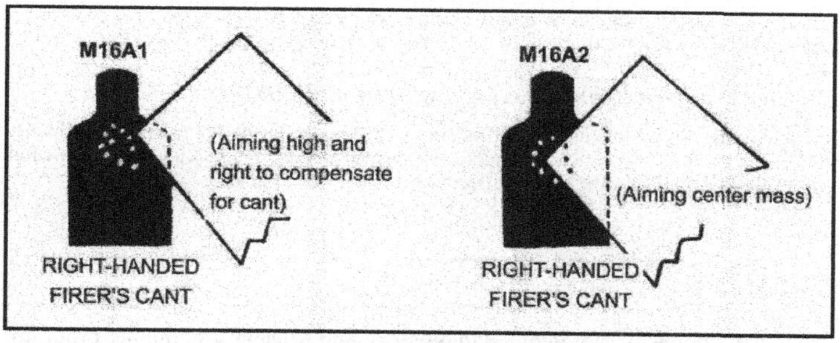

Figure 7-12. Engagement of 175-meter target.

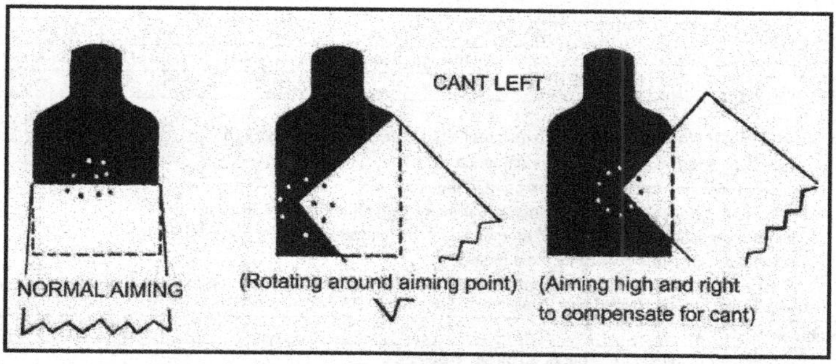

Figure 7-13. Engagement of 300-meter target.

7-106. Although bullet strike is displaced when using a cant, individual differences vary; center-of-mass aiming should be used until the individual knows what aiming adjustment is needed. When distant targets are missed, a right-handed firer should usually adjust his point of aim to the right and high; a left-handed firer should adjust to the left and high. Actual displacement of the point of aim must be determined by using downrange feedback targets positioned more than 75 meters from the firer.

Breath Control

7-107. Moving when encumbered by MOPP equipment requires more physical effort, and breathing is restricted while wearing the protective mask. Physical exertion can produce labored breathing and make settling into a normal breathing rhythm much more difficult, which, in turn, can increase the breath rate. All of these factors make holding and controlling the breath to produce a well-aimed shot more energy- and time-consuming. Emphasis must be placed on rapid target engagement during the limited amount of time a firer can control his breath.

Trigger Squeeze

7-108. Consider the following modifications:

- MOPP gloves complicate the act of grasping the pistol grip and squeezing the trigger with the index finger. The action of the trigger finger is restricted, and the fit of the glove may require the release of the swing-down trigger guard.
- Because the trigger feels different, control differs from that used in barehanded firing. This difference cannot be accurately predicted.

7-109. Dry-fire training using dime/washer exercises or simulators such as the EST 2000 or LMTS may be necessary to ensure that the firer knows the changes he will encounter during live-fire.

MISSION-ORIENTED PROTECTIVE POSTURE EQUIPMENT TRAINING

7-110. Repeated dry-fire training and LFXs are the most efficient means to prepare a Soldier wearing MOPP Level 4 equipment for successful target engagements at any range. The Soldier must follow these procedures and applications to be combat-effective in a CBRN environment.

Conduct

NOTE: See Table 7-9 for the current training program.

Table 7-9. Chemical, Biological, Radiological, and Nuclear fire training program.

CBRN FIRE TRAINING PROGRAM
Instructional Intent
• Develop the Soldier's confidence and ability to engage targets while wearing any level of MOPP equipment.
Special Instructions
Ensure that— • The M16A2/A3/A4 rifle's or M4 carbine's rear sight is set on the 0-2 aperture. • The M16A1's rear sight is set on the unmarked aperture. • Soldiers have inserted lenses before firing, if required. • Soldiers have a proper seal on the mask to prevent fogging and loss of visibility. • Each Soldier is issued 20 rounds of ammunition to be loaded in two magazines (10/10). • Each Soldier engages 20 targets at 50 meters. • Target exposures consist of 10 from the right and 10 from the left from the foxhole supported firing position using Table 1 of the Record Fire Qualification firing table (DA Form 3595-R).
Observables
• Each Soldier obtains 11 hits out of 20 target exposures.

Dry-Fire Exercises

7-111. As with all marksmanship training, Soldiers must start at the basics in order to become proficient at CBRN fire. Modified fundamentals can be taught anywhere, and Soldiers must learn them before participating in an LFX. The dry-fire exercises used during CBRN training are the same as those used during initial rifle marksmanship (dime/washer exercise, target box, SPORTS, EST 2000, and LMTS). Soldiers can also practice by using MILES equipment during force-on-force training.

NOTE: Soldiers should wear MOPP Level 4 when participating in dry-fire exercises; training at the highest degraded level allows Soldiers to adjust their shooting technique to increase their marksmanship ability in a CBRN environment.

Downrange Feedback

7-112. CBRN downrange feedback gives Soldiers the confidence, knowledge, and skills required to consistently deliver accurate, well-aimed fire against combat targets as far away as 300 meters. On a KD range, the Soldier will perform the following scenario:

- The Soldier will be issued six magazines.
  - The first and second magazine will have 5 rounds each.
  - The third and fourth magazine will have 10 rounds each.
  - The fifth and sixth magazine will have 5 rounds each.

- The Soldier engages the appropriate targets.
  - The Soldier engages the 75-meter target with the first and second 5-round magazine from the foxhole, standing, or kneeling supported position.
  - The Soldier engages the 175-meter target with both 10-round magazines from the foxhole, standing, or kneeling supported firing position.
  - The Soldier engages the 300-meter target with the remaining two 5-round magazine from the foxhole, standing, or kneeling supported firing position.
- In order to receive a GO, the Soldier must obtain a minimum of—
  - 8 hits out of 10 shots on the 75-meter target.
  - 14 hits out of 20 shots on the 175-meter target.
  - 5 hits out of 10 shots on the 300-meter target.

NOTES: 1. DA Form 5789-R is used to record the firer's results.

2. Soldiers can use ammunition allocated by DA PAM 350-38 for advanced skill training for the CBRN downrange feedback scenario.

50-Meter Live-Fire Exercise

7-113. The basic CBRN LFX allows all Soldiers to gain confidence in their CBRN firing abilities. Practice and proficiency firing can be conducted on any range; however, when a Remote Electronic Target System (RETS) range is used for this exercise, the two 50-meter mechanisms are used. For CBRN LFXs, Soldiers perform the following scenario.

- Each Soldier is issued 20 rounds of ammunition to be loaded in two 10-round magazines.
- On the command "GAS–GAS–GAS," each Soldier engages twenty 50-meter targets. Target exposures consist of 10 from the right and 10 from the left from the foxhole supported or the kneeling firing position.
- Each Soldier must achieve 11 hits out of 20 target exposures.
- This is a GO/NO GO exercise.

Alternate Fire Exercise

7-114. The CBRN alternate fire course uses the 25-meter scaled silhouette timed-fire target. This silhouette can be used on any 25-meter range, and the target provides feedback to the firer on where the strike of the round impacts the target. This exercise is conducted in the same manner as the 25-meter alternate course record fire.

- Each Soldier will be in MOPP Level 4.
- Each Soldier will be issued 20 rounds of ammunition to be loaded two 10-round magazines.
- Each Soldier will engage each silhouette with two rounds from the foxhole supported or kneeling position using Table I of DA Form 5790-R.
- Each Soldier must achieve 11 hits out of 20 shots.
- This is a GO/NO GO exercise.

NOTE: At the commander's discretion, Table II of DA Form 3595-R may be used as an alternate course of fire. This table allows Soldiers to engage targets from 50 to 300 meters.

SECTION IV. NIGHT FIRE TRAINING

Soldiers should be able to fire their weapons effectively in total darkness, in bright sunlight, and under all conditions between these two extremes. Scheduled marksmanship training should provide a variety of night and limited visibility conditions. The battlefield may be illuminated by ground flares, handheld flares, M203 flares, mortar and artillery illumination, aerial flares, searchlights, exploding rounds, and burning vehicles, or obscured by smoke, fog, and various environmental conditions. The well-trained Soldier should have experienced a number of these conditions and be confident that he can effectively employ his weapon when required. This section provides guidance on training Soldiers to be effective in total darkness without using iron sights and in limited visibility when using iron sights.

NOTE: See Appendix C for more detailed information about night fighting.

UNASSISTED NIGHT FIRE TRAINING

7-115. All Soldiers must be able to effectively employ their weapons during limited visibility. Soldiers must experience the various conditions of night combat—from total darkness to the many types of artificial illumination.

NOTE: See Table 7-10 for the current training program.

Table 7-10. Unassisted night fire training program.

UNASSISTED NIGHT FIRE TRAINING PROGRAM
Instructional Intent
• Develop the Soldier's confidence in his ability to hit targets when he cannot see through his weapon sights and does not have night vision capability.
Special Instructions
Ensure that— • The M16A2/A3/A4 rifle's or M4 carbine's rear sight is set on the 0-2 aperture. • The M16A1's rear sight is set on the unmarked aperture. • Each Soldier is given two 15-round magazines with 10 rounds of ball ammunition and 5 rounds of tracer ammunition in each magazine. • Each Soldier engages the 50-meter F-type silhouette target from the prone supported or foxhole supported fighting position with one magazine. • Each Soldier engages the 50-meter F-type silhouette target from the prone unsupported fighting position or kneeling position with the second magazine.
Observables
• Each Soldier achieves 7 hits out of 30 target exposures.

MODIFICATIONS FOR NIGHT FIRE

7-116. During limited visibility, a firer cannot effectively use his iron sights in most situations, and without artificial illumination, the sights block his field of vision. Trainers must consider the impact of limited visibility on the Soldier's ability to properly apply combat firing skills and the fundamentals of marksmanship.

Operation and Maintenance of the Weapon

7-117. Many actions are affected by nighttime conditions: handling the weapon, performing operation and function checks, loading and unloading, and maintenance. Consider the following modifications:

- Because combat conditions and enforcement of noise and light discipline restrict the use of illumination, Soldiers must be trained to operate, service, and clean their weapons in total darkness.
- Movements are slower, tasks take longer to complete, vision is impaired, and equipment is more easily misplaced or lost.

NOTE: Although initial practice of these tasks should occur during daylight to facilitate control and error correction, repeated practice during actual nighttime conditions should be integrated with other training. Only through repeated practice and training can the Soldier be expected to perform all tasks efficiently.

Immediate Action

7-118. Under normal conditions, a Soldier should clear a stoppage in three to five seconds. After dark, this task usually takes longer. Identifying the cause of the stoppage may be difficult and frustrating for the Soldier. A dry-fire practice (applying SPORTS) under these conditions using dummy or blank rounds is necessary to reduce time and build confidence. Once the Soldier is confident in applying immediate action in darkness, he can perform such actions rapidly during actual firing.

NOTES: 1. Training should be practiced first during daylight for better control and error correction by the trainer.

2. To learn the hands-only technique of identifying a stoppage, the firer must practice applying immediate action with his eyes closed. This is a technique that Soldiers should be able to master with practice.

Firing Positions

7-119. The firing position recommended for use during limited visibility is the supported firing position.

NOTE: This unassisted night fire supported position differs slightly from the supported position taught in earlier periods of instruction because the firer cannot use his sights during limited visibility.

7-120. To effectively engage targets during limited visibility—

(1) Assume a supported firing position.
(2) Establish a raised stock weld (look 2 to 3 inches above the sights, level with the barrel).
(3) Point the weapon at the target.
(4) Fire in the semiautomatic fire mode.

NOTE: To obtain optimum results, keep the eyes open, and move the head, arms, and weapon as one unit.

Marksmanship Fundamentals

7-121. The four marksmanship fundamentals apply to night firing, but some modifications are needed depending on the conditions.

Steady Position

7-122. When applying unassisted night fire, the Soldier must change his head position or stock weld, especially when using weapon/target alignment techniques. Consider the following modifications:

- Position the head high so that the weapon is aligned on the target.
- Look just over the iron sights.
- Keep the cheek in contact with the stock.

7-123. Repeated dry-fire practice, followed by live-fire training, is necessary to learn and refine these modifications and still achieve the steadiest position.

Aiming

7-124. Modifications to the aiming process vary. Consider these modifications:

- When firing unassisted, use off-center vision instead of pinpoint focus.
- Open both eyes to gather the maximum available light.
- Focus both eyes downrange.

Breathing

7-125. This fundamental is not affected by unassisted night fire conditions.

Trigger Squeeze

7-126. This fundamental is not affected by unassisted night fire conditions. The objective is to not disrupt weapon's alignment with the target when squeezing the trigger.

Target Detection

7-127. Trying to detect a target during the day is difficult enough, but at night, it becomes even more so. The firer must detect and engage targets without artificial illumination or NVDs. Consider these modifications:

- Scan potential target areas and when a target is detected, engage it using a modified quick fire position.
- Take a few seconds to improve weapon/target alignment by pointing slightly low to compensate for the usual tendency to fire high (Figure 7-14).
- Apply the three principles of night vision:
  - Dark adaptation.
  - Off-center vision.
  - Scanning.

7-128. Tracer ammunition may provide feedback on the line of trajectory and facilitate any adjustments in the weapon/target alignment.

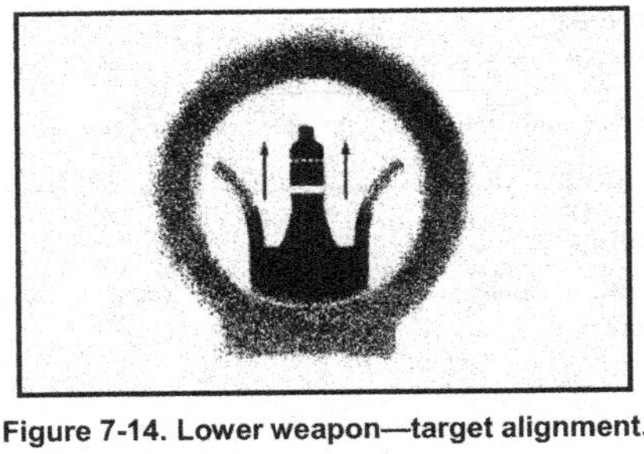

Figure 7-14. Lower weapon—target alignment.

Dark Adaptation

7-129. This process conditions the eyes to see under low levels of illumination. The eyes of the average person take about 30 minutes to acquire 98 percent night vision in a completely darkened area. Moving from illuminated to darker areas will decrease night vision until the eyes have adjusted to the surrounding area again.

Off-Center Vision

7-130. When an individual looks at an object during the daytime, he looks directly at it. However, if he did this at night he would only see the object for a few seconds. To see this object for any length of time, he must look 6 to 10 degrees from this object (Figures 7-15 and 7-16) while concentrating his attention on the object. This allows the light sensitive area of the eye to detect faint light sources or reflection.

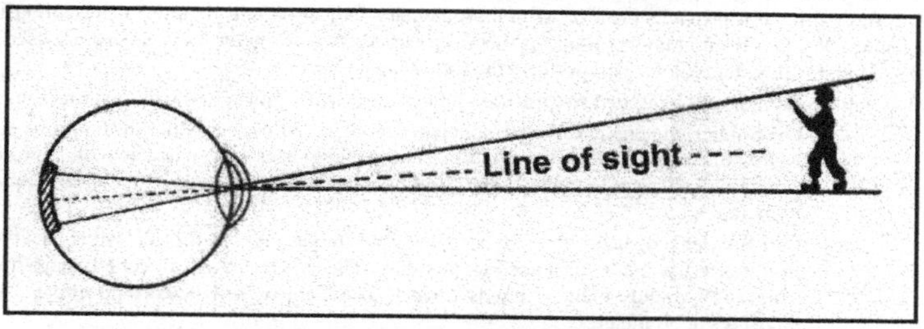

Figure 7-15. Daytime field of view using pinpoint focus.

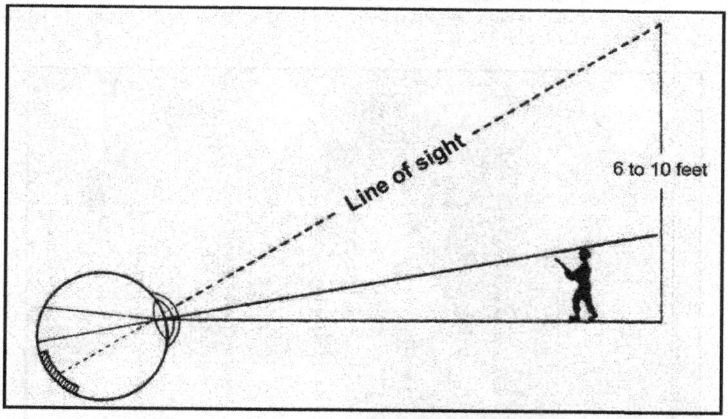

Figure 7-16. Nighttime field of view using off-center vision.

Scanning

7-131. The act of scanning relates to the short, abrupt, irregular movement of the firer's eyes every 4 to 10 seconds around an object or area. Be aware that scanning ranges vary according to levels of darkness.

NIGHT FIRE TRAINING

7-132. All units must include basic unassisted night fire training annually in their unit marksmanship programs.

7-133. Combat units should conduct tactical night fire training at least quarterly. This tactical training should include MILES, during force-on-force training, as well as live-fire training.

Live-Fire Exercises

7-134. The basic unassisted LFX allows all Soldiers to apply night fire principles and to gain confidence in their ability to effectively engage targets out to 50 meters. Practice and proficiency firing can be conducted on any range equipped with mechanical lifters and muzzle flash simulators (Figure 7-17). The muzzle flash simulator provides the firer with a momentary indication that a target is presenting itself for engagement. When a RETS range is used for this exercise, the two 50-meter mechanisms are used. For the unassisted night LFX, the Soldier will perform the following scenario:

- Each Soldier will be issued two 15-round magazines with a tracer and ball combination.
  - The Soldier engages the F-type silhouette target at 50 meters while in the prone supported or foxhole supported firing position. The Soldier uses one magazine of 15 rounds (10 rounds ball; 5 rounds tracer). The Soldier will detect and engage 15 target exposures at 50 meters.
  - The Soldier engages the F-type silhouette target at 50 meters while in the prone unsupported position or kneeling position. The Soldier uses a second magazine of 15 rounds (10 rounds ball; 5 rounds tracer). The Soldier will detect and engage 15 target exposures at 50 meters.
- Each Soldier must achieve 7 hits out of 30 target exposures.
- When the automated range is used, the Soldier's performance is recorded in the tower. If automatic scoring is not available, a coach can observe and score the number of target hits the firer achieves using NVDs.

Figure 7-17. Night-fire target.

ARTIFICIAL ILLUMINATION TRAINING

7-135. Artificial illumination allows for better target detection and long-range accuracy than unassisted night vision. When the artificial light is gone, time must be spent regaining night vision and adapting. Only when the level drops enough so that the target cannot be seen through the iron sights should the firer resume short-range scanning, looking just over the sights.

NOTE: See Table 7-11 for the current training program.

Table 7-11. Artificial illumination training program.

ARTIFICIAL ILLUMINATION TRAINING PROGRAM
Instructional Intent
• Develop the Soldier's confidence in his ability to locate, mark, prioritize and engage targets at night using artificial illumination.
Special Instructions
Ensure that—
• M16A2/A3/A4 or M4-series weapon's rear sight is set on the 0-2 aperture.
• M16A1 rear sight is set on the unmarked aperture.
• Each Soldier is given two 15-round magazines with 10 rounds of ball and 5 rounds of tracer ammunition.
• Each Soldier engages 15 target exposures from the foxhole supported fighting position with the first 15-round magazine.
• Each Soldier engages 15 target exposures from the prone unsupported fighting position with the second 15-round magazine.
Observables
• Each Soldier achieves 15 hits out of 30 target exposures.

EFFECTIVENESS AND CONTROL OF ARTIFICIAL ILLUMINATION

7-136. When artificial illumination is used, the eyes lose most of their night adaptation, and off-center vision is no longer useful. Artificial illumination allows the firer to use the iron sights with the 0-2 rear sight aperture, as he does during the day.

7-137. To preserve night vision while artificial illumination is being used, the Soldier closes his firing eye and scans his sector for enemy targets with his nonfiring eye. This allows the Soldier to have night vision in at least one eye to keep scanning his sector for enemy targets after the artificial illumination has burned out. However, keeping one eye closed to preserve its night vision results in a drastically altered sense of perception when both eyes are opened.

NOTE: Repeated dry-fire training and target detection practice are the keys to successful engagement of targets out to 250 meters or more during live-fire under artificial illumination.

LIVE-FIRE EXERCISE

7-138. The unassisted LFX with artificial illumination allows all Soldiers to apply night fire principles and to gain confidence in their abilities to effectively detect and engage targets 150 meters away and beyond when using artificial illumination. Soldiers use the night record fire scenario:

- Each Soldier is issued two 15-round magazines with the appropriate tracer and ball combination.
  - During the night, each Soldier engages 15 E-type silhouette target exposures from 50 to 250 meters with one magazine of 15 rounds (10 rounds ball, 5 rounds tracer) while in the foxhole supported or prone supported firing position.
  - During the night, each Soldier engages 15 E-type silhouette target exposures from 50 to 250 meters with the second magazine of 15 rounds (10 rounds ball, 5 rounds tracer) while in the prone unsupported or kneeling firing position.
- Each Soldier must achieve 15 hits out of 30 shots.

SECTION V. MOVING TARGET ENGAGEMENTS

In combat situations, enemy Soldiers do not stand still; they rush from one covered or concealed position to another. While making the rush, enemy Soldiers present rapidly moving targets. Moving targets are open to aimed fire at two points in the rush: as the target begins to gain speed (at the beginning) and as it slows down to a new position.

NOTE: See Table 7-12 for the current training program.

Table 7-12. Moving target engagement training program.

MOVING TARGET ENGAGEMENT TRAINING PROGRAM
Instructional Intent
• Soldiers learn to detect and engage moving and stationary targets with the M16 rifle or M4 carbine.
Special Instructions
Ensure that— • The M16A2/A3/A4 rifle's or M4 carbine's rear sight is set on the 0-2 aperture. • The M16A1's rear sight is set on the unmarked aperture. • Soldiers get into a proper semisupported firing position. • Soldiers understand and apply lead guidance rules. • Each Soldier receives two magazines with 25 rounds each of 5.56-millimeter ball ammunition. • Each Soldier engages 34 moving target exposures at ranges from 35 to 185 meters, and 15 stationary target exposures at ranges from 50 to 300 meters.
Observables
• Each Soldier achieves 18 hits out of 50 target exposures.

MODIFICATIONS FOR MOVING TARGET ENGAGEMENTS

7-139. Soldiers in combat do not know if their next target will be stationary or moving; they must fire immediately at whatever target presents itself. Trainers should consider the following when conducting moving target engagement instruction:

- More dispersion and erratic shots are expected when Soldiers are trained to hit moving targets.
- Considering the environment and the variables of the weapon and ammunition, well-trained Soldiers should be able to hit 300-meter stationary silhouette targets.
- When the target is moving laterally, well-trained Soldiers may only hit 150-meter targets 7 out of 10 times. This is considered an acceptable performance.

7-140. Further, trainers should consider modifications to the fundamentals for engaging stationary targets:

- Steady position.
- Aiming.
- Breath control.
- Trigger squeeze.

STEADY POSITION

7-141. When firing at moving targets, firers should assume the standard supported firing position, but be flexible so that they can track any target in the sector. When a target is moving directly at the firer, directly away from the firer, or at a slight angle, the Soldier engages the target without changing his firing position. Consider the following aspects of firing at moving targets:

- When targets are moving laterally, only minor changes are needed to allow effective target engagement.
- Most moving targets are missed in the horizontal plane (firing in front of or behind the target) and not in the vertical plane (firing too low or too high).

7-142. The Soldier must make other adjustments, as highlighted in Table 7-13.

Table 7-13. Modifications for a steady position when firing at moving targets.

PART OF BODY	MODIFICATION
Nonfiring hand	Grip the weapon more tightly with the nonfiring hand, and apply more pressure to the rear. This helps to maintain positive control of the weapon and steady it for rapid trigger action.
Nonfiring-side elbow	Lift the nonfiring-side elbow from the support position only to maintain a smooth track.
Firing hand	Apply more rearward pressure to the pistol grip to steady the weapon during trigger squeeze.
Firing-side elbow	Lift the firing-side elbow from support only to help maintain a smooth track. NOTE: The weapon pocket in the shoulder and the stock weld are the same as for stationary targets.

AIMING

7-143. When aiming at moving targets, Soldiers must apply precise lead rules, and in turn, Soldiers must accurately estimate speed, angle, and range to the target to apply precise lead rules. Then, he must apply the single-lead rule in order to place effective fire on combat targets.

7-144. The procedures used to engage moving targets vary as the angle and speed of the target vary. For example, when a target is moving directly at the firer, stationary target procedures apply. However, for a close, fast-moving target at a 90-degree angle, the weapon and firer's entire upper body must be free from support so the target can be tracked.

Lead Requirements

7-145. Aiming directly at a 300-meter target moving 8 miles per hour at a 90-degree angle would result in missing it; this type of target covers 4 ½ feet while the bullet is traveling toward him. To hit the target, the Soldier must apply target lead (Figure 7-18) and understand how target lead and bullet speed relate to the range, angle, and speed of the target. To hit the target depicted, the Soldier must aim and fire at position D when the target is at position A.

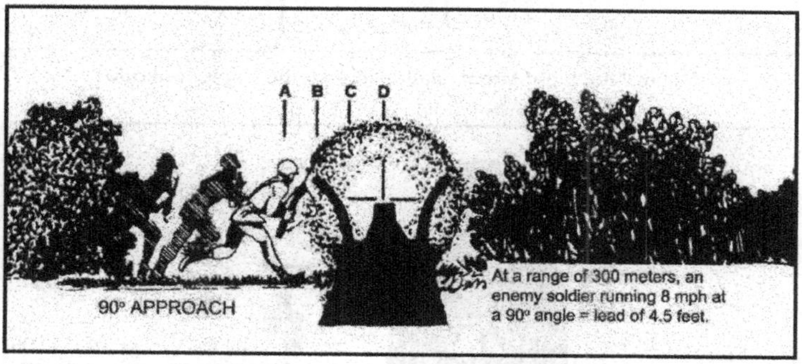

Figure 7-18. Lead requirement based on distance and approach angle.

Single-Lead Rule

7-146. The single-lead rule says: To hit a target moving laterally, place the trailing edge of the front sightpost at the target's center (Figure 7-19). This rule also provides that the lead increases as the range to the target increases (Figure 7-20).

NOTE: At 100 meters, the rule begins to break down for targets moving at slight and large angles.

> **EXAMPLE**
>
> As Figure 7-20 depicts, the front sightpost covers about 1.6 inches at 15 meters and about 16 inches at 150 meters. Since the center of the front sightpost is the actual point of aim, placing the trailing edge of the front sightpost at the target's center provides a .8-inch lead on a 15-meter target and an 8-inch lead on a target at 150 meters.
>
> This provides a dead-center hit on a 15-meter target moving at 7 miles per hour at a 25-degree angle because the target moves .8 inches between the time that the weapon is fired and the time that the bullet arrives at the target. A 150-meter target moving at 7 miles per hour at a 25-degree angle moves 8 inches between the time the weapon is fired and the bullet arrives.

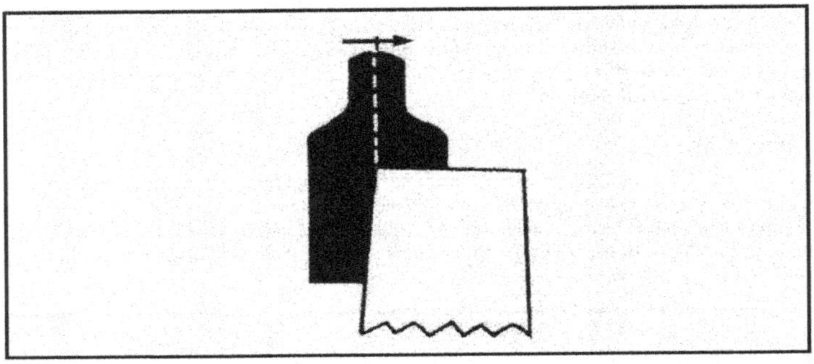

Figure 7-19. Sight-target relationship for the single-lead rule.

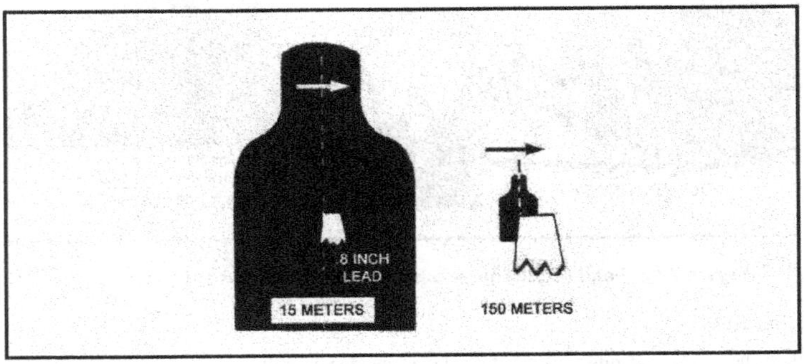

Figure 7-20. Lead increasing at greater ranges.

Target Speed

7-147. Figure 7-21 reflects the differences in lateral speeds for various angles of target movement for a target traveling at 8 miles per hour at a distance of 150 meters from the firer. The angle of target movement is the angle between the target/firer line and the target's direction of movement. An 8-mile-per-hour target moves 24 inches during the bullet's flight time. If the target is moving on a 15-degree angle, it moves 6 inches (the equivalent of 2 miles per hour).

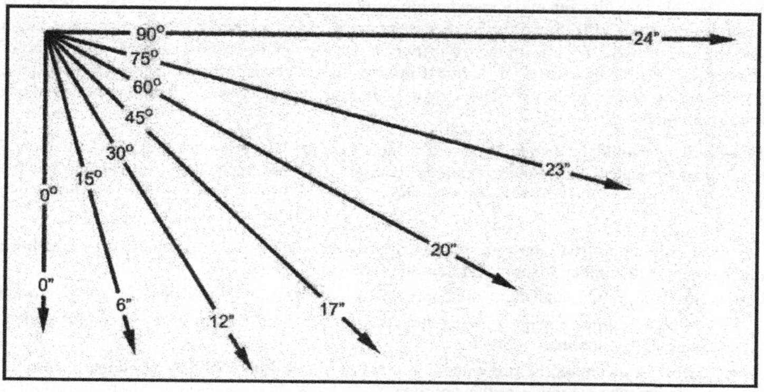

Figure 7-21. Target movement (distance) at various angles.

7-148. Since the target lead is half the perceived width of the front sightpost, at 100 meters, the standard sight provides 5.4 inches of lead for M16A1/2/3/4 rifles and M4 carbines (Table 7-14).

Table 7-14. Angle of target movement.

ANGLE OF TARGET MOVEMENT (Degrees)	RANGE: 100 m		
	(STANDARD SIGHT) TARGET SPEED		
	4 mph	6 mph	8 mph
5	+4.9"	+4.5"	+4.3"
10	+4.1"	+3.5"	+2.7"
15	+3.5"	+2.5"	+1.5"
20	+2.8"	+1.5"	+.2"
25	+2.2"	+.7"	-1.0"
30	+1.7"	-.2"	-2.0"
35	+1.1"	-1.1"	-3.2"
40	+.6"	-1.9"	-4.3"
45	0"	-2.7"	-5.4"
50	-.4"	-3.3"	-6.2"
55	-.8"	-4.0"	-7.0"
60	-1.2"	-4.5"	-7.7"
65	-1.5"	-4.9"	-8.4"
70	-1.7"	-5.3"	-8.8"
75	-1.9"	-5.6"	-9.2"
80	-2.0"	-5.9"	-9.6"
85	-2.1"	-5.9"	-9.7"
90	-2.1"	-6.0"	-9.8"
NOTE: Plus (+) indicates bullet strike in the direction of movement; minus (-) indicates bullet strike behind the target's center.			

Target Distance

7-149. The front sightpost covers only a small part of close-in targets, providing hits on close targets moving at any angle and any speed. However, if the lead rule is applied on more distant targets moving at a slight angle—for example, 5 degrees at 100 meters—the bullet strikes about 4 inches forward of the target's center. Soldiers must be taught to fire at targets as though they are stationary until lateral movement is observed (15 degrees).

7-150. The rule provides for many speed/angle combinations that place the bullet within 2 inches of the target's center (Table 7-14). Since the Soldier is expected to fire a 12-inch group on moving targets at 100 meters, the rule provides for hits on the majority of targets. Even the worst case (a 90-degree target moving at 8 miles per hour) would result in the shot group's center being located 9.8 inches behind the target's center. If bullets were evenly distributed in a 12-inch group, this would result in hitting the target 40 percent of the time.

7-151. Soldiers should be taught to increase their lead if they miss the target, which increases their probability of hitting all targets. For example, if target exhibits much lateral movement and the Soldier feels that he has missed the target by applying the lead rule and firing fundamentals, he should increase his lead.

7-152. The training program must be simple and provide Soldiers with information relevant to improving their performance in combat. It should address the following topics:

- Soldiers should understand and apply the single-lead rule in the absence of more information.
- Soldiers should engage moving targets coming toward them or on a slight angle (0 to 15 degrees) as stationary targets.
- Information should be presented and practice allowed on applying additional lead to targets for Soldiers who demonstrate this aptitude.

Target Angle

7-153. The single-lead rule does not apply to targets moving at small and large angles (Table 7-15). A walking enemy Soldier at 250 meters is hit dead center when he is moving at 40 degrees. Hits can be obtained if he is moving on any angle between 15 and 75 degrees. When he is running, a center hit is obtained when the target is on an angle of 18 degrees; misses occur when he exceeds an angle of 30 to 35 degrees.

7-154. The information provided in Figure 7-21 and Table 7-15 is designed to enhance instructor understanding so proper concepts are presented during instruction. For example, a target at 100 meters moving at 6 miles per hour receives a center hit when moving at 29 degrees. When moving at an angle less than 29 degrees, the bullet strikes somewhat in front of the target's center. When moving at an angle of more than 29 degrees, the bullet strikes somewhat behind the target's center.

Table 7-15. Target angle when dead center; hits occur using the single-lead rule.

STANDARD SIGHT			
RANGE (m)	4 mph	6 mph	8 mph
25	48°	30°	22°
50	47°	30°	22°
100	45°	29°	21°
150	44°	28°	20°
200	41°	27°	19°
250	40°	26°	18°
300	33°	21°	16°
350	38°	24°	18°
400	35°	22°	17°
450	33°	21°	16°

Tracking

7-155. Tracking is a more accurate technique of engaging targets used by experienced firers. It involves establishing and maintaining a point of aim in relationship to the target and moving with the target to maintain that sight picture, while squeezing the trigger. As the target moves, this technique puts the firer in position for a second shot if the first one misses.

Trapping

7-156. Trapping involves setting up a point of aim forward of the target and along the target path. The trigger is squeezed as the target comes into the sights. This technique works on targets with slow lateral movement. It does not require tracking skills, but the firer must know precisely when the weapon is going to fire.

NOTE: Soldiers who can squeeze the trigger without reacting to the weapon firing may fire better using this technique.

Breath Control

7-157. This fundamental is unchanged.

Trigger Squeeze

7-158. To use proper trigger squeeze—

- Apply rearward pressure on the handguard and pistol grip to hold the weapon steady while applying pressure to the trigger.
- Squeeze the trigger quickly (almost a controlled jerk).
- Apply heavy pressure on the trigger (at least half of the pressure it takes to make the weapon fire) before squeezing.

MOVING TARGET LIVE-FIRE EXERCISE

7-159. Soldiers engage in a firing scenario once for practice and then for qualification. Soldiers who fail to qualify on the initial day of qualification receive only one refire on the same day.

SECTION VI. SHORT-RANGE MARKSMANSHIP TRAINING

SRM training allows Soldiers to quickly and effectively engage targets at ranges less than 50 meters. Although normally associated with UO, SRM techniques are also used during operations in restrictive terrain, such as clearing a trench line, the final assault across an objective during an attack or raid, or when fighting in dense vegetation or during periods of limited visibility. SRM instruction consists of four components:

- Phase I—Reflexive Fire Training (blank-fire day and night).
- Phase II—Target Discrimination (blank-fire day and night).
- Phase III—Short-Range Marksmanship Qualification (day and night live-fire) and Barricade Transition Qualification.
- Phase IV—Shotgun and Automatic or Burst Firing Familiarization.

CONDUCT OF SHORT-RANGE MARKSMANSHIP TRAINING

7-160. SRM requires individual Soldiers to be trained to standard on reflexive firing, target discrimination, and all necessary BRM fundamentals prior to semiannual qualification.

NOTE: An explanation of the base level proficiency requirements is provided with each course of fire.

7-161. At a minimum, Soldiers should be qualified on their individual weapon within the previous six months. Shotgun and automatic firing is required for annual familiarization only. Reflexive MILES dry-fire drills are an essential part of the training process and should be conducted by the team leader or squad leader during troop-leading procedures and before any SRC or SRM training.

NOTE: See Table 7-16 for the current training program.

Table 7-16. Short-range marksmanship training program.

SHORT-RANGE MARKSMANSHIP TRAINING PROGRAM
Instructional Intent
• Soldiers gain confidence and knowledge in SRM fundamentals.
Special Instructions
Ensure that—
• The M16A2/A3/A4 rifle's or M4 carbine's rear sight is set on the 0-2 aperture.
• The M16A1's rear sight is set on the unmarked aperture.
• Each Soldier is given one 20-round magazine of 5.56-millimeter ball ammunition for marksmanship qualification.
• Each Soldier is given one 28-round magazine and three 4-round magazines of 5.56-millimeter ball ammunition for barricade transition firing.
NOTE: The round must impact within the lethal zone to be scored a hit.
Observables
• Each Soldier achieves 20 target hits during the day marksmanship qualification.
• Each Soldier achieves 16 target hits during the night marksmanship qualification.
• Each Soldier achieves 14 target hits during the day marksmanship qualification while wearing a protective mask.
• Each Soldier achieves 12 target hits during the night marksmanship qualification while wearing a protective mask.

NOTE: All SRC and SRM training should begin with a review of the principles of safe weapon handling: Assume the weapon is always loaded, and never point the weapon at anything that you do not intend to shoot.

FUNDAMENTALS OF SHORT-RANGE MARKSMANSHIP

7-162. During SRC, there is little margin for error. Too slow a shot at the enemy, too fast a shot at a noncombatant, or inaccurate shots can all be disastrous for the Soldier. Further, the risk of fratricide or noncombatant casualties is greatest during SRC. To survive and accomplish missions in close quarters, Soldiers must master four fundamentals of SRM:

- Firing stance, weapon ready positions, and movement techniques.
- Aiming technique.
- Point of aim.
- Trigger manipulation.

FIRING STANCE, WEAPON READY POSITIONS, AND MOVEMENT TECHNIQUES

7-163. Regardless of the weapon ready position used, Soldiers must always assume the correct firing stance to ensure stability and accuracy when engaging targets.

Firing Stance

7-164. To assume the correct fighting stance—

- Keep the feet approximately shoulder-width apart.
- Point the toes straight to the front (direction of movement).
- Stagger the firing-side foot slightly to the rear of the nonfiring-side foot.
- Bend the knees slightly.

- Lean the upper body slightly forward.
- Square the shoulders and pull them back; don't roll them over or slouch.
- Keep the head up and both eyes open.

7-165. When engaging targets—

- Hold the weapon with the buttstock firmly against the shoulder.
- Hold the firing-side elbow close against the body.

7-166. Although short-range engagements generally take place from the standing position, a Soldier may be required to engage targets from the kneeling position. The kneeling position is generally used when correcting a weapon malfunction.

Weapon Ready Positions

7-167. There are two weapon ready positions: high ready and low ready.

High Ready Position

7-168. The high ready position (Figure 7-22) is best suited for the lineup outside of a building, room, or bunker entrance.

7-169. To hold the weapon in the high ready position—

(1) Hold the weapon's buttstock under the armpit, with the barrel pointed slightly up so that the top of the front sightpost is just below the line of sight, but still within peripheral vision.

(2) Grasp the handguards toward the front sling swivel with the nonfiring hand. Keep the trigger finger outside of the trigger well and the thumb of the firing hand on the selector lever.

7-170. To engage a target from the high ready position—

(1) Push the weapon forward, as if to bayonet the target.

(2) Bring the buttstock firmly against the shoulder as it slides up the body.

Figure 7-22. High ready position.

Low Ready Position

7-171. The low ready position (Figure 7-23) is best suited for movement inside of buildings.

7-172. To hold the weapon in the low ready position—

(1) Place the weapon's buttstock firmly in the pocket of the shoulder, with the barrel pointed down at a 45-degree angle.

(2) Grasp the handguards toward the front sling swivel. Keep the trigger finger outside of the trigger well and the thumb of the firing hand on the selector lever.

7-173. To engage a target from the low ready position, bring the weapon up until the proper sight picture is achieved.

Figure 7-23. Low ready position.

Movement Techniques

7-174. Soldiers must practice moving with their weapons up until they no longer look at the ground, but concentrate on their sectors of responsibility and move without stumbling over their own feet.

NOTE: The low ready position is the best position to use when moving or turning.

7-175. To engage a target to the left—

(1) Step toward the target with the left foot.

(2) Move the right foot even with the left, assuming a proper standing firing position.

7-176. To engage a target to the right—

(1) Step toward the target with the right foot.

(2) Move the left foot even with the right, assuming a proper standing firing position.

7-177. To turn to the rear—

(1) Position the firing-side foot forward.

(2) Place the body weight on the firing-side foot.

(3) Pivot the body, similar to the drill movement "rear march."

Aiming Techniques

7-178. Four aiming techniques are used during SRC:

- Slow aimed fire.
- Rapid aimed fire.
- Aimed quick kill.
- Instinctive fire.

7-179. Each has advantages and disadvantages, and the Soldier must understand when, how, and where to use each technique.

Slow Aimed Fire

7-180. Slow aimed fire is the slowest, but most accurate, technique. When using this technique, Soldiers take a steady position, properly align the sight picture, and squeeze off rounds. This technique should only be used to engage targets more than 25 meters away, when good cover and concealment is available, or when the need for accuracy overrides the need for speed.

Rapid Aimed Fire

7-181. The rapid aimed fire technique utilizes an imperfect sight picture. When using this technique, the Soldier focuses on the target and raises his weapon until the target is obscured by the front sightpost. Elevation is less critical than windage when using this technique. This aiming technique is extremely effective on targets 0 to 15 meters away.

Aimed Quick Kill

7-182. The aimed quick kill technique is the quickest and most accurate method of engaging targets up to 12 meters away. As Soldiers become more experienced at using this technique, they may use it at greater ranges. When using this technique, the Soldier aims over the rear sight, down the length of the carrying handle, and places the top ½ to ¾ of an inch of the front sightpost on the target.

Instinctive Fire

7-183. Instinctive fire is the least accurate technique and should only be used in emergencies. It relies on instinct, experience, and muscle memory. To use this technique, the firer concentrates on the target and points the weapon in the general direction of the target. While gripping the handguards with the nonfiring hand, he extends the index finger to the front, automatically aiming the weapon on a line toward the target.

Point of Aim

7-184. Most short-range engagements will be decided by who hits his target with the first round. During this type of engagement, it is more important to put the target down as quickly as possible than it is to kill him immediately.

7-185. Soldiers must aim at the lethal zone (center of mass) of the body. Although shots to the center of the body may prove to be eventually fatal, they may not immediately incapacitate the target. A shot that does not immediately incapacitate the target may be no better than a clean miss. Because of this, and the possible presence of military equipment or protective vests, Soldiers must also be able to engage targets with incapacitating shots.

Lethal Shot Placement

7-186. The target's lethal zone (Figure 7-24) is its center of mass, between the waist and the chest. Shots in this area maximize the hydrostatic shock of the shot pellets. Due to the nature of SRC, Soldiers must continue to engage targets until they go down.

Figure 7-24. Lethal zone.

Incapacitating Shot Placement

7-187. Only one shot placement guarantees immediate and total incapacitation: roughly centered in the face, below the middle of the forehead and the upper lip, and from the eyes in. Shots to the side of the head should be centered between the crown of the skull and the middle of the ear opening, from the center of the cheekbones to the middle of the back of the head (Figure 7-25).

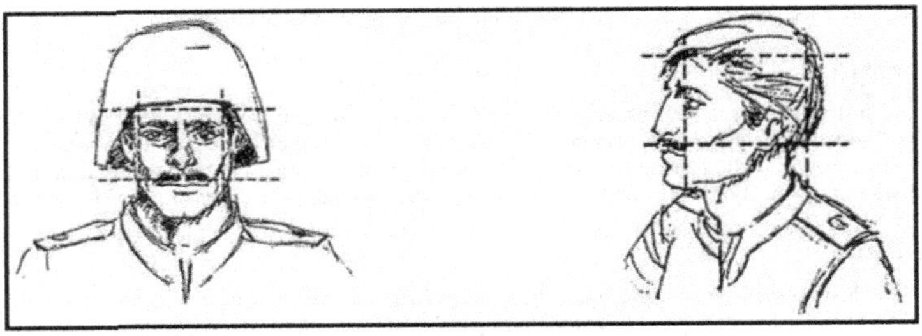

Figure 7-25. Incapacitation zone.

TRIGGER MANIPULATION

7-188. SRC engagements are usually quick, violent, and deadly. Due to the reduced reaction time, imperfect sight picture, and requirement to effectively place rounds into targets, Soldiers must fire multiple rounds during each engagement in order to survive. Multiple shots may be fired either through the use of a controlled pair or automatic weapon fire.

Controlled Pair

7-189. A controlled pair is two rounds fired in rapid succession. Controlled pairs should be fired at single targets until they go down. When multiple targets are present the Soldier must fire a controlled pair at each target, and then reengage any targets left standing. To fire a controlled pair—

(1) Fire the first round, and allow the weapon to move in its natural arc without fighting the recoil.

(2) Rapidly bring the weapon back on target, and fire a second round.

7-190. Soldiers must practice firing the controlled pair until it becomes instinctive.

Automatic Fire

7-191. While rapid, aimed, semiautomatic fire is the most accurate method of engaging targets during SRC and controlled three-round bursts are better than automatic fire, automatic weapon fire may be necessary to maximize violence of action or gain fire superiority when gaining a foothold in a room, building, or trench. When properly trained, Soldiers should be able to fire six rounds (two three-round bursts) in the same time it takes to fire a controlled pair. With practice, the accuracy of engaging targets can be equal to that of semiautomatic fire at 10 meters.

NOTE: The key to burst or automatic firing is to squeeze the trigger, not jerk it.

7-192. For the majority of Soldiers, fully automatic fire is rarely effective and can lead to unnecessary noncombatant casualties or fratricide. Not only is fully automatic fire inaccurate and difficult to control, but it also rapidly empties ammunition magazines. A Soldier who finds himself out of ammunition with an armed, uninjured enemy Soldier during SRC will become a casualty unless a fellow Soldier intervenes.

Failure Drill

7-193. To make sure that a target is completely neutralized, Soldiers should be trained to execute the failure drill. The firer will fire a controlled pair into the lethal zone, followed by a third round placed into the incapacitation zone. This type of target engagement is particularly useful when engaging targets wearing body armor.

PRELIMINARY SHORT-RANGE MARKSMANSHIP INSTRUCTION

7-194. As with all other forms of marksmanship training, preliminary SRM instruction must be conducted to establish a firm foundation. Soldiers must be taught, and must understand, the fundamentals of SRM. Blank-fire drills are conducted to ensure a thorough understanding of the fundamentals, as well as to provide the trainers with valuable feedback about each Soldier's level of proficiency.

NOTE: To maximize safety during training and in combat situations, it is important to emphasize muzzle awareness and selector switch manipulation during preliminary SRM instruction. The risk of fratricide or noncombatant casualties is greatest during SRC.

7-195. Table 7-17 outlines the tasks that preliminary SRM instruction should include (at a minimum).

Table 7-17. Preliminary SRM tasks and explanation.

TASK	EXPLANATION
Weapon ready positions and firing stance	Ensure that each Soldier understands and can properly carry his weapon.
Magazine changing drills	Have Soldiers perform magazine changes using the two methods: • Parallel method. • L-shaped method. NOTE: See the paragraph below for more information about magazine changing drills.
Moving with a weapon	Ensure that Soldiers can walk, run, and turn left, right, and to the rear, as well as move from the standing to kneeling firing position and the kneeling back to the standing firing position.
Weapon malfunction drills	Ensure that Soldiers instinctively drop to the kneeling firing position, clear a malfunction using SPORTS, and continue to engage targets. This drill can be performed by issuing each Soldier a magazine loaded with six to eight rounds of blank ammunition with one expended blank round.
Target engagement drills	Teach Soldiers to move from the weapon ready position to the firing stance, emphasizing speed and precision. Soldiers must be observed to ensure that the finger is outside of the trigger well and that the selector switch remains on SAFE until the weapon is raised to the firing position. NOTE: This is a force protection issue and must be drilled until all Soldiers can perform to standard.

MAGAZINE CHANGING DRILLS

NOTE: Before performing magazine changing drills, have Soldiers configure their LBE or magazine racks to allow for the easiest access for the magazine change. The time limit to perform any of the magazine changing drills is 8 seconds.

7-196. There are two methods of changing the magazine:

- Parallel method.
- L-shaped method.

Parallel Method

7-197. When shooting from the right side, perform the right-side parallel magazine changing method (Figure 7-26):

(1) Engage the target until the magazine is empty.

(2) Take a knee.

(3) Place the weapon at a 45-degree angle, with the magazine well facing in, the muzzle pointed in a safe direction, and the buttstock in the crease of the right elbow.

(4) Slide the nonfiring hand down the handguard to the receiver, and press the magazine release catch.

(5) Secure a full magazine with two fingers and the thumb of the nonfiring hand.

(6) Bring the full magazine (top of the magazine toward the magazine well) next to and parallel with the empty magazine.

(7) Slide the empty magazine out with the nonfiring hand and insert the full magazine.

(8) Hit the bolt release.

(9) Stow the empty magazine in the ammunition pouch.

Figure 7-26. Right-side parallel magazine changing method.

7-198. When shooting from the left side, perform the left-side parallel magazine changing method (Figure 7-27):

(1) Engage the target until the magazine is empty.
(2) Take a knee.
(3) Place the weapon at a 45-degree angle, with the magazine well facing in, the muzzle pointed in a safe direction, and the buttstock in the crease of the left elbow.
(4) Secure a full magazine with two fingers and the thumb of the nonfiring hand.
(5) Bring the full magazine (top of the magazine toward the magazine well) next to and parallel with the empty magazine.
(6) Press the magazine release catch with the trigger finger.
(7) Slide the empty magazine out with the nonfiring hand, and insert the full magazine.
(8) Hit the bolt release.
(9) Stow the empty magazine in the ammunition pouch.

Figure 7-27. Left-side parallel magazine changing method.

L-Shaped Method

7-199. When shooting from the right side, perform the right-side L-shaped magazine changing method (Figure 7-28):

(1) Engage the target until the magazine is empty.
(2) Take a knee.
(3) Place the weapon at a 45-degree angle, with the magazine well facing in, the muzzle pointed in a safe direction, and the buttstock in the crease of the right elbow.
(4) Slide the nonfiring hand down the handguard to the receiver, and press the magazine release catch.
(5) Secure a full magazine with two fingers and the thumb of the nonfiring hand.
(6) Bring the full magazine (top of the magazine facing to the rear) next to and perpendicular to the empty magazine, forming an L-shape.
(7) Slide the empty magazine out with the nonfiring hand, rotate the full magazine so the top of the magazine faces the magazine well, and insert the full magazine.
(8) Hit the bolt release.
(9) Stow the empty magazine in the ammunition pouch.

Figure 7-28. Right-side L-shaped magazine changing method.

7-200. When shooting from the left side, perform the left-side L-shaped magazine changing method (Figure 7-29):

(1) Engage the target until the magazine is empty.
(2) Take a knee.
(3) Place the weapon at a 45-degree angle, with the magazine well facing in, the muzzle pointed in a safe direction, and the buttstock in the crease of the left elbow.
(4) Press the magazine release catch with the trigger finger.
(5) Secure a full magazine with two fingers and the thumb of the nonfiring hand.
(6) Bring the full magazine (top of the magazine facing to the rear) next to and perpendicular to the empty magazine, forming an L-shape.
(7) Slide the empty magazine out with the nonfiring hand, rotate the full magazine so the top of the magazine faces the magazine well, and insert the full magazine.
(8) Hit the bolt release.
(9) Stow the empty magazine in the ammunition pouch.

Figure 7-29. Left-side L-shaped magazine changing method.

PHASE I—REFLEXIVE FIRE TRAINING

7-201. Reflexive fire training involves the practical application of all four of the fundamentals of SRM. All Soldiers must receive a GO on the task, Conduct Reflexive Firing, before proceeding with training. This is a perishable skill that must be constantly reinforced, and reflexive fire training should be conducted frequently as refresher training to ensure that the Soldiers' skills are always at the highest level possible.

REFLEXIVE FIRING TARGETS

7-202. Targets can be purchased locally (FBI style) or manufactured by the unit (bowling pin targets). E-type silhouettes may be painted as shown in Figure 7-30.

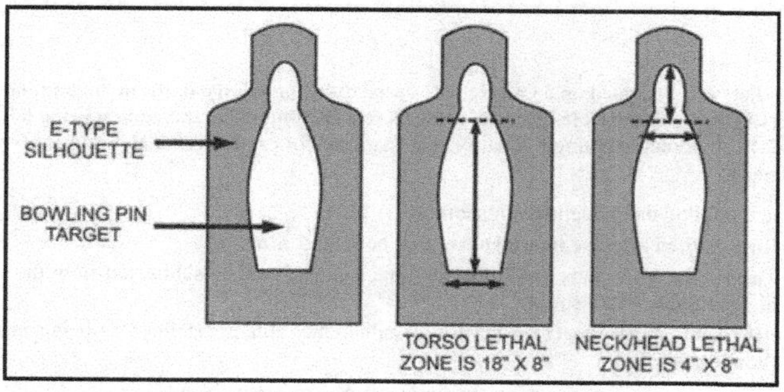

Figure 7-30. Dimensions and placement of bowling pin targets.

Range Setup

7-203. The range must be at least 25 meters in length, with identification marks at the 5-, 10-, 15-, and 25-meter distances. Each lane should be marked in a way that prevents crossfiring between lanes. A lane safety coach is assigned to each lane to observe and evaluate the Soldiers' performance, as well as ensure the safe conduct of firing. All firing cues are given by the tower or line safety.

Conduct of Training

NOTE: Unit commanders should conduct training continually to first establish and then sustain levels of proficiency in reflexive firing.

7-204. Each Soldier conducts a dry-fire exercise and a blank-fire exercise prior to conducting the LFX. The dry-fire and blank-fire exercises give the Soldier the repetition needed to successfully engage targets quickly and accurately.

7-205. During the dry-fire exercise, Soldiers identify and engage the proper targets at ranges from 5 to 25 meters from the stationary position, while turning and walking. The tower or line safety gives all firing cues. All tables are fired at night, with and without protective masks, using automatic fire for familiarization, and while using NVDs. The dry-fire exercise proceeds as follows:

(1) The Soldier starts at the 25-meter line at the low ready position, facing the targets.

(2) The Soldier is then told the engagement position (for example, facing left, turns right) and, once in position, is given the cue to fire.

(3) On cue, the Soldier assumes the proper firing position and stance, places the selector lever on SEMI, uses the correct aiming technique for the target's distance, and engages the target.

(4) After engaging the target, the Soldier continues to cover the target to reinforce firing until the threat is eliminated.

NOTE: If Soldiers will be engaging targets using lasers, optics, or the protective mask, they should complete all steps using the same equipment. Do not have Soldiers familiarize with iron sights and then fire the LFX with optics.

Evaluation

7-206. Soldiers are evaluated on a GO/NO GO basis, based on the standards in the training and evaluation outline (T&EO) and scoring table. Soldiers must score a GO on the familiarization firing tables (Table 7-18 and Table 7-19) before attempting to qualify. The standard for protective mask firing is 60 percent day and 50 percent night.

7-207. Use the following guidelines for scoring:

- Rounds fired after the time standard will be scored as a miss.
- The number of rounds fired after the time standard will be subtracted from the total number of hits the Soldier has scored.
- All rounds must impact on the E-type silhouette. Hits are defined as being in the lethal zone (bowling pin).

Table 7-18. Familiarization (stationary).

POSITION	ROUNDS FIRED	DISTANCE (m)	METHOD	TIME STANDARD	LETHAL ZONE HIT STANDARD
Straight ahead	4	25	Single shot	None	3
Straight ahead	4	10	Single shot	None	3
Straight ahead	4	25	Controlled pair	None	3
Straight ahead	4	10	Controlled pair	None	3

Table 7-19. Familiarization (moving).

POSITION	ROUNDS FIRED	DISTANCE (m)	METHOD	TIME STANDARD	LETHAL ZONE HIT STANDARD
Facing left; turning right	4	25	Controlled pair	None	3
Facing right; turning left	4	25	Controlled pair	None	3
Straight ahead walking	4	10, start at 15	Controlled pair	None	3
Straight ahead	4	5	Controlled pair	None	3

PHASE II—TARGET DISCRIMINATION TRAINING

7-208. During SRC, there is little or no margin for error. Too slow a shot at the enemy, too fast a shot at a noncombatant or friendly Soldier, or inaccurate shots can all be disastrous for the Soldier. Target discrimination is the act of distinguishing between threat and nonthreat targets during SRC. Target discrimination is an inescapable responsibility and must be stressed in all situations regardless of mission.

NOTE: It is essential that target discrimination training be aimed at instilling fire control and discipline in Soldiers. The first priority is always the safety of a Soldier.

TARGET DISCRIMINATION TARGETS

7-209. Targets used to conduct target discrimination training include—

(1) Two or more E-type silhouettes with bowling pins painted on each side of the silhouette (for example, brown side and green side).

(2) Multiple E-type silhouettes with different painted shapes (for example, squares, triangles, and circles).

(3) A series of 3-inch circles on E-type silhouettes.

(4) Pop-up targets (for example, electrical or pull targets).

(5) Target's hands and center of the uniform.

Two or More E-Type Silhouettes with Bowling Pins Painted on Each Side

7-210. Target discrimination is best taught using two or more E-type silhouettes with bowling pins painted on each side of the silhouette. To conduct training using this type of target—

(1) The instructor calls out a color for the firer to identify.

(2) On the command "READY, UP" or at the whistle blast, the firer quickly scans all targets for the color and engages the correct target using a controlled pair.

7-211. This exercise trains Soldiers to accomplish missions under the expected ROE. During force-on-force training, the OPFOR wears distinctive uniforms; this exercise prepares Soldiers to eliminate threats based on enemy uniforms and reduces the chances of a Soldier hesitating and becoming a casualty. Using realistic targets displaying threat and nonthreat personnel is another variation.

Multiple E-Type Silhouettes with Different Painted Shapes

7-212. To conduct training using this type of target—

(1) The instructor calls out a shape for the firers to identify.

(2) On the command "READY, UP" or at a whistle blast, the firers quickly scan all targets, searching for the shape, and engage the correct target using the controlled pair technique.

7-213. This is repeated until one shape is mastered. Subsequently, a sequence of shapes is announced, and the firers engage accordingly.

Series of 3-Inch Circles on E-Type Silhouettes

7-214. To conduct training using this type of target—

(1) Instructors call out which circle to engage (for example, top left).

(2) Firers react accordingly.

7-215. Marksmanship is emphasized using this technique.

Target's Hands and Center of the Uniform

7-216. If a target is a threat, the first and most obvious indicator is a weapon in the target's hands. A good technique for teaching Soldiers target discrimination is to have them focus on the target's hands and the center of the uniform. To conduct this exercise, the Soldier must visualize the entire target because an armed target could be a fellow Soldier or other friendly. This level of target discrimination should not be trained until Soldiers are thoroughly proficient in basic SRC and SRM tasks.

Range Setup

7-217. The range must be at least 25 meters in length, and each lane should be at least 5 meters wide. Each lane should have target holders and should be marked in a way that prevents crossfiring between lanes. A coach/safety is assigned to each lane to observe and control the Soldier's performance. The tower, lane safety, or senior instructor gives all firing commands.

Conduct

7-218. Regardless of the type of target used, the exercise proceeds as follows:

NOTE: Each Soldier must complete a dry-fire exercise and a blank-fire exercise before moving to the live-fire portion.

(1) The Soldier faces away from the target.

NOTE: This action requires the Soldier to identify and discriminate targets, and reinforces skills used during reflexive firing training.

(2) The Soldier is given a target description.

(3) On the command "READY," the Soldier begins to scan for the target.

(4) On the command "UP," voice command, or whistle blast, the Soldier turns toward and engages the target.

NOTES: 1. Instructors should vary commands and targets so that the Soldier does not fall into a pattern. Intermixing "NO FIRE" commands will add to realism.

2. The first priority is always the safety of the Soldier.

7-219. All Soldiers must receive a GO on this task before SRM qualification. Targets must be scored and marked after each firing distance. A Soldier will be scored as a NO GO if he fails to engage a target or engages a target other than the one called for by the instructor. Soldiers will also receive a NO GO if, at any time, they point their weapon at another Soldier or they fail to keep their weapon on SAFE before acquiring and engaging the targets.

NOTE: Initial training and sustainment training may be conducted by changing the uniform in the standards statement.

Evaluation

7-220. Table 7-18 will be used to score this exercise.

PHASE III—SHORT-RANGE MARKSMANSHIP QUALIFICATION

7-221. Soldiers should conduct SRM qualification semiannually. In addition to qualification, commanders should conduct familiarization using the same qualification standards while altering the conditions. Firing the qualification tables in protective masks and during periods of limited visibility should be included; Soldiers should train as they fight—with all assigned equipment.

NOTE: Although the qualification is intended to be fired with open sights only, iterations using laser aiming devices, CCOs, and NVDs are highly encouraged.

Record and Practice Fire

Conduct

NOTES: 1. Soldiers must complete a blank-fire iteration of the qualification tables before conducting live-fire qualification.

2. Each Soldier will have a coach to ensure that he acquires the target, keeps the weapon on SAFE until time to engage the target and is then places it back on SAFE, and maintains muzzle awareness throughout the exercise.

3. If Soldiers will be engaging targets with lasers, optics, or the protective mask, they should complete all steps using the same equipment. Do not have Soldiers familiarize with iron sights and fire the exercise while wearing the protective mask.

7-222. Each Soldier engages the target IAW the firing table (Table 7-20).

Table 7-20. Record and practice fire.

POSITION	ROUNDS FIRED	DISTANCE (m)	METHOD	TIME STANDARD
Straight ahead	2	4	Controlled pair	3 sec from command "UP"
Left turn	2	7	Controlled pair	3 sec from command "UP"
Right turn	2	10	Controlled pair	3 sec from command "UP"
Walk straight	2	5, start at 10	Controlled pair	3 sec from command "UP"
Walk straight	2	10, start at 15	Controlled pair	3 sec from command "UP"
Run, stop, and shoot	2	10, start at 20	Controlled pair	3 sec from command "UP"
Straight ahead	2	25	Controlled pair	3 sec from command "UP"
Run, stop, kneel, and shoot	2	15, start at 25	Controlled pair	3 sec from command "UP"
Walk laterally right	2	7	Controlled pair	3 sec from command "UP"
Walk laterally left	2	7	Controlled pair	3 sec from command "UP"

Evaluation

NOTE: For scoring purposes, a hit is a round that impacts within the lethal zone.

7-223. Each Soldier scores 16 hits day and night. The standard when wearing a protective mask is 14 day and 12 night. In addition to achieving a qualifying score, all 20 rounds must hit the E-type silhouette in order to qualify.

BARRICADE TRANSITION QUALIFICATION FIRING

Range Setup

7-224. The range must be at least 50 meters in length, with tall barricades at the 35- and 50-meter distances and short barricades at the 25- and 40-meter distances. Each lane should be marked in a way that prevents crossfiring between lanes. A lane safety coach is assigned to each lane to observe and evaluate the Soldier's performance, as well as ensure that firing is conducted safely. The tower or line safety gives all firing cues.

Conduct

NOTE: Soldiers must complete a dry-fire and a blank-fire iteration of the qualification tables before conducting live-fire qualification. The dry-fire and blank-fire exercises gives the Soldier the repetition needed to successfully engage targets quickly and accurately.

7-225. The exercise proceeds as follows:

(1) The Soldier receives one 28-round and three 4-round magazines.

(2) The Soldier loads one 28-round magazine.

(3) The Soldier assumes the low ready position at the 50-meter line, facing the targets.

(4) The Soldier engages the targets as shown in Table 7-21.

(5) On cue, the Soldier assumes the proper firing position and stance, places the selector lever on SEMI, uses the correct aiming technique for the target's distance, and engages the target.

NOTE: If a Soldier is having difficulty during the blank-fire exercise, he must be retrained before continuing with the qualification.

Evaluation

7-226. Use the following guidelines for scoring:

- All rounds must impact on the E-type silhouette.
- Hits are defined as being in the lethal zone (bowling pin).

Table 7-21. Barricade transition fire.

BARRICADE/DISTANCE	POSITION	ROUNDS FIRED	METHOD
Start with 28-round magazine in the open.			
Open/50 m	Standing	2	Controlled pair
	Kneeling	2	Controlled pair
	Prone	2	Controlled pair
	Sitting	2	Controlled pair
Tall/50 m	Left	2	Controlled pair
	Right	2	Controlled pair
Short/40 m	Left	2	Controlled pair
	Top	2	Controlled pair
	Right	2	Controlled pair
Tall/35 m	Left	2	Controlled pair
	Right	2	Controlled pair
Short/25 m	Left	2	Controlled pair
	Top	2	Controlled pair
	Right	2	Controlled pair
Transition back to the tall 35-m barricade. Perform parallel or L-shaped magazine change. Insert 4-round magazine.			
Tall/35 m	Left	2	Controlled pair
	Right	2	Controlled pair
Transition back to short 40-m barricade. Perform parallel or L-shaped magazine change. Insert 4-round magazine.			
Short/45 m	Left	2	Controlled pair
	Right	2	Controlled pair
Transition back to tall 50-m barricade. Perform parallel or L-shaped magazine change. Insert 4-round magazine.			
Tall/50 m	Left	2	Controlled pair
	Right	2	Controlled pair

PHASE IV—SHOTGUN AND AUTOMATIC OR BURST FIRING FAMILIARIZATION

7-227. Shotgun and automatic or burst firing familiarization is no different for SRM than for BRM.

NOTE: Publications for shotgun firing familiarization are currently being developed.

SECTION VII. SQUAD DESIGNATED MARKSMAN TRAINING

Advances in technology have led to the development of weapon systems that are increasingly more accurate and able to engage targets at much longer ranges. Conversely, today's rifleman is trained to engage targets only out to 300 meters. This 300-meter limit is well short of the weapon/ammunition combination's capability. Snipers engage targets at 600 meters and beyond. The SDM engages targets with direct small arms fire in the gap between the engagement range of the average combat Soldier and the sniper. Possessing the ability to estimate range, detect targets, and place effective, well-aimed fire on intermediate range targets, the SDM plays a vital role on the modern battlefield.

MISSION OF THE SQUAD DESIGNATED MARKSMAN

7-228. The SDM program provides the squad with a designated marksman that has been trained to engage targets from 300 to 500 meters. He will operate and maneuver as a rifleman, but will have the added responsibility of engaging targets out to 500 meters with effective, well-aimed fires. The SDM is a vital member of his individual squad, not a squad sniper. The SDM has neither the equipment nor training to engage targets at extended ranges with precision fires while operating individually or in a small team. He can also be used to help direct the fires of other squad members into enemy positions. Due to the increased skill level required for his position, the SDM must maintain a high level of proficiency through continued training of the required skills. The SDM must possess a thorough understanding and mastery of—

- The fundamentals of rifle marksmanship.
- Ballistics.
- Elevation and windage.
- Hold-off (adjusted points of aim).
- Sight manipulation.
- Range estimation.

SELECTION

7-229. The platoon sergeant and squad leaders must take special consideration in selecting the SDM. The SDM must have—

- A solid marksmanship performance.
- A clear understanding of the fundamentals.
- The ability to apply these fundamentals consistently during dry-fire and live-fire training.

SQUAD DESIGNATED MARKSMAN SKILLS PROGRESSION

7-230. The skills progression program assesses the Soldier's ability to apply the fundamentals of marksmanship and trains and assesses the Soldier's proficiency in several key areas. Firing events will serve to both reinforce and assess these areas.

CONDUCT

7-231. While conducting the skills progression program, instructors-trainers will adhere to the following guidelines:

- The skills progression program for the SDM is based on the M16-/M4-series weapon system and a 98-rounds-per-man ammunition requirement.
- Soldiers will use their assigned weapon during the training.
- The firing events will be conducted with the iron sights or BUISs only.
- The firing events will be conducted on a KD range that enables firing out to 600 meters at a minimum.

QUALIFICATION

7-232. SDM qualification requires the completion of five phases:

(1) Position evaluation.
(2) Dry-fire training.
(3) Range estimation and sight manipulation.
(4) Hold-off.
(5) Field fire.

7-233. Each phase stresses marksmanship fundamentals and specific skill areas required to perform as an SDM. To continue training, Soldiers must receive a GO in each phase. Soldiers who fail in any area should be removed from training.

NOTE: If an optic is issued for use, the phases dealing with hold-off and field record fire will be removed and relevant optics training and testing will be substituted.

PHASE I—POSITION EVALUATION

7-234. Phase I of the training consists of demonstrating the ability to consistently assume proper firing positions. The foxhole supported and prone unsupported firing positions will provide the Soldier with the smallest target exposure to the enemy and will be used during this training cycle. The prone supported position can be substituted for the foxhole supported position dependent on range configurations. Prior to this phase of training, trainers ensure that the—

- Weapon is cleared and that no ammunition is loaded prior to training.
- Weapon is zeroed prior to training.
- Soldier is able to assume a steady firing position.

Foxhole Supported

7-235. The Soldier must be able to successfully assume a proper supported position while firing from a foxhole. The trainer must ensure that the Soldier has a good steady position. After the Soldier has assumed a good supported position in the foxhole—

(1) The trainer uses DA Form 7650-R (Squad Designated Marksman—Position Evaluation) to evaluate his position and take notes on all of the following characteristics:

NOTE: See the end of this publication for a blank, reproducible copy of DA Form 7650-R.

- Eye relief.
- Trigger finger.
- Elbows.
- Nonfiring hand.
- Legs.

NOTE: The main areas that will differ between the foxhole supported and the prone supported positions are in the placement of the elbows, legs, and nonfiring hand. These body positions will be similar to those of the prone unsupported position.

(2) After all characteristics have been noted, the trainer has the Soldier lay his weapon down, relax, and then assume another supported position in the foxhole.

(3) The trainer evaluates this position by comparing his notes from the original supported position. The Soldier should maintain the same characteristics in the second evaluation as he did in the first.

(4) Once the trainer is satisfied that the Soldier has demonstrated the proper position and is able to show it in two consecutive attempts, the Soldier moves to the unsupported prone position.

Eye Relief

7-236. To evaluate the Soldier's eye relief, ensure that the Soldier—

- Demonstrates a consistent eye relief by checking the placement of the Soldier's cheek on the weapon's buttstock.
- Places his eye is the same distance from the rear sight each time he is evaluated.

Trigger Finger

7-237. To evaluate the placement of the Soldier's trigger finger, ensure that the Soldier—

- Uses his own style; not all Soldiers place their finger on the trigger in the same place.
- Places his finger on the trigger the same way each time he is evaluated.

Elbows

7-238. To evaluate the placement of the Soldier's elbows, ensure that the Soldier—

- Places his elbows firmly a comfortable distance apart on the outside edge of the foxhole.
- Uses a sandbag, and not his arms, to support the weapon's weight.
- Assumes a stable position each time he is evaluated by slightly nudging him.

Nonfiring Hand

7-239. To evaluate the placement of the Soldier's nonfiring hand, ensure that the Soldier—

- Places the nonfiring hand in a position that is comfortable and provides the best weapon stability and support. Show the Soldier different ways this can be done.
- Is supporting the weapon properly by nudging him after the weapon has been stabilized.

Legs

7-240. To evaluate the placement of the Soldier's legs, ensure that the Soldier—

- Places the legs inside the foxhole while firing.
- Plants the legs firmly for a stable position while firing. Slightly nudge the Soldier to make sure that his legs are firmly planted in the foxhole.

Prone Unsupported

7-241. The Soldier must be able to successfully assume a proper unsupported firing position. The trainer must ensure that the Soldier has a good steady position. After the Soldier has assumed a good unsupported firing position —

(1) Taking special care to observe the positioning of the elbows, the nonfiring hand, and the legs, the trainer uses DA Form 7650-R (Squad Designated Marksman—Position Evaluation) to evaluate the same characteristics as with the supported firing position with the exception of the—

- Elbows.
- Nonfiring hand.
- Legs.

NOTE: See the end of this publication for a blank, reproducible copy of DA Form 7650-R.

---

(2) Once the trainer has noted the Soldier's position, he has the Soldier lay his weapon down, stand up, relax, and then get back down into another unsupported prone position.

(3) The trainer evaluates this position by comparing his notes from the original position. The Soldier should maintain the same characteristics in the second evaluation as he did in the first evaluation.

(4) The trainer lets the Soldier hold this firing position for approximately 15 seconds to check for shaking. If the Soldier starts to shake, have him relax and reposition himself.

(5) Once the trainer is satisfied that the Soldier has demonstrated the proper position and is able to accomplish it in two consecutive attempts, the Soldier moves on to the next phase of training.

Elbows

7-242. To evaluate the placement of the Soldier's elbows, ensure that the Soldier—

- Places the elbows a comfortable distance apart on the ground.
- Uses the bone, not the muscles, to support the weapon's weight. This will prevent any unnecessary muscle fatigue and will allow for a steadier firing position.
- Assumes a stable position. Slightly nudge the Soldier to ensure that his position is stable.

Nonfiring Hand

7-243. To evaluate the placement of the Soldier's nonfiring hand, ensure that the Soldier—

- Places his nonfiring hand in a comfortable position on the handguards.
- Does not support his nonfiring hand on the ground, sandbag or anything that would create a supported position.

Legs

7-244. To evaluate the placement of the Soldier's legs, ensure that the Soldier—

- Positions his legs in such a way that he has a stable position; not all Soldiers position their legs the same way while shooting from the prone position.
- Spreads his legs a comfortable distance apart, with the heels on the ground or as close as possible without causing strain.

Zero Confirmation

7-245. After completing Phases I and II, the Soldier conducts a firing event (Table 7-22) to zero or confirm the zero on his weapon and reinforce the fundamentals of marksmanship. This firing event will be conducted on a 25-meter range. If the Soldier cannot zero within 18 rounds, the trainer recommends re-training, retesting, or possible removal from the course. After the weapon is zeroed, any additional rounds will be fired and the coach will observe the Soldier for deficiencies in his marksmanship fundamentals.

Table 7-22. Zero/zero confirmation firing event.

FIRING EVENT	ROUNDS	TARGET RANGE (m)
Zero/Zero Confirmation	18	25

PHASE II—DRY-FIRE TRAINING

7-246. SDMs must have a solid grasp on the fundamentals to successfully engage targets at longer ranges. During this phase of training, the Soldier must demonstrate that he can apply the fundamentals of marksmanship correctly. If the Soldier does not receive a GO in this phase of training, he will be dropped from the course. The components of this phase of training are—

- Follow-through.
- Borelight exercise.
- Target box exercise.
- Dime/washer drill.
- Zero confirmation.

7-247. Prior to this phase of training trainers ensure that the—

- Weapon is cleared and no ammunition is loaded prior to training.
- Weapon is zeroed prior to training.
- Soldier is able to consistently apply the fundamentals of marksmanship.

Follow-Through

7-248. Follow-through involves applying all of the marksmanship fundamentals while and after the weapon fires. A good follow-through ensures that the weapon is allowed to fire and recoil naturally. The Soldier/weapon combination reacts as a single unit to such actions. It consists of the following:

- Keeping the cheek in firm contact with the stock (stock weld).
- Keeping the finger on the trigger all the way to the rear.
- Continuing to look through the rear aperture.
- Keeping muscles relaxed.
- Avoiding reaction to recoil or noise.
- Releasing the trigger only after the recoil has stopped.

Borelight Exercise

7-249. The borelight dry-fire exercise provides evaluation of the Soldier throughout the integrated act of firing.

NOTE: If a borelight is not available, the target box exercise will be used.

7-250. To perform a borelight exercise—

(1) The trainer attaches a 25-meter zero target to a flat surface.

(2) The trainer positions the Soldier 10 meters away, facing the target.

(3) The Soldier assumes a good prone supported firing position with the borelight inserted in the weapon's barrel and with the weapon placed in the dry-fire mode.

(4) The trainer uses DA Form 7650-R (Squad Designated Marksman—Position Evaluation) to evaluate the Soldier's position:

NOTE: See the end of this publication for a blank, reproducible copy of DA Form 7650-R.

(5) The Soldier aims at the silhouette's center of mass on the 25-meter zero target and squeezes the trigger.

(6) The borelight is activated as the trigger is fired. The laser is seen on the 25-meter zero target.

(7) The trainer marks the 25-meter zero target exactly were the borelight laser hit the target.

(8) The Soldier gets out of position and then back into a prone supported firing position.

7-251. This process will be done until a three-round shot group has been achieved. The Soldier will do the same from the prone unsupported firing position. To receive a GO, the Soldier must place a three-round shot group in a 3-centimeter circle from both prone positions.

Target Box Exercise

7-252. The target box exercise checks the consistency of aiming and placement of three-round shot groups in a dry-fire environment. To conduct the exercise—

(1) The target man places the silhouette on a plain sheet of paper 25 meters away from the firer and moves the silhouette target as directed by the Soldier.

(2) When the Soldier establishes proper aiming, he squeezes the trigger.

(3) The target man marks through the silhouette with a pen or pencil at the target's center of mass.

(4) The target man moves the silhouette to another spot on the paper and tells the firer to repeat the process twice more to obtain a shot group.

NOTE: A simulated shot group covered within a 1/2-centimeter circle indicates consistent aiming.

7-253. Since no weapon or ammunition variability is involved and shot groups do not have to be placed in a certain location, a 1/2-centimeter standard may be compared to obtaining a 4-centimeter shot group on the 25-meter live-fire zero range.

Dime/Washer Drill

7-254. The dime/washer drill is an effective way of measuring the Soldier's trigger squeeze. To conduct the exercise—

(1) The Soldier takes aim and squeezes the trigger.

(2) If the dime or washer remains in place, he has successfully squeezed the trigger.

7-255. The Soldier must successfully obtain five out of five consecutive shot groups within 1 centimeter and without allowing the dime or washer to drop. The trainer evaluates the Soldier's performance and gives the Soldier a GO or NO GO. If the Soldier receives a NO GO, the trainer recommends retraining, retesting, or possible removal from the course.

Zero Confirmation

7-256. Zero confirmation is conducted as shown in Table 7-6.

PHASE III—RANGE ESTIMATION AND SIGHT MANIPULATION 100 TO 500 METERS

7-257. SDMs must use range estimation methods to determine the distance between their position and the target. Trainers ensure that the—

- Weapon is cleared and no ammunition is loaded prior to training.
- Weapon is zeroed prior to training.
- Soldier knows how to adjust for wind and gravity.
- Soldier can manipulate the rear sight for different ranges.

7-258. The trainer sets up a range estimation course using E-type silhouettes at ranges from 100 meters to 700 meters. Soldiers practice on this course until they find the method that works best for them.

7-259. Once the Soldiers have had time to practice, trainers test their ability to estimate range. The Soldier is given six targets. He must estimate the range within 50 meters of the actual range to receive a GO. The Soldier must estimate range correctly six out of six targets to move on to the next portion of this phase.

Methods of Range Determination

7-260. SDMs can use five different methods of range determination:

- 100-meter unit-of-measure method.
- Range card method.
- Front sightpost method.
- Appearance of objects method.
- Combination method.

100-Meter Unit-of-Measure Method

7-261. To use this method, the SDM must be able to visualize a distance of 100 meters on the ground.

- For ranges up to 500 meters, the SDM determines the number of 100-meter increments between the two objects he wishes to measure.
- Beyond 500 meters, he must select a point halfway to the object, determine the number of 100-meter increments to the halfway point, and then double the number.

NOTES: 1. See Chapter 6 for more information about this method of range determination.

2. For example, terrain with much dead space limits the accuracy of the 100-meter method.

Range Card Method

7-262. SDMs use a range card to quickly determine ranges throughout the target area. Once a target is detected, the SDM determines its location on the card and then reads the proper range to the target.

Front Sightpost Method

7-263. Using the front sightpost as a scale is another method of estimating range. This method can be used for a quick on-the-spot estimation and engagement.

- If a man-sized target is approximately ½ of the width of the front sightpost, he is approximately 300 meters away.
- If the target is 1/2 the width of the front sightpost, then the target is approximately 600 meters away.

NOTE: See Chapter 6 for more information about this method of range determination.

Appearance of Objects Method

7-264. This method of range determination is based on the size and visible characteristics of an object. To use this method with any degree of accuracy, the SDM must be familiar with the appearance and visible detail of an object at various ranges. Some common guidelines can be used to determine the range of a human target:

- At 200 meters, a human target is clear and details can be seen.
- At 300 meters, the target is still clear, but no details can be seen.
- At 400 meters, the target's outline is clear; however, the target itself is blurry.
- At 500 meters, the body tapers and the head disappears.
- At 600 meters, the body resembles a wedge shape.

NOTE: See Chapter 6 for more information about this method of range determination.

Combination Method

7-265. In a combat environment, perfect conditions rarely exist. Therefore, only one method of range estimation may not be enough for the SDM's specific mission. By using a combination of two or more methods to determine an unknown range, an experienced SDM should arrive at an estimated range close to the true range.

Factors Affecting Range Estimation

7-266. Three factors affect range estimation:

- Nature of the target
- Nature of the terrain.
- Light conditions.

Nature of the Target

7-267. The nature of the target affects its perceived range:

- An object with a regular outline, such as a house, appears closer than one with an irregular outline, such as a clump of trees.
- A target that contrasts with its background appears to be closer than it actually is.
- A partly exposed target appears more distant than it actually is.

Nature of the Terrain

7-268. The contour of the terrain affects the observer's ability to estimate range:

- As the observer's eye follows the contour of the terrain, he tends to overestimate distant targets.
- Observing over smooth terrain, such as sand, water, or snow, causes the observer to underestimate distant targets.
- Looking downhill, the target appears farther away.
- Looking uphill, the target appears closer.

Light Conditions

7-269. Light conditions affect range estimation:

- The more clearly a target can be seen, the closer it appears.
- When the sun is behind the observer, the target appears to be closer.
- When the sun is behind the target, the target is more difficult to see and appears to be farther away.

Elevation Knob Training

7-270. Elevation knob training involves nothing more than being able to adjust the rear elevation knob (front sightpost on the M16A4) for the various ranges that the SDM must engage. With this knowledge, he can better determine his range settings for the different distances between the 100-meter adjustments.

7-271. The rear elevation knob adjusts the point of aim—

- From 300 to 800 meters on the M16A2.
- From 300 to 600 meters on the M16A4 and M4.

Conduct

7-272. During elevation knob training, the Soldier determines the number of adjustments (clicks) between the different range settings on his rear elevation adjustment knob (Tables 7-23 and 7-24). Once the Soldier understands how to set the proper point of aim for his target using his rear elevation knob, the instructor/trainer has him conduct another range estimation course. This time, the instructor/trainer has him estimate the range and set the rear elevation for the range that he has estimated. The Soldier must estimate range and set his rear elevation knob properly six out of six times to receive a GO. If the Soldier receives a NO GO, the trainer recommends retraining, retesting, or possible removal from the course.

7-273. Once the Soldier has an understanding of range estimation and sight manipulation, he is able to begin the live-fire training exercise (Table 7-25). The Soldier will be given 20 rounds in which to engage 20 targets at ranges from 100 to 500 meters using mechanical sight adjustments.

Table 7-23. Elevation knob, M16A2/3 and front sightpost, M16A4.

DISTANCE (m)	DISTANCE ONE CLICK WILL ADJUST THE POINT OF IMPACT		
	FRONT SIGHTPOST	WINDAGE KNOB	ELEVATION WHEEL
25	.83 cm (3/8 in)	.33 cm (1/8 in)	.5 cm (1/4 in)
50	1.50 cm (5/8 in)	.5 cm (1/4 in)	1.5 cm (1/2 in)
75	2.50 cm (1 in)	1.0 cm (3/8 in)	2.0 cm (3/4 in)
100	3.50 cm (1 3/8 in)	1.5 cm (1/2 in)	2.75 cm (1 in)
150	5.00 cm (2 in)	2.0 cm (3/4 in)	4.0 cm (1 1/2 in)
175	6.00 cm (2 3/8 in)	2.25 cm (7/8 in)	5.0 cm (2.0 in)
200	6.50 cm (2 5/8 in)	2.5 cm (1 in)	5.5 cm (2 1/4 in)
250	8.50 cm (3 3/8 in)	3.5 cm (1 1/4 in)	7.0 cm (2 3/4 in)
300	10.0 cm (4 in)	4.0 cm (1 1/2 in)	8.5 cm (3 1/4 in)
400	13.5 cm (5 3/8 in)	5.5 cm (2 1/4 in)	11.0 cm (4 1/2 in)
500	17.0 cm	6.5 cm (2 1/2 in)	14.0 cm (5 1/2 in)
600	20.5 cm	8.0 cm (3 1/8 in)	16. 75 cm (6 1/2 in)
700	24.0 cm	9.0 cm (3 5/8 in)	19.5 cm (7 1/2 in)
800	27.5 cm	10.5 cm (4 1/8 in)	22.5 cm (8 3/4 in)
NOTE: All values were rounded off.			

Table 7-24. Elevation knob, M4/M4A1 and windage, M16A4.

DISTANCE (m)	DISTANCE ONE CLICK WILL ADJUST THE POINT OF IMPACT		
	FRONT SIGHTPOST	WINDAGE KNOB	ELEVATION WHEEL
25	1.2 cm (1/2 in)	.5 cm (1/4 in)	.5 cm (1/4 in)
50	2.4 cm (1 in)	1.5 cm (1/2 in)	1.5 cm (1/2 in)
75	3.6 cm (1 1/2 in)	2.0 cm (3/4 in)	2.0 cm (3/4 in)
100	4.8 cm (1 7/8 in)	2.75 cm (1 in)	2.75 cm (1 in)
150	7.2 cm (2 7/8 in)	4.0 cm (1 1/2 in)	4.0 cm (1 1/2 in)
175	8.4 cm (3 3/8 in)	5.0 cm (2.0 in)	5.0 cm (2.0 in)
200	9.6 cm (3 3/4 in)	5.5 cm (2 1/4 in)	5.5 cm (2 1/4 in)
250	12.0 cm (4 3/4 in)	7.0 cm (2 3/4 in)	7.0 cm (2 3/4 in)
300	14.4 cm (5 3/4 in)	8.5 cm (3 1/4 in)	8.5 cm (3 1/4 in)
400	19.2 cm (7 1/2 in)	11.0 cm (4 1/2 in)	11.0 cm (4 1/2 in)
500	24.0 cm (9 1/2 in)	14.0 cm (5 1/2 in)	14.0 cm (5 1/2 in)
600	28.8 cm (11 1/4 in)	16. 75 cm (6 1/2 in)	16.75 cm (6 1/2 in)
NOTE: All values were rounded off.			

Table 7-25. Known distance (mech. adj.) firing event.

FIRING EVENT	ROUNDS	TARGET RANGE (m)
Known Distance (Mech. Adj.)	20	100 to 500

PHASE IV—HOLD-OFF 100 TO 500 METERS

7-274. Hold-off is shifting the point of aim to achieve a desired point of impact. Hold-off is used only when the SDM does not have time to change his sight setting. Certain situations, such as multiple targets at varying ranges and rapidly changing winds, do not allow proper windage and elevation adjustments. Familiarization and practice of elevation and windage hold-off techniques prepares the SDM to meet these situations. The SDM rarely achieves pinpoint accuracy when holding off; a minor error in range determination or a lack of a precise point of aim might cause the bullet to miss the desired point.

7-275. Prior to this phase of training, trainers ensure that the—

- Weapon is cleared and that no ammunition is loaded prior to training.
- Weapon is zeroed prior to training.
- Soldier knows how to adjust for wind and gravity.
- Soldier can manipulate the rear sight for different ranges.

Elevation

7-276. When a Soldier aims directly at a target at ranges greater than the set range, his bullet will hit below the point of aim. At lesser ranges, his bullet will hit higher than the point of aim. If the SDM understands this and knows about trajectory and bullet drop, he will be able to hit the target at ranges other than that for which the weapon was adjusted.

7-277. For example, the SDM adjusts the weapon for a target located 500 meters downrange, and another target appears at a range of 600 meters. The hold-off would be 25 inches; that is, the SDM should hold off 25 inches above the center of visible mass in order to hit the center of mass of that particular target. If another target were to appear at 400 meters, the SDM would aim 14 inches below the center of visible mass in order to hit the center of mass.

7-278. The chart in Figure 7-31 shows the projectile's trajectory when fired from the M4 carbine and the M16A2 rifle. This demonstrates the drop of the round at various ranges.

NOTE: This diagram will assist the trainer in teaching vertical hold-off during this phase.

7-279. As the chart in Figure 7-31 shows, the hold-off at 400 meters is about half the height of the standard E-type silhouette; to hold-off at 400 meters, the firer aims half the height of the target over the target to hit it. The drop at 500 meters is considerably larger, so holding off will not be practical. The firer will have to adjust his rear elevation knob to get the proper point of aim for that distance.

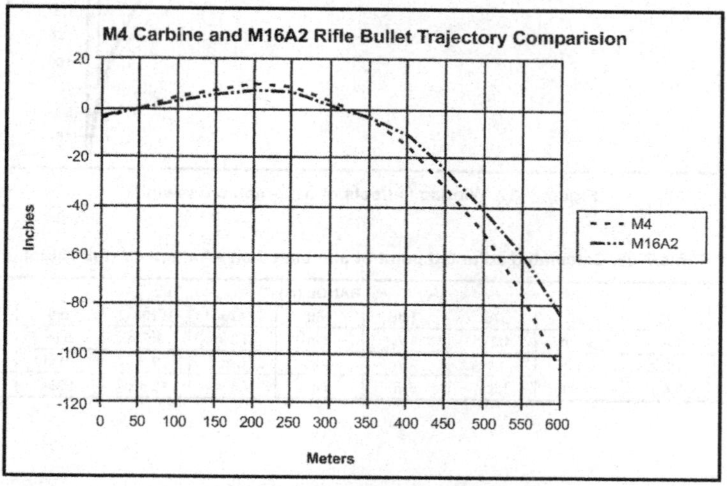

Figure 7-31. Bullet trajectory comparison.

Windage

7-280. When firing during windy conditions and there is no time to make sight adjustments, the SDM must use hold-off to adjust for windage (Figure 7-32).

- When holding off, the SDM aims into the wind. If the wind is moving from the right to left, his point of aim is to the right. If the wind is moving from left to right, his point of aim is to the left.
- If the SDM misses the target and the point of impact of the round is observed, he notes the lateral distance of his error and re-fires, holding off that distance in the opposite direction.

7-281. Table 7-26 shows calculated adjusted points of aim based on wind speed.

7-282. Table 7-27 shows the drift for a 10-mph wind using 5.56-millimeter M855 ball ammunition fired in a M16A2 rifle with a 300-meter battlesight zero.

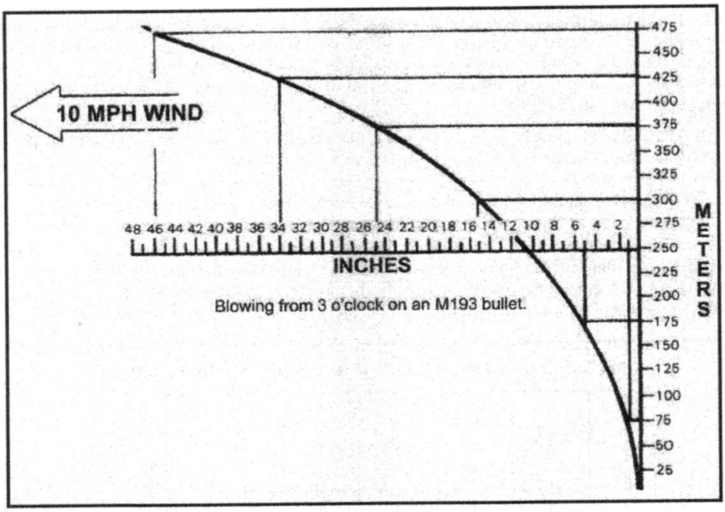

Figure 7-32. Windage effects of a 10-mph crosswind.

Table 7-26. Calculated adjusted point of aim based on wind speed (full value).

WIND SPEED	RANGE (m)								
	25	50	75	100	150	175	200	250	300
5 mph	1/4 in	3/8 in	1/2 in	1 in	2 in	2.5 in	3.5 in	5 in	7.5 in
10 mph	1/2 in	3/4 in	1 in.	2 in	4 in	5 in	7 in	10 in	15 in
15 mph	3/4 in	1 1/8 in	1.5 in	3 in	6 in	7.5 in	10.5 in	15 in	22.5 in

Table 7-27. Drift for 10-mph wind using M855 ammunition when fired from M16A2 rifle with 300-meter battlesight zero.

RANGE (m)	VELOCITY (fps)	TRAJECTORY (in)	DROP (in)	DRIFT (in)
0	3,100	-2.5	0.0	0.0
100	2,751	4.4	-2.3	1.1
200	2,420	5.8	-10.2	4.9
300	2,115	0.0	-25.3	11.8
400	1,833	-15.0	-49.5	22.4
500	1,569	-42.9	-86.7	38.0
600	1,323	-88.2	141.3	59.5
700	1,106	-156.1	-220.9	88.4
800	1,010	-267.7	-339.2	124.9

7-283. Firers use the M15 sighting device to demonstrate that they understand holding off. To do so, the firer aligns the sights on the silhouette on the proper adjusted point of aim.

NOTE: See Appendix A for information about the M15 sighting device.

7-284. Once the firer has an understanding of elevation and windage hold-off, he can begin the live-fire training exercise (Table 7-28). The firer will be given 20 rounds in which to engage 20 targets at ranges from 100 to 500 meters using elevation and windage hold-off.

Table 7-28. Firing event, known distance (hold off).

FIRING EVENT	ROUNDS	TARGET RANGE (m)
Known Distance (Hold Off)	20	100 to 500

PHASE V—FIELD FIRE 100 TO 500 METERS

7-285. The field fire events (Table 7-29) will test the individual's marksmanship, range estimation, and target detection skills. Field fire will consist of both a Record Fire I and a Record Fire II course. The Record Fire I course requires the individual to use mechanical elevation and windage adjustments. The Record Fire II course requires the individual to use elevation and windage hold-off (adjusted points of aim).

Conduct

7-286. To complete this course, Soldiers fire two firing events: Record Fire I and Record Fire II.

7-287. During each event, each Soldier will engage a total of 20 targets with 20 rounds. Soldiers must attain a total of 14 hits out of 20 targets on each record fire to pass. Table 7-29 depicts these two events and provides related information, such as number of rounds that must be fired, position that must be used, and the distance away from the firer that the target must be placed.

Table 7-29. Firing event, Record Fire I and II.

FIRING EVENT	POSITION	ROUNDS	NUMBER OF TARGETS	TARGET RANGE (m)
Record Fire I	Foxhole supported or prone supported position and the prone unsupported firing position	20	20	100 to 500
Record Fire II	Foxhole supported or prone supported position and the prone unsupported firing position	20	20	100 to 500

NOTE: If the SDM is issued an optic, the Record Fire II course will substitute use of that optic instead of using adjusted points of aim.

7-288. Prior to training, ensure that—

- The weapon is zeroed prior to training.
- Each Soldier assumes a steady firing position.
- Each Soldier consistently applies the fundamentals of marksmanship.
- Each Soldier knows how to adjust for wind and gravity.
- Each Soldier manipulates the rear sight for different ranges.

NOTE: See Appendix B for a sample completed form and the end of this publication for a blank, reproducible copy.

CERTIFICATION

7-289. Once the firer has successfully completed the SDM program, he is designated as an SDM and will be able to perform all duties and responsibilities set forth by these guidelines.

NOTE: SDM skills are highly perishable, and sustainment training should be conducted to ensure retention of the skills. At a minimum, sustainment training should be conducted semiannually.

Chapter 8

Advanced Optics, Lasers, and Iron Sights

BRM teaches Soldiers how to effectively engage targets with their weapons using the iron sights. ARM adds other marksmanship situations that a combat Soldier may encounter. This chapter discusses how to enhance marksmanship skills through proper training, using the Army's newest optics and lasers to ensure that Soldiers can fight as well at night as they can during the day. This chapter implements new night qualification standards to complement current Army training strategies.

SECTION I. BORELIGHT

The borelight is an accurate means of zeroing weapons and most aided-vision equipment without the use of ammunition. The time and effort required to ensure a precise boresight will, in turn, save time and ammunition. Table 8-1 outlines weapon/aided-vision device combinations that can be zeroed using the borelight.

NOTE: See Table 8-2 for the current training program.

Table 8-1. Weapon/aided-vision device combinations.

WEAPON / AIDED-VISION DEVICE	M16A2	M16A3/A4	M4/MWS
IRON SIGHT	X	N/A	N/A
BUIS	N/A	X	X
AN/PAQ-4B/C	X	X	X
AN/PEQ-2A/B	X	X	X
AN/PAS-13B/C/D	X	X	X
M68 CCO	X	X	X
ACOG	X	X	X

NOTES:
1. Precise boresighting of a laser will allow direct engagement of targets without a 25-meter zero. If a borelight is not available, a 25-meter zero must be conducted to zero the laser.
2. All optics must be 25-meter zeroed; a borelight only aids in zeroing.

Table 8-2. Borelight training program.

BORELIGHT
Instructional Intent
• Soldiers align the optic, laser, or iron sight to the bore of the weapon to reduce or eliminate the time and ammunition it requires to live-fire zero.
Special Instructions
Ensure that Soldiers— • Zero the borelight. • Use only approved 10-meter boresighting targets from Picatinny Arsenal. • Use the proper 10-meter boresighting target for weapon configuration. • Boresight 10 meters from the end of the barrel. • Stabilize both the weapon and the target (or the boresight will not be accurate). • Install filters for aiming lasers to reduce blooming.
Observables
Soldiers confirm that the— • Borelight spins on itself when zeroed at 10 meters. • Weapon configuration is boresighted using the official and proper targets only. • Target and weapon do not move during the boresighting procedure. • Borelight is centered on the circle on the target. • Aiming device is aiming at the center of the crosshair on the offset.

CONCEPT

8-1. Boresighting is a simple procedure that saves time and ammunition. The visible laser of the borelight is aligned with the barrel of a designated weapon. Then, using a 10-meter boresighting target, the weapon can be boresighted with any optic, laser, or iron sight that the Soldier is assigned to fire.

ZEROING THE BORELIGHT

WARNINGS

Before using the borelight, ensure that the weapon is clear and on SAFE, and that the bolt is locked in the forward position.

When rotating the borelight to zero it, ensure that the mandrel is turning counterclockwise (from the firer's point of view) to avoid loosening the borelight from the mandrel.

DANGERS

DO NOT STARE INTO THE VISIBLE LASER BEAM.

DO NOT LOOK INTO THE VISIBLE LASER BEAM THROUGH BINOCULARS OR TELESCOPES.

DO NOT POINT THE VISIBLE LASER BEAM AT MIRROR-LIKE SURFACES.

DO NOT SHINE THE VISIBLE LASER BEAM INTO OTHER INDIVIDUALS' EYES.

8-2. Before boresighting the weapon system, the borelight must first be zeroed to the weapon. To zero the borelight to the weapon—

(1) Stabilize the weapon by placing it in a rifle box rest or by laying two rucksacks side by side and placing another rucksack on top of the weapon.

NOTE: The weapon does not have to be perfectly level with the ground when boresighting.

---

(2) Align the visible laser with the weapon's barrel.

> **CAUTION**
>
> Do not over-adjust the laser. Do not point the laser at Soldiers or reflective material.

(3) Attach the 5.56-millimeter mandrel to the borelight.

(4) Insert the mandrel into the weapon's muzzle.

NOTE: The borelight is seated properly when the mandrel cannot be moved any further into the muzzle and the mandrel spins freely.

---

(5) Measure 10 meters with the 10-meter cord that comes with the borelight, or pace off eleven paces.

(6) Draw a zeroing mark (small dot) on a piece of paper or tree bark, or use the borelight reference point on the 10-meter boresighting target (Figure 8-1).

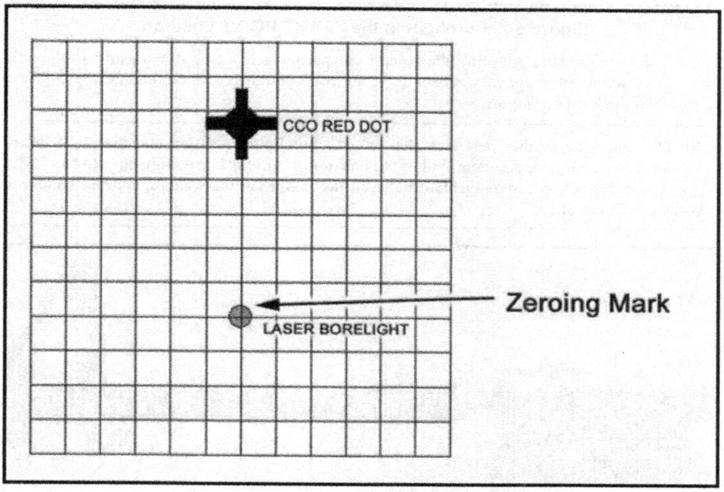

Figure 8-1. Example of a zeroing mark.

(7) Rotate the borelight until the battery compartment is facing upward and the adjusters are on the bottom (Figure 8-2).

NOTE: Once the Soldier performs step (7), the borelight's position and the place where the visible laser is pointing are identified as the start point. The command "START POINT" is given to ensure clear communication between the Soldier at the weapon and the Soldier at the boresighting target.

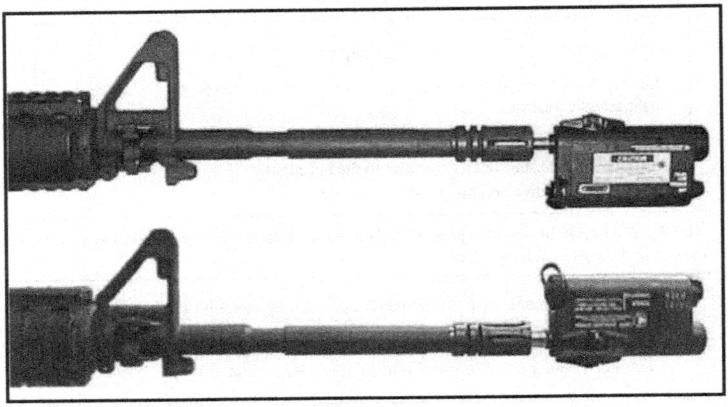

Figure 8-2. Borelight in the START POINT position.

(8) Rotate the borelight until the battery compartment is facing downward and the adjusters are on top to allow for easy access to the adjusters and help with communication and stabilization of the weapon (Figure 8-3).

NOTE: Once the Soldier performs step (8), the borelight's position and the place where the visible laser is pointing are identified as the half-turn position. The command "HALF-TURN" is given to ensure clear communication between the Soldier at the weapon and the Soldier at the boresighting target.

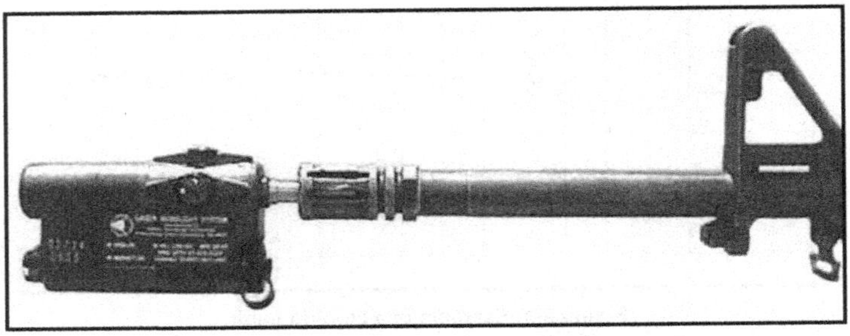

Figure 8-3. Borelight in the HALF-TURN position.

(9) Identify the point approximately halfway between the start point and the half-turn point. This is the reference point (Figure 8-4).

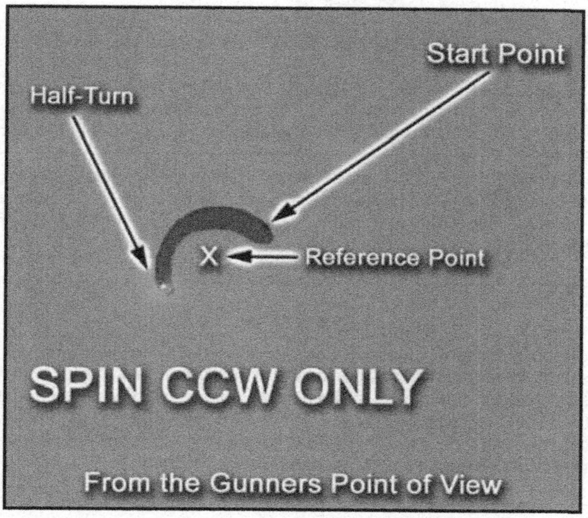

Figure 8-4. Example of a start point, half-turn, and reference point.

(10) Turn the borelight on and spin it until it is in the start point position.

(11) Place the zeroing mark approximately 10 meters from the end of the barrel so that the visible laser strikes the zeroing mark.

(12) Slowly rotate the borelight 180 degrees while watching the visible laser made by the borelight.

NOTE: If the visible laser stops on the zeroing mark, the borelight is zeroed to the weapon. If the borelight does not stop on the zeroing mark, elevation and windage adjustments must be made to the borelight.

---

(13) From the start point, realign the zeroing mark with the visible laser, rotate the borelight 180 degrees to the half-turn position, and identify the reference point.

(14) Using the adjusters on the borelight, move the visible laser to the reference point.

(15) Rotate the borelight back to the start point; move the zeroing mark to the visible laser.

NOTE: If the visible laser cannot be located when the borelight is spun to the half-turn position, start this procedure at 2 meters instead of 10 meters. When the visible laser is adjusted to the reference point at 2 meters, restart the procedure at 10 meters.

---

(16) Repeat steps (13) through (15) until the visible laser spins on itself.

NOTE: Every barrel is different, so steps (10) through (15) must be performed with every weapon to ensure that the borelight is zeroed to that barrel. If the borelight is zeroed, go directly to the boresighting procedures.

BORESIGHTING

8-3. Two Soldiers are required to properly boresight a weapon: a firer and a target holder. Their duties are as follows:

- The firer's primary duty is to zero the borelight and make all adjustments on the aided-vision device being used.
- The target holder secures the 10-meter boresighting target straight up and down 10 meters from the borelight and directs the firer in making necessary adjustments to the aiming device. The target holder must wear night vision goggles when boresighting IR aiming lasers.

NOTE: Appendix F shows the most current 10-meter boresighting target. The 10-meter boresighting target grids are 1x1-centimeter squares, unlike those on 25-meter zeroing targets. Contact the proponent of this publication (C Co, 2/29 IN, Fort Benning, GA) for information about the availability of boresighting targets.

CAUTION

Weapon stability is crucial in boresighting. The weapon should be in the bolt forward position and must not be canted left or right during boresighting procedures. If the weapon is boresighted using field-expedient methods (for example, sandbags, rucksacks) and the weapon is laid on its side for stability, ensure that the boresighting target is also oriented in the same manner.

BORESIGHTING THE WEAPON USING THE BACKUP IRON SIGHTS

8-4. The BUIS can be boresighted to a new user to expedite 25-meter zeroing. To boresight the weapon using the BUIS—

(1) Align the BUIS with the Canadian bull on the 10-meter boresighting target.

(2) Make adjustments to the windage and elevation of the BUIS until the borelight is centered with the circle on the boresighting target.

BORESIGHTING THE WEAPON USING THE M68 CLOSE COMBAT OPTIC

NOTES:

1. Before boresighting, ensure that the borelight has been zeroed to the weapon. The more accurate the boresight of the M68 CCO to the assigned weapon, the closer to a battlesight zero the weapon will be.
2. 25-meter zeroing must be conducted to ensure that the M68 CCO is properly zeroed.
3. The M68 CCO is a parallax-free sight beyond 50 meters; boresighting is conducted at 10 meters. To get a solid boresight, ensure that the red dot is centered within both the CCO and the target's center of mass.

8-5. To boresight the weapon using the M68 CCO—

(1) Select the proper 10-meter boresighting target for the weapon/M68 configuration.

(2) With the help of an assistant, place the boresighting target 10 meters in front of the weapon.

(3) Turn the M68 CCO to the desired setting by turning it clockwise to position 10, turning it counterclockwise one click at a time until you can no longer see the dot, and then turning it clockwise two clicks.

NOTE: The desired setting may need to be adjusted throughout the day depending on ambient light conditions.

(4) Get behind the weapon in a stable supported firing position, and look through the M68 CCO.
(5) Aim the red dot of the M68 CCO on the crosshair located on the 10-meter boresighting target.
(6) Make adjustments to the M68 CCO until the visible laser of the borelight is centered on the borelight circle on the 10-meter boresighting target.
(7) Turn the borelight off.
(8) Move the weapon off of the crosshair, realign the M68 CCO's red dot on the crosshair, and turn the borelight back on.
(9) If the borelight is on the circle and the M68 CCO's red dot is on the crosshair, the weapon system is boresighted.
(10) Turn the laser off, and carefully remove the borelight and the mandrel from the weapon so that the borelight device is not damaged.

NOTE: If the firer does not get the same sight picture after the second realignment, he more than likely has a fundamentals problem with his firing position and sight picture. To save time on the range, a coach should troubleshoot the Soldier before continuing to boresight the M68 CCO.

BORESIGHTING THE ADVANCED COMBAT OPTICAL GUNSIGHT

NOTES: 1. Before boresighting, ensure that the borelight has been zeroed to the weapon. The more accurate the boresight of the ACOG to the assigned weapon, the closer to a battlesight zero the weapon will be.

2. 25-meter zeroing must be conducted to ensure that the ACOG is properly zeroed.

8-6. To boresight the ACOG—

NOTE: Use the 10-meter boresighting target that is used for boresighting the M68 CCO.

(2) With the help of an assistant, place the boresighting target 10 meters in front of the weapon.
(4) Get behind the weapon in a stable supported firing position, and look through the ACOG.
(5) Aim the tip of the 300-meter post on the crosshair located on the 10-meter boresighting target.
(6) Make adjustments to the ACOG until the visible laser of the borelight is centered on the borelight circle on the boresighting target.
(7) Turn the borelight off.
(8) Move the weapon off of the crosshair, realign the tip of the 300-meter post on the crosshair, and turn the borelight back on.
(10) If the borelight is on the circle and the tip of the 300-meter post is on the crosshair, the weapon system is boresighted.
(10) Turn the laser off, and carefully remove the borelight and the mandrel from the weapon so that the borelight device is not damaged.

Boresighting the Thermal Weapon Sight

NOTES: 1. Before boresighting the TWS, ensure that the borelight has been zeroed to the weapon. The more accurate the boresight of the TWS to the assigned weapon, the closer to a battlesight zero the weapon will be.

2. 25-meter zeroing must be conducted to ensure that the TWS is properly zeroed.

3. Both the narrow field of view (NFOV) and wide field of view (WFOV) must be boresighted and zeroed.

8-7. To boresight the TWS—

(1) Select the proper 10-meter boresighting target for the weapon/TWS configuration.
(2) With the help of an assistant, place the boresighting target 10 meters in front of the weapon.
(3) Ensure that the weapon's reticle is displayed.
(4) Get behind the weapon in a stable supported firing position, and look through the TWS.
(5) Place a finger on each oval on the 10-meter boresighting target.
(6) Aim between the fingers with the 300-meter point of aim.
(7) Make adjustments to the TWS until the visible laser of the borelight is centered on the borelight circle on the 10-meter boresighting target.
(8) Move off of the aiming block, realign the TWS to the center of the heated block, and then turn the borelight back on.
(9) If the boresight is properly aligned, the weapon system is boresighted; otherwise, you will need remedial training on your sight picture.
(10) Change the field of view (FOV) on the sight by rotating the FOV ring, and repeat steps (1) through (9).
(11) Turn the laser off, and carefully remove the borelight and the mandrel from the weapon so that the borelight device is not damaged.

Boresighting the AN/PAQ-4B/C

NOTE: Before boresighting the AN/PAQ-4B/C, make sure that the borelight has been zeroed to the weapon. The more accurate the boresight of the AN/PAQ-4B/C to the assigned weapon, the closer to a battlesight zero the weapon will be.

DANGERS

DO NOT STARE INTO THE IR LASER BEAM WITH THE NAKED EYE OR THROUGH BINOCULARS OR TELESCOPES.

DO NOT POINT THE IR LASER BEAM AT MIRROR-LIKE SURFACES OR OTHER INDIVIDUALS' EYES.

ALTHOUGH THIS LASER IS EYE-SAFE, IT IS A SAFE PRACTICE TO TREAT ALL LASERS AS NOT EYE-SAFE.

3x EXTENDERS MAGNIFY THE LASER AIMING LIGHT. WHEN USING THE 3x EXTENDERS, THE AN/PAQ-4B/C IS NOT CONSIDERED EYE-SAFE AT ANY DISTANCE.

DO NOT STORE THE AN/PAQ-4B/C WITH THE BATTERIES INSTALLED.

8-8. To boresight the AN/PAQ-4B/C —

(1) Select the proper 10-meter boresighting target for the weapon/AN/PAQ-4B/C configuration.
(2) With the help of an assistant, place the boresighting target 10 meters in front of the weapon.
(3) Install the borelight filter, and turn the AN/PAQ-4B/C on.
(4) Align the 10-meter boresighting target with the visible laser of the borelight.
(5) Adjust the adjusters on the AN/PAQ-4B/C until the IR laser is centered on the crosshair located on the 10-meter boresighting target.

NOTE: For windage and elevation, one click equals 1 centimeter at 25 meters. A click is defined as the sound or feel of the positive detent movement.

CAUTION

Do not turn the adjustment screws too much. Over-turning the adjustment screws will cause them to break.

NOTES: 1. The boresighting target and zeroing mark must be kept stable during the boresighting procedure.

2. Regardless of the mounting location, the adjuster that is on top or bottom will always be the elevation adjuster, and the one on the side will always be the windage adjuster.

BORESIGHTING THE AN/PEQ-2A/B

NOTES: 1. Before boresighting the AN/PEQ-2A/B, ensure that the borelight has been zeroed to the weapon. The more accurate the boresight of the AN/PEQ-2A/B to the assigned weapon, the closer to a battlesight zero the weapon will be.

2. The boresighting target and zeroing mark must be kept stable during the boresighting procedure.

DANGERS

THE AN/PEQ-2A/B EMITS INVISIBLE LASER RADIATION. AVOID DIRECT EXPOSURE TO THE BEAM.

DO NOT STARE INTO THE IR LASER BEAM WITH THE NAKED EYE OR THROUGH BINOCULARS OR TELESCOPES.

DO NOT POINT THE IR LASER BEAM AT MIRROR-LIKE SURFACES OR OTHER INDIVIDUALS' EYES.

EYE DAMAGE CAN OCCUR IF CARELESS HANDLING OF THE LASER OCCURS. EYE-SAFE DISTANCE IS BEYOND 25 METERS (IN DUAL LO MODE) IN TRAINING MODE AND BEYOND 220 METERS IN TACTICAL MODE.

3x EXTENDERS MAGNIFY THE LASER AIMING LIGHT. WHEN USING THE 3x EXTENDERS, THE AN/PEQ-2A/B IS NOT CONSIDERED EYE-SAFE AT ANY DISTANCE.

DO NOT STORE THE AN/PEQ-2A/B WITH THE BATTERIES INSTALLED.

8-9. To boresight the AN/PEQ-2A/B—

(1) Select the proper 10-meter boresighting target for the weapon/AN/PEQ-2A/B configuration.

(2) With the help of an assistant, place the boresighting target 10 meters in front of the weapon.

(3) Install the filter on the aiming laser, and turn the AN/PEQ-2A/B on.

(4) Align the 10-meter boresighting target with the visible laser of the borelight.

(5) Adjust the adjusters on the AN/PEQ-2A/B until the IR laser is centered on the crosshair located on the 10-meter boresighting target.

NOTE: For windage and elevation, one click equals 1 centimeter at 25 meters. A click is defined as the sound or feel of the positive detent movement.

CAUTION

Do not turn the adjustment screws too much. Over-turning the adjustment screws will cause them to break.

NOTES: 1. Each click of elevation and windage is 1 centimeter and each square of the 25-meter zeroing target is .9 centimeter. For ease of use, round up to one square per click of elevation. This will have a negligible effect on the zero unless a large adjustment is needed.

2. Regardless of the mounting location, the adjuster that is on top or bottom will always be the elevation adjuster, and the one on the side will always be the windage adjuster.

(6) Adjust the illuminator in the same manner.

(7) Turn the laser off, and carefully remove the borelight and the mandrel from the weapon so that the borelight device is not damaged.

BORESIGHTING THE AN/PVS-4

NOTES: 1. Before boresighting, ensure that the borelight has been zeroed to the weapon. The more accurate the boresight of the AN/PVS-4 to the assigned weapon, the closer to a battlesight zero the weapon will be.

2. 25-meter zeroing must be conducted to ensure that the AN/PVS-4 is properly zeroed.

(1) Select the proper 10-meter boresighting target for the weapon/AN/PVS-4 configuration.

(1) With the help of an assistant, place the boresighting target 10 meters in front of the weapon.

(2) Ensure that the M16 reticle is displayed.

(3) Get behind the weapon in a stable supported firing position, and look through the AN/PVS-4.

(4) Turn the borelight laser on.

(5) Align the borelight laser with the circle on the 10-meter target offset.

(6) Keeping the laser in place, adjust the windage and elevation until the reticle of the AN/PVS-4 is aligned with the circular crosshair.

NOTE: If there is not enough ambient light to see the 10-meter target offset circular crosshair, shine a flashlight indirectly at the target to provide ambient light.

(7) Turn the borelight off.

(8) Move the reticle off of the circular crosshair, and then realign on the target.

(9) Turn the borelight laser back on.

(9) If the borelight is in the circle, then the AN/PVS-4 is boresighted.

(10) Turn the laser off, and carefully remove the borelight and mandrel from the weapon so that the borelight device is not damaged.

SECTION II. TRAINING STRATEGIES AND QUALIFICATION STANDARDS

Before beginning a night marksmanship program, Soldiers must qualify on their assigned weapons during the day, as outlined in the previous chapters of this manual. Commanders should follow these training strategies and abide by the qualification standards set forth. Although some courses of fire may seem redundant or inappropriate, numerous tests show that these training strategies work, and the qualification standards are achievable if the strategies are followed.

BACKUP IRON SIGHT

8-10. The BUIS is a semipermanent flip-up sight equipped with a rail-grabbing base. The BUIS provides a backup capability effective out to at least 600 meters and can be installed on M16A4 rifles and M4 carbines.

NOTE: See Table 8-3 for the current training program.

Table 8-3. Backup iron sights training program.

BACKUP IRON SIGHTS
Instructional Intent
• Soldiers zero and qualify with the BUIS.
Special Instructions
Ensure that Soldiers— • Apply the marksmanship fundamentals. • Place the BUIS in the full vertical position and lock them prior to firing. • Install the plastic insert in the BUIS during boresighting and zeroing.
Observables
• Soldiers attain the same day standards for zeroing and qualification as with the standard iron sights (See Chapter 5).

CONCEPT

8-11. The BUIS is adjusted for a 300-meter battlefield zero to provide backup in the event an optic or laser device fails to function. The BUIS is zeroed on the M4/M4A1 target on the backside of the M16A2 zeroing target (NSN 6920-01-395-2949).

NOTE: The 25-meter zeroing procedures are the same as for conventional rear sight assembly on M16 rifles and M4 carbines.

CONDUCT OF TRAINING

NOTES: 1. The BUIS training strategy is the same as the iron sight training strategy.

2. All procedures for the BUIS are the same as with standard iron sights.

8-12. BUIS equipment training should familiarize the Soldier with the proper operation and characteristics of the BUIS.

Boresight the Backup Iron Sights

8-13. Optional.

Zero

8-14. The zeroing standards for the BUIS are the same as with iron sights.

- To zero the BUIS for M4 carbines, set the range selector to 300 meters.
- To zero the BUIS to an M16A4, place the range selector on the white line below the 300-meter mark.

Target Detection

8-15. Target detection procedures for the BUIS are the same as with standard iron sights.

Practice Qualification

8-16. A practice qualification must always precede an actual qualification. Practice qualification allows the Soldier to practice and refine the skills needed to succeed during qualification. Practice qualification standards for the BUIS are the same as with standard iron sights. If the Soldier qualifies during the practice qualification, it may be counted as the record qualification.

Record Qualification

8-17. Qualification with the BUIS is conducted on a standard record fire range, and the standards for qualification are the same as the record fire day standards.

M68 CLOSE COMBAT OPTIC

8-18. The M68 CCO is a reflex (nontelescopic) sight. It uses a red aiming reference (collimated dot) and is designed for the two eyes open method of sighting. The dot follows the horizontal and vertical movement of the firer's eye, while remaining fixed on the target. No centering or focusing is required.

NOTE: See Table 8-4 for the current training program.

Table 8-4. M68 close combat optic training program.

M68 CLOSE COMBAT OPTIC
Instructional Intent
• Soldiers qualify with the M68 CCO.
Special Instructions
Ensure that Soldiers— • Are proficient with the M68 CCO. • Install the half-moon spacer when using the M68 CCO with the M4, M16A4, and MWS. • Do not install the half-moon spacer when using the M68 CCO with the M16A1/A2. • Use the proper offset during boresighting procedures. • Confirm 10-meter boresight with a 25-meter zero. • Retighten the rail grabber after the initial 3 rounds are fired. • Sets the M68 CCO's dot for best sight picture. • Apply the marksmanship fundamentals. • Zero and qualify with the same sight picture (one eye or two eyes open method). • Zero on the M16A2 25-meter zeroing target. Ensure that the designated impact zone is 1.4 centimeters down from the center of mass of the 300-meter silhouette on the 25-meter zeroing target.
Observables
• Soldiers zero the M68 CCO to the same standards as with iron sights. • Rounds impact in the 4x4-centimeter square designated impact zone. • Soldiers achieve the same practice qualification and qualification standards as with day record fire.

CONCEPT

8-19. Soldiers must qualify on their assigned weapons during daylight conditions, as outlined in this manual. The integrated act of firing with the M68 CCO is identical to the iron sights except for the change in the sight picture.

CONDUCT OF TRAINING

NOTES: 1. The M68 CCO training strategy is the same as the iron sight training strategy.

2. All procedures for the M68 are the same as with standard iron sights.

8-20. M68 equipment training should familiarize the Soldier with the proper operation and characteristics of the M68 CCO IAW TM 9-1240-413-13&P.

MODIFIED FUNDAMENTALS

8-21. M68 CCO operation requires modifications to the fundamentals of marksmanship:

- Steady position.
- Aiming.
- Breath control.
- Trigger squeeze.

Steady Position

8-22. When operating the M68 CCO, Soldiers no longer need a good stock weld to get a good sight picture after the M68 is zeroed at 25 meters. The M68's reflexive sight allows the Soldier to fire the weapon with his cheek in a comfortable position, but since the CCO is parallax-free beyond 50 meters only, the Soldier must zero and fire using the same cheek position.

Aiming

8-23. When using the M68 CCO, the two eyes open method is the preferred method of aiming.

NOTE: The aiming method used to zero must also be used to engage targets. When using the M68 CCO, the weapon must not be canted during aiming or firing.

Two Eyes Open Method (Preferred)

8-24. This method allows a much greater field of view and makes scanning for targets much easier; however, getting accustomed to the two eyes open method takes practice. The Soldier must keep the rifle and M68 in a vertical alignment each time he fires.

8-25. To use the two eyes open method—

(1) Position the head so that one eye can focus on the red dot and the other eye can scan downrange.

(2) Place the red dot on the target's center of mass, and engage.

One Eye Open Method

8-26. To use the one eye open method—

(1) With the nonfiring-side eye closed, look through the M68 CCO to ensure that the red dot can be seen clearly.

(2) Place the red dot on the target's center of mass, and engage.

Breath Control

8-27. This fundamental does not change.

Trigger Squeeze

8-28. This fundamental does not change.

ZEROING WITHOUT A BORELIGHT

8-29. To zero without a boresight—

(1) Begin with a securely installed and live-fire zeroed BUIS.

(2) Mount the CCO to the front of the receiver rail or to the top ARS, as preferred.

(3) Adjust windage and elevation on the reflex sight until the center of the aiming dot is at the tip of the front sightpost when viewed through the BUIS while assuming a normal firing position.

25-METER ZEROING PROCEDURES

NOTE: Conduct zeroing only on the M16A2 25-meter target.

8-30. When zeroing the CCO at 25 meters, a designated point of impact zone must be identified on the 25-meter zeroing target (Figure 8-5). To zero the CCO at 25 meters—

(1) Starting from center of mass of the 300-meter silhouette on the 25-meter zeroing target, count down 1½ squares or 1.4 centimeters. This is now the point of impact when zeroing the M68 CCO.

(2) Continue to aim at the center of mass of the 300-meter silhouette, and make adjustments to the M68 CCO so that the rounds impact in the secondary 4x4-centimeter circular box, 1½ squares or 1.4 centimeters down from the point of aim.

8-31. Other procedures are the same as standard iron sight procedures:

NOTE: A click is defined as the sound or feel of the positive detent movement.

- For windage and elevation, two clicks equal 1 centimeter at 25 meters.
- For elevation, one clockwise click moves the bullet strike down.
- For windage, one clockwise click moves the bullet strike left.

NOTES: 1. At ranges of 50 meters and beyond, the effects of parallax are minimal. However, at ranges of 50 meters and closer, parallax exists and the firer must ensure that the red dot is centered while zeroing.

2. The aiming method (two eyes open or one eye open) used to zero must be used to engage targets.

---

25 METER ZEROING TARGET

M16A2/3/4

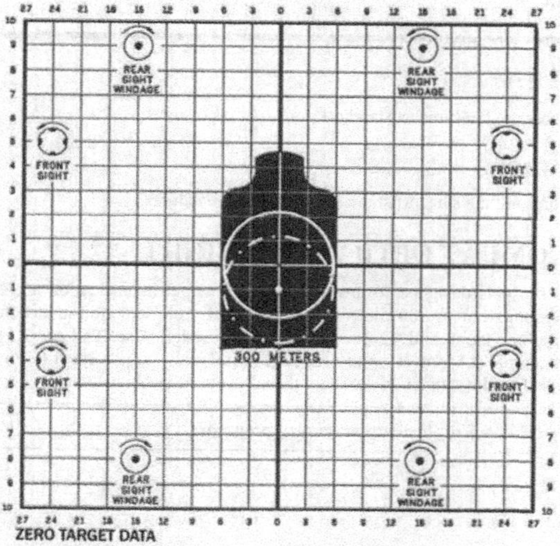

ZERO TARGET DATA

1- ROTATE REAR SIGHT ELEVATION KNOB TO THE 8/3 SETTING. THEN UP (RIGHT) ONE CLICK PAST THE 300 MARK FOR ZEROING AT 25 METERS.

2- AIM AT TARGET CENTER, ADJUST SIGHTS TO MOVE SHOT GROUP CENTER AS CLOSE AS POSSIBLE TO THE WHITE DOT IN THE CENTER OF THE TARGET.

3- AFTER COMPLETION OF THE 25 METER ZERO, ROTATE THE REAR SIGHT ELEVATION KNOB BACK ONE CLICK TO THE 300/800 METER MARK. THE WEAPON WILL BE ZEROED FOR 300 METERS.

4- 1.4 CENTIMETERS OR 1.5 SQUARES STARTING FROM CENTER MASS OF THE 300-METER SILHOUETTE ON THE 25-METER ZERO TARGET WILL BE THE POINT OF IMPACT FOR THE M68, CCO.

Figure 8-5. Close combat optic, 25-meter zeroing target.

TARGET DETECTION

8-32. Target detection procedures for the M68 are the same as with standard iron sights.

> **WARNING**
>
> **In Positions 4 and above, the red dot is visible through the front of the sight. For night vision operations—**
>
> - **Close the front lens cover before turning the rotary switch clockwise to Positions 2 and 3.**
> - **Check the light for proper intensity before opening the front lens cover.**
> - **Close the front lens cover before turning the rotary switch counterclockwise to the OFF position.**
>
> **Failure to follow this warning could reveal your position to the enemy.**

PRACTICE QUALIFICATION

8-33. The procedures are the same as standard iron sight procedures.

RECORD QUALIFICATION

8-34. The procedures are the same as standard iron sight procedures.

ADVANCED COMBAT OPTICAL GUNSIGHT

8-35. The ACOG scope is designed to provide enhanced target identification and hit probability for the M16/M4 series weapons out to 800 meters. It is designed with dual illuminated technology using fiber optics for daytime illumination and tritium for night and low-light use. The ACOG scope is a lightweight rugged, fast and accurate 3.5 power optic scope. It is internally adjustable to allow the shock of rough handling to be carried by the scope body and not the adjustment mechanism.

NOTE: See Table 8-5 for the current training program.

Table 8-5. Advanced combat optical gunsight training program.

ADVANCED COMBAT OPTICAL GUNSIGHT
Instructional Intent
• Soldiers qualify with the ACOG.
Special Instructions
Ensure that Soldiers— • Are proficient with the ACOG. • Use the proper offset during boresighting procedures. • Confirm the 10-meter boresight with 100-meter zero (or 25-meter zero if the 100-meter range is not available). • Retighten the rail grabber after the initial 3 rounds are fired. • Apply the marksmanship fundamentals. • Zero and qualify with the same sight picture. Ensure that— • The point of aim and point of impact are the same at 100 meters. • The designated impact zone is 1 centimeter down from the center of mass of the 300-meter silhouette on the 25-meter zeroing target.
Observables
• Soldiers zero the ACOG to the same standards as with iron sights. • Soldiers achieve the same practice qualification and qualification standards as with day record fire.

CONCEPT

8-36. Soldiers must qualify on their assigned weapons during the day, as outlined in this manual. The integrated act of firing with the ACOG is identical to the iron sights except for the change in sight picture.

CONDUCT OF TRAINING

NOTES: 1. The ACOG training strategy is the same as the iron sight training strategy.

2. All procedures for the ACOG are the same as with standard iron sights.

8-37. ACOG equipment training should familiarize the Soldier with its proper operation and characteristics.

OPERATION

8-38. The ACOG scope is internally adjustable. Adjustment is made using the adjuster mechanisms located inside the adjuster caps on the top and right-hand side of the scope. Adjustment can be made with a small screwdriver, coin, or other hard object that fits the adjustment screws. The caps are very tight to ensure a waterproof seal with the O-rings inside. The caps should only be off the scope when adjustments are being made.

CAUTION

The ACOG contains an internal adjustment mechanism to allow zeroing. Adjustments to the extreme ends of the range can result in damage to the internal prism assembly. Do not continue to adjust the windage and elevation mechanisms if you encounter resistance.

8-39. The ACOG scope is shipped with a pre-centered setting. Normally this means that only small adjustments are necessary. Do not adjust the scope to the extremes. It is possible that over-adjustment will damage the precise alignment of the prism assembly inside the rifle scope.

8-40. The ACOG scope is adjusted at the factory to be as parallax-free as possible at 100 meters. This in no way affects the accuracy of the scope.

8-41. The reticle patterns in the scope have been designed to provide many features while retaining simplicity of operation. The user does not need to make any manual adjustments between shots at different ranges. Ranging capability is built into the reticle patterns.

8-42. The outside legs of the chevron reticle in the ACOG correspond to 19 inches (average width of a man's shoulders) at 300 meters.

8-43. The widths of the horizontal hash marks on the BDC reticles in all ACOG scopes correspond to the width of a .5 meter (19 inches) silhouette (man-size) at that range (Figure 8-6).

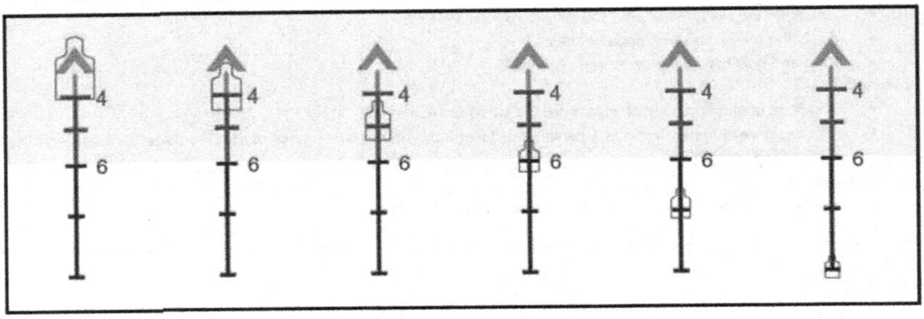

Figure 8-6. Width of horizontal hash marks.

Zeroing the Advanced Combat Optical Gunsight

8-44. The adjustment increments for the ACOG are 1/4 inch per click at 100 meters. A click is defined as the sound or feel of the positive detent movement. This means that 4 clicks are required to move the bullet approximately one inch on the target at 100 meters. At 25 meters, 16 clicks move the bullet one inch. Turning the screw in the direction of the arrow moves the bullet impact in the direction marked (up/right).

NOTE: To ensure a consistent zero, tap the scope with the palm of your hand to stabilize the adjustment mechanism after an adjustment has been made, and then fire a three-shot group on the target.

8-45. The ACOG scope with chevron reticle is designed to be zeroed at 100 meters, using the tip of the chevron reticle for POA/POI (Figure 8-7).

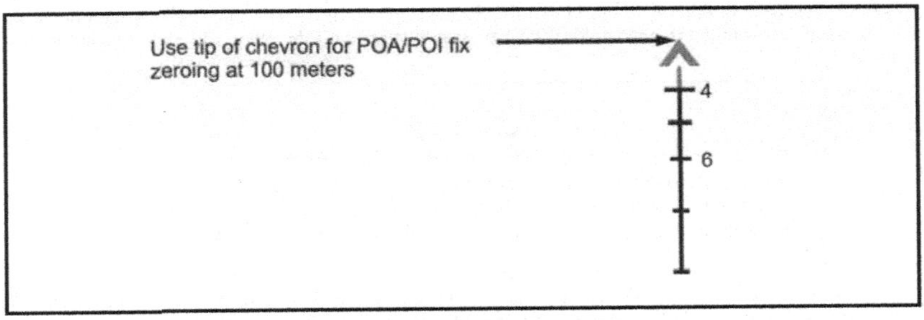

Figure 8-7. Advanced combat optical gunsight reticle point of aim at 100 meters.

8-46. For a quick combat zero, the scope may be zeroed at 25 meters, using the tip of the 300-meter post as POA/POI. A 25-meter zero is less precise than a 100-meter zero and should be verified at longer distances once time and a range is available (Figure 8-8).

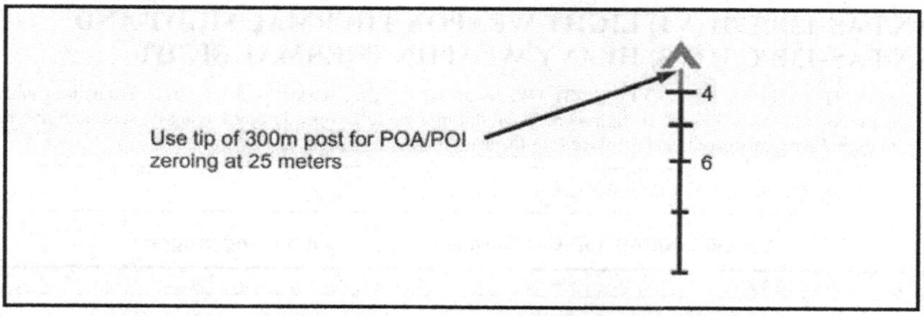

Figure 8-8. Advanced combat optical gunsight reticle point of aim at 25 meters.

MODIFIED FUNDAMENTALS

8-47. ACOG operation requires modifications to the following fundamentals of marksmanship:

- Steady position.
- Aiming.
- Breath control.
- Trigger squeeze.

Steady Position

8-48. This fundamental does not change.

Aiming

8-49. When the TA11F ACOG is attached to the carrying handle of an M16A1/A2/A3 rifle, the BDC reticle will be off slightly at extended distances. Beginning at 400 meters, the BDC reticle will be approximately one MOA off to the extent of the BDC reticle, once properly zeroed.

Example: At 500 meters, the impact will be 5 inches high; at 600 meters, the impact will be 6 inches high.

Once the target is ranged beyond 400 meters, hold slightly low.

Breath Control

8-50. This fundamental does not change.

Trigger Squeeze

8-51. This fundamental does not change.

TARGET DETECTION

8-52. Target detection procedures for the ACOG are the same as with standard iron sights.

PRACTICE QUALIFICATION

8-53. The procedures are the same as standard iron sight procedures.

Record Qualification

8-54. The procedures are the same as standard iron sight procedures.

AN/PAS-13B/C/D (V1) LIGHT WEAPON THERMAL SIGHT AND AN/PAS-13B/C/D (V3) HEAVY WEAPON THERMAL SIGHT

8-55. The AN/PAS-13B/C/D (V1, V3) TWS is an IR imaging sensor used for target acquisition under conditions of low visibility. IR light is received through the telescope, detected by an IR sensor, converted to digital data, processed, and displayed for the user.

NOTE: See Table 8-6 for the current training program.

Table 8-6. AN/PAS-13B/C/D thermal weapon sight training program.

AN/PAS-13B/C/D TWS
Instructional Intent
• Soldiers qualify with the ANPAS-13B/C/D TWS.
Special Instructions
Ensure that Soldiers— • Are proficient with the TWS. • Use the spacer with the M4, M16A4, and MWS. • Use the proper 10-meter boresighting target during boresighting procedures. • Boresight both FOVs. • Confirm the 10-meter boresight with a 25-meter zero. • Use the M16A2 zeroing target with a 4x4-centimeter square cut out of the center of the silhouette. • Thermalize the zero range and qualification range (inspect for targets that aren't thermalized). • Use every other lane during zero and qualification.
Observables
• Soldiers zero the TWS to the same standards as with iron sights. • Soldiers achieve the same practice qualification and qualification standards as with day record fire.

WARNING

Ensure that the weapon is not loaded and is on SAFE before installing the TWS on the weapon. A loaded weapon may accidentally discharge, causing severe injury or death.

Concept

8-56. The AN/PAS-13B/C/D TWS training strategy is much the same as that of aiming lights. The TWS does not require the use of NVDs. The course of fire for the TWS is the same scenario as the day qualification tables, with the same standards of fire as for current day standards. Qualification standards are the same for day and night.

Conduct of Training

8-57. AN/PAS-13B/C/D equipment training should familiarize the Soldier with the proper operation and characteristics of the TWS IAW the TM.

MODIFIED FUNDAMENTALS

8-58. AN/PAS-13B/C/D TWS operation requires modifications to the fundamentals of marksmanship:

- Steady position.
- Aiming.
- Breath control.
- Trigger squeeze.

Steady Position

8-59. This fundamental slightly changes due to the height of the sight. Consider the following modifications:

- Soldiers must adjust their position so they can properly look through the sight.
- In most cases, the cheek-to-stock weld no longer exists.

Aiming

8-60. To properly aim with the TWS, Soldiers must ensure that the correct reticle is selected in the sight.

NOTE: Refer to TM 11-5855-312-10 for reticle selection and point of aim for use with the TWS.

Breath Control

8-61. This fundamental is not affected by night firing conditions when using the TWS.

Trigger Squeeze

8-62. This fundamental of marksmanship does not change during night firing.

25-METER ZEROING PROCEDURES

NOTE: Refer to TM 11-5855-312-10 for target preparation.

8-63. To zero the AN/PAS-13B/C/D TWS at 100 meters, use the same procedures and standards as with iron sights, along with the following:

- At the 25-meter range, each incremental adjustment to the azimuth or elevation setting moves strike of the round as follows:
  - 1 ½ centimeters for LWTS on WFOV.
  - ½ centimeter for LWTS on NFOV.
  - ¾ centimeter for HWTS on WFOV.
  - ¼ centimeter for HWTS on NFOV.
- Retighten the rail grabber after firing the first three rounds.
- Zero both FOVs (Figure 8-9).

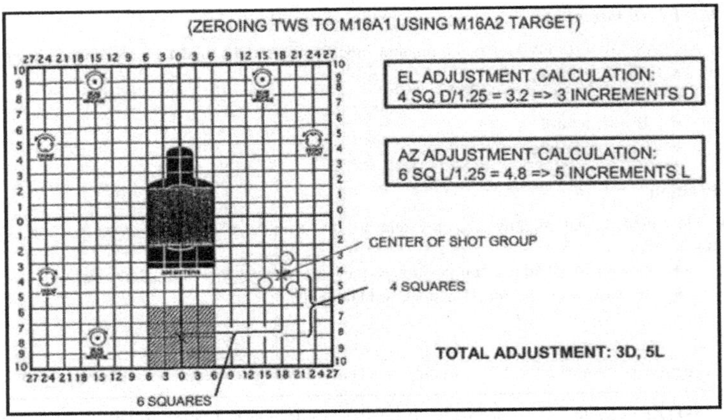

Figure 8-9. Example of thermal weapon sight zeroing adjustments.

TARGET DETECTION

8-64. With NVDs, the field of view is much smaller, scanning becomes much more deliberate, and, with the TWS, camouflage becomes less of a factor. Even though NVDs greatly enhance the Soldier's ability to acquire a target at night, increased awareness of target detection must be trained to allow the Soldier to recognize the visual cues of IR imagery.

Select Position

8-65. The TWS is a large device; selecting a position that allows for good fields of view, but does not silhouette the Soldier and his equipment, might be a challenge. Since the TWS detects thermal energy (heat) emitted from an object, a position near an object emitting a vast amount of thermal energy (for example, a vehicle with the engine running or a fire) may affect the Soldier's ability to acquire a target.

Scanning

8-66. With earlier versions of the TWS, scanning too fast causes a stuttering on the screen, which might cause the Soldier to miss or overlook a target. With these versions, Soldiers must scan slowly in order to maintain a good thermal image on the screen.

8-67. With the newer version, this stuttering is not as obvious. One advantage of the TWS is its two FOVs—wide and narrow. Each has advantages and disadvantages:

- The NFOV increases magnification but decreases the field of view.
- The WFOV decreases magnification but increases the field of view.

Target Indicators

8-68. While scanning the sector or lane with the TWS, the Soldier should recognize thermal cues that allow him to detect and identify targets. The engine compartment, exhaust, and tires of a vehicle that has been moving are all examples of thermal cues. Adjusting the brightness, contrast, and polarity helps enhance the thermal cues of a target, allowing for quicker detection and identification.

Sound

8-69. Use the same techniques outlined in day and night target detection.

Movement

8-70. Thermal cues become much more obvious on a moving object than on an object standing still. A good example is a vehicle's tires. When the vehicle is not moving, the tires are cold. On a moving vehicle, the friction between the road and the tires causes the tires to heat up and become prominent when observed through the TWS. The same is true with the human body—a moving person generates more heat than someone standing still.

Camouflage

8-71. The TWS allows Soldiers to see thermal signatures not completely masked by camouflage, such as paint, foliage, and camouflage netting, thereby increasing both day and night target detection abilities.

WARNING

If the TWS is operated with the eyecup removed, light emitting from the eyepiece may be visible to the enemy's NVDs.

PRACTICE QUALIFICATION

8-72. Practice qualification with the TWS is the same as day practice qualification with iron sights. Dry-fire exercises are performed to allow Soldiers to make adjustments to the TWS. Every other firing lane should be used so that the Soldier engages only the targets in his lane.

RECORD QUALIFICATION

8-73. Record qualification with the TWS is the same as day record qualification with iron sights.

NOTES:

1. Record qualification with the TWS can be done day and or night. Regardless of the qualification, the standard day record fire for the iron sights will be used. The standards for qualification with the TWS, either day or night, are 23 out of 40.
2. During practice qualification and qualification, the Soldier chooses his own polarity and FOV.

AN/PAQ-4B/C AND AN/PEQ-2A/B INFRARED AIMING LASERS

8-74. The newest IR aiming lasers greatly increase the night firing accuracy of all infantry weapons. IR aiming lasers complete the transition from day optics to night optics. Their effectiveness is limited by the capability of the image-intensifier (I2) sight with which they are used.

NOTE: See Table 8-7 for the current training program.

Table 8-7. AN/PAQ-4B/C or AN/PEQ-2A/B infrared aiming laser training program.

AN/PAQ-4B/C OR AN/PEQ-2A/B INFRARED AIMING LASERS
Instructional Intent
• Soldiers qualify with the AN/PAQ-4B/C or AN/PEQ-2A/B IR aiming laser.
Special Instructions
Ensure that Soldiers— • Are proficient with the AN/PAQ-4B/C or AN/PEQ-2A/B IR aiming laser. • Use the proper 10-meter boresighting target during boresighting procedures. • Use the borelight filter. • Set the AN/PEQ-2A/B IR aiming laser to AIM LO. • Boresight the illuminator on the AN/PEQ-2A/B. • Use the M16A2 25-meter zeroing target for 25-meter zero. Ensure that a 3x3-centimeter hole is cut in the center of the 25-meter zeroing target and E-type silhouette.
Observables
• Soldiers conduct either 10-meter boresight or a 25-meter zero. • Soldiers display good scanning, IR discipline, and IR walking technique. • Soldiers achieve at least 17 target hits out of 40 target exposures.

CONCEPT

8-75. Two training strategies have been devised to adequately train Soldiers in the use of AN/PAQ-4B/C and AN/PEQ-2A/B IR aiming lasers: the night initial training strategy and the night sustainment training strategy. Units should always review the night initial training strategy prior to the night sustainment training strategy.

Night Initial Training Strategy

8-76. The night initial training strategy is used for Soldiers who have little or no previous experience with night vision goggles, or for units beginning a night training program.

Night Sustainment Training Strategy

8-77. The night sustainment training strategy is for Soldiers who are familiar with night vision goggles, and for units that have already implemented a night training program.

CONDUCT OF TRAINING

8-78. AN/PAQ-4B/C and AN/PEQ-2A/B IR aiming laser equipment training should familiarize Soldiers with the proper operation and characteristics of AN/PAQ-4B/C and AN/PEQ-2A/B IR aiming lasers IAW the TM.

MODIFIED FUNDAMENTALS

8-79. NVD operation requires modifications to the fundamentals of marksmanship:

- Steady position.
- Aiming.
- Breath control.
- Trigger squeeze.

Steady Position

8-80. Consider the following modifications:

- Realize that a good cheek-to-stock weld is not possible with NVGs mounted on the head.
- Ensure that the weapon's buttstock is firmly pulled into the pocket of the shoulder to prevent the laser from wobbling.
- When ready to fire, plant the elbows firmly on the ground to prevent the laser from wobbling excessively.

Aiming

8-81. Consider the following modifications:

- Practice raising the head just enough to clear the weapon with the NVGs.
- Acquire a good sight picture by walking the laser onto the target and then aiming at the center of mass.

Breath Control

8-82. This fundamental is not modified for night firing conditions.

Trigger Squeeze

8-83. Do not disrupt the laser/target alignment by jerking the trigger.

25-Meter Zeroing Procedures

8-84. If the borelight is not available, a 25-meter zero must be conducted. A 25-meter zeroing target is shown in Figure 8-10.

AN/PAQ-4B/C Infrared Aiming Laser

8-85. The 25-meter zeroing procedures for the AN/PAQ-4B/C IR aiming laser are as follows:

NOTE: The zero standards are the same as with iron sights.

---

(1) Set the adjusters to their zero preset position.

NOTE: See TM 11-5855-301-12&P for more information about setting the adjusters to their zero preset position.

---

(2) Prepare a 25-meter zeroing target by cutting a 3x3-centimeter square out of the center of the silhouette.

(3) Adjust for windage and elevation.

- For windage and elevation, one click equals 1 centimeter at 25 meters.
- For elevation, one clockwise click moves the bullet strike up.
- For windage, one clockwise click moves the bullet strike left.

(4) Retighten the rail grabber after the first three rounds are fired.

NOTE: When cutting the 3x3-centimeter square out of the target, some of the strike zone may be cut out. Take care when annotating the impact of the rounds. When the weapon is close to being zeroed, some of the shots may be lost through the hole in the target.

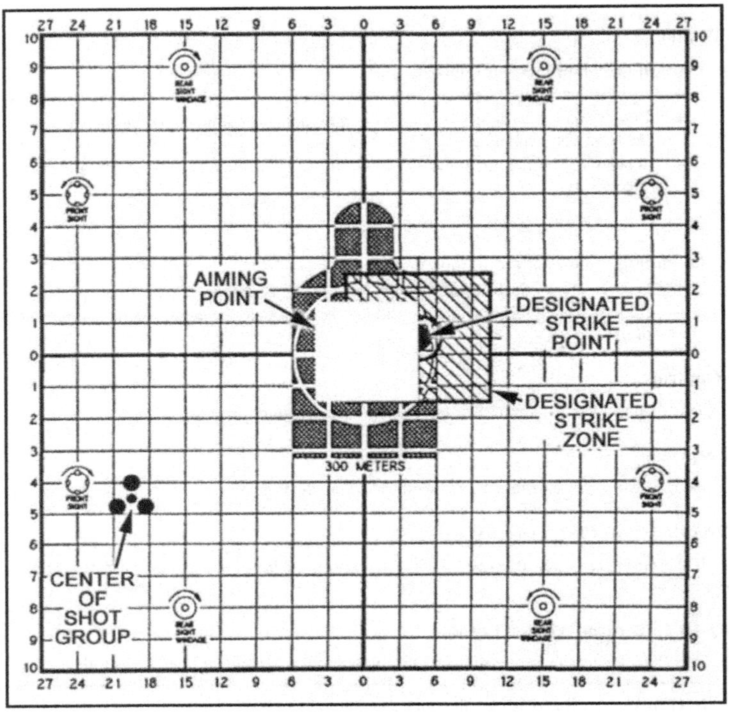

Figure 8-10. Example of shot group adjustment with strike zone.

AN/PEQ-2A/B Infrared Aiming Laser

8-86. The 25-meter zeroing procedures for the AN/PEQ-2A/B IR aiming laser are as follows:

NOTE: The zero standards are the same as with iron sights.

---

(1) Set the adjusters to their zero preset position.

NOTE: See TM 11-5855-308-12&P for more information about setting the adjusters to their zero preset position.

---

(2) Prepare the 25-meter zeroing target by cutting out a 3x3-centimeter square in the center of the target and E-type silhouette.

(3) Turn the aiming beam on in the low power setting (AIM LO).

(4) Install the aim point filter to eliminate excessive blooming.

(5) Adjust the AN/PEQ-2A/B (top-mounted) aiming point as follows:

- For windage and elevation, one click equals 1 centimeter or one square at 25 meters.
- For elevation, one clockwise click moves the bullet strike up.
- For windage, one clockwise click moves the bullet strike right.

(6) Adjust the AN/PEQ-2A/B (top-mounted) target illuminator as follows:
- For windage and elevation, one click equals 1 centimeter or one square at 25 meters.
- For elevation, one clockwise click moves the bullet strike down.
- For windage, one clockwise click moves the bullet strike right.

(7) Retighten the rail grabber and the AN/PEQ-2A/B.

(8) Once the aiming beam is zeroed, rotate the selector knob to the DUAL LO, DUAL LO/HI, or DUAL HI/HI mode to observe both aiming and illumination beams.

(9) Rotate the illumination beam adjusters to align the illumination beam with the aiming beam.

NOTES: 1. Failure to fully tighten the mounting brackets and AN/PEQ-2A/B thumbscrew may cause zero retention problems. Confirm that equipment is tight prior to zeroing.

2. To retain zero, remove the TPIAL and rail grabber as a whole assembly and place back onto the same notch as removed.

TARGET DETECTION

8-87. Soldiers should receive in-depth instruction on the proper use and fit of night vision goggles, to include characteristics and capabilities, maintenance, and mounting procedures. Extensive testing has proven that the average Soldier does not properly use NVDs. Unit leaders must be proficient in the train-the-trainer strategy. At night, Soldiers should conduct a terrain walk to become more familiar and build confidence using the night vision goggles.

Scanning for Targets

8-88. NVDs have a 40-degree field of view, which causes the average firer to miss easy targets of opportunity. Soldiers must be trained to aggressively scan their sectors of fire for targets.

8-89. The art of target detection at night is only as good as the Soldier practices. Regular blinking during scanning relieves some of the eyestrain that Soldiers experience when trying to spot distant targets. Regular blinking must be reinforced during training. After Soldiers have mastered the art of scanning, they will find that targets are more easily detected by acknowledging the flicker or movement of a target.

Infrared Discipline

8-90. Soldiers must be taught that what they can see downrange or on the battlefield through NVGs, the enemy can also see. Soldiers must train to activate the laser at the base of the target and engage the target as soon as the target is detected. After the target has been engaged, the laser is deactivated.

8-91. When a Soldier uses proper IR discipline while scanning for targets, he must keep his weapon oriented within his sector of fire. When the target is detected, the Soldier orients his weapon around the base of the target, activates his laser, and walks the laser to the target's center of mass for engagement.

FIELD FIRE

8-92. During dry-fire exercises, Soldiers acquire a sight picture on all exposed silhouette targets before conducting the field fire scenario. This allows Soldiers to focus on the targets at range. The procedures for field fire include the following:

- Conduct a dry-fire exercise prior to conducting a live-fire.
- Conduct Field Fire in the same manner as Field Fire II.
- Engage targets at 50, 150, and 250 meters.
- Fire 36 rounds:
  - 18 rounds from the supported firing position.
  - 18 rounds from the prone unsupported firing position.

Practice Qualification

8-93. The procedures for practice qualification include the following:

- Conduct a dry-fire exercise.
- Use coaches.
- Fire 40 rounds:
  - 20 rounds from the prone supported firing position.
  - 10 rounds from the unsupported firing position.
  - 10 rounds from the kneeling firing position.
- Engage targets from 50 to 250 meters.
- Meet the standards (17 hits out of 40 target exposures).

Record Qualification

8-94. The procedures for record qualification include the following:

- Conduct a dry-fire exercise.
- Fire 40 rounds:
  - 20 rounds from the prone supported firing position.
  - 10 rounds from the unsupported firing position.
  - 10 rounds from the kneeling firing position.
- Engage targets from 50 to 250 meters.
- Meet the standards (17 hits out of 40 target exposures).

AN/PVS-4 NIGHT VISION DEVICE

8-95. The AN/PVS-4 NVD is a portable, battery-operated electro-optical instrument used for observation and aimed fire of weapons at night. It amplifies reflected light, such as moonlight, starlight, and sky glow, so that the viewed scene becomes clearly visible to the operator. It can be mounted on the M16A2/A3/A4 rifle and M4/M4 MWS. Mounting brackets are provided for each type of weapon.

NOTE: See Table 8-8 for the current training program.

Table 8-8. AN/PVS-4 night vision device training program.

AN/PVS-4 NIGHT VISION DEVICE
Instructional Intent
• Soldiers qualify with the AN/PVS-4 NVD.
Special Instructions
Ensure that Soldiers— • Are proficient with the AN/PVS-4 NVD. • Install the spacer and Picatinny rail grabber when mounting on the MWS. • Use the proper 10-meter boresighting target during boresighting procedures. • Use the proper reticle. • Confirm 10-meter boresight with a 25-meter zero.
Observables
• Soldiers zero the AN/PVS-4 to the same standard as with the iron sight. • Soldiers achieve the same practice and qualification standards as with day record fire.

CONCEPT

8-96. Training strategy for the AN/PVS-4 is much the same as that for aiming lights, and the course of fire and qualifications standards are the same as those for aiming lasers.

CONDUCT OF TRAINING

8-97. This training should familiarize the Soldier with the proper operation and characteristics of the AN/PVS-4 IAW the TM.

MODIFIED FUNDAMENTALS

8-98. NVD operation requires modifications to the fundamentals of marksmanship:

- Steady position.
- Aiming.
- Breath control.
- Trigger squeeze.

Steady Position

8-99. This fundamental slightly changes due to the height of the sight. Consider the following modifications:

- Adjust your position so you can properly look through the sight.
- Recognize that, in most cases, the cheek-to-stock weld no longer exists.

Aiming

8-100. Consider the following modifications:

- To properly aim the AN/PVS-4, ensure that the proper reticle is inserted in the sight.
- Place the aiming point on the target's center of mass.

NOTE: Refer to TM 11-5855-213-10 for more information about inserting the proper reticle.

Breath Control

8-101. This fundamental of marksmanship does not change during night firing.

Trigger Squeeze

8-102. This fundamental of marksmanship does not change during night firing.

25-METER ZEROING PROCEDURES

8-103. To zero the AN/PVS-4 at 25 meters, use the same procedures and standards as with the iron sights, along with the following:

- At 25-meter range, each increment of azimuth or elevation setting moves the strike of the round .63 centimeters or ¼ mil.
- Two clicks of the windage or elevation will move the strike of the round approximately one square on the M16A2 zeroing target.
- Retighten the thumbscrew on the rail grabber after firing the initial three rounds.

NOTE: If there is not enough ambient light to see the boresight mark at ten meters or the silhouette on the zeroing target during boresighting or zeroing procedures, shine a flashlight indirectly at the target to provide ambient light.

Target Detection

8-104. Target detection with the AN/PVS-4 is very similar to target detection with the night vision goggles. The AN/PVS-4 has a 14.5-degree field of view, causing the average firer to miss easy targets of opportunity, more commonly the 50-meter left or right.

8-105. Soldiers must be trained to aggressively scan their sectors of fire for targets. The art of target detection at night is only as good as the Soldier practices. Regular blinking during scanning relieves some of the eyestrain that Soldiers experience when trying to spot distant targets. Regular blinking must be reinforced during training. After Soldiers have mastered the art of scanning, they will find that targets are more easily detected by acknowledging the flicker or movement of a target.

Field Fire

8-106. During dry-fire exercises, Soldiers acquire a sight picture on all exposed silhouette targets before conducting the field fire scenario. This allows Soldiers to focus on the targets at range. The procedures for field fire include the following:

- Conduct a dry-fire exercise prior to conducting a live-fire.
- Conduct Field Fire in the same manner as Field Fire II.
- Engage targets at 50, 150, and 250 meters.
- Fire 36 rounds:
  - 18 rounds from the supported firing position.
  - 18 rounds from the prone unsupported firing position.

Practice Qualification

8-107. The procedures for practice qualification include the following:

- Conduct a dry-fire exercise.
- Use coaches.
- Fire 40 rounds:
  - 20 rounds from the prone supported firing position.
  - 10 rounds from the unsupported firing position.
  - 10 rounds from the kneeling firing position.
- Engage targets from 50 to 250 meters.
- Meet the standards (17 hits out of 40 target exposures).

Record Qualification

8-108. The procedures for record qualification include the following:

- Conduct a dry-fire exercise.
- Fire 40 rounds:
  - 20 rounds from the prone supported firing position.
  - 10 rounds from the unsupported firing position.
  - 10 rounds from the kneeling firing position.
- Engage targets from 50 to 250 meters.
- Meet the standards (17 hits out of 40 target exposures).

Appendix A

Training Aids, Devices, and Exercises

Training aids and devices must be included in a marksmanship program. This appendix lists those available and provides information on how to obtain them for marksmanship training.

SECTION I. ENGAGEMENT SKILLS TRAINER 2000

The EST 2000 (Figure A-1) supports realistic and comprehensive gated rifle marksmanship instruction, identifies Soldiers' needs by requiring them to satisfy gate requirements in order to progress, and facilitates any necessary remedial training prior to qualification. This unit/institutional, indoor, multipurpose, multilane, small arms, crew-served, and individual antitank training simulator is used to—

- Train and evaluate individual marksmanship training for initial entry Soldiers (BCT/OSUT).
- Provide unit sustainment training for active and reserve components in preparation for qualification on individual and crew small arms live-fire weapons.
- Provide unit collective tactical training for static dismounted infantry, scout, engineer, military police squads, and sustainment elements.
- Simulate training events which lead to live-fire individual and crew weapon qualification.
- Simulate training events that contribute to increased weapon, crew, fire team, and squad combat effectiveness.
- Simulate squad collective, defensive, ambush, gunnery, and tactical tasks.
- Train leaders of fire teams and squads in the command, control, and distribution of fires.
- Save ammunition, travel time, transportation costs, and other range support resources.
- Support functional gunnery training strategies and standards in weapon training.

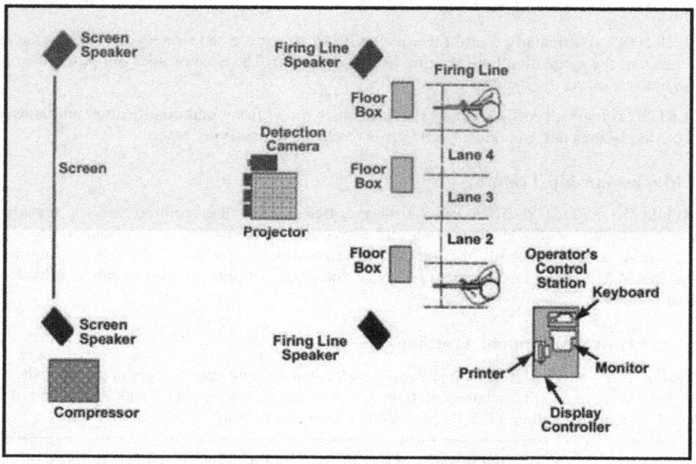

Figure A-1. Engagement skills trainer (five-lane configuration).

GENERAL CHARACTERISTICS

A-1. The EST 2000 replicates eleven weapons, including the rifle, carbine, pistol, grenade launcher, all machine guns, MK19, shotgun, and AT4. It has three modes of training:

- Marksmanship training.
- Tactical collective training.
- Shoot/don't shoot rules of engagement training.

A-2. It also includes a scenario editor for generating or tailoring new scenarios.

MARKSMANSHIP TRAINING

A-3. When used in the marksmanship training mode, the EST 2000—

- Uses Army standard courses of fire for all small arms weapons.
- Accurately simulates live-fire ranges in daylight and limited visibility conditions using precision-scaled targets, high-resolution imagery, and the essential weapon's system accuracy to compensate for errors (drift, parallax).
- Isolates, captures, and displays shots with replay that highlights firer's errors in the application of the fundamentals of marksmanship. Replay of the point of aim trace (before the shot, during the shot, and after the shot) diagnoses firer problems with aiming, breathing, steady hold, trigger control, and shot recovery for on-the-spot corrections. Cant sensors visually indicate firer-induced right or left cant, possibly resulting in missed shots.

Basic Rifle Marksmanship Training

NOTE: The marksmanship core scenarios can be found in the EST 2000 Operator's Manual, TD-07-6910-702-10.

A-4. The EST 2000 begins training the fundamentals of marksmanship right from the beginning, before the Soldier has a chance to develop bad habits. Using EST 2000 technology, Soldiers and units can reduce their rate of marksmanship failures and increase Soldiers' confidence in being able to fire their assigned weapons.

A-5. EST 2000 is particularly useful for teaching BRM, where a gated strategy requires the Soldier to pass certain gates in the simulation before firing live ammunition. The Soldier does not proceed or pass a gate scenario until he meets the standard.

A-6. EST 2000 marksmanship training provides basic range firing and qualification and is accomplished in 5-, 10-, or 15-lane configurations. Each firer is restricted to one firing lane.

Remedial Marksmanship Training

A-7. While the EST 2000 BRM gated strategy often reduces the requirements for remedial live-fire training, it is highly useful in diagnosing and correcting problems before the Soldier fires live rounds. Using the EST 2000 technology of rifle cant, trigger pressure, and before-the-shot, during-the-shot, and after-the-shot AARs, trainers can quickly identify and correct problems, thus raising confidence and first-time qualifications.

Individual and Unit Sustainment Training

A-8. Sustainment training and prequalification refresher training can be conducted with the EST 2000. During individual and unit sustainment training, grouping, zeroing, and Field Fire I and II should be conducted to standard in the EST 2000 prior to live-fire qualification.

NOTE: EST 2000 will not be used in lieu of live-fire qualification except for those outlined in DA Pam 350-38.

TACTICAL COLLECTIVE TRAINING

A-9. When used in the tactical collective training mode, the EST 2000—

- Presents tactical collective exercises that train squad, team, and element leaders in fire distribution and control.
- Provides fully articulated interactive targets with variable outcomes based on a squad's action or inaction.
- Uses realistic 3D-modeled battlefield terrain and variable environmental effects (for example, day/night and dawn/dusk, variable weather conditions, and illumination).
- Uses other special effects, such as a weapon's effects, explosions, and vehicle damage, to enhance the static eye point of the battlefield.
- Allows trainers and units to build scenarios as they would fight.
- Provides shot-by-shot feedback that is tied to each firer's lane of fire.
- Will soon include an entry-level indirect fire capability (product improvement).

NOTE: Tactical collective training core scenarios can be found in the EST 2000 Operator's Manual, TD-07-6910-702-10.

A-10. Tactical collective training is conducted on two networked five-lane subsystems. This configuration can support up to 11 weapons, including tandem weapons.

NOTE: The tandem weapons capability is available in collective training only. This capability allows the use of an extra weapon connected to the fifth lane (port 6) in the third floor box, allowing a firer to manage two weapons. For further instructions, refer to the EST 2000 Operator's Manual.

SHOOT/DON'T SHOOT RULES OF ENGAGEMENT TRAINING

A-11. When used in the shoot/don't shoot rules of engagement training mode, the EST 2000—

- Can be configured to enhance special operations and counterterrorism training.
- Is the premier training simulation for stability and support operations training.
- Uses video-based graphic overlays with multiple escalation or de-escalation points that require the firer to justify his actions based on his situational awareness.

NOTE: Shoot/don't shoot core scenarios can be found in the EST 2000 Operator's Manual, TD-07-6910-702-10.

A-12. Shoot/don't shoot rules of engagement training is conducted on a five-lane subsystem. This training uses video-based graphic overlays that provide important clues (for example, facial expressions and body language) for the firer to cue on. Multiple escalation or de-escalation points require the firer to justify his actions based on his situational awareness.

SCENARIO EDITOR

A-13. The scenarios currently available in the EST 2000 meet 90 percent of a unit's training requirements. As a unit's mission changes, weapons and TTPs are changed, or additional training requirements occur, the unit can use the scenario editor to generate or tailor new scenarios. The scenario editor can be used to enhance the individual Soldier's skills and, collectively, the squad's ability to engage and destroy an enemy threat.

NOTE: Detailed instructions on how to create and modify scenarios can be found in the EST 2000 training support package.

WEAPON SAFETY

A-14. The following general safety precautions should be adhered to:

- Fire simulated weapons only if they are pointed downrange.
- Post warning signs at all entry doors.
- Do not allow personnel to stand downrange from the firing line.

WARNING

No one should be allowed beyond the firing line.

- Instruct weapon handlers never to look directly into a barrel.
- Take the weapon off-line for testing and service at the first indication of malfunction and refer to the troubleshooting procedures.

DANGER

Each simulated weapon has the same appearance as a fully functioning weapon, with the exception of the trainer-peculiar umbilical cable. Under certain circumstances, especially in the subdued light of a training room, it is possible to mistake a live firearm for a simulated weapon. This creates the potential for personal injury or damage to property. To avoid confusion, no live or blank ammunition and no live weapons should be allowed in the training room.

Simulated weapons will not accept live or blank ammunition. Any attempt to chamber a live or blank round may damage the simulated weapon and create an unsafe situation.

LASER SAFETY

A-15. The lasers used in the simulated weapons meet ANSI Standard Z136.1-1993 Class I Standards for single laser pulse power. This classification is commonly referred to by the industry rating of "eye-safe." However, even eye-safe lasers may be dangerous under extraordinary circumstances. To ensure personnel safety, weapon handlers should not stare directly down a simulated weapon's barrel. Serious eye injury could result if a laser malfunctioned while a user was staring into the weapon's muzzle.

WARNING

Laser light is used in the operation of this equipment. Injury may result if personnel fail to observe safety precautions.

Never stare into the laser beam, look down the barrel of the simulated weapon, or directly view the laser beam with optical instruments.

Avoid direct eye exposure.

The instructor should ensure that all persons entering the training room are aware that laser radiation is present.

SECTION II. LASER MARKSMANSHIP TRAINING SYSTEM

The LMTS supports training with a Soldier's own weapon without the use of live ammunition. It is not designed to replace live-fire training or to eliminate the need for knowledgeable instructors. Major components include a battery-powered laser transmitter mounted to a mandrel inserted in the rifle barrel or affixed to the front sight with a mounting bracket, and a variety of laser-sensitive targets. The exercise is performed in the same manner as an LFX, except that the ammunition is a laser beam. The target senses shot locations, which are shown on a laptop screen. Unit commanders should expect and require the following:

- A training process that focuses on the four fundamentals of marksmanship (steady position, sight alignment and picture, breath control, and trigger squeeze).
- Opportunity for experienced marksmen to "test out" and serve as peer trainers or return to other duties.
- Real-time feedback.
- All-season training.
- Soldiers trained on their assigned weapon throughout the process.

Using LMTS technology, units can consistently reduce the rate of first-time marksmanship failures and increase new Soldiers' confidence in their ability to fire their basic weapon.

EQUIPMENT

A-16. Software enhancements continue to optimize the training process and minimize computer requirements by enabling an instructor to control up to 10 targets with only one computer. This feature reduces overall system costs and provides maximum throughput with a minimum number of instructors.

A-17. Minimum LMTS systems consist of a basic laser transmitter with a rod to fit the weapon and a laser target. Systems can be expanded to include a variety of components.

NOTE: Table A-1 provides a complete component list.

Table A-1. Laser marksmanship training strategy parts list.

PART	INCLUDES
110 system (110v or 220v)	TR-700 target, LT100C laser, transmitter rod, mask set, AC power adapter, and user's manual
330A system (110v or 220v)	TR-900 target, LT100C laser, transmitter rod, cable, software, AC power adapter, and user's manual
360 system (available in 3-, 4-, and 5-target array)	TR-900 target, LT100 laser, transmitter rod, cable, software, AC power adapter, control box, and user's manual
430 system (110v or 220v)	TR-900 target, MP400 laser, 556C rod, software, AC power adapter, and user's manual
Mini-range	TR-700 target, LT100C laser, transmitter rod, transceiver unit with RS-232 cable, software, E-tat, AC power adapter, and user's manual
Sound and recoil replicator system	M16A2 upper receiver M4 upper receiver CO_2-powered weapon simulator (M4 or M16)
Borelight kit	MP400 laser, LTA-556C transmitter rod, and carrying case
Targets	TR-700 electronic TR-900 electronic 25-m zero reflective (small) 25-m zero reflective (large) Other reflective targets available
Laser transmitters	LT-100 laser transmitter LT-500 in-barrel laser transmitter

Table A-1. Laser marksmanship training strategy parts list (continued).

PART	INCLUDES
System software	330A system 360 system Mini-range system
Instruction booklets	110 system 330A system 360 system
Laser transmitter rods	LTA-1200 Type B 12-gauge LTA-190 Type B Cal .177 LTA-220 Type B Cal .22 LTA-240 Type B Cal .25 LTA-310-2 Type A Cal .30, 2" barrel LTA-380 Type A Cal .38 and .357 LTA-380-2 Type A Cal .38 and .357, 2" barrel LTA-410 Type A Cal .40 and .41 LTA-440 Type A Cal .44 LTA-450-2 Type A Cal .45, 2" barrel LTA-500 Type A Cal .50 LTA-556C Cal 5.56-mm (special order) LTA-762C Cal 7.62-mm (special order)
Mask sets for TR-700 targets	Various E-type silhouettes; military mask set to simulate 300 m, 450 m, and 600 m for training conducted at 25 m
System components and accessories	Training vest for TR-700 target Sound and Recoil Replicator Consumable—5.56-mm, 7.62-mm, and 9-mm Notebook computer Safety rod CO2 4 oz., 7.5 oz., and 20 lbs. AC-600 100v adapter for TR-700 target AC-610 220v adapter for TR-700 target AC-910 110v adapter for TR-900 target AC-920 220v adapter for TR-900 target RC-260 remote cable BL-265 daisy chain cable BL-640 long cable BL-654 daisy chain cable RS-232 15-m cable extension CB-440 control boxTR-700 RF control (E-tag) Mini-range transceiver unit (tag-receiver)
Carrying cases	110 system, 330A system, and 360 system (per 3 targets) M16/M4 sound and recoil replicator Borelight kit

MARKSMANSHIP TRAINING

A-18. For initial skill development (for example, IET), Exercises 1 through 4 should be conducted sequentially. After grouping and zeroing standards, the Soldier moves to the LMTS alternate course C target where the course of fire replicates the live-fire course, except that the ammunition is a laser beam. Failure to meet the standards for this course of fire identifies the Soldier as a candidate for remedial training.

REMEDIAL TRAINING

A-19. Failure to achieve the standards set forth in this manual identifies the Soldier as a candidate for remedial training. Using the LMTS technology, trainers can quickly identify and correct problems, significantly raising qualification rates after subsequent attempts at qualification. After remedial training, the Soldier moves to the LMTS alternate course C target where the course of fire replicates that of the live-fire course, except that the ammunition is a laser beam.

SUSTAINMENT TRAINING

A-20. The training model in the exercises provides commanders and unit trainers with a sustainment training system that can be employed year-round, ideally as integrated concurrent training to cause the least disruption to other planned training.

A-21. Soldiers are administered a skill test at a regular frequency (current training guidance recommends quarterly). The results of this test allow commanders to focus training efforts on those Soldiers least able to demonstrate the minimum skills required.

A-22. For quarterly sustainment training, Soldiers should be pretested to determine the extent of training required. The pretest should begin with the grouping exercise (from Exercise 3) followed by the electronic alternate course C or mini-RETS (Exercise 4). Soldiers unable to meet pretest standards are given refresher training in the four fundamentals of rifle marksmanship, followed by completion of Exercises 1 through 4.

EXERCISES

A-23. LMTS exercises define procedures for using LMTS equipment to train and sustain basic marksmanship fundamentals. They may be conducted as independent stations or combined on a single station as appropriate for the training scenario. Trainers should employ LMTS equipment in a manner that accounts for:

- Space and time available at the training site.
- Unit size and composition.
- Remedial training requirements.
- Equipment availability.

NOTE: Check the LMTS operator's manual for specific information about equipment setup and operation.

A-24. Training in Exercises 1 through 3 should be conducted using the Soldier's own service rifle in the dry-fire mode. Exercise 4 may be conducted in the dry-fire mode, but an optional sound and recoil replicator should be employed for added realism.

NOTE: Sound and recoil replicators provide nearly 100 percent of the recoil felt with full rifle function. They require the Soldier to properly load magazines and enable the trainer to cause the rifle to misfeed or misfire to verify a Soldier's ability to perform immediate action procedures to reduce a stoppage.

A-25. If LMTS training immediately precedes a live-fire grouping and zeroing exercise and time permits, trainers may wish to use the system's prezeroing capability during Exercise 3 by using calibrated or "spun" lasers Using the calibrated lasers, Soldiers make adjustments to their own rifle sights during Exercise 3, resulting in a savings of time and ammunition on the grouping and zeroing range. All LMTS-based zeros must be confirmed by live-fire. If no live-firing is planned, calibrated lasers need not be used, and adjustments are made to the laser in Exercise 3.

NOTE: See the LMTS operator's manual for a description of the laser calibration process.

EXERCISE 1: REFLECTIVE TARGET EXERCISE

Table A-2. Action, conditions, and standards for a reflective target exercise.

ACTION	Demonstrate the four fundamentals of rifle marksmanship while using the LMTS reflective zeroing target.
CONDITIONS	Given an M16-/M4-series weapon, laser transmitter with mandrel, and reflective target.
STANDARDS	Demonstrate the four fundamentals of marksmanship by: • Achieving a good steady position. • Applying the proper sight alignment and sight picture. • Applying proper breath control. • Applying proper trigger squeeze.

A-26. Exercise 1 introduces Soldiers to the four fundamentals of marksmanship, outlines how to diagnose and correct firer problems, and reinforces proper application of the fundamentals.

A-27. This exercise requires a high degree of instructor involvement, but one instructor may effectively train up to 20 lanes. Decreased trainer/firer ratio will result in decreased efficiency and effectiveness. One trainer per 10 lanes is the optimum ratio.

A-28. The exercise requires little time to complete; to retain group integrity, it should be combined with Exercise 2 to allow more advanced firers to progress while problem firers receive remedial training.

A-29. A reflective zeroing target with MP400 laser/mandrel provides a simple but effective tool for remedial training during LFXs. Problem firers should be sent to a remedial station for a quick check of the application of the fundamentals and remedial training.

Exercise Performance

A-30. To perform Exercise 1, use the following procedures:

(1) The Soldier assumes a proper supported position using sandbags. The trainer inserts the MP-400/LTA-556C assembly into the rifle barrel and uses laser windage and elevation adjustments to achieve a bold sight adjustment with laser spot on front sight (Figure A-2). Then, the Soldier turns the laser to the ON position.

NOTE: Soldiers should become familiar with both supported and unsupported firing positions.

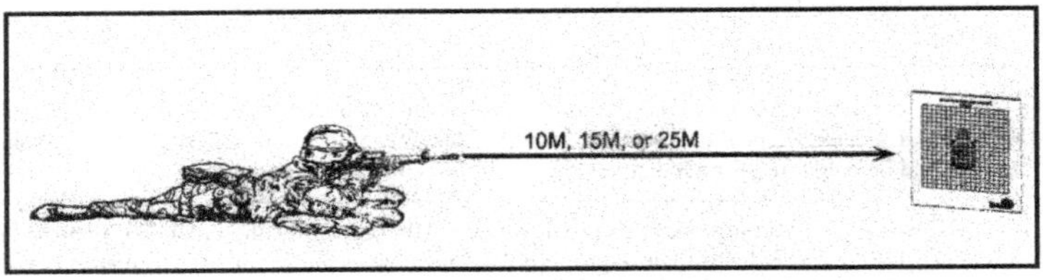

Figure A-2. Exercise 1.

(2) Under trainer supervision, the Soldier establishes a proper sight alignment and sight picture on a reflective zeroing target set at 10 meters, 15 meters, or 25 meters. With the MP-400 laser turned to ON, the trainer or coach blocks the beam with his finger.

NOTE: Use the appropriate target with the corresponding distance.

---

(3) When the Soldier is confident with the sight alignment and picture, the trainer removes his finger and observes the location of the red laser dot on the target.

(4) If the laser dot is in the 4-centimeter circle, proceed to Step 5. If the laser dot is outside of the 4-centimeter circle, the trainer instructs the Soldier regarding correct aiming techniques to bring the dot inside the circle, and repeats Step 2.

NOTE: If the trainer is reasonably certain that the laser and sights are aligned, the visible laser dot may be used to help the firer understand correct sight picture and alignment. The firer should be instructed to bring the laser dot to the target silhouette's center of mass, and then observe the relationship of the front and rear sights to the target.

(5) With the MP-400 in constant ON mode, the trainer or coach uses the red dot trace to confirm steady hold and proper breathing and trigger control.

NOTE: This trace can also be used to show the effects of improper steady position breath control and trigger control, and reinforce proper techniques.

(6) The trainer or coach turns the MP-400 to the training (TRN) mode and instructs the Soldier to fire six shots into the target's center of mass. The trainer or coach observes the laser hits to confirm proper application of the four fundamentals of rifle marksmanship.

NOTE: Failure to achieve this standard provides an early indication of the need for more intense instruction in the fundamentals of marksmanship or remedial training.

Exercise 2: Interactive Dry-Fire Exercise

Table A-3. Action, conditions, and standards for an interactive dry-fire exercise.

ACTION	Demonstrate the integrated act of firing while using the LMTS 130-target system.
CONDITIONS	Given an M16-/M4-series weapon, laser transmitter with mandrel, and TR-700 targets with military masks.
STANDARDS	• Achieve 8 hits out of 10 shots two times on an open-faced target from the prone unsupported position. • Achieve 8 hits out of 10 shots two times on a 300-meter masked target from the supported position.

A-31. This exercise provides Soldiers with an opportunity for practicing the four fundamentals of rifle marksmanship in the integrated act of firing and may easily be conducted concurrently with Exercise 1 on the same station. TR-700 targets may be used both indoors and outdoors in various environments and arrangements to meet the training requirement.

Exercise Performance

A-32. To perform Exercise 2, use the following procedures:

(1) The Soldier assumes a proper firing position using sandbags for a supported position (Figure A-3).

NOTE: Sleeping mats should be used on hard floors.

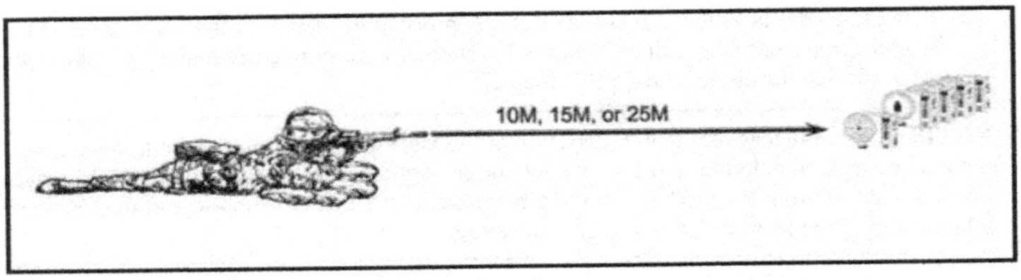

Figure A-3. Exercise 2.

(2) The Soldier applies the four fundamentals of marksmanship to engage a TR-700 open-faced target with 10 shots from the prone unsupported position. The Soldier cocks the rifle after each shot, forcing a break and reestablishing a proper stock weld to build muscle memory.

NOTE: The TR-700 open-faced target at 25 meters equals a doublewide E-silhouette target at 300 meters.

---

(3) The trainer inspects the target score for the number of hits. If the number is less than eight, the trainer should perform a visual laser/sight alignment check. If the laser/sight alignment is correct, the trainer reconfirms the Soldier's understanding of the four fundamentals of marksmanship and directs the Soldier to repeat Step 2. If the number of hits is less than eight after several tries, the Soldier reports for remedial training. If the number of hits is eight or more, the Soldier repeats Step 2 to confirm, and then proceeds to Step 4.

(4) After the Soldier completes Step 3, a 300-meter scaled E-silhouette mask is installed over the face of a TR-700 target. The Soldier repeats Step 2 from the supported position and repeats Step 4 to confirm.

(5) OPTIONAL: As time allows, the Soldier increases the number of shots to 20 and/or installs smaller masks for an additional skill challenge. Additional firing positions may also be reinforced, if needed.

NOTES:
1. The largest mask presents a 300-meter E-target size scaled for 25 meters.
2. The middle mask presents a 300-meter E-target size scaled for 15 meters or a 450-meter E-target size scaled for 25 meters.
3. The smallest mask presents a 300-meter E-target size scaled for 10 meters, a 450-meter E-target size scaled for 15 meters, or a 600-meter E-target size scaled for 25 meters.

EXERCISE 3: GROUPING AND ZEROING EXERCISE

Table A-4. Action, conditions, and standards for a grouping and zeroing exercise.

ACTION	Group and zero an M16-/M4-series weapon using the TR-900 target system with military mask.
CONDITIONS	Given an M16-/M4-series weapon, laser transmitter with mandrel, and TR-900 target system with military mask.
STANDARDS	From the supported firing position— • Grouping. Fire up to 27 shots (dry-fire) in three-round shot groups and achieve two consecutive shot groups within a 4-centimeter circle (25 meters), 2.4-centimeter circle (15 meters), or 1.6-centimeter circle (10 meters). • Battlesight Zero. Adjust the sights so that five out of six rounds in two consecutive shot groups strike within the zeroing circle in the silhouette on the zeroing target.

A-33. This exercise evaluates a Soldier's ability to apply the four fundamentals of rifle marksmanship in the integrated act of firing through shot grouping. The exercise is conducted in the same manner as live-fire grouping and zeroing exercises and can make those exercises more efficient and effective. All normal range commands should be used to reinforce training in proper range procedures. Up to 10 targets may be grouped together for scoring on one computer. This exercise is most efficient with one trainer to run the control and scoring console, plus one trainer for every five lanes. Training distance must correspond to the distance used in Exercise 4.

NOTE: Whenever this exercise is conducted prior to an LFX, calibrated lasers should be used to support prezeroing. Adjustments to the rear sight of an M16A2/A3/A4 or M4-series weapon, associated optics, and the front sight of an M16A1 must be made when training at 10 meters or 15 meters to compensate for parallax error. See the LMTS operator's manual for a detailed description of these adjustments.

Exercise Performance

A-34. To perform Exercise 3, use the following procedures:

(1) From the supported firing position (Figure A-4), the Soldier fires three-round shot groups at the target overlay's center of mass, continuing until two consecutive groups fall within a 4-centimeter circle anywhere on the target (maximum 27 shots). Trainers should provide feedback to the Soldier between each shot group. If the Soldier is unable to achieve the standard within 27 shots, the trainer attempts remedial actions or sends the Soldier to the remedial training station.

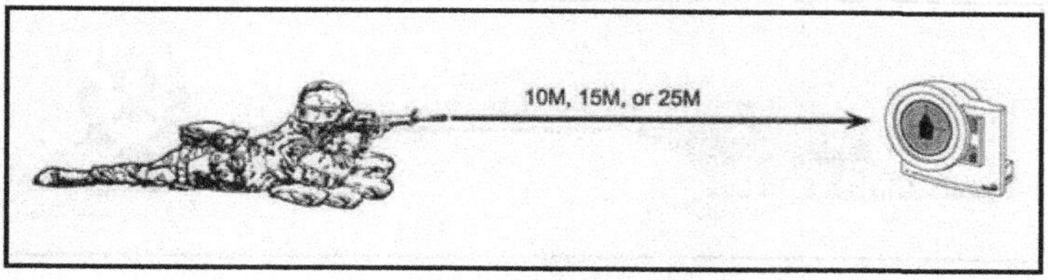

Figure A-4. Exercise 3.

(2) When the grouping standard is met, the Soldier makes appropriate sight changes as instructed by the trainer, who begins a new session for the zeroing process. The Soldier continues to fire three-round shot groups, adjusting the sights as instructed by the trainer to bring the shot groups (maximum 18 shots) within the zeroing circle on the target silhouette. When a shot group falls within the zeroing circle, the Soldier fires an additional shot group for confirmation. Five of six shots must fall within the zeroing circle.

NOTE: Failure to achieve the standard identifies the Soldier as a candidate for remedial training. The Soldier should not progress to Exercise 4 until the standard is met.

Exercise 4: Laser Marksmanship Training Strategy Prequalification Exercise

Table A-5. Action, conditions, and standards for a laser marksmanship training strategy prequalification exercise.

ACTION	Engage 10-, 15-, or 25-meter alternate course C scaled silhouettes with an M16-/M4-series weapon.
CONDITIONS	Given an M16-/M4-series weapon, laser transmitter with mandrel, and electronic alternate C target system with ten 10-, 15-, or 25-meter scaled silhouettes. Engage each silhouette with two shots from the supported position and two shots from the prone unsupported position.
STANDARDS	Without assistance, the Soldier engages 10 target silhouettes using the M16-/M4-series weapon with laser transmitter and achieves a minimum of 30 hits out of 40 shots.

A-35. Exercise 4 is used as a skill test to determine the need for training or the results of training, and serves as an accurate predictor of live-fire alternate course C performance.

NOTE: Soldiers failing to meet the standards of this exercise should receive remedial training prior to live-fire qualification.

A-36. The exercise may be conducted in the dry-fire mode, with Soldiers recocking the rifle between shots. When performing the dry-fire method, Soldiers should use a magazine with the follower and spring removed. Another option for the dry-fire mode involves removing the charging handle and attaching a piece of cord (looped on both ends with the free end about 3 inches from the end of the stock) to the bolt. The coach can recock the rifle between shots by pulling the cord directly to the rear.

A-37. The sound and recoil replicator options include the M16A2 Blazer and the alternate laser-mounting bracket used with a standard blank firing attachment (BFA). Both options offer added realism by providing full rifle function, sound, and recoil. When using the special safe Blazer blanks, the Blazer option may be employed indoors without hearing protection. Standard M200 blanks may only be used with the BFA while outdoors and while wearing hearing protection. When available, the LMTS mini-RETS range should be employed for added training realism and to prepare Soldiers for firing on pop-up targets.

Exercise Performance

A-38. To perform Exercise 4, use the following procedures:

(1) The Soldier assumes the proper supported firing position using sandbags (Figure A-5).

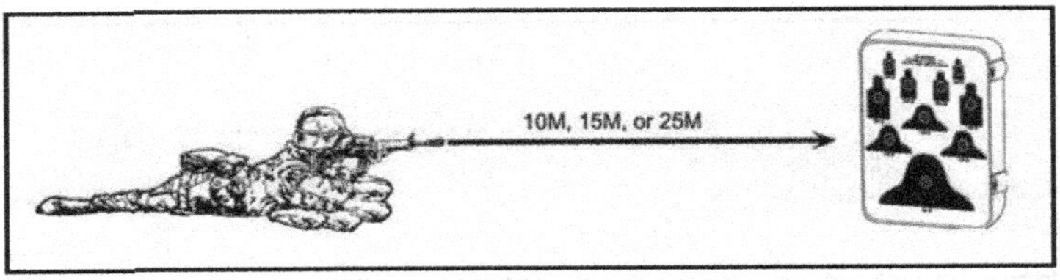

Figure A-5. Exercise 4.

(2) The trainer prepares the electronic 25-meter alternate C target and computer control station for the prequalification skill test. The course may be conducted at 25 meters, 15 meters, or 10 meters using the appropriate target overlay.

NOTE: The trainer should use appropriate range commands and enforce all range safety procedures.

---

(3) To fire the alternate course C, the Soldier applies the four fundamentals of rifle marksmanship using the service rifle (laser-zeroed) during Exercise 3.

(4) The computer automatically times the test and can print a score sheet.

SOUND AND RECOIL REPLICATOR

A-39. To add realism to the training, a special upper receiver sound and recoil replicator provides full live-fire functionality (without the projectile). It supplies nearly 100 percent of the recoil with 50 percent of the noise using a specially-designed nontoxic theatrical blank.

SECTION III. OTHER TRAINING DEVICES AND EXERCISES

This section provides the classification and nomenclature for training aids, devices, and targets.

TRAINING SUPPORT CENTERS

A-40. Training support centers (TSCs) are located throughout the world to provide training aids and devices. Each TSC provides training aid services to customers (for example, active Army units and schools, reserve components, and ROTC units) in their geographic area of support.

NOTE: For more information concerning TSC operations, write:

Commander
United States Army Training Support Center
ATTN: ATIC-DM
Fort Eustis, VA 23604

SELECTION OF TRAINING AIDS AND DEVICES

A-41. After training requirements have been established, appropriate training aids and devices can be selected from the TSC. Table A-6 lists many of those aids and devices available and their identification numbers.

Table A-6. Training aids and devices.

TYPE/NOMENCLATURE	IDENTIFICATION NUMBER
Rifle Marksmanship Trainer (Weaponeer)	DVC 07-57/DVC 07/57B
M15A2 Blank Firing Attachment	Supply Item (see TM)
Chamber Block (M16A1/A2)	TAD-0001
M16 Sighting Device (A1 0r A2) (Left and Right)	DVC-T 7-84
Target Box Paddle	DVC-T 7-86
Riddle Device	DVC-T 7-87
M16 Rifle Brass Deflector	DVC-T 23-30
M15A1 Aiming Card	DVC-T 07-26
M16A1 Display Mat (canvas)	TAD-0034 (locally)
Rifle Rest (for target box exercise)	TAD-12 (locally)
Front and Rear Sight, M16 Rifle	TAD-26 (locally)
Front and Rear Sight, M16A2 Rifle	TAD-0026A
Graphic Training Aids (GTA)	
M16A1 Disassembly Mat (paper)	GTA 09-06-43
Rifle, 5.56-mm, M16A1 Mechanical	GTA 7-1-26
Rifle, M16 Disassembly (M16A1)	GTA 9-6-43
M16A1 Rifle Malfunction	GTA 9-6-44
M16A1 Rifle Maintenance Card	GTA 21-1-3
Training Films	
*Rifle, M16A1 Part I, Care, Cleaning, Lubrication	TF 21-3907
*Rifle, M16A1 Part II Field Expedients	TF 21-3908
*Also available in videotape.	
Videotapes	
Engagement of Moving Personnel Targets with the M16A1 Rifle Team from the Foxhole Position	2E/010-071-1271-B
Cycle of Functioning M16A1 Rifle	2E/010-071-0444-B
Overview of BRM Training	2E/010-071-0086-B
TVT 7-13 (Feb 87)	2E/010-071-0725-B
TVT 7-1 Teaching Rifle Marksmanship: Part I	
TVT 7-2 Teaching Rifle Marksmanship: Part II	

TARGET ORDERING NUMBERS

A-42. Table A-7 lists the description and NSN to use when ordering marksmanship targets.

Table A-7. Target ordering numbers.

DESIGNATION	DESCRIPTION	NSN
D Prone	Full-length face with V through two scoring areas	6920-00-922-7450
D Prone	Repair center with V through two scoring areas	6920-00-922-7451
E-Silhouette	Full-length face, solid color paper	6920-00-600-6874
E-Silhouette	Full-length, pop-up, solid color plastic	6920-00-071-4780
E-Silhouette	Full-length face, cardboard, kneeling	6920-00-079-1806
F-Silhouette	Short-length face, solid color paper	6920-00-610-9086
F-Silhouette	Short-length, pop-up, solid color plastic	6920-00-071-4589
F-Silhouette	Short-length face, pasteboard	6920-00-795-1807
25-Meter Alternate Course Scaled Qualification Target	50- to 300-meter scaled silhouette target	6920-01-167-1398
15-Meter Battlesight Zeroing Target (.22 Caliber RFA)	250-meter scaled silhouette target (50-foot indoor range)	6920-01-167-1393
15-Meter Alternate Course C (.22 Caliber RFA)	50- to 300-meter scaled silhouette target (50-foot indoor range)	6920-01-167-1396
25-Meter M16A1 Zeroing Target	250-meter scaled silhouette target	6920-01-167-1392
25-Meter M16A2 Zeroing Target	300-meter scaled silhouette target	6920-01-253-4005
25-Meter M16A1 Slow Fire Target	75- to 300-meter scaled silhouette target	6920-01-167-1391
25-Meter M16A1 Timed Fire Target	50- to 300-meter scaled silhouette target	6920-01-167-1397
75-Meter M16A1 Feedback Target	75-meter scaled F-type silhouette	6920-01-169-6921
75-Meter M16A2 Feedback Target	75-meter scaled F-type silhouette	6920-01-253-4006
175-Meter M16A1 Feedback Target	175-meter scaled E-type silhouette	6920-01-167-1395
175-Meter M16A2 Feedback Target	175-meter scaled E-type silhouette	6920-01-167-1395
Pasters, Black		6920-00-165-6354
Pasters, Buff		6920-00-172-3572
Landscape target		6920-00-713-8253
Spindle, Target Spotter, Wood		6920-00-713-8257
Spotters, 1 1/2 inches in diameter		6920-00-789-0869
Spotters, 3 inches in diameter		6920-00-713-8255
Spotters, 5 inches in diameter		6920-00-713-8254
Thermal Blankets		6920-01-516-9912

TRAINING DEVICES

A-43. Several marksmanship training devices are available to aid in PMI and sustainment training. They are beneficial when ammunition is limited for training or during practice exercises, such as field firing on the EST 2000 or zeroing and qualifying with short-range training ammunition (SRTA). Some training devices are complex, costly, and in limited supply, while others are relatively simple, cheap, and in large supply. Individuals or squads can sustain and practice basic marksmanship skills and fundamentals using devices and aids alone or in combinations.

AIMING CARD

A-44. The M15A1 aiming card (Figure A-6) determines if the Soldier understands how to aim at a target's center of mass, how to adjust the point of aim, how to allow for gravity, and how to engage a moving target.

Figure A-6. M15A1 aiming card.

A-45. To use the aiming card—

(1) The card is misaligned, the Soldier is instructed to establish the correct point of aim, and a trainer checks it. Each Soldier demonstrates six out of six of the points of aim.

NOTE: The sight/target relationship on the card is the same visual perception that the Soldier should have when he is zeroing on a standard silhouette target.

---

(2) The Soldier shows the side alignment technique three times—place the front sightpost on the left or right edge of the target and bring the front sightpost to the target's center of mass.

(3) The Soldier shows the bottom-up alignment technique—place the front sightpost at the bottom of the target and bring the front sightpost to target's center of mass.

Riddle Sighting Device

A-46. The Riddle sighting device (Figure A-7) indicates if the Soldier understands the aiming process while using the rifle. It is a small plastic plate with a magnet and a drawing of an E-type silhouette target. A two-man team is required for its use.

A-47. To use the Riddle sighting device—

(1) The Soldier assumes a supported or prone firing position. The assistant places the Riddle device on the front sight assembly and adjusts the plastic plate in the direction of the firer until he reports the proper sight picture.

(2) Without disturbing the plastic plate, the trainer or coach aims through the sights to determine if the Soldier has aligned the target and sight properly.

A-48. Many sightings are conducted, and the trainer may include variations to ensure that the Soldier understands the process. Each Soldier demonstrates six out of six points of aim, starting with the plastic plate offset to the front sightpost.

NOTE: This device has a small metal clip that slips over the front sight assembly to allow a smooth surface for attachment of the magnet. The device may also be used without the metal clip.

This device is provided with a small metal clip that slips over the front sight assembly. It all a smoother surface for attachment of the magnet; however, the device can be used without the metal clip.

Figure A-7. Riddle sighting device.

M16 SIGHTING DEVICE

A-49. The M16 sighting device (Figure A-8) is made of metal, with a tinted square of glass placed at an angle. When the device is attached to the rear of the M16A1 carrying handle, an observer can look through the sight to see what the firer sees.

NOTE: The M16 sighting device can be mounted on the M16A2 rifle, but the charging handle must be pulled to the rear first. Then, the M16 sighting device is mounted on the rear of the carrying handle, and the charging handle is returned forward.

A-50. When using the M16 sighting device, observe the following recommendations:

- The M16 sighting device can be used in a dry-fire or live-fire environment, but a brass cartridge deflector must be used during live-fire.
- The observer must practice with the sight to be effective. For example, if the observer looks at a reflected image and the Soldier is aiming to the right, it appears left to the observer.
- The device must be precisely positioned on the rifle (it may need to be bent to stay on).
- The observer's position must remain constant.
- The observer must talk with the firer to ensure a correct analysis of the aiming procedures.

A-51. The Soldier must achieve six out of six proper sight alignment drills.

NOTE: The M16 sighting device is made for left- and right-handed firers.

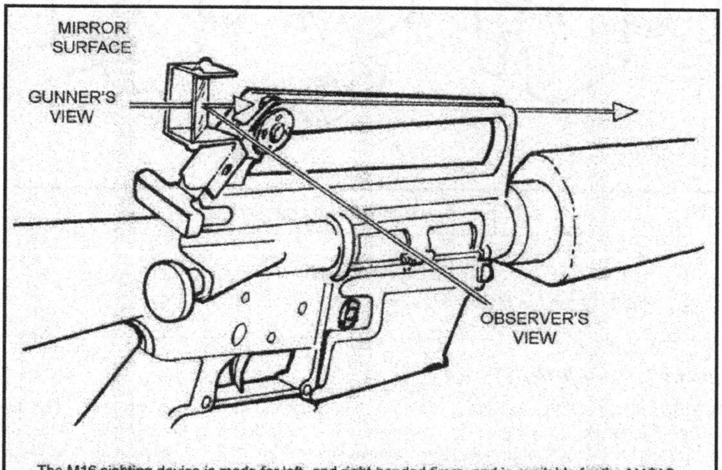

The M16 sighting device is made for left- and right-handed firers, and is available for the M16A2.

Figure A-8. M16 sighting device.

M15A2/M23 BLANK FIRING ATTACHMENT

A-52. The M15A2/M23 BFA (Figure A-9) attaches to the muzzle of M16-/M4-series weapons. It is designed to keep sufficient gas in the weapon's barrel to allow for semiautomatic, automatic, or burst firing with blank ammunition (M200 only).

A-53. When using the M15A2/M23 BFA, observe the following recommendations:

- After firing 50 rounds, check the attachment for a tight fit.
- Continuous blank firing results in a carbon buildup in the bore, gas tube, and carrier key. If this occurs, follow the cleaning procedures outlined in TM 9-1005-249-10.

NOTE: The M15A2 is painted red and is used on the M16-series weapons. The M23 is painted yellow and is used on the M4-series weapons. For identification, the M23 is stamped "M4 Carbine Only."

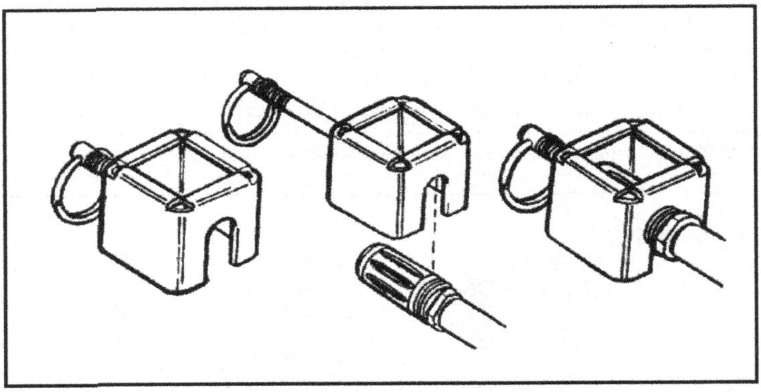

Figure A-9. Blank firing attachment.

CAUTION

Hand tighten the M15A2/M23 only.

LOCATION OF MISS-AND-HIT SYSTEM

A-54. The LOMAH system is a range aid used during downrange feedback exercises. The device uses acoustical triangulation to compute the exact location of a supersonic bullet as it passes through a target. The bullet impact is displayed instantly on a video monitor at the firing line. More importantly, it shows the location of a bullet miss, allowing the firer to make either a sight adjustment or a hold-off for subsequent shots.

NOTE: LOMAH, like other devices, is only an aid. When the Soldier uses LOMAH, he must understand the weapon and firing techniques and have a coach/instructor.

A-55. LOMAH ranges have been fielded in USAREUR and Korea. In locations where KD ranges are not available and restrictions prohibit walking downrange, the LOMAH system is a practical alternative to essential downrange feedback.

NOTE: Requests for LOMAH devices should be sent to:

Commander
US Army Training Support Center
ATIC-DM
Fort Eustis, VA 23604.

M261 .22-Caliber Rimfire Adapter

A-56. The .22-caliber RFA can contribute to a unit's marksmanship program when 5.56-millimeter ammunition is not available or when ranges that allow firing 5.56-millimeter ammunition are not available. The RFA can be useful for marksmanship training such as night fire, quick fire, and assault fire, but is not recommended for primary marksmanship training.

Training Considerations

A-57. When service ammunition is in short supply, the RFA can be used to complement a unit's training program.

Rifle Performance

A-58. The RFA cannot be depended on to fire in the same place as 5.56-millimeter ammunition. The RFA and its .22-caliber ammunition cannot replicate the exact ballistics of the 5.56-millimeter ammunition, but efforts to match RFAs with specific rifles can result in reasonable replication.

NOTE: Under ideal training conditions, the RFA should be used with dedicated rifles. It is not necessary for the Soldier to use his own weapon during RFA training.

A-59. Finding the right match of RFA and rifle can eliminate some variability. A trial-and-error technique can match RFAs to rifles, which results in weapons that fire well.

Rifle Zero

A-60. RFA ammunition differs from 5.56-millimeter ammunition in the following ways:

- RFA ammunition will not usually group in the same location as 5.56-millimeter ammunition at 25 meters and cannot be used for weapon zero.
- RFA ammunition normally fires a slightly larger shot group than 5.56-millimeter ammunition.

A-61. When a Soldier uses an RFA in his rifle, he must be careful not to lose his 5.56-millimeter zero. This can be accomplished by using hold-off while firing .22-caliber ammunition or keeping a record of sight changes so the sights can be moved back.

NOTES:
1. The .22-caliber round approximates the 5.56-millimeter trajectory out to 25 meters.
2. The correct zeroing target or appropriate scaled-silhouette targets can be used for practice firing exercises at 15 meters (50 feet) or 25 meters.

Advantages and Disadvantages

A-62. If the RFA is used as a training aid, the advantages and disadvantages must be considered during training.

Advantages

A-63. The .22-caliber ammunition is cheaper and may be available in larger quantities than 5.56-millimeter ammunition. It can be fired on all approved indoor ranges and in other close-in ranges where 5.56-millimeter ammunition is prohibited. RFA training can be used to sustain marksmanship skills during periods when full-caliber 5.56-millimeter ammunition training cannot be conducted.

Disadvantages

A-64. Some negative training aspects exist because of differences in the weapon's functioning when using the RFA. These differences include—

- The forward assist does not work.
- The bolt does not lock to the rear after the last round is fired.
- More malfunctions can occur with the RFA than with 5.56-millimeter ammunition.
- Immediate action procedures are different.

SHORT-RANGE TRAINING AMMUNITION

A-65. SRTA is a plastic practice cartridge (M862) that enables a unit to conduct realistic firing training at shorter distances with reduced danger areas. The M862 has a maximum range of 250 meters. The blue plastic projectile reduces the risk of overpenetration and ricochet, which makes it ideal for UO training.

A-66. To fire the M862 SRTA from an M16-/M4-series weapon, the standard bolt and bolt carrier must be replaced by the M2 practice bolt. The M2 practice bolt consists of a bolt carrier, which is a fixed bolt. The practice bolt changes the weapon from a gas-operated action to a blow-back action that permits cyclic fire with the lower-powered M862.

NOTES: 1. Because of the design of the M2 practice bolt, standard 5.56-millimeter rounds cannot be fired from the weapon while it is installed.

2. See TM 9-6920-746-12&P for more information on the M862 SRTA and the M2 practice bolt.

WEAPONEER

A-67. The Weaponeer is an effective rifle marksmanship training device that simulates the live-firing of the M16-series rifle. The system can be used for developing and sustaining marksmanship skills, diagnosing and correcting problems, and assessing basic skills.

NOTE: The EST 2000 has replaced the Weaponeer, but some units still use the Weaponeer.

Operation

A-68. Figure A-10 shows the Weaponeer in the standing supported firing position. With the exception of smoke and cartridge ejection, the rifle operates normally and has the same weight and balance as the standard weapon. An IR aiming sensor simulates round trajectory and hit point to an accuracy of better than one MOA. The recoil rod that attaches at the rifle's muzzle end simulates recoil.

NOTE: Recoil is provided in both semiautomatic and automatic modes of fire and is adjustable from no-net force to 30 percent more than that of a live M16.

A-69. Sound is provided through headphones and is adjustable from 115 to 135 decibels. Special magazines are used. One magazine simulates a continuous load; the other is used to train rapid magazine change and can be loaded with 1 to 30 simulated rounds. Selectable misfire can be used to detect gun-shyness and drill immediate action. The front and rear sights are zeroed the same as standard rifles.

Figure A-10. Weaponeer set up in the standing supported position.

A-70. The Weaponeer range can be raised or lowered to accommodate all firing positions. The target assembly contains four targets: a scaled 25-meter zeroing target and three pop-up targets. E-type and F-type silhouettes at ranges from 75 meters can be used on the Weaponeer. Known distance and various other types of targets displayed in fixed or random sequences can be used. Target exposure times may be set to unlimited or from 1 to 30 seconds. The fall-when-hit mode can be selected with the KILL button.

A-71. The operator's console contains the system control buttons, graphics printer, and video feedback monitor. The back of the console has counters that total rounds and hours, and a storage bin for storing magazines, printer paper and ribbon, headphones, two wrenches for assembling the Weaponeer, and a small hex-head wrench for aligning the rifle sensor. A remote control, which attaches to the back of the console, enables a trainer or firer to operate select functions away from the console.

Feedback

A-72. The Weaponeer provides feedback to help trainers to teach and Soldiers to learn marksmanship skills. The tools used for feedback include—

- Fall-when-hit mode.
- Real-time aiming point display.
- Immediate shot impact display.
- Replay.
- Shot groups.
- Printer.

Fall-When-Hit Mode

A-73. Lighting the KILL button enables the fall-when-hit mode. When the button is activated, targets fall when hit. This feedback provides the same hit-or-miss information as a train-fire (RETS) range.

Real-Time Aiming Point Display

A-74. When a firer aims on or near a target, his aiming point relative to the target is continuously displayed on the video screen. The aiming point display allows the trainer to teach and verify aiming techniques, and to continuously monitor the firer's steadiness, techniques, time on target, trigger squeeze, and recovery from recoil.

Immediate Shot Impact Display

A-75. When a shot is fired, its impact relative to the target is immediately displayed on the video screen as a blinking white dot (Figure A-11, left target).

Replay

A-76. After a shot is fired, a real-rate display of how the firer engaged the target can be replayed on the video screen.

A-77. The target to the right in Figure A-11 shows the type of information that can be replayed on the video screen after a series of shots are fired. To show the sequence, the dots have been numbered.

A-78. To show a replay, the firer—

(1) Selects the shot he wishes to replay by operating the EACH SHOT button.

(2) Presses the REPLAY button.

NOTE: Some Weaponeers record and store replays for only the first three shots.

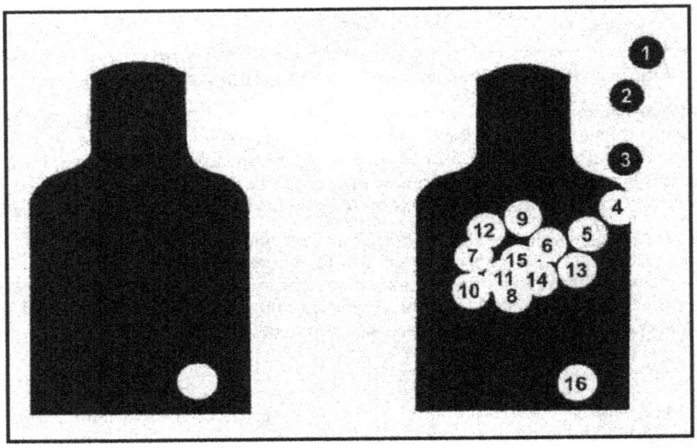

Figure A-11. Replay of shot.

Shot Groups

A-79. The impact location of up to 32 shots is automatically stored in the Weaponeer's memory and displayed on the video screen. Each impact is indicated by a white dot, and the last shot blinks for indication. All 32 shots can be fired and displayed on a single target or split among a combination of targets. The CLEAR button erases all shots from the Weaponeer's memory.

Printer

A-80. A hard copy printer is provided for postfiring analysis, for firer progress tracking, and for record keeping. Pressing the PRINT button causes the target displayed on the video to print (Figure A-12). Some Weaponeers can print the three pop-up targets at the same time by holding in the REPLAY button and pressing the PRINT button.

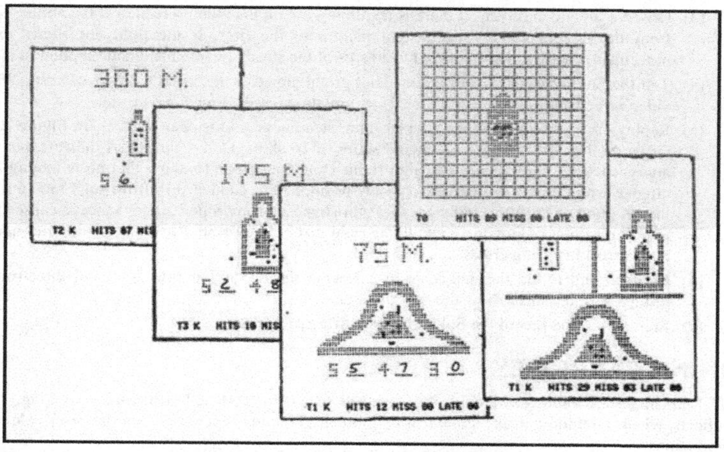

Figure A-12. Weaponeer printouts.

Use of the Weaponeer

A-81. The Weaponeer is used to evaluate the firer's ability to apply the four fundamentals. It is used throughout the program to help diagnose and remediate problems. In the unit, the Weaponeer should be used much like it is used in BRM.

A-82. When training Soldiers on the Weaponeer—

- Proceed at a relaxed pace, and emphasize accuracy before speed.
- If possible, train with small groups, allowing each Soldier several 10- to 15-minute turns on the device.
- For remedial training, try to relax the Soldier. A nervous Soldier will have trouble learning and gaining confidence in his marksmanship skills. For sustainment training, encourage competition between individuals or units.

Diagnosis of Firing Problems

A-83. The Weaponeer assists in the diagnosis of firing problems. The seven-step program is recommended as a guide. These seven steps are designed to diagnose and show the Soldier his firing errors. This could be enough to correct the error.

NOTES: 1. Diagnosis needs to be followed up with remedial exercises with the Weaponeer, target box exercise, or dime/washer exercise.

2. Depending on the extent of the firer's problems and time constraints, the number of shots may be increased.

A-84. To diagnose firing problems—

(1) Tell the Soldier to assume a good firing position, aim at a target, and hold steady (supported and prone unsupported positions).

(2) Visually check the firer's position and correct any gross errors.

(3) Observe the video screen. If there is no aiming dot on the video screen or if the aiming dot is far from the target's center, teach sight picture to the firer. If the light dot shows excessive movement, check and correct the techniques of the steady position and natural point of aim.

(4) Tell the Soldier to fire a three-round shot group aimed at the target's center of mass. Watch the video screen and Soldier as he fires. Note violations of the four fundamentals.

(5) Replay each shot to show the firer his aim, steadiness, and trigger squeeze. In Figure A-11, the target on the right shows a numbered series of 16 shots. Dots 1 through 4 indicate that the firer approached the target from the high right. Dots 5 through 15 show that he is aiming near the target's center, but does not have a steady position. The sudden shift from dot 15 to 16 (dot 16 is the hit point of the shot) indicates that gun-shyness or improper trigger squeeze caused the firer to pull his point of aim down and to the right just before firing. Replay helps the firer understand and correct his firing errors.

(6) Confirm and refine the diagnosis by allowing the Soldier to fire additional three-round shot groups. Use replay to show the firer his faults.

(7) Summarize and record the Soldier's basic firing problems.

TRAINING EXERCISES

A-85. Marksmanship training exercises are designed to aid in PMI and sustainment training. They are beneficial when ammunition is limited for training. Training exercises can be used alone or in combinations.

TARGET BOX EXERCISE

A-86. The target box exercise checks the consistency of aiming and placement of three-round shot groups in a dry-fire environment.

A-87. To conduct the exercise (Figure A-13)—

(1) The target man places the silhouette anywhere on the plain sheet of paper and moves the silhouette target as directed by the firer.

NOTE: The two positions (separated by 15 yards or 25 meters) must have already been established so the rifle is pointed at some place on the paper.

---

(2) When the firer establishes proper aiming, he signals the target man to "Mark."

NOTE: Only hand signals are used since voice commands would be impractical when training several pairs of Soldiers at one time.

Figure A-13. Target box exercise.

(3) The target man places the pencil through the hole in the silhouette target and makes a dot on the paper. Then, he moves the silhouette to another spot on the paper and indicates to the firer that he is ready for another shot.

(4) When the three shots are completed, the target man triangulates the three shots and labels it "Shot Group 1." The firer and instructor view the shot group.

(5) The Soldier fires several shot groups. After two or three shot groups are completed in one location, the rifle, paper holder, or paper is moved so shots fall on a clean section of the paper.

NOTE: Any movement of the rifle or paper between the first and third shots of a group voids the exercise. Two devices are available to hold the rifle. Rifle-holding devices (Figures A-14 and A-15) are positioned on level ground, or are secured by sandbags or stakes to ensure the rifle does not move during the firing of the three shots. Movement of the paper is eased by using a solid backing (Figure A-16). Any movement of either is reflected in the size of the shot group. Several varieties of wooden target boxes have been locally fabricated.

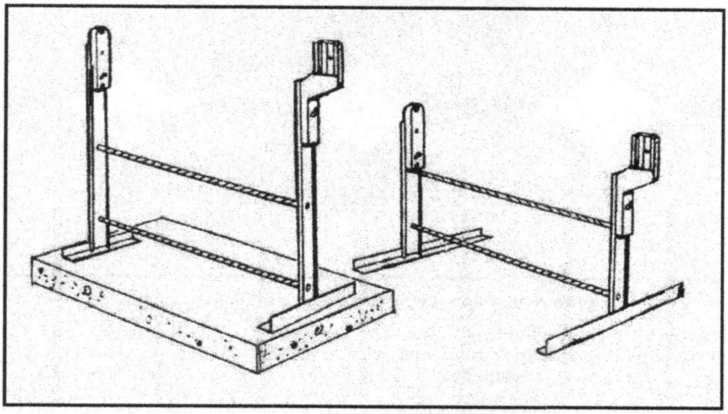

Figure A-14. Rifle-holding device (TA-G-12A).

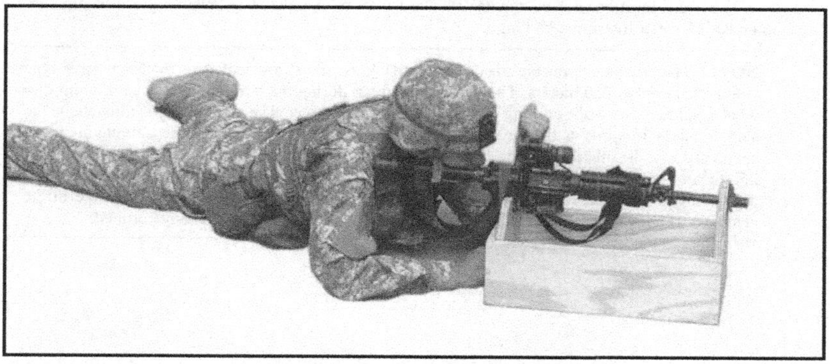

Figure A-15. Staked rifle holding box.

Figure A-16. Paper being placed on a stationary object.

A-88. Each Soldier dry-fires the exercise until he has demonstrated six out of six of the points of aim within the plastic target box paddle's 4-centimeter template. The exercise should be repeated as many times as necessary to achieve two consecutive shot groups that will fit into the same 2-centimeter circle.

A-89. A simulated shot group covered with a 1-centimeter (diameter) circle indicates consistent aiming. Since no rifle or ammunition variability is involved and since there is no requirement to place the shot group in a certain location, a 1-centimeter standard may be compared to obtaining a 4-centimeter shot group on the 25-meter live-fire zero range.

NOTE: The silhouettes on the plastic paddle (Figure A-17) are scaled to represent an E-type silhouette target at 250 meters. The visual perception during the target box exercise is similar to what a Soldier sees while zeroing on a standard zeroing target. The small E-type silhouette is the same scale at 15 yards as the larger silhouette is at the 25-meter range (some training areas are set up at 15 yards; others are set up at 25 meters). While there are some benefits to representing a 250-meter target, the main benefit of this exercise can be obtained at any distance. In place of the paddle, a standard zeroing target can be used at 25 meters by placing a small hole in the center (dot), moving the target sheet over the paper, and marking as previously outlined.

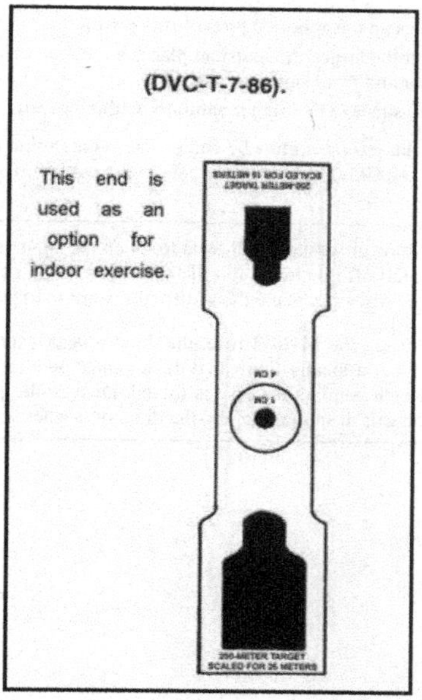

Figure A-17. Target box paddle (DVC-T-7-86).

A-90. The shot group exercise provides a chance for the trainer to critique the Soldier on his aiming procedures, aiming consistency, and placement of shot groups. Assuming the rifle and paper remain stationary and the target man properly marks the three shots, the only factor to cause separation of the dots on the paper is error in the Soldier's aiming procedure. When the Soldier can consistently direct the target into alignment with the sights on this exercise, he should be able to aim at the same point on the zero range or on targets at actual range.

BALL AND DUMMY EXERCISE

A-91. This exercise is conducted on a live-fire range and is used to detect if the firer is anticipating the shot or using improper trigger squeeze.

A-92. To perform this exercise—

(1) The coach or designated assistant inserts a dummy round into a magazine of live rounds. The firer must not know when a dummy round is in the magazine.

(2) When the hammer falls on a dummy round (which the firer thought was live), the firer and his coach may see movement. The firer anticipating the shot or using improper trigger squeeze causes this. Proper trigger squeeze results in no movement when the hammer falls.

A-93. The Soldier demonstrates the ability to properly utilize the fundamentals of marksmanship six consecutive times.

DIME/WASHER EXERCISE

A-94. This dry-fire technique is used to teach or evaluate the skill of trigger squeeze and is effective when conducted from an unsupported position. To perform this exercise—

(1) The Soldier cocks the weapon.

(2) The Soldier assumes an unsupported prone firing position.

(3) The Soldier aims at the target. An assistant places a dime or washer on the rifle's barrel between the flash suppressor and front sightpost assembly.

(4) The Soldier tries to squeeze the trigger naturally without causing the dime or washer to fall off.

A-95. Several repetitions of this exercise must be conducted to determine if the Soldier has problems with trigger squeeze. The Soldier is a GO if he can dry-fire six of six consecutive shots without causing the dime or washer to fall.

NOTES: 1. If the dime or washer is allowed to touch the sight assembly or flash suppressor, it may fall off due to the jolt of the hammer. The strength of the hammer spring on some rifles can make this a difficult exercise to perform.

2. When using the M16A2 rifle, the dime/washer exercise is conducted the same, except that a locally fabricated device must be attached to the weapon. A piece of 3/4-inch bonding material is folded into a clothespin shape and inserted in the weapon's flash suppressor and the dime or washer is placed on top of it.

Appendix B

Scorecards

During live-fire events, a Soldier's hit-and-miss performance is recorded to facilitate the instructor/trainer's critiques or to indicate where more training is needed. The following are examples of completed scorecards.

EXAMPLES OF COMPLETED SCORECARDS

B-1. Figures B-1 through B-8 show examples of completed scorecards.

REPRODUCIBLE FORMS

B-2. Blank copies of the following can be found at the end of this publication:

- DA Form 3595-R (Record Fire Scorecard).
- DA Form 3601-R (Single Target—Field Fire I Scorecard).
- DA Form 5239-R (100-, 200-, and 300-Meter Downrange Feedback Scorecard).
- DA Form 5241-R (Single and Multiple Targets—Field Fire II Scorecard).
- DA Form 5789-R (Record Fire Scorecard—Known-Distance Course).
- DA Form 5790-R (Record Fire Scorecard—Scaled Target Alternate Course).
- DA Form 7489-R (Record Night Fire Scorecard).
- DA Form 7649-R (Squad Designated Marksman—Record Fire I and II).
- DA Form 7650-R (Squad Designated Marksman—Position Evaluation).

NOTE: These forms, scorecards, and position evaluation sheets are not available through the normal supply channels. You may reproduce them locally on 8 1/2- x 11-inch paper or download them from the Army Publishing Directorate at http://www.apd.army.mil/.

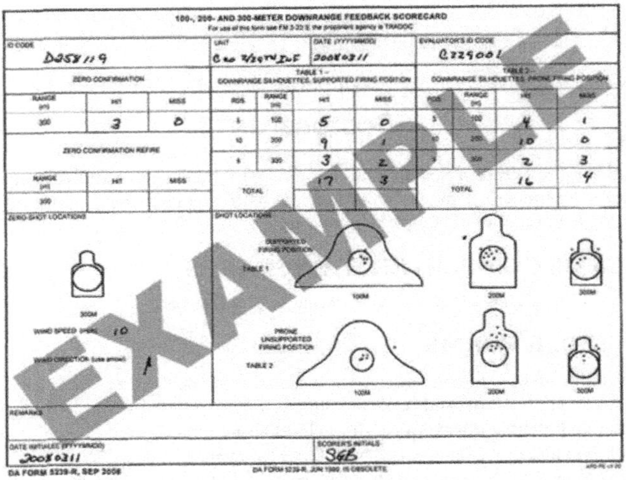

Figure B-1. Example of completed DA Form 5239-R (100-, 200-, and 300-Meter Downrange Feedback Scorecard).

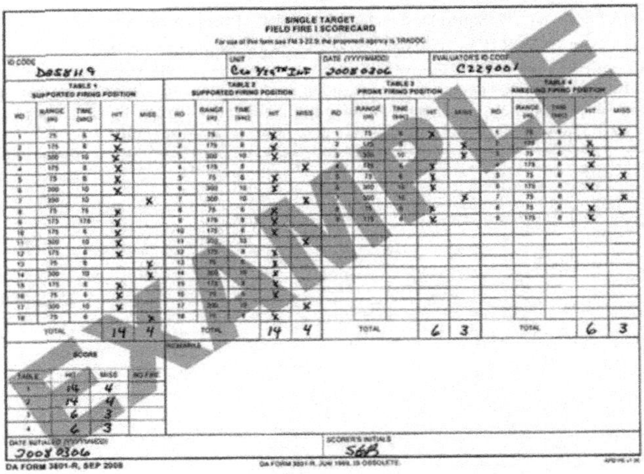

Figure B-2. Example of completed DA Form 3601-R (Single Target—Field Fire I Scorecard).

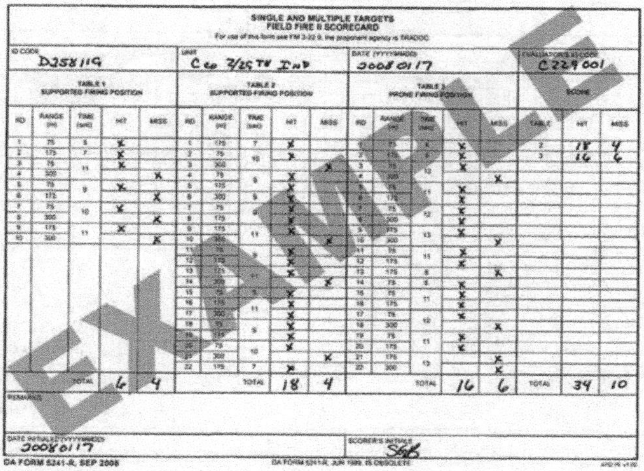

SINGLE AND MULTIPLE TARGETS
FIELD FIRE II SCORECARD
For use of this form see FM 3-22.9; the proponent agency is TRADOC

ID CODE	UNIT	DATE (YYYYMMDD)	EVALUATOR'S ID CODE
D258119	C Co 3/25TH INF	20080117	C229001

	TABLE 1 SUPPORTED FIRING POSITION	TABLE 2 SUPPORTED FIRING POSITION	TABLE 3 PRONE FIRING POSITION
TOTAL HIT	6	18	16
TOTAL MISS	4	4	6

SCORE TABLE	HIT	MISS
2	18	4
3	16	6
TOTAL	34	10

REMARKS

DATE INITIALED (YYYYMMDD): 20080117 — SCORER'S INITIALS: SGB

DA FORM 5241-R, SEP 2008 — DA FORM 5241-R, JUN 1989, IS OBSOLETE

Figure B-3. Example of completed DA Form 5241-R
(Single and Multiple Targets—Field Fire II Scorecard).

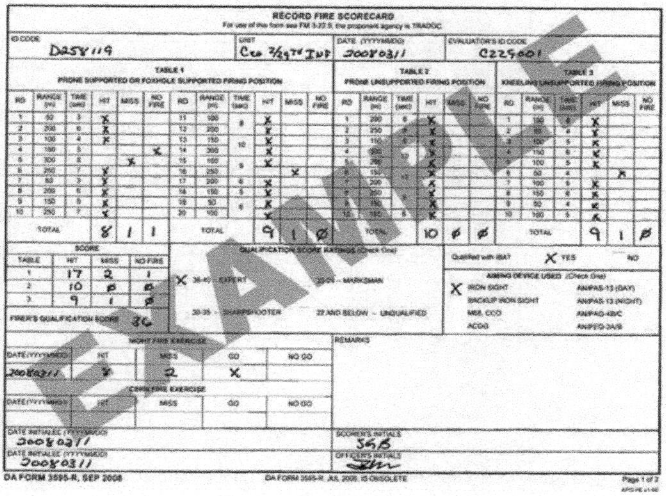

RECORD FIRE SCORECARD
For use of this form see FM 3-22.9; the proponent agency is TRADOC

ID CODE	UNIT	DATE (YYYYMMDD)	EVALUATOR'S ID CODE
D258119	C Co 2/29TH INF	20080311	C229001

	TABLE 1 PRONE SUPPORTED OR FOXHOLE SUPPORTED FIRING POSITION (RD 1–10)	TABLE 1 (RD 11–20)	TABLE 2 PRONE UNSUPPORTED FIRING POSITION	TABLE 3 KNEELING UNSUPPORTED FIRING POSITION
TOTAL HIT	8	9	10	9
TOTAL MISS	1	1	ø	1
TOTAL NO FIRE	1	ø	ø	ø

SCORE TABLE	HIT	MISS	NO FIRE
1	17	2	1
2	10	ø	ø
3	9	1	ø

FIRER'S QUALIFICATION SCORE: 36

QUALIFICATION SCORE RATINGS (Check One)
X 36-40 – EXPERT; 23-29 – MARKSMAN; 30-35 – SHARPSHOOTER; 22 AND BELOW – UNQUALIFIED

Qualified with IBA? X YES / NO

AIMING DEVICE USED (Check One)
X IRON SIGHT; BACKUP IRON SIGHT; M68, CCO; ACOG; AN/PAS-13 (DAY); AN/PAS-13 (NIGHT); AN/PAQ-4B/C; AN/PEQ-2A/B

NIGHT FIRE EXERCISE

DATE (YYYYMMDD)	HIT	MISS	GO	NO GO
20080311	[illegible]	2	X	

CBRN FIRE EXERCISE

DATE (YYYYMMDD)	HIT	MISS	GO	NO GO

REMARKS

DATE INITIALED (YYYYMMDD): 20080311 — SCORER'S INITIALS: SGB
DATE INITIALED (YYYYMMDD): 20080311 — OFFICER'S INITIALS: [illegible]

DA FORM 3595-R, SEP 2008 — DA FORM 3595-R, JUL 2005, IS OBSOLETE — Page 1 of 2

Figure B-4. Example of completed DA Form 3595-R
(Record Fire Scorecard).

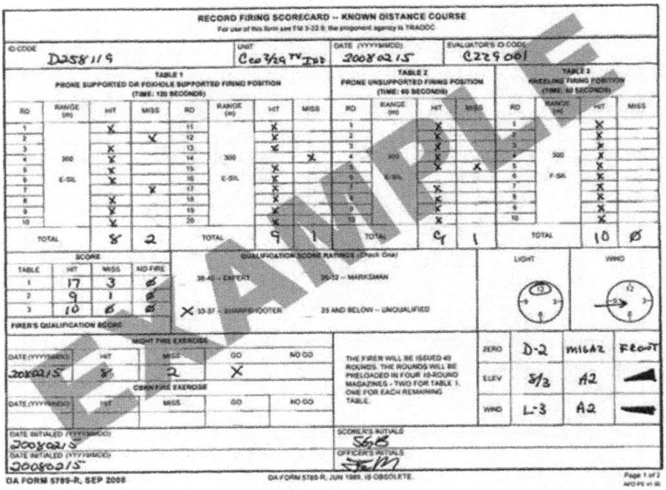

RECORD FIRING SCORECARD -- KNOWN DISTANCE COURSE

For use of this form see FM 3-22.9; the proponent agency is TRADOC

ID CODE	UNIT	DATE (YYYYMMDD)	EVALUATOR'S ID CODE
D258119	Cco 3/29 TH INF	20080215	C229001

TABLE 1 PRONE SUPPORTED OR FOXHOLE SUPPORTED FIRING POSITION (TIME: 120 SECONDS) — TOTAL HIT 8, MISS 2; TOTAL HIT 9, MISS 1

TABLE 2 PRONE UNSUPPORTED FIRING POSITION (TIME: 60 SECONDS) — TOTAL HIT 9, MISS 1

TABLE 3 KNEELING FIRING POSITION (TIME: 60 SECONDS) — TOTAL HIT 10, MISS Ø

TABLE	HIT	MISS	NO FIRE
1	17	3	Ø
2	9	1	Ø
3	10	Ø	Ø

QUALIFICATION SCORE RATINGS (Check One): 36-40 – EXPERT; 26-32 – MARKSMAN; X 33-37 – SHARPSHOOTER; 25 AND BELOW – UNQUALIFIED

FIRER'S QUALIFICATION SCORE

NIGHT FIRE EXERCISE

DATE (YYYYMMDD)	HIT	MISS	GO	NO GO
20080215	8	2	X	

CBRN FIRE EXERCISE

DATE (YYYYMMDD)	HIT	MISS	GO	NO GO

THE FIRER WILL BE ISSUED 40 ROUNDS. THE ROUNDS WILL BE PRELOADED IN FOUR 10-ROUND MAGAZINES - TWO FOR TABLE 1, ONE FOR EACH REMAINING TABLE.

ZERO	D-2	M16A2	FRONT
ELEV	5/3	A2	
WIND	L-3	A2	

DATE INITIALED (YYYYMMDD) 20080215 — SCORER'S INITIALS SGB

DATE INITIALED (YYYYMMDD) 20080215 — OFFICER'S INITIALS

DA FORM 5789-R, SEP 2008 — DA FORM 5789-R, JUN 1989, IS OBSOLETE. — Page 1 of 2

Figure B-5. Example of DA Form 5789-R
(Record Fire Scorecard—Known Distance Course).

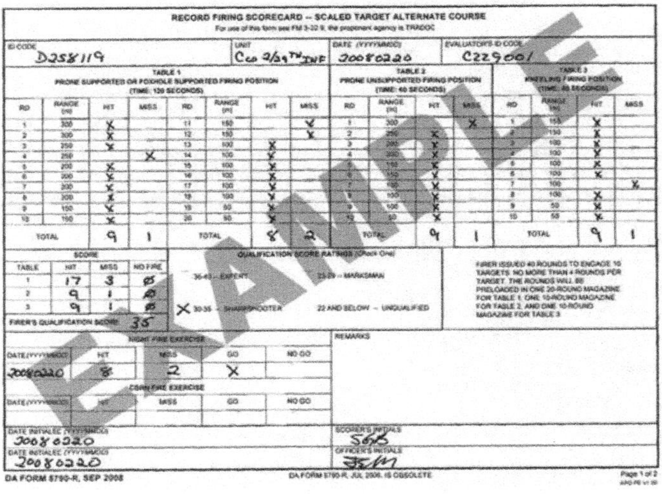

RECORD FIRING SCORECARD -- SCALED TARGET ALTERNATE COURSE

For use of this form see FM 3-22.9; the proponent agency is TRADOC

ID CODE	UNIT	DATE (YYYYMMDD)	EVALUATOR'S ID CODE
D258119	Cco 2/29 TH INF	20080220	C229001

TABLE 1 PRONE SUPPORTED OR FOXHOLE SUPPORTED FIRING POSITION (TIME: 120 SECONDS) — TOTAL HIT 9, MISS 1; TOTAL HIT 8, MISS 2

TABLE 2 PRONE UNSUPPORTED FIRING POSITION (TIME: 60 SECONDS) — TOTAL HIT 9, MISS 1

TABLE 3 KNEELING FIRING POSITION (TIME: 60 SECONDS) — TOTAL HIT 9, MISS 1

TABLE	HIT	MISS	NO FIRE
1	17	3	Ø
2	9	1	Ø
3	9	1	Ø

FIRER'S QUALIFICATION SCORE 35

QUALIFICATION SCORE RATINGS (Check One): 36-40 – EXPERT; 23-29 – MARKSMAN; X 30-35 – SHARPSHOOTER; 22 AND BELOW – UNQUALIFIED

FIRER ISSUED 40 ROUNDS TO ENGAGE 10 TARGETS. NO MORE THAN 4 ROUNDS PER TARGET. THE ROUNDS WILL BE PRELOADED IN ONE 20-ROUND MAGAZINE FOR TABLE 1, ONE 10-ROUND MAGAZINE FOR TABLE 2, AND ONE 10-ROUND MAGAZINE FOR TABLE 3

NIGHT FIRE EXERCISE

DATE (YYYYMMDD)	HIT	MISS	GO	NO GO
20080220	8	2	X	

CBRN FIRE EXERCISE

DATE (YYYYMMDD)	HIT	MISS	GO	NO GO

REMARKS

DATE INITIALED (YYYYMMDD) 20080220 — SCORER'S INITIALS SGB

DATE INITIALED (YYYYMMDD) 20080220 — OFFICER'S INITIALS

DA FORM 5790-R, SEP 2008 — DA FORM 5790-R, JUL 2006, IS OBSOLETE — Page 1 of 2

Figure B-6. Example of completed DA Form 5790-R
(Record Fire Scorecard—Scaled Target Alternate Course).

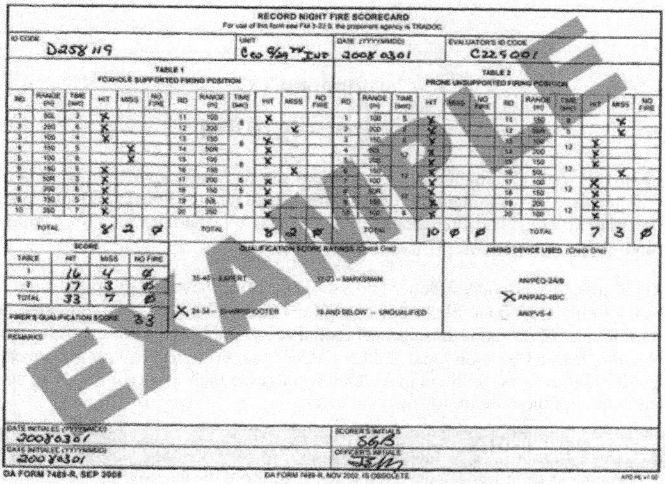

RECORD NIGHT FIRE SCORECARD

For use of this form see FM 3-22.9; the proponent agency is TRADOC

ID CODE	UNIT	DATE (YYYYMMDD)	EVALUATOR'S ID CODE
D258119	C Co 3/29TH INF	20080301	C225001

TABLE 1 — FOXHOLE SUPPORTED FIRING POSITION

RD	RANGE (m)	TIME (sec)	HIT	MISS	NO FIRE
1	50L	3	X		
2	200	6	X		
3	100	4	X		
4	150	5		X	
5	100	6		X	
6	150	5	X		
7	50R	3	X		
8	200	6	X		
9	150	5	X		
10	250	7	X		
TOTAL			8	2	Ø

RD	RANGE (m)	TIME (sec)	HIT	MISS	NO FIRE
11	100	8	X		
12	200			X	
13	150	8	X		
14	50R		X		
15	100	8	X		
16	150			X	
17	200	6	X		
18	150	5	X		
19	50L	8	X		
20	250		X		
TOTAL			8	2	Ø

TABLE 2 — PRONE UNSUPPORTED FIRING POSITION

RD	RANGE (m)	TIME (sec)	HIT	MISS	NO FIRE
1	100	5	X		
2	200	[illegible]	X		
3	150	[illegible]	X		
4	50L	[illegible]	X		
5	200	[illegible]	X		
6	150	[illegible]	X		
7	100	[illegible]	X		
8	50R	[illegible]	X		
9	150	[illegible]	X		
10	100	5	X		
TOTAL			10	Ø	Ø

RD	RANGE (m)	TIME (sec)	HIT	MISS	NO FIRE
11	150	[illegible]		X	
12	50R	5		X	
13	100	12	X		
14	200		X		
15	150	12	X		
16	50L			X	
17	100	12	X		
18	150		X		
19	200	12	X		
20	100		X		
TOTAL			7	3	Ø

SCORE

TABLE	HIT	MISS	NO FIRE
1	16	4	Ø
2	17	3	Ø
TOTAL	33	7	Ø

FIRER'S QUALIFICATION SCORE 33

QUALIFICATION SCORE RATINGS (Check One)

35-40 – EXPERT
X 24-34 – SHARPSHOOTER
17-23 – MARKSMAN
16 AND BELOW – UNQUALIFIED

AIMING DEVICE USED (Check One)

AN/PEQ-2A/B
X AN/PAQ-4B/C
AN/PVS-4

REMARKS

DATE INITIALED (YYYYMMDD)	SCORER'S INITIALS
20080301	SGB
20080301	OFFICER'S INITIALS: JBM

DA FORM 7489-R, SEP 2008 — DA FORM 7489-R, NOV 2002, IS OBSOLETE.

Figure B-7. Example of completed DA Form 7489-R (Record Night Fire Scorecard).

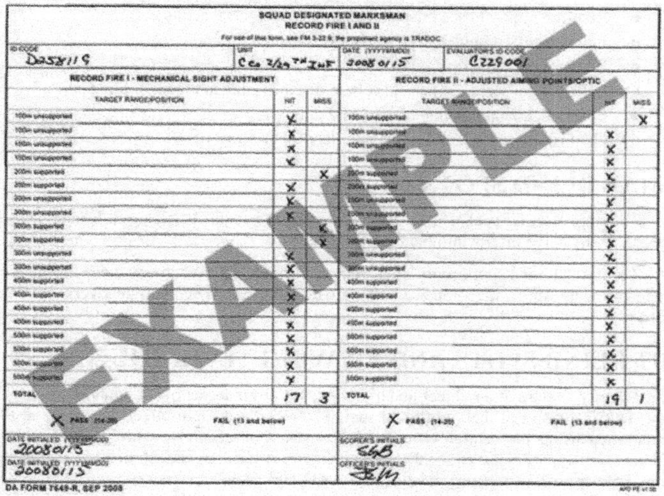

SQUAD DESIGNATED MARKSMAN
RECORD FIRE I AND II

For use of this form, see FM 3-22.9; the proponent agency is TRADOC

ID CODE	UNIT	DATE (YYYYMMDD)	EVALUATOR'S ID CODE
D258119	C Co 3/29TH INF	20080115	C229001

RECORD FIRE I - MECHANICAL SIGHT ADJUSTMENT

TARGET RANGE/POSITION	HIT	MISS
100m unsupported	X	
100m unsupported	X	
100m unsupported	X	
100m unsupported	X	
200m supported		X
200m supported	X	
200m unsupported	X	
200m unsupported	X	
300m supported		X
300m supported		X
300m unsupported	X	
300m unsupported	X	
400m supported	X	
400m supported	X	
400m supported	X	
400m supported	X	
500m supported	X	
500m supported	X	
500m supported	X	
500m supported	X	
TOTAL	17	3

X PASS (14-20) FAIL (13 and below)

RECORD FIRE II - ADJUSTED AIMING POINTS/OPTIC

TARGET RANGE/POSITION	HIT	MISS
100m unsupported		X
100m unsupported	X	
100m unsupported	X	
100m unsupported	X	
200m supported	X	
200m supported	X	
200m unsupported	X	
200m unsupported	X	
300m supported	X	
300m supported	X	
300m unsupported	X	
300m unsupported	X	
400m supported	X	
400m supported	X	
400m supported	X	
400m supported	X	
500m supported	X	
500m supported	X	
500m supported	X	
500m supported	X	
TOTAL	19	1

X PASS (14-20) FAIL (13 and below)

DATE INITIALED (YYYYMMDD)	SCORER'S INITIALS
20080115	SGB
20080115	OFFICER'S INITIALS: JBM

DA FORM 7649-R, SEP 2008

Figure B-8. Example of completed DA Form 7649-R (Squad Designated Marksman—Record Fire I and II).

Appendix C
Night Fighting

With the Army's emphasis on night operations, commanders must be sure that riflemen are lethal at night as well as during the day. That lethality depends largely on whether riflemen can fire effectively with today's technology: NVGs, aiming lights, and TWSs.

This appendix provides a better understanding of how eyes adapt to the night, as well as information about night devices. The information has been provided through continuous observation during operational testing, the Army Research Institute, and lessons learned by individual Soldiers across the Army. Be it with the naked eye or with night devices, Soldiers must learn to increase their ability not only to see better at night, but their ability to "own the night."

SECTION I. UNAIDED NIGHT VISION

Although operating at night has definite advantages, it is also difficult. Eyes do not work as well as they do during the day, yet they are crucial to effective performance; 80 percent of sensory input comes through them. Soldiers need to be aware of the constraints that their eyes place upon them at night. These constraints include—

- Reduced ability to see crisp and clear images (visual acuity).
- Inability to distinguish colors under certain conditions.
- Reduced depth perception.
- Difficulty in seeing objects at certain distances due to the night blind spot.
- Lost dark adaptation due to exposure to light.
- Confusion (your eyes may seem to play tricks on you).

NORMAL BLIND SPOTS

C-1. The normal blind spot is always present, day and night. It is caused by the lack of light receptors where the optic nerve inserts into the back of the eye.

C-2. The normal blind spot occurs when you use just one eye. When you close the other eye, objects 12 to 15 degrees away from where you are looking will disappear. When you uncover your eye, the objects will reappear.

NIGHT BLIND SPOTS AND VIEWING TECHNIQUES

C-3. When you stare at an object under starlight or lower levels of illumination, it can disappear or fade away. This is a result of the night blind spot. The night blind spot affects both eyes at the same time and occurs when using the central vision of both eyes. Consequently, when looking directly at an object, Soldiers miss larger objects as the distances increase. A hand grenade 2 meters away might not be seen; an enemy Soldier at 50 meters may be missed. An M1 tank at 300 meters can even be missed.

Avoiding Night Blind Spots

C-4. To avoid night blind spots—

- Look to all sides of objects you are trying to find or follow.
- **DO NOT STARE.**

C-5. Diamond viewing, a technique similar to the off-center vision technique taught in rifle marksmanship, is a good technique for peripheral viewing. To use the diamond viewing technique—

- Move your eyes just slightly, a few degrees, in a diamond pattern around the object you wish to see.
- Do not move your head; use your peripheral vision.

DARK ADAPTATION

C-6. In order for your visual system to work efficiently at night, you need to dark-adapt, or get your eyes used to seeing things under low light conditions. Dark adaptation is similar to walking into a movie theater when it's very dark; you can't see things at first, but as your eyes gradually adapt, you can see better.

C-7. It takes about 30 to 45 minutes to fully dark-adapt when going from a brightly lighted area into the dark. However, people dark-adapt at varying rates. People who are older, people who smoke, or people who are not in great physical shape will take longer to dark-adapt.

C-8. There are three stages of dark adaptation:

- Daylight vision.
- Twilight vision.
- Night vision.

Daylight Vision

C-9. Daylight vision occurs under maximum lighting conditions, such as when the sun is shining or in a well-lit room. Under these conditions, Soldiers have—

- Optimal visual acuity. In daylight conditions, Soldiers use both their central and peripheral vision, which provides optimal visual acuity—20/10, 20/15, and 20/20 vision.
- Optimal color vision. Colors look most vivid under daylight conditions.
- Quickest reaction time.

Twilight Vision

C-10. Twilight vision occurs during many military night operations and when driving around in a car at night. It occurs at dawn and dusk, down to full moonlight; when there is artificial illumination; and when snow is on the ground at night. It can occur in the daytime with several layers of jungle canopy. Under these conditions, Soldiers have—

- Poorer visual acuity. Visual acuity can be as poor as 20/100. Under twilight conditions, optimal visual acuity is between 20/50 and 20/100.
- Poorer color vision. Colors will not be as vivid.
- Slower reaction times.

NIGHT VISION

C-11. Night vision occurs under starlight, as well as on moonless and cloudy nights when there are no stars or cultural lighting. Remember, there is a night blind spot, as discussed earlier. Under these conditions, Soldiers have—

- The worst visual acuity—from 20/200 to 20/400 and possibly much worse. Soldiers can recognize silhouettes, but not details of the objects. This is why knowing the silhouettes of vehicles and critical natural and man-made objects is important.
- Poor color vision. Soldiers cannot see colors, only various shades of gray. The longer wavelengths of light, such as the reds and oranges, are hard to see and appear dark. Unless a dark color is bordered by two lighter colors, it becomes totally invisible. On the other hand, greens and blues appear brighter, but Soldiers may not be able to determine their color.

NOTE: Reds are almost invisible at night. Red crosses are on white backgrounds on tents or vehicles so they can be seen more easily at night.

PROTECTING (BEFORE OPERATION)

C-12. Soldiers must protect their eyes before night operations so they can dark-adapt in an efficient manner. To efficiently dark-adapt, use the following suggestions:

- Don't smoke before night operations. Not smoking four to six hours before night operations will aid in dark adaptation.
- Wear sunglasses if you plan to spend time in the sun. Without sunglasses, it will take longer to dark-adapt.
- Watch what you eat. Maintain adequate levels of Vitamin A.
- Use dim white lighting or red lighting before night operations.

PROTECTING (DURING OPERATION)

C-13. Once a Soldier has dark-adapted, it is important to maintain that dark adaptation. To maintain dark adaptation, use the following suggestions:

- Minimize your use of unnecessary lighting.
- Close one eye before being flashed by flares and other bright lights.

ILLUSION (APPARENT MOVEMENT OF LIGHT)

C-14. The illusion of movement, which a static light exhibits when stared at in the dark, is related to the loss of surrounding visual references that normally serve to stabilize visual perceptions. Consequently, very small eye movements are perceived by the brain as movement of the light. To avoid illusions of movement—

- Begin a scan pattern, and control the eye movement.
- Use large movements, and scan.
- Try to find another light, and shift your gaze back and forth between the lights.

SECTION II. AIDED NIGHT VISION

There are three devices available to dismounted Soldiers that will help increase their lethality at night: night observation devices (NODs), aiming lasers, and TWSs. Each provides the dismounted Soldier with different views of the IR spectrum. Before Soldiers can fully operate these devices, they must receive training on how the systems work within the IR range and the electromagnetic (light) spectrum. Soldiers must also know what constraints and advantages each piece of equipment provides so that they can determine when to employ each device.

ELECTROMAGNETIC (LIGHT) SPECTRUM

C-15. To understand why some night devices cannot be used in conjunction with others, Soldiers must understand the electromagnetic spectrum. The electromagnetic spectrum is a spectrum of energy (light) containing x-rays, gamma rays, radio waves, cosmic rays, and ultraviolet rays. Also within this spectrum is visible light, light visible with the naked eye. Just beyond red visible light is IR light. IR light is broken down into three different ranges: near IR, middle IR, and far IR.

C-16. Two types of night devices will increase the Soldier's vision into the IR range:

- Image-intensification devices.
- Thermal sight.

IMAGE-INTENSIFICATION DEVICES

C-17. Image-intensification devices rely on ambient light and energy within the near IR range, such as light emitted from natural and artificial sources (for example, moonlight or starlight). Image-intensification devices include the AN/PVS-4, PVS-5, PVS-7A/B/C/D, and PVS-14s. The Army also has aiming devices that emit near IR energy in a colliminated beam, such as the AN/PAQ-4B/C and the AN/PEQ-2A/B. Since image-intensification devices and aiming lasers work within the same range of near IR energy, they can be used in conjunction with each other.

THERMAL WEAPON SIGHT

C-18. In the past, thermal technology has been solely reserved for tanks, fighting vehicles, and antiarmor specialists (for example, TOW and Dragon gunners). These devices were bulky, heavy, and impractical for the dismounted Soldier. Now, the Army has a thermal device that can be mounted on a dismounted Soldier's weapon or handheld. The TWS operates within the middle/far IR range. It can detect IR light emitted from friction, combustion, or objects radiating natural thermal energy. Since the TWS and other thermal devices operate within the middle/far IR range, they cannot be used in conjunction with image intensifiers or other image-intensification devices.

IMAGE-INTENSIFICATION DEVICES

C-19. As the name implies, image-intensification devices are designed to amplify light. To be effective, some degree of light must be available. When light enters the image-intensification tube, the light releases electrons, which the tube accelerates until the light is much brighter. Under optimal conditions, second-generation devices, such as the PVS-5-series, intensify ambient light up to about 1,500 times. Third-generation devices, such as the PVS-7/14-series NODs, double that level of intensification.

ADJUSTMENTS

C-20. To acquire and engage a target at night, image-intensification devices must be adjusted properly.

NOTE: Soldiers will not be able to obtain the acuity level that they have during the day. Under optimal night conditions, a Soldier with 20/20 daytime vision can expect no better than 20/50 vision with second-generation NODs and 20/40 vision with third-generation NODs. To approach these levels of acuity, you must be able to adjust your NOD for optimal clarity.

C-21. To adjust the NOD—

(1) Mount the NOD. Mount the head mount or helmet mount IAW the appropriate TM. If using the helmet mount, ensure that the tilt is adjusted until you have a comfortable viewing angle. Use the nape strap to maintain proper acuity with the NODs.

NOTE: If the mounting bracket is permanently attached to the helmet, ensure that the nape strap rear bracket is also permanently attached. (See TM 11-5855-306-10 for more information about attaching the nape strap rear bracket.) Use of the nape strap will prevent the NOD's weight from pulling the helmet downward, causing the NOD to rest on the bridge of your nose. The nape strap allows for proper acuity of the sight and allows you to engage targets with more ease and accuracy.

(2) Set the eye-relief. Move the goggles so that the eyecups cover the eye, but not so close that the eyepiece touches your eyelashes or glasses.

(3) Turn the goggles on.

(4) Set the inner-pupillary distance (AN/PVS-7 series). Move each eyepiece until they are centered over each eye. Close one eye, and make adjustments until the eye that is open is viewing a complete circle and not an oval. Continue to make adjustments to the other eye.

(5) Adjust the diopter focus ring. Close one eye, and with the eye that is open, turn the diopter focus ring in one direction until the diopter is totally out of focus. Then, turn the diopter focus ring the opposite direction until the display is focused to your eye. Follow the same procedures for the other eye if using the AN/PVS-7 series. No further adjustments should have to be made to the diopter focus ring.

NOTE: Adjust the diopter focus ring before adjusting the objective focus ring. The diopter focus ring focuses the display lens to your eye, while the objective focus ring focuses the target. You cannot focus the sight to the target without your eye being focused to the display first.

(6) Adjust the objective focus ring. While looking at an object, turn the objective focus ring until the objective lens is out of focus and then slowly turn the objective focus ring in the opposite direction until the object becomes as clear as possible.

NOTE: Adjustments will have to be made for targets at different ranges using the objective focus ring.

(7) Adjust the variable gain control (AN/PVS-14 only). Turn the variable gain control to the point that both eyes are almost receiving the same amount of light.

NOTE: The AN/PVS-14 has a variable gain control that controls the illumination input to the eye. Keeping the variable gain turned up will cause your brain to form two separate images, one darker and one very bright. With the variable gain turned down to the point that both eyes are almost receiving the same amount of light, the brain will produce one image, making it seem like both eyes are looking through the same sight.

AIMING LASERS

C-22. AN/PAQ-4-series and the AN/PEQ-2A/B aiming lasers operate within the electromagnetic spectrum, specifically near the IR range, and are seen through image-intensification devices. Aiming lasers emit a highly colliminated beam of IR energy that allows for quick "point and shoot" capability at night. Even though aiming lasers provide a quick and easy means of engaging the enemy at night, special attention must be given to the following:

- Proper adjustments to the image intensifiers.
- 10-meter boresighting procedures or 25-meter zeroing procedures.
- Scanning.
- Walking.
- IR discipline.

SCANNING

C-23. NVDs have a 40-degree field of view, leaving the average firer to miss easy targets of opportunity, more commonly targets 50 meters left or right.

C-24. Soldiers must train to aggressively scan their sectors of fire for targets. Training must reinforce regular blinking during scanning, which relieves some of the eyestrain from spotting far targets. After Soldiers master the art of scanning, they find that targets are easier to detect by acknowledging the flicker or movement of a target.

WALKING

C-25. Once a Soldier has located a target, he must be aware of the placement of the aiming laser. If he activates his laser while it is pointing over the target and into the sky, he wastes valuable time trying to locate exactly where the laser is pointing and increases his chances of being detected and fired upon by the enemy. Walking the laser to the target is a quick and operationally secure means of engaging the enemy with the aiming laser. To walk the laser to a target—

(1) Aim the laser at the ground just in front of the target.

(2) Walk the aiming laser along the ground and up to the target's center of mass.

(3) Engage the target.

INFRARED DISCIPLINE

C-26. To exercise proper IR discipline—

- While on the range, actively scan for targets with the laser off.
- Once a target is located, walk the laser to the target, and engage.
- Once a target has been located and engaged with the aiming laser, deactivate the laser.

THERMAL WEAPON SIGHT

C-27. To understand how the TWS is able to convert energy into an image suitable for viewing, Soldiers must understand the electromagnetic spectrum and the range of IR light in which the TWS operates. The TWS absorbs all available light into the lens, and then filters out all light except for middle/far IR (thermal) light. Then, the TWS converts the thermal light into an image and creates a video that is displayed on the raster.

C-28. The TWS can convert thermal energy that is reflected, radiated, or generated from an object. All objects, such as trees, metal, plastic, and living creatures, display a quality that allows them to be seen with thermal technology. How well the objects display these qualities will determine how well they are seen.

Absorption

C-29. During the day, all inanimate objects absorb thermal energy from the sun to varying degrees. Metal objects have a much higher rate of absorption than wood, leaves, or grass; therefore, a metal object sitting in the sun will stand out more than the grass surrounding it when viewed through the TWS.

Exposure

C-30. The amount of time an object is exposed to thermal energy determines how well that object will be seen. Naturally, an object with a long exposure time will have absorbed more thermal energy than an object exposed to the same thermal energy for a shorter period of time.

Emissivity

C-31. Emissivity is the rate at which an object emits the thermal energy it has absorbed or generates. Usually, objects with a high absorption rate will have a high emissivity factor. Although the human body does not have a high absorption rate, it has a high emissivity factor due to the fact that it generates a high amount of thermal energy. An object that has a high emissivity factor will be much hotter, and, therefore, when seen through the TWS, much easier to see and recognize.

Reflection

C-32. Items such as glass and water have virtually no absorption rate. Instead, they reflect thermal energy, which makes it very difficult to see objects through glass and water.

C-33. Snow and ice have the same effect, especially during the day with no clouds present. Snow and ice reflect most of the thermal energy from the sun, so it is difficult to acquire a good thermal image on objects that are close to the ground.

Diurnal Cycle

C-34. There are two times during the day when motionless objects that do not generate their own thermal energy, such as trees, rocks, and man-made objects, become the same temperature as the surrounding air: once in the morning and once in the evening. This is known as the diurnal cycle. The specific times that this cycle will take effect are based on the time of year, but it usually occurs shortly after sunrise and shortly after sunset.

C-35. The diurnal cycle contains crossover points. During the day, a motionless object will absorb thermal energy from the sun; the crossover point is the time when that object stops absorbing thermal energy (day) and starts radiating thermal energy (night). As the night goes on, that same object will come to a point where it stops radiating thermal energy and will once again start absorbing thermal energy (day).

Adjustments

C-36. Rain, snow, fog, smoke, and the diurnal cycle are just a few environmental or combat situations that may affect your thermal image. These conditions may cause objects to be difficult to see, so adjustments must be made to the TWS in order to refine the thermal image.

C-37. To allow Soldiers to maximize the capability of the sight, the TWS is equipped with—

- A diopter focus ring.
- Two FOVs.
- An objective focus ring.
- A brightness knob.
- An auto and manual contrast switch.
- A polarity switch.

Diopter Focus Ring

C-38. When making adjustments to the sight, Soldiers begin with the diopter focus ring. The diopter focus ring focuses the display screen (raster) to the eye. To adjust the diopter focus ring, adjust the diopter focus ring until everything on the display screen is clear and easily read.

NOTE: This is best done with the objective lens cover closed.

C-39. Once you have adjusted the diopter focus to your eye, no other adjustments to the diopter focus ring should be necessary.

Fields of View

C-40. The TWS has two operating FOVs—wide and narrow. The Soldier should be allowed to select the FOV that suits him best. Through use, Soldiers learn to use the appropriate FOV under different combat situations.

NOTE: When selecting an FOV, make sure that the FOV ring is turned completely to the left or to the right. If the FOV ring is turned only halfway, you will not be able to see through the sight.

Wide Field of View

C-41. The WFOV has the least magnification, but a greater FOV and is great for scanning.

Narrow Field of View

C-42. The NFOV has greater magnification, but less degrees of FOV.

Objective Focus Ring

C-43. The objective focus ring focuses the sight to the target. Adjustments to the objective focus ring are based on the range of the object being viewed.

NOTES:
1. Over-adjustment to the objective focus ring will lock the FOV ring to the point that the FOV cannot be changed.
2. Make adjustments to the objective focus ring only after focusing the diopter focus ring.

Brightness Knob

C-44. The dual-function brightness knob is used to turn on the TWS, adjust the brightness of the raster, and refine the thermal image. Used in conjunction with the contrast knob, it helps to combat the effects of the diurnal cycle and other conditions that might require fine-tuned adjustment to the thermal image.

Contrast Switch

C-45. The dual-function contrast switch has automatic and manual contrast modes. Used in conjunction with the brightness knob, the contrast switch allows Soldiers to obtain the best possible thermal image.

Automatic Contrast Mode

C-46. The automatic contrast mode is used under normal operating conditions.

Manual Contrast Mode

C-47. The manual contrast is used under conditions other than normal (for example, during 10-meter boresighting or 25-meter zeroing; during rain, fog, smoke, or snow; during the diurnal cycle; or when trying to obtain as much detail of a target as possible).

Polarity Switch

C-48. The polarity switch has two modes: white hot or black hot. Use of the polarity switch is a user's preference. Through continued use, Soldiers decide which polarity setting works best under different combat or environmental conditions.

White Hot Mode

C-49. When in the white hot mode, hotter objects appear white, while cooler objects have shades of gray to black.

Black Hot Mode

C-50. When using the black hot mode, hotter objects appear black, while the cooler objects are shades of gray to white.

Appendix D

Range Safety and Risk Management

All personnel training on a rifle range should be briefed on the safety and local requirements for that range. This briefing fulfills the minimum requirements for a rifle range safety briefing. Information may be added to conform to local requirements and safety regulations. ARs 385-10 and 385-63 and DA Pam 385-63 should be reviewed by all range personnel before operating any range.

RECOMMENDED BRIEFING

D-1. The first priority on any range is training, but safety must be at the forefront of the training program. The safety program prescribes the safety precautions necessary to minimize accidents when firing and using ammunition during training and range operations. The safety program should identify—

- Surface danger zones (SDZs), as described in AR 385-63.
- The location of medical personnel.
- Left and right limits of the range. Firers never fire outside of these limits.
- The designated smoking area (if applicable).

D-2. Further, the safety program should include the following information:

- Inspect for objects located near the weapon's muzzle before firing, especially during unassisted night fire.
- When not on the firing line, ensure that the weapon's selector lever is on SAFE and the bolt is locked to the rear.
- Ensure that firers enter and exit the firing line at the entry or exit point.
- Before occupying a firing position, inspect it for wildlife or obstructions.
- Always keep the weapon's muzzle pointed downrange when on the firing line, and keep the finger outside of the trigger housing area.
- Never touch a weapon while personnel are downrange or in front of the firing line.
- Load the weapon only on command from the tower or control point.
- Never fire without using hearing protection.
- Ensure that left-handed firers attach left-handed brass deflectors to their weapon before firing.
- When entering or exiting the firing line, clear the weapon with a cleaning rod.
- Consider the rifle loaded at all times, even in break areas. Never point the weapon at anyone.
- Anyone observing an unsafe act will immediately call "CEASE FIRE," place the weapon on SAFE, place it in the V-notched stake or lay it on the sandbags, and give the verbal and visual command of cease fire.
- Once cleared off of the firing line, firers report to the ammunition point and turn in all brass and ammunition.
- No one will leave the range until they have been inspected for live ammunition and brass.
- In case of an electrical storm, personnel will be directed to lock and clear all weapons, ground their equipment (except wet weather gear), and disperse into a predetermined area.
- Eating and drinking are not permitted on the firing line unless the tower operator permits drinking from the canteen.

NOTE: Drink water often to prevent heat injuries.

PERSONNEL AND DUTIES

D-3. To provide a safe and efficient range operation and effective instruction, certain duties may be required of personnel. The personnel may include—

- OIC.
- RSO.
- NCOIC.
- Ammunition detail.
- Unit armorer.
- Assistant instructor.
- Medical personnel.
- Control tower operators.
- Maintenance detail.

OFFICER IN CHARGE

D-4. The OIC is responsible for the overall operation of the range before, during, and after live-firing.

RANGE SAFETY OFFICER

D-5. The RSO—

- Is responsible for the safe operation of the range.
- Conducts a safety orientation before each scheduled LFX.
- Ensures that a brass and ammunition check is made before the unit leaves the range.
- Ensures that all personnel comply with the safety regulations and procedures prescribed for the conduct of an LFX.
- Ensures that all left-handed firers use left-handed firing devices.

NOTE: This officer should not be assigned duties other than those of the safety officer.

NONCOMMISSIONED OFFICER IN CHARGE

D-6. The NCOIC assists the OIC and safety officer by performing duties as required; for example, he might supervise enlisted personnel who are supporting the LFX.

AMMUNITION DETAIL

D-7. This detail is composed of one or more ammunition handlers. The ammunition detail—

- Breaks down, issues, receives, accounts for, and safeguards live ammunition.
- Collects expended ammunition casings and other residue.

UNIT ARMORER

D-8. The unit armorer—

- Repairs rifles.
- Replaces parts.

ASSISTANT INSTRUCTOR

D-9. One assistant instructor is assigned for each one to ten firing points. Each assistant instructor—

- Ensures that all firers observe safety regulations and procedures.
- Assists firers having problems.

Medical Personnel

D-10. Medical personnel provide medical support, as required by regulations governing LFXs.

Control Tower Operators

D-11. Control tower operators—

- Raise and lower the targets.
- Time the exposures.
- Sound the audible signal.
- Give the fire commands.

NOTE: If possible, two men should be chosen to perform these functions.

Maintenance Detail

D-12. This detail should be composed of two segments: one to conduct small arms repair and one to perform minor maintenance on the target-holding mechanisms.

AMMUNITION POSITIONING AND ISSUANCE

D-13. To provide a safe and operational range, the following are recommended procedures for handling ammunition:

- Locate all ammunition for the weapons involved at firing sites outside of the backblast area (when applicable). Store ammunition at a position that minimizes the potential for ignition, explosion, or rapid burning.
- Issue ammunition to firing units immediately before scheduled training exercises. Distribute small arms ammunition to troops only when they are on the ready line or firing line.
- Cover all ammunition to protect it from the elements and direct rays of the sun. For proper ventilation, provide air circulation between the ammunition and cover.
- Limit the unpacking of ammunition at the firing line to the minimum number of rounds needed. Retain packaging material until firing is complete. Units do not burn wooden containers or indiscriminately fire ammunition to preclude return to a storage site.

COMPOSITE RISK MANAGEMENT

D-14. Composite risk management (CRM) is a decision-making process used to mitigate risks associated with all hazards that have the potential to injure or kill personnel, damage or destroy equipment, or otherwise impact mission effectiveness. The guiding principles of CRM are as follows:

- Integrate CRM into all phases of the mission and operations.
- Make risk decisions at the appropriate level.
- Accept no unnecessary risk.
- Apply the process cyclically and continuously.
- Do not be risk averse.

D-15. CRM is a five-step process:

(1) Identify hazards.
(2) Assess hazards to determine risk.
(3) Develop controls, and make risk decisions.
(4) Implement controls.
(5) Supervise and evaluate.

NOTE: Steps 1 and 2 are assessment steps; Steps 3 through 5 are management steps.

STEP 1–IDENTIFY HAZARDS

D-16. Hazards may arise from any number of areas. They can be associated with enemy activity, accident potential, weather or environmental conditions, health, sanitation, behavior, and/or materiel or equipment. Mission, enemy, terrain and weather, troops and support available, time available, civil considerations (METT-TC) factors serve as a standard format for identifying hazards, on- or off-duty.

D-17. CRM does not differentiate between the sources of the hazard. The loss of personnel, equipment, or materiel due to any hazard has the same disruptive impact on readiness or mission capabilities, no matter what the source.

STEP 2–ASSESS HAZARDS TO DETERMINE RISK

D-18. During mission analysis, course of action (COA) development, or analysis, rehearsal, and execution steps of the military decision-making process (MDMP), commanders assess hazards and assign risk in terms of probability and severity of adverse impact. During their assessment, they must consider both mission- and non-mission-related aspects that may have an impact. The end result of this assessment is an initial estimate of risk for each identified hazard as determined from the standardized application of the risk assessment matrix.

D-19. There are three substeps in this step:

(1) Assess the probability of an event or occurrence.

(2) Estimate the expected result or severity of an event or occurrence.

(3) Determine the level of risk using the standard risk assessment matrix.

Assess the Probability of an Event or Occurrence

D-20. Probability is the likelihood an event will occur based on prior experience. The probability levels estimated for each hazard are based on the mission, COA, or frequency of a similar event. There are five levels of probability (Table D-1):

- Frequent.
- Likely.
- Occasional.
- Seldom.
- Unlikely.

Table D-1. Five levels of probability.

LEVEL OF PROBABILITY	EXPLANATION
Frequent	Occurs very often, known to happen regularly.
Likely	Occurs several times, a common occurrence.
Occasional	Occurs sporadically, but is not common.
Seldom	Remotely possible, could occur at some time.
Unlikely	Can assume will not occur, but not impossible.

Estimate the Expected Result or Severity of an Event or Occurrence

D-21. Severity is the degree to which an incident will impact combat power, mission capability, or readiness. The degree of severity estimated for each hazard is based on the results of similar events. Severity is addressed in the four levels used on the risk assessment worksheet (Table D-2):

- Catastrophic.
- Critical.
- Marginal.
- Negligible.

Table D-2. Four levels of severity.

LEVEL OF SEVERITY	EXPLANATION
Catastrophic	• Complete mission failure or loss of ability to accomplish a mission. • Death or permanent total disability. • Loss of major or mission-critical systems or equipment. • Major property or facility damage. • Severe environmental damage. • Mission-critical security failure. • Unacceptable collateral damage.
Critical	• Severely degraded mission capability or unit readiness. • Permanent partial disability or temporary total disability exceeding three months time. • Extensive major damage to equipment or systems. • Significant damage to property or the environment. • Security failure. • Significant collateral damage.
Marginal	• Degraded mission capability or unit readiness. • Minor damage to equipment or systems, property, or the environment. • Lost days due to injury or illness, not exceeding three months. • Minor damage to property or the environment.
Negligible	• Little or no adverse impact on mission capability. • First aid or minor medical treatment. • Slight equipment or system damage, but fully functional or serviceable. • Little or no property or environmental damage.

Determine the Level of Risk Using the Standard Risk Assessment Matrix

D-22. Using the standard risk assessment matrix (Table D-3), commanders convert probability and severity for each identified hazard into a specified level of risk. All accepted residual risk must be approved at the appropriate level of command.

NOTE: This assessment is an estimate, not an absolute. It may or may not be indicative of the relative danger of a given operation, activity, or event.

Table D-3. Risk assessment matrix.

RISK ASSESSMENT MATRIX						
Severity		Probability				
		Frequent A	Likely B	Occasional C	Seldom D	Unlikely E
Catastrophic	I	E	E	H	H	M
Critical	II	E	H	H	M	L
Marginal	III	H	M	M	L	L
Negligible	IV	M	L	L	L	L
E – Extremely High		H – High		M – Moderate	L – Low	

D-23. Risk is addressed in the four levels listed in the lower left corner of the matrix (Table D-4):

- Extremely high risk.
- High risk.
- Moderate risk.
- Low risk.

Table D-4. Four levels of risk.

LEVEL OF RISK	EXPLANATION
Extremely High Risk	Loss of ability to accomplish the mission if hazards occur. In the example noted in Table D-3, a frequent or likely probability of catastrophic loss (IA or IB) or a frequent probability of critical loss (IIA) exists. This implies that the risk associated with this mission, activity, or event may have severe consequences. The decision to continue must be weighed carefully against the potential gain to be achieved by continuing this COA.
High Risk	Significant degradation of mission capabilities (in terms of the required mission standard), inability to accomplish all parts of the mission, or inability to complete the mission to standard will result if hazards occur during the mission. Occasional to seldom probability of catastrophic loss (IC or ID) exists. A likely to occasional probability of a critical loss (IIB or IIC) exists. Frequent probability of marginal losses (IIIA) exists. This implies that if a hazardous event occurs, serious consequences will occur. The decision to continue must be weighed carefully against the potential gain to be achieved by continuing this COA.
Moderate Risk	Expected degraded mission capabilities (in terms of the required mission standard) will result if hazards occur during the mission. An unlikely probability of catastrophic loss (IE) exists. The probability of a critical loss is seldom (IID). Marginal losses occur with a likely or occasional probability (IIIB or IIIC). A frequent probability of negligible (IVA) losses exists.
Low Risk	Expected losses have little or no impact on accomplishing the mission. The probability of critical loss is unlikely (IIE), while that of marginal loss is seldom (IIID) or unlikely (IIIE). The probability of a negligible loss is likely or less (IVB through (IVE). Expected losses have little or no impact on accomplishing the mission. Injury, damage, or illness are not expected, or may be minor and have no long-term impact or effect.

Step 3 – Develop Controls, and Make Risk Decisions

D-24. In this step, commanders develop and apply controls, reassess the hazard to determine a residual risk, and make risk decisions. This process continues until an acceptable level of risk is achieved or until all risks are reduced to a level where benefits outweigh the potential cost. This step is accomplished during the COA development, COA analysis, COA comparison, and COA approval of the MDMP.

Step 4 – Implement Controls

D-25. Leaders and staffs ensure that controls are integrated into SOPs, written and verbal orders, mission briefings, and staff estimates. The critical check for this step is to ensure that controls are converted into clear and simple execution orders. This step includes coordination and communication with—

- Appropriate superior, adjacent, and subordinate units, organizations, and individuals.
- Logistics Civil Augmentation Program (LOGCAP) organizations and civilian agencies that are part of the force or may be impacted by the activity, hazard, or its control.
- The media and nongovernmental organizations (NGO) when their presence impacts or is impacted by the force.

Step 5 – Supervise and Evaluate

D-26. This step involves implementing risk controls and enforcing them to standard, and validating the adequacy of selected control measures in supporting the objectives and desired outcomes. This continuous process provides the ability to identify weaknesses and to change or adjust controls based on performance, changing situations, conditions, or events.

RESPONSIBILITIES

D-27. CRM responsibilities are spread across three levels:

- Commander.
- Leaders.
- Individual.

COMMANDER

D-28. During implementation of the CRM process, the commander—

- Ensures that warfighting functions (WFF) are performed to standard to minimize human error, materiel failure, and environmental effects.
- Establishes a force protection policy and publishes a safety philosophy with realistic safety goals, objectives, and priorities.
- Ensures that his training assessment considers the WFF's ability to protect the force. Selects long-, short-, and near-term control actions and ensures implementation to improve force protection.
- Ensures that his staff integrates risk management into the planning and execution of training and operational missions.
- Makes risk decisions. Selects, monitors, and enforces implementation of controls for hazards most likely to result in loss of combat power. After implementing controls, if risk remains above the tolerance level established by higher command, he must elevate the risk decision to the appropriate command level.
- Ensures that the CRM process is evaluated during all AARs.
- Determines if unit performance meets force protection guidance. Determines effectiveness of hazard controls and necessary changes to guidance and controls. Ensures that these changes are fed back into the training management cycle and guidance for operational missions, including unit SOPs.

LEADERS

D-29. During implementation of the CRM process, leaders—

- Enforce METL task performance to standard. Adopt the crawl—walk—run approach in planning and executing training.
- Make use of automated on- and off-duty CRM tools and surveys available from the United States Army Combat Readiness Center (USACRC).
- Execute risk reduction controls selected by the commander by developing and implementing supporting leader level controls. Apply risk management procedures to each.

INDIVIDUAL

D-30. All Soldiers must understand how to use the CRM process to enhance mission success and to reduce or eliminate loss.

D-31. During implementation of the CRM process, Soldiers—

- Support commanders and leaders in the rapid identification and communication of hazards and associated risks that may impact on the mission.
- Provide immediate feedback to the leader as the mission progresses and hazards are encountered. Use short written messages, hand and arm signals, or radio transmissions to communicate first-hand information to leaders.
- In extreme situations, act alone or make risk decisions within the context of orders.

DA FORM 7566 (COMPOSITE RISK MANAGEMENT WORKSHEET)

D-32. DA Form 7566 (Composite Risk Management Worksheet, Figure D-1a and D-1b) provides a starting point to logically track the CRM process. It can be used to document risk management steps taken during planning, preparation, and execution of training and combat missions.

NOTE: For detailed CRM procedures, see FM 5-19.

COMPOSITE RISK MANAGEMENT WORKSHEET

For use of this form, see FM 5-19; the proponent agency is TRADOC.

1. MSN/TASK	2a. DTG BEGIN	2b. DTG END	3. DATE PREPARED *(YYYYMMDD)*

4. PREPARED BY

a. LAST NAME	b. RANK	c. POSITION

5. SUBTASK	6. HAZARDS	7. INITIAL RISK LEVEL	8. CONTROLS	9. RESIDUAL RISK LEVEL	10. HOW TO IMPLEMENT	11. HOW TO SUPERVISE (WHO)	12. WAS CONTROL EFFECTIVE?

Additional space for entries in Items 5 through 11 is provided on Page 2.

13. OVERALL RISK LEVEL AFTER CONTROLS ARE IMPLEMENTED *(Check one)*

☐ LOW ☐ MODERATE ☐ HIGH ☐ EXTREMELY HIGH

14. RISK DECISION AUTHORITY

a. LAST NAME	b. RANK	c. DUTY POSITION	d. SIGNATURE

DA FORM 7566, APR 2005 Page 1 of 2

APD V3.00

EXAMPLE

Figure D-1a. Sample DA Form 7566 (Composite Risk Management Worksheet).

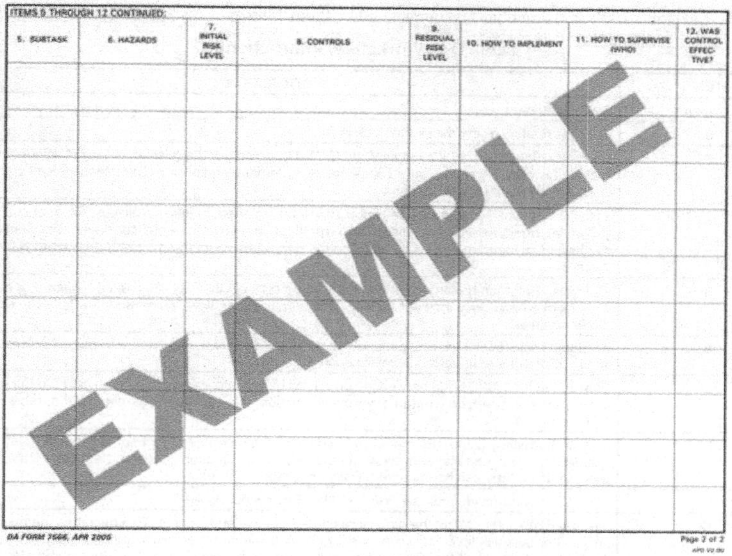

ITEMS 5 THROUGH 12 CONTINUED:

5. SUBTASK	6. HAZARDS	7. INITIAL RISK LEVEL	8. CONTROLS	9. RESIDUAL RISK LEVEL	10. HOW TO IMPLEMENT	11. HOW TO SUPERVISE (WHO)	12. WAS CONTROL EFFEC-TIVE?

DA FORM 7566, APR 2005

Page 2 of 2

APD V2.00

Figure D-1b. Sample DA Form 7566 (Composite Risk Management Worksheet) (page 2).

D-33. See Table D-5 for instructions on completing DA Form 7566.

Table D-5. Worksheet instructions

ITEM	INSTRUCTION
1 through 4	Self-explanatory.
5	Subtask relating to the mission or task in Block 1.
6	Hazards–Identify hazards by reviewing METT-TC factors for the mission or task. Additional factors include historical lessons learned, experience, judgment, equipment characteristics and warnings, and environmental considerations.
7	Initial Risk Level–Factors include historical lessons learned, intuitive analyses, experience, judgment, equipment characteristics and warnings, and environmental considerations. Determine the initial risk for each hazard by applying the risk assessment matrix. Enter the risk level for each hazard.
8	Controls– For each hazard, develop one or more controls that will eliminate the hazard or reduce the risk of a hazardous incident. Specify who, what, where, why, when, and how for each control. Enter controls.
9	Residual Risk Level–Determine the residual risk for each hazard by applying the risk assessment matrix. Enter the residual risk level for each hazard.
10	How to Implement–Decide how each control will be put into effect or communicated to the personnel who will make it happen (written or verbal instruction; tactical, safety, garrison SOPs; rehearsals). Enter controls.
11	How to Supervise (Who)–Plan how each control will be monitored for implementation (continuous supervision, spot-checks) and reassess hazards as the situation changes. Determine if the controls worked and if they can be improved. Pass on lessons learned.
12	Was Control Effective–Indicate "Yes" or "No." Review during AAR.
13	Overall Risk Level–Select the highest residual risk level and circle it. This becomes the overall mission or task risk level. The commander decides whether the controls are sufficient to accept the level of residual risk. If the risk is too great to continue the mission or task, the commander directs development of additional controls or modifies, changes, or rejects the COA.
14	Risk Decision Authority–Signed by the appropriate level of command.

Appendix E

Range Procedures and Range Operations Checklist

This appendix contains the procedures used to conduct live-fire training exercises. These procedures support Army regulations (TCs 7-9 and 25-1), local range regulations, and established unit training SOPs. Conduct of the training exercise should clearly define and establish details and equipment needed to open and operate the range so that these actions do not adversely impact Soldiers' training time. The procedures outlined in this appendix should be followed in order to open the range and conduct effective training.

RANGE PROCEDURES

E-1. Before beginning an LFX, all personnel must receive an orientation on range operations. The orientation should outline the procedures for conducting the exercise, to include the duties of the nonfiring orders.

PERSONNEL AND DUTIES

D-2. To provide a safe and efficient range operation and effective instruction, certain duties may be required of personnel. The personnel may include—

- OIC.
- RSO.
- NCOIC.
- Ammunition detail.
- Unit armorer.
- Assistant instructor.
- Medical personnel.
- Control tower operators.
- Maintenance detail.

Officer In Charge

D-3. The OIC is responsible for the overall operation of the range before, during, and after live-firing.

Range Safety Officer

D-4. The RSO—

- Is responsible for the safe operation of the range.
- Conducts a safety orientation before each scheduled LFX.
- Ensures that a brass and ammunition check is made before the unit leaves the range.
- Ensures that all personnel comply with the safety regulations and procedures prescribed for the conduct of an LFX.
- Ensures that all left-handed firers use left-handed firing devices.

NOTE: This officer should not be assigned duties other than those of the safety officer.

Noncommissioned Officer In Charge

D-5. The NCOIC assists the OIC and safety officer by performing duties as required; for example, he might supervise enlisted personnel who are supporting the LFX.

Ammunition Detail

D-6. This detail is composed of one or more ammunition handlers. The ammunition detail—

- Breaks down, issues, receives, accounts for, and safeguards live ammunition.
- Collects expended ammunition casings and other residue.

Unit Armorer

D-7. The unit armorer—

- Repairs rifles.
- Replaces parts.

Assistant Instructor

D-8. One assistant instructor is assigned for each one to ten firing points. Each assistant instructor—

- Ensures that all firers observe safety regulations and procedures.
- Assists firers having problems.

Medical Personnel

D-9. Medical personnel provide medical support, as required by regulations governing LFXs.

Control Tower Operators

D-10. Control tower operators—

- Raise and lower the targets.
- Time the exposures.
- Sound the audible signal.
- Give the fire commands.

NOTE: If possible, two men should be chosen to perform these functions.

Maintenance Detail

D-11. This detail should be composed of two segments: one to conduct small arms repair and one to perform minor maintenance on the target-holding mechanisms.

FIRING ORDER LINEUP

E-12. After the range cadre have given the safety and range briefings, they assemble the Soldiers in firing orders in correlation with the number of firing points on that range. After the firing order has been determined, firers have their weapons rodded, move to the firing line, and proceed to their assigned firing points, keeping their weapons pointed up and downrange at all times.

TOWER COMMANDS

E-13. Simple, standard fire commands are needed to avoid confusion and misunderstanding during LFXs.

GENERAL COMMANDS

E-14. The following general commands may be altered when necessary:

- "Firers, assume the ________ position."
- "Issue the firer ________ rounds of ammunition."
- "Coach, secure ________ rounds of ammunition."
- "Lock one round. Load."
- "Ready on the right?"
- "Ready on the left?"
- "Ready on the firing line?"
- "Commence firing when your targets appear."
- "Cease firing. Lock and clear all weapons."

Grouping Commands

E-15. Grouping commands include the following:

- "Firers, assume a good supported prone position."
- "Lock one of three single rounds. Load."
- "Ready on the right?"
- "Ready on the left?"
- "The firing line is ready."
- "Place your selector lever on SEMIAUTOMATIC."
- "Commence firing."
- "Cease fire. Lock and clear your weapons."
- "Clear on the right?"
- "Clear on the left?"
- "The firing line is clear."
- "Move down to your targets, and triangulate your shot group."
- "After all personnel have triangulated their targets, move back to the firing line."
- "At this time, make adjustments to your sights."
- "Repeat all firing commands until grouping standards are met."

ZERO COMMANDS

E-16. Tower commands are the same as grouping commands. Repeat all firing commands until zeroing standards are met.

Field Firing Exercises

E-17. Simple, standard fire commands are needed to avoid confusion during field firing exercises. Commands for exercises from stationary positions are as follows:

- "Firers, assume a good ________ position."
- "Lock one magazine of ________ rounds. Load."
- "Ready on the right?"
- "Ready on the left?"
- "The firing line is ready."
- "Place your selector lever on SEMIAUTOMATIC."
- "Scan your sector."
- "Cease fire. Lock and clear your weapon." (Place the selector lever in the SAFE position.)

E-18. The range officer relays his commands to the pit NCOIC by radio or telephone so he can keep abreast of the conduct of fire. Before each firing exercise, the range officer informs the pit NCOIC of the next exercise and any special instructions for target operation (for example, "The next firing will be for zero. Mark targets after each three-round shot group," or for slow fire, "The next firing will be ten rounds. Slow fire. Mark targets after each shot.").

E-19. Radio telephone operators (RTOs) relay commands to the pit and pass on special instructions to target operators as requested by assistant instructors.

NOTES: 1. Radio telephone operators never identify a firer on a particular firing point.

2. The command "MARK TARGET NUMBER ________" indicates that the target has been fired upon but has not withdrawn for marking.

Practice Record Fire and Record Fire

E-20. Simple, standard fire commands are needed to avoid confusion and misunderstanding during practice record fire and record fire.

Practice Record Fire

E-21. Practice record fire commands include the following:

- "Firers, assume a good supported (prone unsupported) position."
- "Scorers, point out the limits of your lane."
- "Firers, lock your first magazine. Load."
- "Scan your sector."
- "Cease fire."
- "Lock and clear all weapons."

Record Fire

E-22. Record fire commands include the following:

- "Firers, assume a good supported (prone unsupported) position."
- "Scorers, point out the limits of your sector."
- "Firers, lock your first 20-round magazine. Load."
- "Scan your sector."
- "Cease fire."
- "Lock and clear all weapons."

Rapid Fire Exercises

E-23. The following commands are used for rapid fire exercises:

- "Lock and clear all weapons."
- "Clear on the right?"
- "Clear on the left?"
- "The firing line is clear."
- "Firers, assume the ________ position."
- "Assistants, secure two magazines of five rounds each." ("Issue the firer one magazine of five rounds.")
- "Lock one magazine. Load."
- "Ready on the right?"
- "Ready on the left?"
- "Ready on the firing line?"
- "Watch your targets." ("Firers, assume the appropriate firing position and commence firing when the targets are presented.")

E-24. When all of the targets are withdrawn, the range officer checks for slow firers or malfunctions and then allows them to fire.

E-25. The pit NCOIC organizes, orients, and provides safety for the pit detail.

E-26. The success of KD firing depends on efficient operation of the targets and the close coordination between the pit NCOIC and range officer. All operators must know the proper procedure for operating and marking the target.

Marking Targets for Zeroing and Slow Fire

E-27. Targets are marked quickly after each shot or group of shots without command. During slow fire, the firer has a time limit of one minute for each shot. Twenty seconds is considered the maximum time limit for marking. A marker (spotter) is placed in the hit regardless of its location on the target. Each time the target is marked, the marker is removed from the previous hit, and the hole is pasted.

NOTE: 3-inch markers are used for 100, 200, and 300 meters; 5-inch markers are used for 500 meters.

Using Disk Markers

E-28. The target markers are painted black on one side and white on the opposite side. They are available in three dimensions:

- 1 1/2 inches (NSN 6920-00789-0864).
- 3 inches (NSN 6920-00-713-8255).
- 5 inches (NSN 6920-00-713-8254).

E-29. The disk spindle (NSN 6920-00-7138257) may also be procured through the supply channels.

RANGE OPERATIONS CHECKLIST

E-30. This checklist consists of nine sections, each covering a topic relating to range operations:

- Mission analysis.
- Double check.
- Become an expert.
- Determine requirements.
- Determine available resources.
- Foolproof.
- Occupy the range and conduct training.
- Close the range.
- KD range.

NOTE: The checklist should be modified to include local policy changes to regulations or SOPs.

E-31. The person responsible for the training must answer the questions in each section. To answer these questions, he should use the following procedures:

- Ask each question in order.
- Record a "Yes" answer by checking the GO column.
- Record a "No" or "Don't know" answer by checking the NO GO column.

- Refer to the checklist to find the GO and NO GO columns.
- After all of the questions in a section are answered, analyze the NO GOs.
  - Contact the people who reported them and ask if they have corrected each problem. If so, change the answer to GO.
  - If any NO GO remains, analyze it and implement a countermeasure for the shortfall. Afterward, check to ensure that the countermeasures work.
  - Before range operations start, be sure that a workable countermeasure is implemented for each safety hazard presented by a NO GO answer.

MISSION ANALYSIS

E-32. Mission analysis includes identifying the following:

- Who will be firing on the range? ________
- Number of personnel________ Units________
- What weapons and course will be used? Weapons________ Course________
- Where will the training be conducted? Range________
- When is the range scheduled for operations? Date________ Opens________ Closes________

DOUBLE CHECK

	GO	NO GO	REMARKS
1. Has sufficient ammunition been requested for the number of personnel?			
2. Are the range facilities adequate for the type of training to be conducted?			
3. Has enough time been scheduled to complete the training?			
4. Have conflicts that surfaced been resolved?			

BECOME AN EXPERT

	GO	NO GO	REMARKS
1. Review TMs and FMs on the weapons to be fired.			
2. Talk with the armorer and other personnel experienced with the weapons to be fired.			
3. Review AR 385-63.			
4. Visit range control and read the installation's range instructions.			
5. Reconnoiter the range (preferably while it is in use).			
6. Check ARTEPs to see if training tasks can be integrated into the range training plan.			

DETERMINE REQUIREMENTS

E-33. Determining the requirements involves identifying the equipment and personnel necessary to conduct the training.

Personnel

E-34. The following personnel are required for range operations:

- OIC.
- Safety officer.
- Assistant safety officer.
- NCOIC.
- Ammunition NCO.

- Ammunition personnel (determined by type of range).
- Target detail and target operators.
- Tower operator.
- Concurrent training instructors.
- Assistant instructors.
- Radio telephone operators.
- Guards (range requirements).
- Medic(s).
- Air guard.
- Armorer.
- Truck driver (range personnel and equipment).
- Mechanic for vehicles.

E-35. Further, the following must be checked:

- Have you overstaffed your range?

Equipment

E-36. The following equipment is required for range operations:

- Range packet and clearance form.
- Safety fan and diagram (if applicable).
- Other safety equipment (for example, aiming circle, compass).
- Publications pertaining to the training that will be conducted.
- Lesson plans, status reports, and reporting folder.
- Range flag and light (night firing).
- Radios.
- Field telephone and wire.
- 292 antenna (if necessary).
- PA set with backup bullhorn(s).
- Concurrent training markers.
- Training aids for concurrent training stations.
- Sandbags.
- Tentage (briefing tent, warm-up tent).
- Space heaters (if needed).
- Colored helmets for control personnel.
- Safety paddles and vehicle flag sets or lights.
- Ambulance or designated vehicle.
- Earplugs.
- Water for drinking and cleaning.
- Scorecards.
- Master score sheet.
- Armorer's tools and cleaning equipment for weapons.
- Brooms, shovels, and other cleaning supplies and equipment.
- Tables and chairs (if needed).
- Target accessories.
- Fire extinguishers.
- Tarp, stakes, and rope to cover the ammunition.
- Toilet paper.
- Spare weapons and repair parts (as needed).

- Tow bar and slave cables for vehicles.
- Fuel and oil for vehicles and target mechanisms.

Determine Available Resources

E-37. Determining available resources involves the following:

- Fill personnel spaces.
- Keep unit integrity.
- Utilize NCOs.
- Coordinate with supporting organizations:
  - Ammunition.
  - Transportation.
  - Training aids.
  - Medics.
  - Weapons.
  - Other equipment.

Foolproof

E-38. To foolproof range operations—

- Write an overall lesson plan for the range.
- Organize a plan for firing:
  - Determine the range organization.
  - Outline courses of fire to be used.
  - Have fire commands typed for use on the range.
  - Set the rotation of stations.
- Rehearse concurrent training instructors and assistants.
- Brief the RTO on unique range control radio procedures.
- Brief and rehearse the reporting NCO on range operation and all of his duties.
- Collect and concentrate equipment for use on the range in one location.
- Obtain training aids.
- Pick up targets from the range warehouse (if required).
- Report to range control for a safety briefing (if required) and sign for any special items.
- Publish the LOI:
  - Uniform of range and firing personnel (helmets and earplugs).
  - Mode of transportation, departure times, and places.
  - Methods of messing to be used.
  - Any special requirements being placed on units.

Occupy the Range and Conduct Training

E-39. Certain actions must be performed to properly occupy the range and conduct training.

Occupy the Range

E-40. When occupying the range, perform the following actions:

- Request permission to occupy the range.
- Establish good communications.
- Have designated areas prepared:
  - Parking.
  - Ammunition point.

  - Medical station.
  - Water point.
  - Concurrent training.
  - Mess.
  - Helipad.
  - Armorer.
- Inspect the range for operational condition.
- Raise the flag when occupying or firing according to the local SOP.
- Check the ammunition to ensure that it is the correct type and quantity.
- Ensure that range personnel are in the proper uniform and the equipment is in position.
- Receive firing units.
- Conduct safety checks on weapons.
- Check for clean, fully operational weapons.
- Conduct a safety briefing (to include administrative personnel on the range).
- Organize personnel into firing orders (keep unit integrity, if possible).
- Request permission to commence firing from range control.

Conduct of Firing

E-41. When conducting firing, the following must be checked:

- Are communications to range control satisfactory?
- Are commands from the tower clear and concise?
- Are range areas policed?
- Is ammunition accountability maintained?
- Is the master score sheet updated?
- Is personnel accountability maintained?
- Are vehicles parked in the appropriate areas?
- Is the air guard on duty and alert?
- Are personnel in proper uniform?
- Are earplugs in use?
- Are troops responding properly to commands?
- Are on-the-spot corrections being made when troops use poor techniques or fail to hit the target?
- Is conservation of ammunition being enforced?
- Are weapons cleared before they are taken from the firing line?
- Are personnel checked for brass or ammunition before they leave the range?
- Is anyone standing around not involved in training or support?

CLOSE THE RANGE

E-42. To properly close the range, the following must be performed:

- Close downrange according to the local SOP.
- Remove all equipment and ammunition from the range.
- Police the range.
- Repaste and resurface targets as required by range instructions.
- Perform other maintenance tasks as required by the local SOP.
- Request a range inspector from range control when ready to be cleared.
- Submit an after-action report to headquarters.
- Report any noted safety hazards to the proper authorities.

KNOWN DISTANCE RANGE

E-43. Events held on the KD range require additional personnel and equipment.

Personnel

E-44. In addition to the personnel previously identified, the KD range requires—

- NCOIC of pit detail.
- Assistant safety officer for the pit area.

Equipment

E-45. In addition to the equipment previously identified, the KD range requires—

- A sound set for the pit area.
- Positive communication from the firing line to the pit area.
- Pasters.
- Glue and brushes for resurfacing targets.
- Lubricant for target frames.
- Proper targets mounted in target frames.
- Briefing on how to operate a KD range.
- Procedure for marking targets.
- Procedure for pit safety.

COMPUTER CONTROLLED RANGES

E-46. Modern computer-controlled ranges allow trainers to develop scenarios and control targets and battlefield simulation devices. This permits Soldiers and units to practice mission-essential tasks in a stressful environment. Computerized systems also provide performance feedback. Using data recorded during training, the computer generates AARs. This accurate feedback allows commander to assess the unit's performance and leaders to assess their unit's mission status and design training programs to overcome the identified shortcomings. The performance feedback highlights positive actions to reinforce correct procedures and to foster Soldiers' confidence, enabling Soldiers and leaders to recognize and correct their shortcomings.

RANGE SELECTION

E-47. This section identifies primary and alternate ranges used for training and qualification, with specific weapon systems and weapons based on applicable FMs. The following table lists primary and alternate ranges to satisfy weapon system training events.

Table E-1. Primary/alternate range selection.

WEAPON SYSTEM	TRAINING EVENT	RANGE	
		PRIMARY	ALTERNATE
M16-Series and M4	Zero	25-m range	AFF/ARF/MRF with 25m boots
	Sustainment/Record	AFF/ARF/MRF	25-m range
	Qualification	ARF/MRF/QTR	25-m range
	Night	AFF/ARF/MRF Night	25-m range
	CBRN Conditions	AFF/ARF/MRF	25-m range

AFF: Automated Field Fire Range
ARF: Automated Record Fire Range
MRF: Modified Record Fire Range
QTR: Qualification Training Range

Rifle/Machine Gun Zero Range (17801)

E-48. This range (Figure E-1) is used to train individual Soldiers on the skills necessary to align the sights and practice basic marksmanship techniques against stationary targets. The range is designed for conducting shot-grouping and zeroing exercises with the M16- and M4-series weapons, as well as crew-served machine guns. Its primary features include—

- 32 target frames at 25 meters.
- 16 target frames at 10 meters.
- 32 foxholes.

E-49. This range requires no automation. All targets are fixed at 25 meters from the firing line for M16- and M4-series weapons and 10 meters for machine gun.

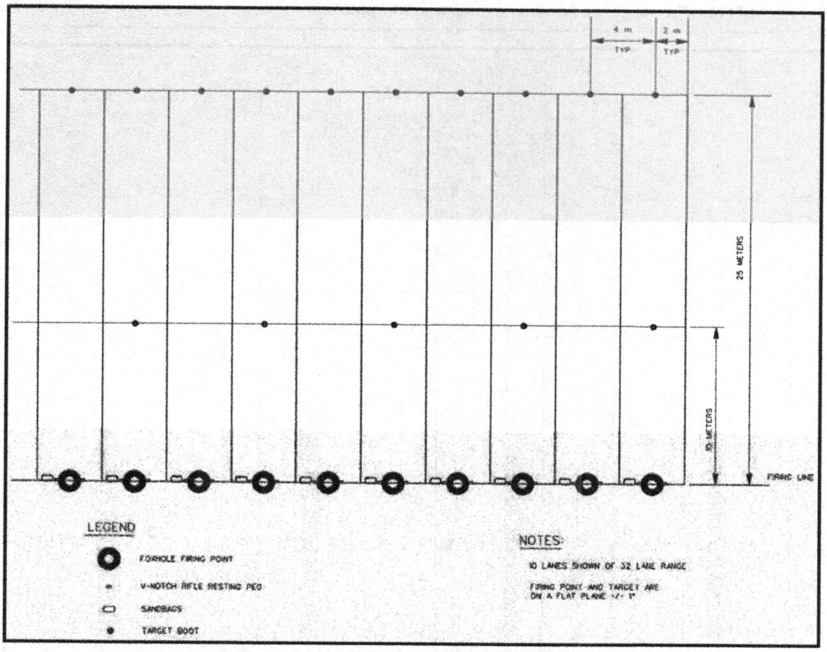

Figure E-1. Rifle/machine gun zero range (17801).

Automated Field Fire Range (17803)

E-50. The AFF range (Figure E-2) is used to train and familiarize Soldiers on the skills necessary to identify, engage, and hit stationary Infantry targets with M16- and M4-series weapons. Its primary features include—

- 96 stationary Infantry targets.
- 32 foxhole positions.

E-51. All targets are fully automated, and the event-specific target scenario is computer-driven and scored from the range operations center. The range operating system is fully capable of providing immediate performance feedback to the participants.

NOTE: This range can be used for automatic rifle practice.

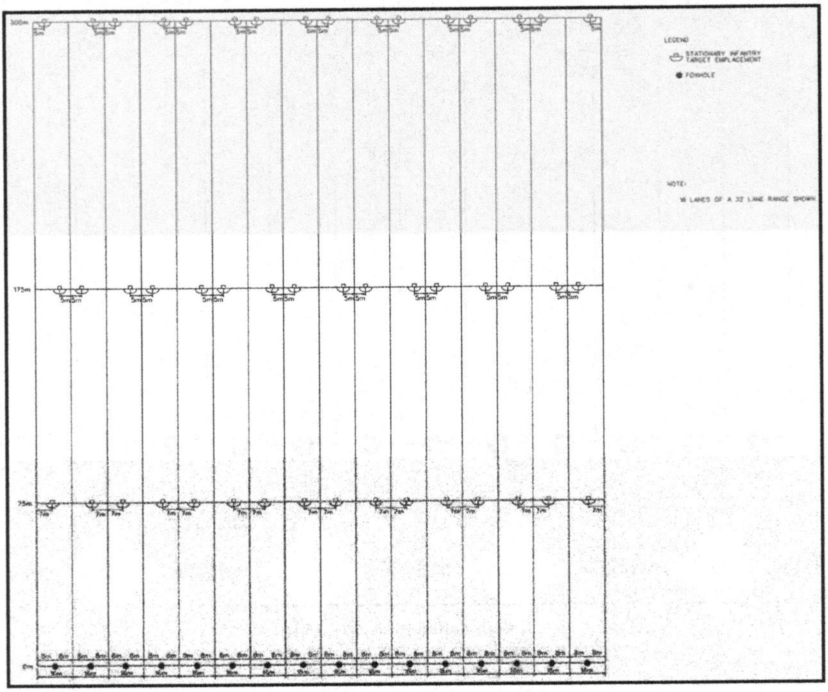

Figure E-2. Automated field fire range (17803).

Automated Record Fire Range (17805)

E-52. The ARF range (Figure E-3) is used to train and test individual Soldiers on the skills necessary to identify, engage, and defeat stationary Infantry targets for day/night qualification requirements with M16- and M4-series weapons. Its primary features include—

- 112 stationary Infantry targets.
- 16 foxholes.

E-53. All targets are fully automated, and the event-specific target scenario is computer-driven and scored from the range operations center. The range operating system is fully capable of providing immediate performance feedback to the participants.

NOTE: 1. To perform night fire, Soldiers fire at 50-meter targets from the night-fire line and baseline.

2. Range Set-up: Replace one of the 50m F-type silhouettes with an E- type silhouette. Low light illumination capability is required in both of the 50-meter target emplacements. Engage the F- type from the 25-meter night-firing line, and the E- type from the baseline.

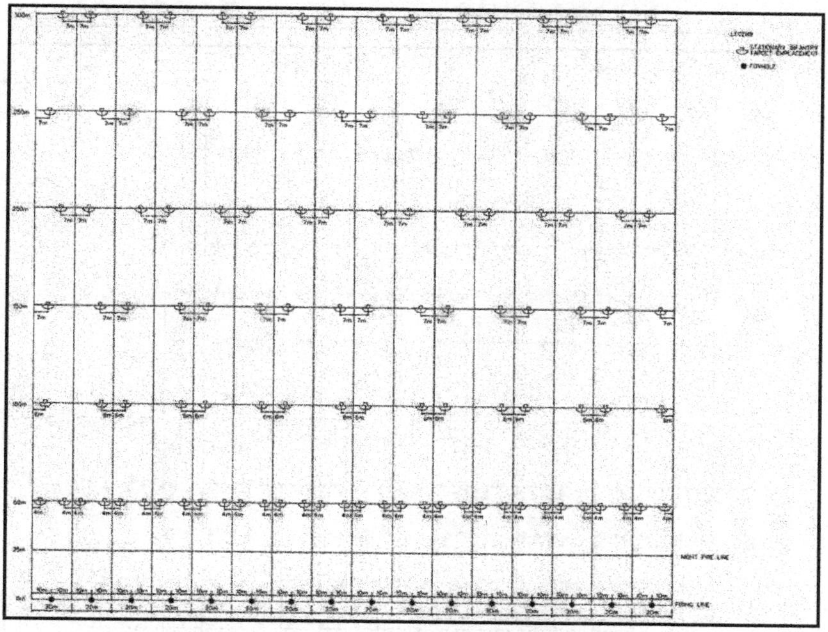

Figure E-3. Automated record fire range (17805).

Modified Record Fire Range (17806)

E-54. The modified record fire (MRF) range (Figure E-4) is used to train and test individual Soldiers on the skills necessary to identify, engage, and defeat stationary Infantry targets for day/night qualification requirements with M16- and M4-series weapons. This range combines the capabilities of the AFF (17803), ARF (17805), and automated night fire (17808) ranges to reduce land and maintenance requirements and increase efficiencies. Its primary features include—

- 144 stationary Infantry targets.
- 16 foxholes

E-55. All targets are fully automated, and the event-specific target scenario is computer-driven and scored from the range operations center. The range operating system is fully capable of providing immediate performance feedback to the participants.

NOTE: 1. To perform night fire, Soldiers fire at 50-meter targets from the night-fire line and baseline.

2. Range Set-up: Replace one of the 50m F-type silhouettes with an E- type silhouette. Low light illumination capability is required in both of the 50-meter target emplacements. Engage the F- type from the 25-meter night-firing line, and the E- type from the baseline.

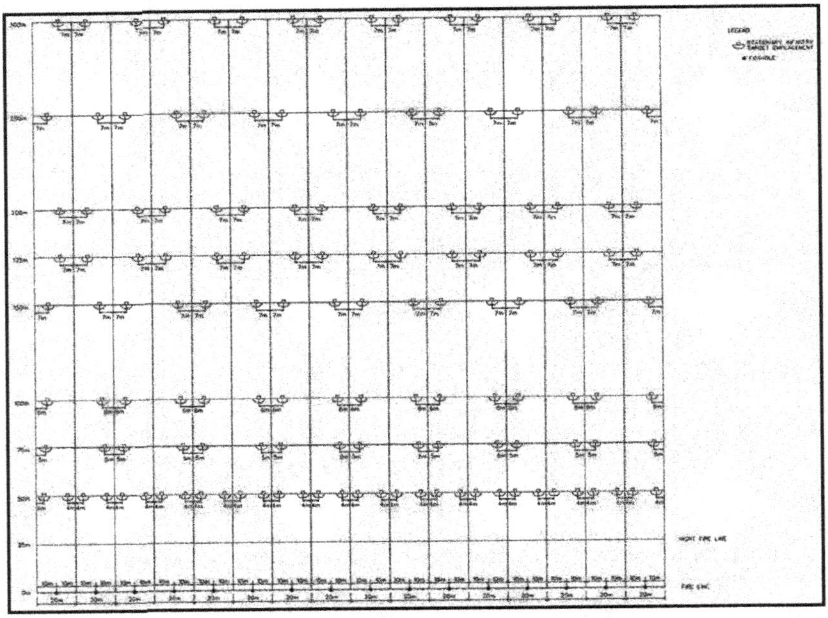

Figure E-4. Modified record fire range (17806).

Qualification Training Range (17809)

E-56. The qualification training (QTR) range (Figure E-5) is used to train and test Soldiers on the skills necessary to detect, identify, engage, and defeat stationary and moving Infantry targets in a tactical array with their prescribed weapons. This range combines the capabilities of the MRF range (17806), automated sniper field fire range (17812), combat pistol/military police firearms qualification course (17822), and multipurpose machine gun range (17833) to centralize training and reduce land, maintenance, and unit overhead requirements. Its primary features include—

- 429 stationary Infantry targets.
- 20 stationary armor targets.
- 20 moving Infantry targets.
- 10 stationary Infantry emplacements.
- 32 lanes rifle/machine gun zero.
- 15 lanes combat pistol qualification.
- 10 lanes sniper field fire.
- 16 lanes modified record fire.
- 10 lanes multipurpose machine gun.
- 2 target mechanisms each (400-meter and 700-meter).

E-57. All targets are fully automated, and the event-specific target scenario is computer-driven and scored from the range operations center. The range operating system is fully capable of providing immediate performance feedback to the participants.

NOTE: This range enhances throughput capability for units with multiple weapon densities by consolidating unit efforts to operate one training facility.

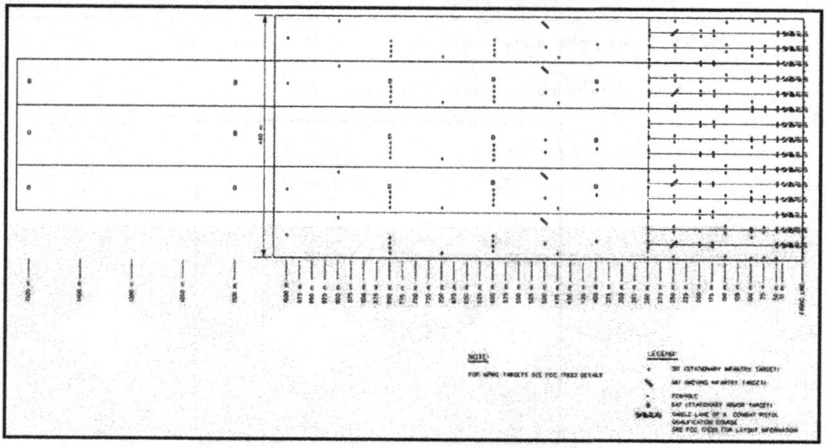

Figure E-5. Qualification training range (17809).

Appendix F

10-Meter Target Offsets and 25-Meter Zero Offsets

This section provides the 10-meter target offsets and the 25-meter zero offsets for M16- and M4-series weapons mounted with iron sights, optics, MILES, or aiming lasers. A blank, reproducible 10-meter target offset (Figure F-2) and an example of each weapon configuration (Figures F-3 through F-6) are provided. The M16A2 300-meter zeroing target is used for 25-meter zeroing with all weapon configurations.

MARKING 10-METER TARGET OFFSETS

F-1. To mark the proper 10-meter target offsets—

(1) Find the correct template for the weapon configuration.

(2) Starting from the center of the borelight circle on the offset, count the number of squares to the desired point of aim.

EXAMPLE

L2.0, U2.4

Starting from the center of the borelight circle (0.0, 0.0), move left 2 squares and up 2.4 squares.

NOTE: Each template also provides a number formula for the proper offset.

(3) Place the appropriate symbol or mark (Figure F-1).

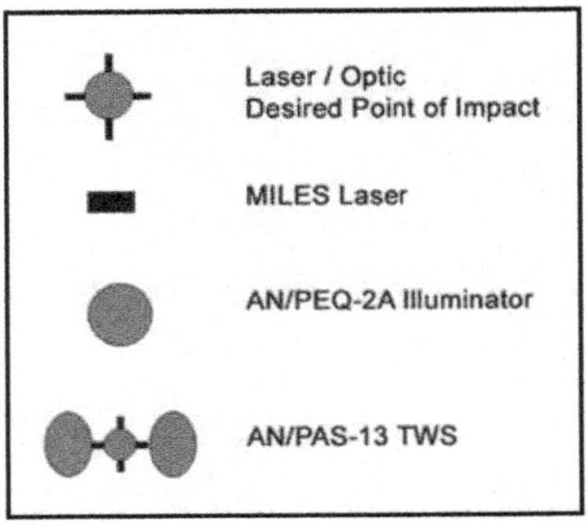

Figure F-1. 10-meter target offset symbols.

MARKING 25-METER ZERO OFFSETS

F-2. To mark the proper 25-meter zero offsets—

(1) Use only a M16A2 300-meter zeroing target.

(2) Find the correct target template for the weapon configuration.

(3) Count the number of squares, starting from the center of the 300-meter zeroing silhouette.

(4) Mark the designated strike point by drawing a small circle at the appropriate number of squares from the center of the 300-meter zeroing silhouette.

(5) Draw a 4-centimeter by 4-centimeter square, keeping the designated strike point at the center.

NOTES: 1. To reproduce the 10-meter target offset, copy the blank 10-meter target offset and place the example of the weapon being used on the back. This reproducible copy can be laminated and used repeatedly.

2. Table F-1 provides offset mounting information for various weapon configurations.

10-METER BORESIGHT TARGET

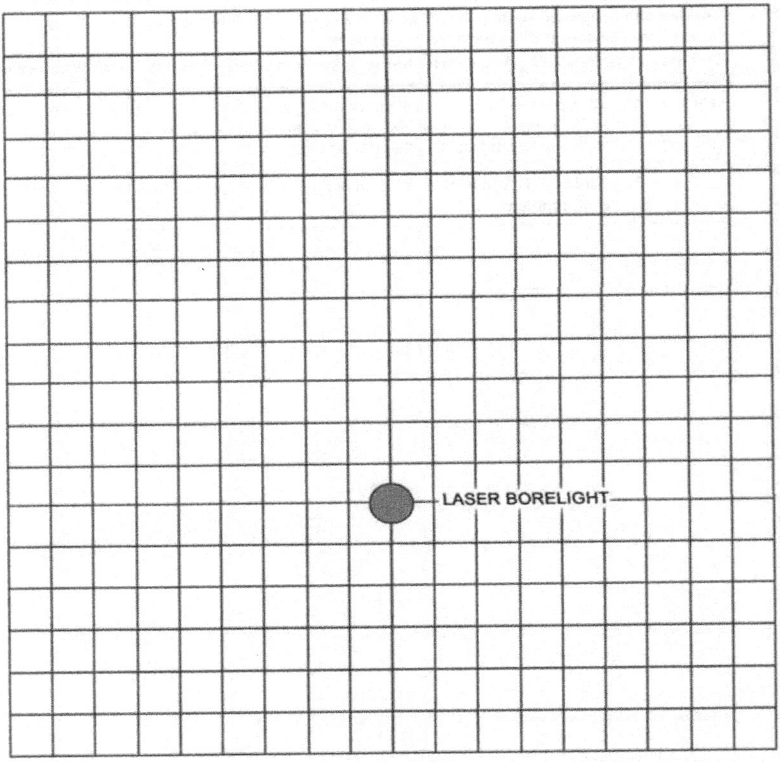

1 - USE THE CORRECT OFFSET FOR THE WEAPON, SIGHT AND LOCATION CONFIGURED
2 - STABILIZE THE WEAPON AND THE OFFSET
3 - ZERO BORELIGHT WHILE INSIDE THE BARREL OF THE WEAPON
4 - ALIGN THE LASER OF THE BORELIGHT WITH THE DOT ON THE 10M OFFSET
5 - ALIGN THE MILES LASER WITH THE MILES RECTANGLE ON THE 10M OFFSET (IF APPLICABLE)
6 - ADJUST AIMING LASER UNTIL CENTERED ON CROSSHAIR
7 - PLACE OPTIC AIMPOINT CENTERED ON CROSSHAIR AND MAKE ADJUSTMENTS ON OPTIC UNTIL THE BORELIGHT LASER IS ALIGNED WITH THE DOT ON THE 10M OFFSET
8 - RECONFIRM ALL DEVICES ARE STILL ALIGNED TO THEIR AIMING MARK

US Army ARDEC
AMSTA-AR-CCL-A
AMSTA-AR-FSF-R
Picatinny Arsenal, NJ 07089

Figure F-2. Blank 10-meter target offset.

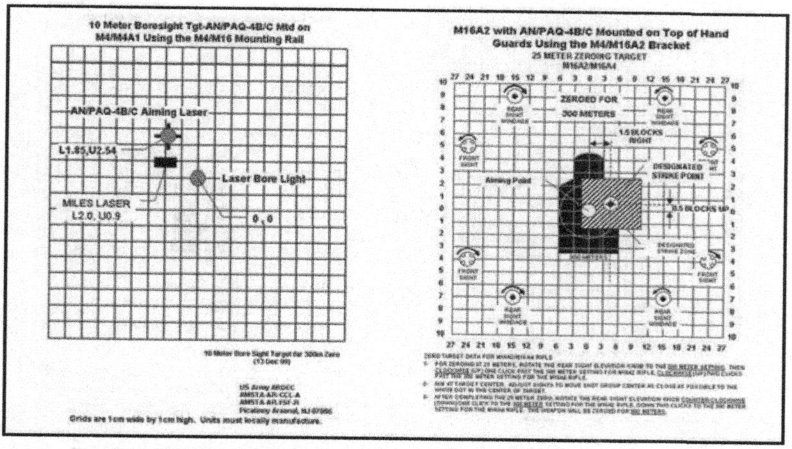

Figure F-3. M16A2 10-meter boresighting target/25-meter zeroing target offsets.

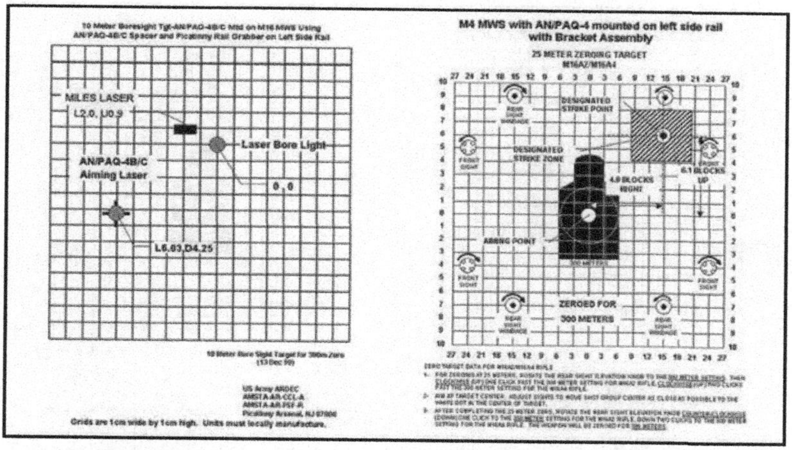

Figure F-4. M4 MWS 10-meter boresighting target/25-meter zeroing target offsets.

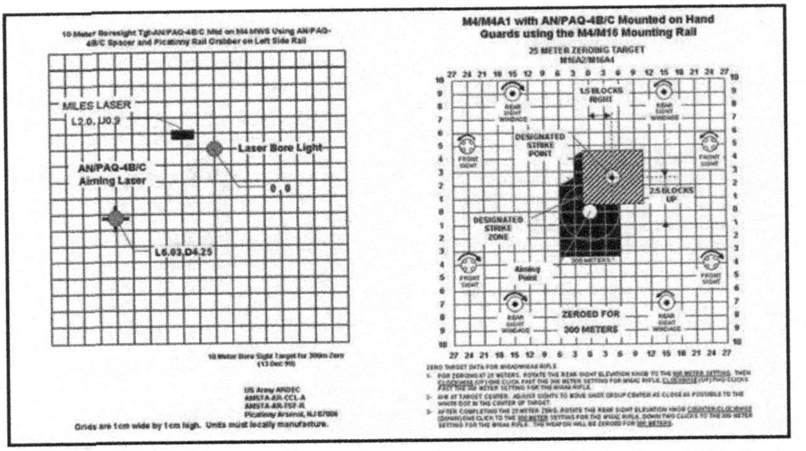

Figure F-5. M4/M4A1 10-meter boresighting target/25-meter zeroing target offsets.

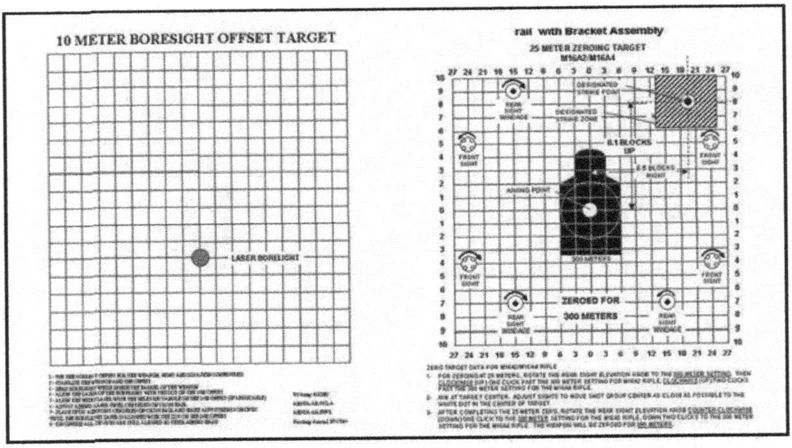

Figure F-6. M16A4 MWS 10-meter boresighting target/25-meter zeroing target offsets.

Table F-1. Offset mounting.

WEAPON	ACCESSORY	RAIL GRABBER	MOUNT	LOCATION	RANGE TO ZERO	ZERO OFFSET	BORESIGHT TARGET	MILES OFFSET
M16A2	Iron sight	N/A	N/A	N/A	300 m	0.0 0.0	0.0 4.2U	2.0L 0.9U
M16A2	M68	N/A	M68 gooseneck bracket	Carrying handle	300 m	0.0 1.4 cm DN	0.0 5.2U	2.0L 2.4U
M16A2	LTWS	TWS	TWS bracket assembly	Carrying handle	300 m	0.0 10D	0.0 13.4U	2.0L 2.4U
M16A2	TWS	N/A	TWS bracket assembly	Carrying handle	300 m	0.0 8.1D	0.0 11.5U	2.0L 2.4U
M16A2	AN/PVS-4	N/A	AN/PVS-4 mounting knob	Carrying handle	300 m	0.0 7.0D	0.0 9.4U	2.0L 0.9U
M16A2	AN/PAQ-4B/C	N/A	M4/M16 bracket	Hand guards	300 m	1.5R 0.5U	1.85L 2.54U	2.0L 0.9U
M16A2	AN/PEQ-2A/B	N/A	M4/M16 bracket	Hand guards	300 m	1.5L 0.5U	1.8R 2.4U	2.0L 0.9U
M16/M203	AN/PAQ-4B/C	N/A	Spacer and AN/PVS-4 mounting knob	Carrying handle	300 m	1.85R 2.6D	1.85L 8.6U	2.0L 3.9U
M16/M203	AN/PVS-4	N/A	AN/PVS-4 mounting bracket assembly	Carrying handle	300 m	4.2R 9.8D	TBD	2.0L 0.9U
M4/M4A1	BUIS	N/A	N/A	Upper receiver	300 m	0.0 0.0	0.0 4.01U	2.0L 0.9U
M4/M4A1	AN/PAQ-4B/C	N/A	M4/M16 bracket	Hand guards	300 m	1.5R 2.5U	1.85L 2.54U	2.0L 0.9U
M4/M4A1	LTWS	TWS	N/A	Upper receiver	300 m	0.0 4.5D	0.0 7.9U	TBD
M4/M4A1	TWS	Picatinny	TWS spacer and rail grabber	Upper receiver	300 m	0.0 5.7D	0.0 9.4U	2.0L 2.4U
M4/M4A1	AN/PEQ-2A/B	N/A	M4/M16 bracket	Hand guards	300 m	1.0L 0.3U	1.8R 2.4U	2.0L 0.9U
M4/M4A1	AN/PVS-4	Picatinny	Spacer and rail grabber	Upper receiver	300 m	0.0 3.4D	0.0 7.6U	2.0L 0.9U
M4/M4A1	M68	M68	Half-moon spacer	Upper receiver	300 m	0.0 1.4 cm DN	0.0 5.63U	2.0L 2.4U
M4/M203	BUIS	N/A	N/A	Upper receiver	300 m	0.0 0.0	0.0 6.01U	2.0L 0.9U
M4/M203	AN/PAQ-4B/C	N/A	Spacer and AN/PVS-4 mounting knob	Carrying handle	300 m	1.3R 1.9D	1.85L 8.6U	2.0L 0.9U
M4/M203	AN/PVS-4	Picatinny	Spacer and rail grabber	Upper receiver	300 m	0.0 3.4D	0.0 9.6U	2.0L 3.9U

NOTE: Target offsets not yet developed are indicated by TBD (to be developed).

Table F-1. Offset mounting (continued).

WEAPON	ACCESSORY	RAIL GRABBER	MOUNT	LOCATION	RANGE TO ZERO	ZERO OFFSET	BORESIGHT TARGET	MILES OFFSET
M4 MWS	BUIS	N/A	N/A	Upper receiver	300 m	0.0 0.0	0.0 4.01U	2.0L 0.9U
M4 MWS	AN/PVS-4	Picatinny	Spacer and rail grabber	Upper receiver	300 m	0.0 3.4D	0.0 7.6U	2.0L 0.9U
M4 MWS	M68	M68	Rail grabber	Upper receiver	300 m	0.0 1.4 cm DN	0.0 5.63U	2.0L 2.4U
M4 MWS	LTWS	TWS	N/A	Upper receiver	300 m	0.0 4.5D	0.0 7.9U	2.0L 2.4U
M4 MWS	TWS	TWS	Spacer	Upper receiver	300 m	0.0 5.7D	0.0 9.4U	2.0L 2.4U
M4 MWS	ANPEQ-2A	Insight	N/A	Left	300 m	TBD	4.5L 1.0D	2.0L 0.9U
M4 MWS	AN/PEQ-2A/B	Insight	N/A	Right	300 m	N/A	5.5R 5.4D	2.0L 0.9U
M4 MWS	AN/PEQ-2A/B	Insight	N/A	Top	300 m	1.5L 0.5D	2.9R 2.3U	2.0L 0.9U
M4 MWS	AN/PEQ-2A/B	Picatinny	Spacer	Top	300 m	N/A	1.95R 4.1U	2.0L 0.9U
M4 MWS	AN/PEQ-2A/B	Picatinny	Spacer	Right	300 m	N/A	6.35R 4.4D	2.0L 0.9U
M4 MWS	AN/PEQ-2A/B	Picatinny	Spacer	Left	300 m	6.9R 2.0U	6.2L 0.60D	2.0L 0.9U
M4MWS	AN/PEQ-2A/B	Insight	Training adapter	Top	300 m	2.0L 1.5D	N/A	2.0L 0.9U
M4 MWS	AN/PAQ-4B/C	Picatinny	AN/PAQ-4B/C bracket adapter	Top	300 m	4.9R 6.1U	1.75L 3.9U	2.0L 0.9U
M4 MWS	AN/PAQ-4B/C	Picatinny	AN/PAQ-4B/C bracket adapter (spacer)	Right	300 m	N/A	6.9R 0.9D	2.0L 0.9U
M4 MWS	AN/PAQ-4B/C	Insight	N/A	Top	300 m	N/A	1.75L 2.15U	2.0L 0.9U
M4MWS	AN/PAQ-4B/C	Insight	N/A	Right	300 m	N/A	4.35R 0.65D	2.0L 0.9U
M4MWS	AN/PAQ-4B/C	Insight	N/A	Left	300 m	N/A	4.30L 4.25D	2.0L 0.9U
M4 MWS M203	BUIS	N/A	N/A	Upper receiver	300 m	0.0 0.0	0.0 6.01U	2.0L 0.9U
M4 MWS M203	AN/PAQ-4B/C	Picatinny	Bracket adapter (spacer	Left	300 m	4.9R 6.1U	6.0L 4.0D	2.0L 3.9U
M4 MWS M203	AN/PVS-4	Picatinny	Spacer	Upper receiver	300 m	0.0 3.4D	0.0 9.6U	2.0L 3.9U
M16A4 MWS	BUIS	N/A	N/A	Upper receiver	300 m	0.0 0.0	0.0 4.01U	2.0L 0.9U
M16A4 MWS	AN/PAQ-4B/C	Picatinny	AN/PAQ-4B/C bracket adapter (spacer)	Left	300 m	6.5R 8.1U	6.03L 4.25D	2.0L 0.9U
M16A4 MWS	TWS	TWs	Spacer	Upper receiver	300 m	0.0 6.0D	0.0 9.4U	2.0L 2.4U
M16A4 MWS	M68	M68	N/A	Upper receiver	300 m	0.0 0.0	0.0 5.63U	2.0L 2.4U
M16A4 MWS	AN/PEQ-2A/B	Insight	N/A	Left	300 m	3.0R 3.0U	4.5L 1.0D	2.0L 0.9U
M16A4 MWS	AN/PVS-4	Picatinny	Spacer	Upper receiver	300 m	0.0 4.6D	0.0 7.6U	2.0L 0.9U

NOTE: Target offsets not yet developed are indicated by TBD (to be developed).

Table F-1. Offset mounting (continued).

WEAPON	ACCESSORY	RAIL GRABBER	MOUNT	LOCATION	RANGE TO ZERO	ZERO OFFSET	BORESIGHT TARGET	MILES OFFSET
M16A4 MWS M203	BUIS	N/A	N/A	Upper receiver	300 m	0.0 0.0	0.0 6.01U	2.0L 0.9U
M16A4 MWS M203	AN/PAQ-4B/C	Picatinny	AN/PAQ-4B/C bracket adapter (spacer	Left	300 m	6.5R 8.1U	6.0L 4.0D	2.0L 3.9U
M16A4 MWS M203	AN/PVS-4	Picatinny	Spacer	Upper receiver	300 m	0.0 4.6D	0.0 9.6U	2.0L 3.9U
M16A4 MWS	AN/PEQ-2A/B	Picatinny	Spacer	Left	300 m	6.0R 2.0U	6.2L 0.60D	2.0L 0.9U
M16A4 MWS	AN/PEQ-2A/B	Picatinny	Spacer	Right	300 m	TBD	6.35R 4.4D	2.0L 0.9U
M16A4 MWS	AN/PEQ-2A/B	Picatinny	Spacer	Top	300 m	TBD	1.95R 4.1U	2.0L 0.9U
M16A4 MWS	AN/PEQ-2A/B	Insight	N/A	Right	300 m	TBD	5.5R 5.4 D	2.0L 0.9U
M16A4 MWS	AN/PEQ-2A/B	Insight	N/A	Top	300 m	1.5L 0.5D	2.0R 2.3U	2.0L 0.9U
M16A4 MWS	AN/PEQ-2A/B	Insight	Training adapter	Top	300 m	2.0L 1.5D	TBD	2.0L 0.9D
M16A4 MWS	AN/PAQ-4B/C	Picatinny	AN/PAQ-4B/C bracket adapter	Top	300 m	4.9R 6.1U	1.75L 3.9U	2.0L 0.9U
M16A4 MWS	AN/PAQ-4B/C	Picatinny	AN/PAC-4B/C bracket adapter	Right	300 m	N/A	6.0R 0.9D	2.0L 0.9U
M16A4 MWS	AN/PAQ-4B/C	Insight	N/A	Top	300 m	N/A	1.75L 2.15U	2.0L 0.9U
M16A4 MWS	AN/PAQ-4B/C	Insight	N/A	Right	300 m	N/A	4.35R 0.65D	2.0L 0.9U
M16A4 MWS	AN/PAQ-4B/C	Insight	N/A	Left	300 m	N/A	4.30L 4.25D	2.0L 0.9U

NOTE: Target offsets not yet developed are indicated by TBD (to be developed).

Glossary

SECTION I – ACRONYMS AND ABBREVIATIONS

AAR	after-action review
ACH	Army combat helmet
ACOG	advanced combat optical gunsight
AFF	automated field fire
AMU	Army Marksmanship Unit
AP	armor-piercing
AR	Army regulation
ARF	automated record fire
ARM	advanced rifle marksmanship
ARS	adapter rail system
ATC	Army training center
ATPIAL	advanced target pointer/illuminator aiming light
BDC	bullet drop compensated
BFA	blank firing attachment
BOLC	Basic Officer Leaders Course
BRM	basic rifle marksmanship
BUIS	backup iron sight
CBRN	chemical, biological, radiological, and nuclear
CCO	close combat optic
COA	course of action
cm	centimeter or centimeters
CRM	composite risk management
DA Pam	Department of the Army pamphlet
EST	Engagement Skills Trainer
FM	field manual
FOV	field of view
fps	feet per second
ft	foot or feet
HWTS	heavy weapon thermal sight
I2	image-intensifier
IBA	interceptor body armor
IET	initial entry training
in	inch or inches
IR	infrared
KD	known distance
lb	pound or pounds
LBE	load-bearing equipment
LFX	live-fire exercise
LMTS	Laser Marksmanship Training System
LOGCAP	Logistics Civil Augmentation Program
LOMAH	location of misses and hits

LWTS	light weapon thermal sight
m	meter or meters
MDMP	military decision-making process
METL	mission-essential task list
METT-TC	mission, enemy, terrain and weather, troops and support available, time available, civil considerations
min	minute or minutes
mm	millimeter or millimeters
MOA	minute of angle
MOPP	mission-oriented protective posture
mph	miles per hour
MRF	modified record fire
MWS	modular weapon system
NCOES	Noncommissioned Officers Education System
NCOIC	noncommissioned officer in charge
NFOV	narrow field of view
NGO	nongovernmental organizations
NOD	night observation devices
NSN	national stock number
NVD	night vision device
OIC	officer in charge
OSUT	one-station unit training
OTV	outer tactical vest
PASGT	personnel armor system for ground troops
PH	probability of hit
PMCS	preventive maintenance checks and services
PMI	preliminary marksmanship instruction
POI	program of instruction
QTR	qualification training
RETS	Remote Electronic Target System
RFA	rimfire adapter
RSO	range safety officer
RTO	radio telephone operator
SAPI	small arms protective insert
SAW	squad automatic weapon
SDM	squad designated marksman
SDZ	surface danger zone
sec	second or seconds
SRC	short-range combat
SRM	short-range marksmanship
SRT	special reaction team
SRTA	short-range training ammunition
STRAC	Standards in Training Commission (DA Pam 350-38)
STX	situation training exercise
TADSS	training aids, devices, simulators, and simulations
TC	training circular
TSC	training support centers
TM	technical manual

TPIAL	target pointer illuminator/aiming light
TWS	thermal weapon sight
UO	urban operations
USACRC	United States Army Combat Readiness Center
WFF	warfighting functions
WFOV	wide field of view

SECTION II – TERMS

Active Army: All Regular Army (RA) forces in the Active Army.

adjusted aiming point: An aiming point that allows for gravity, wind, target movement, zero changes, and MOPP firing.

advanced marksmanship: Normally refers to marksmanship skills taught during ARM.

advanced rifle marksmanship: Normally refers to the formal marksmanship instruction received by infantrymen upon completion of BRM during OSUT.

aiming: A marksmanship fundamental; refers to the precise alignment of the rifle sights with the target.

aiming card: The M15A1 aiming card is a cardboard sleeve with a moveable insert. The rear sight aperture, front sight post, and target are pictured. This training device is used in conjunction with aiming instructions.

aiming point: A place on a target in which the rifle sights are aligned normally the target center of mass.

alibi target: A target or additional target a soldier is allowed to engage during qualification firing when unable to complete a record fire scenario due to circumstances beyond his control; for example, a target mechanism, weapon, or ammunition malfunction.

alternate course: Alternatives to standard qualification courses.

ammunition lot: A quantity of cartridges, each of which is made by one manufacturer under uniform conditions and is expected to work in a uniform manner.

ammunition lot number: Code number that identifies a particular quantity of ammunition from one manufacturer.

aperture: The hole in the rear sight.

armorer: One who services and makes repairs on small arms and performs similar duties to keep small arms ready for use.

Army Training and Evaluation Program: A guide for the training and evaluation of critical unit combat missions – crew/squad through battalion/task force echelon.

Army Training Center: Conducts OSUT and BRM. Locations are Fort Benning, GA; Fort Jackson, SC; Fort Knox, KY.

artificial illumination: Any light from a man-made source.

assault course: An area of ground used for training soldiers in attacking an enemy in close combat.

automatic fire: A firing mode that causes the weapon to continue firing as long as the trigger is held or until all ammunition has been expended.

ball: The projectile; the bullet.

ball ammunition: General-purpose standard service ammunition with a solid core bullet.

ball and dummy: An exercise that substitutes a dummy round for a live round without the firer knowing it. An excellent exercise for identifying and correcting trigger jerks.

ballistics: A science that deals with the motion and flight characteristics of projectiles.

barrel erosion: Wearing away of the surface of the bore due to the combined effects of gas washing, coring, and mechanical abrasion.

basic marksmanship: Fundamental marksmanship skills taught in BRM during IET and OSUT.

basic rifle marksmanship: The formal course of marksmanship instruction received by all soldiers.

battlesight zero: A sight setting that soldiers keep on their weapons. It provides the highest probability of hitting most high-priority combat targets with minimum adjustment to the aiming point, a 250 meter sight setting as on the M16A1 rifle, and a 300 meter sight setting as on the M16A2 rifle.

blank ammunition: A complete cartridge without the bullet used to simulate weapon firing.

blank firing adapter: A device that fits in the muzzle of the rifle; used only with blank ammunition.

brass: An alloy of copper and zinc used to make cartridge cases and bullet jackets. Also, a common name for expended cases.

breath control: The third marksmanship fundamental; refers to the control of breathing to help keep the rifle steady during firing.

bullet: The projectile or ball; the part that goes downrange. It may also be used to refer to the complete cartridge.

bull's-eye target: Any target with a round black circle and scoring rings. Normally used in competitive marksmanship training.

buttplate: Metal or rubber covering of the end of the stock on the rifle.

cadre coach: A trainer with expertise and knowledge exceeding that of the firer.

caliber: Diameter of the bore; for example, the M16-series rifle bore is 5.56mm (.223 inch).

cartridge: A complete round of ammunition.

center of mass: A point that is horizontally (left and right) and vertically (up and down) at the center of the target.

chambering: The step in the cycle of operation that refers too fully seating the round in the chamber of the rifle.

chamber plug: A range safety device that is a small plastic plug designed to fit into the chamber of the M16. A handle extends out the ejection port so safety personnel can see at a glance that the rifle is clear of ammunition.

clock method: Method of calling shots by referring to the figures on an ordinary clock dial assumed to have the target at its center. Also a method of determining the strength and direction of wind.

coach: Any individual who assists firers on the firing line.

coach-and-pupil method: Method of training in which pairs of pupils take turns practicing a procedure explained by the instructor/trainer.

cocking: The step in the cycle of operation that refers to the rearward movement of the bolt riding over the hammer, resetting the weapon for subsequent firing.

collective firing proficiency: Units delivering effective fire in a tactical setting. It requires individual skill plus command and control to engage all targets within an assigned sector.

concurrent training: Training that occurs at the same time that other unit members are using the primary training facilities.

cookoff: A round that fires as a result of a hot chamber without the trigger being pulled. It can occur any time until the weapon is cooled.

crack and thump: A method to determine the general direction and distance to an enemy firer who is shooting at you.

cradle: A vise-like mechanism that holds a weapon in a secure position for test firing.

cross dominance: A soldier with a dominant hand and a dominant eye that are not the same; for example, a right-hander firer with a dominant left eye.

cycle of operation: The eight steps involved in firing a round of ammunition: feeding, chambering, locking, firing, unlocking, extracting, ejecting, and cocking.

cyclic rate of fire: The maximum rate at which a weapon will fire in the automatic mode.

dime-washer exercise: A dry-fire exercise used to practice trigger squeeze.

downrange feedback: Used to describe any training technique that provides precise knowledge of bullet strike (whether hit or miss).

dry fire: A technique used to simulate the firing of a live round with an empty weapon. Any application of the fundamentals of marksmanship without live ammunition may be referred to as dry fire.

dry-fire moving target trainer: A small-motorized scaled target device used to teach the engagement of moving personnel targets.

dummy ammunition: A cartridge without a primer or powder. Primarily used for ball-and-dummy exercises on the live-fire line.

effective wind: The average of all the varying winds encountered.

ejection: The step in the cycle of operation that removes the expended cartridge from the weapon out of the ejection port.

elevation adjustment: Rotating the front sight post to cause the bullet to strike higher or lower on the target.

expert: The highest qualification rating.

external ballistics: What happens to the bullet between the time it leaves the rifle and the time it arrives at the target.

extraction: The step in the cycle of operation that pulls the round from the chamber.

eye relief: The distance from the firing eye to the rear sight. Eye relief is a function of stock weld.

feedback: Obtaining knowledge of performance.

feedback target: Targets designed for use at 75, 175, or 300 meters; includes an overprinted grid similar to a zero target.

feeding: The step in the cycle of operation that is the forward movement of the bolt, stripping the top round from the magazine and moving it toward the chamber.

field firing: Training on the standard field firing range with target banks at 75, 175, and 300 meters.

firing: The step in the cycle of operation that refers to pulling the trigger, releasing the hammer to strike the firing pin, which strikes the primer. The primer ignites and, in turn, ignites the powder charge within the cartridge case.

firing hand: The right hand of a right-handed firer. The left hand of a left-handed firer.

firing pin: Plunger in the bolt of a rifle that strikes the primer.

fleeting target: A moving target remains within observing or firing distance for such a short period that it affords little time for deliberate adjustment and fire against it.

functioning: (See cycle of operation.)

fundamentals of rifle marksmanship: The four essential elements needed to hit targets: steady position, aiming, breath control, and trigger squeeze.

gravity: The natural pull of all objects to the center of the earth.

grouping: A live-fire exercise with the objective of shooting tight shoot groups.

gun bore line: A reference line established by the linear extension of the bore axis of a gun.

headspace: The distance between the face of the bolt(fully closed) and the face of a fully chambered cartridge.

hold-off: (See adjusted aiming point.)

horizontal dispersion: The left-to-right displacement of bullets on a target.

immediate action: procedures applied to rapidly reduce any rifle stoppage without determining its cause.

individual firing proficiency: Individual firing skills; for example, an individual's performance on the record fire course.

Infantry Remoted Target System (IRETS): (See RETS.)

infrared aiming light: A unique night sighting system that uses infrared light to assist in the aiming process.

initial entry training: Indicates the first training received by a new soldier, includes the MOS-producing portion of his training such a one-station unit training (OSUT).

initial pressure: The applications of about half of the total trigger pressure it takes to fire a rifle.

instructor-trainer ratio: The number of soldiers for which each instructor/trainer is responsible.

internal ballistics: What happens to the bullet before it leaves the muzzle of the rifle.

known distance: Describes the older range complexes with large target frames behind a large berm and firing lines at 100 yards or 100-meter increments. (See FM 25-7.)

laser: Light amplification by simulated emission of radiation.

lead: Distance ahead of a moving target that a rifle must be aimed to hit the target.

lead rule: Provides the soldier guidance on how to adjust his aiming point to hit moving targets.

line of sight: A line between the rifle and the aiming point, extending from the firing eye through the center of the rear aperture, across the tip of the front sight post, and onto the target.

location of misses and hits : A projectile location system that provides immediate and precise information to the firer concerning bullet strike (hit or miss).

locking: The step in the cycle of operation that is a counterclockwise rotation of the bolt, securing it into the barrel locking lugs.

long-range sight: The aperture marked L on the M16A1 rifle equipped with standard sights; provides for a zero at 375 meters. The M16A1 rifle equipped with LLLSS has an aperture marked L, but it is a regular sight.

Low-Light Level Sight System (LLLSS): A sighting system for low visibility firing that replaces the standard front and rear sights on the M16A1 rifle.

marksman: The designation given to the lowest qualification rating.

maximum effective range: The greatest distance at which a soldier may be expected to deliver a target hit.

maximum effective rate of fire: The highest rates of fire that can be maintained and still achieve target hits.

maximum range: The longest distance a projectile will travel when fired from a weapon held at the optimum angle.

minute of angle: A angle that would cover 1 inch at a distance of 100 yards, 2 inches at 200 yards, and so on. Each click of sight adjustment on the M16A1 rifle with standard sights is equal to one minute of angle.

Multiple Integrated Laser Engagement System (MILES): A tactile shooting device that uses a low-powered laser to activate detectors placed on people and vehicles.

muzzle velocity: The speed of a projectile as it leaves the muzzle of the weapon.

natural point of aim: The direction of the body/rifle combination is oriented while in a stable, relaxed firing position.

natural respiratory pause: The temporary cessation of breathing between an exhale and inhale.

night firing: Firing performed under all conditions of limited visibility.

nonfiring hand: The opposite of the firing hand.

optical sight: Sight with lenses, prisms, or mirrors used in lieu of iron sights.

Paige sighting device: A device with a small-scaled target that fits into the muzzle of the weapon, allowing the soldier to practice aiming.

pasters: Small white or black gum-backed paper used for covering bullet holes.

peep sight: The rear sight; a sight with a small aperture (hole).

peer coach: A soldier with shooting experience and knowledge equal to that of the firer he is coaching.

pit: The target area behind the large berm of a KD range.

plastic practice ammunition: Ammunition with a plastic projectile, high-muzzle velocity (the light weight causes it to lose velocity rapidly with a maximum range of 250 meters or less) designed for use in close-in training areas; frangible bullet.

point of aim: The exact spot on a target the rifle sights are aligned with.

point of impact: The point that a bullet strikes; usually considered in relation to point of aim.

pop, no kick: A firing condition when the primer ignites and the powder charge does not. This normally results in lodging the bullet inside the barrel.

pop-up target: A silhouette target that is activated remotely so it can suddenly appear and fall when struck by a bullet.

practice record: Firing conducted on a qualification course for practice.

predetermined fire: A technique of aligning the rifle during good visibility so the rifle can be aligned and fired on designated areas when they cannot be seen due to darkness, smoke, or fog.

preparatory marksmanship training: All marksmanship training that takes place before live fire.

primer: A small explosive device in the center base of the cartridge case that is struck by the firing pin to fire the round.

probability of hit: Ranging from 0 to 1.0, it refers to the odds of a given round hitting the target at a given range.

qualification firing: Firing on any authorized course that results in meeting qualification requirements; may also be called record fire. (See record fire.)

quick fire: A technique of fire used to engage surprise targets at close range.

range card: Small chart on which ranges and directions to various targets and other important points in the area under fire are recorded.

rapid semiautomatic fire: A firing procedure that results in an accurate shot being fired every one or two seconds.

receiver: That portion of a firearm that holds the barrel and houses the bolt and firing mechanism.

recoil: The rearward motion or kick of a gun upon firing.

record fire: Any course of fire used to determine if qualification standards are met. The standard record fire course consists of 40 target exposures at ranges between 50 and 300 meters. The standard course requires 23 hits to qualify as marksman, 30 for sharpshooter, and 36 for expert.

reduced range ammunition: Ammunition that is designed to be a ballistic match with service ammunition to an appropriate range for training (may be less than maximum effective range) and a reduced maximum range.

regular rear sight: The M16A1 rifle rear sight that is zeroed for 250 meters (the unmarked aperture on rifles with standard sights and the aperture marked L on rifles equipped with LLLSS).

reinforcement training: Training conducted that is over and above scheduled training.

remedial action: A procedure applied after immediate action has failed to correct a malfunction, which determines the cause of the malfunction.

remedial training: Additional training presented to soldiers who have demonstrated special shooting problems.

Remote Electronic Target System: Range complexes. Some ranges include moving targets.

Reserve Components: Includes Army National Guard and Army Reserve forces.

ricochet fire: Fire in which the projectile glances from a surface after impact.

Riddle sighting device: A small magnetic device with a scaled target that attaches to the front sight assembly, allowing the soldier to practice aiming.

rifle cant: Any leaning of the rifle to the left or right from a vertical position during firing.

rim-fire adapter: The caliber .22-rim fire adapter (M261) consists of a bolt and a magazine insert, which allows standard .22 caliber ammunition to be fired in the M16 rifle.

round: May refer to a complete cartridge or to a bullet.

scaled-silhouette target: Any target that is reduced in size. When it is observed from 25 meters, it looks the same size as though at a greater range.

sector of fire: An area assigned to an individual, weapon, or unit to be covered by fire.

semiautomatic fire: A mode of fire that allows one round to be fired each time the trigger is pulled.

serviceability checks: A technical inspection of the rifle to determine if it is safe to fire and in working condition. (May not ensure accuracy.)

service ammunition: Standard ammunition used by the military. Ammunition designed for combat.

service rifle: The primary rifle of a military force.

service school: Branch schools such as the US Army Infantry School at Fort Benning, Ga. and the Armor School at Fort Knox, Ky.

sharpshooter: The middle rating of qualification.

shot group: A number of shots fired using the same aiming point, which accounts for rifle, ammunition, and firer variability. Three shots are enough, but any number of rounds may be fired in a group.

shot group analysis: A procedure for analyzing the size of shot groups on a target to determine firer error.

sight alignment: Placing the center tip of the front sight post in the exact center of the rear aperture.

sighter rounds: Rounds fired that allow the bullet strike to be observed in relation to the aiming point.

sight picture: Placing correct sight alignment on a selected aiming point on a target.

sight radius: The distance from the front sight post to the rear sight aperture of a rifle.

sighting device (M16): A small metal device with a tinted square of glass that is placed on the carrying handle, allowing a coach to see what the firer sees through the sights.

silhouette target: A target that represents the outline of a man.

spotters: A round cardboard disk placed in bullet holes with a small wooden peg so the bullet strike can be observed from the firing line.

squad automatic weapon: A lightweight, one-man, 5.56mm machine gun.

starlight scope: A weapon scope that amplifies ambient light so targets can be seen and effectively engaged during darkness. The AN/PVS-2 and AN/PVS-4 are used on the M16 rifle.

steady position: The first marksmanship fundamental, which refers to the establishment of a position that allows the weapon to be held still while it is being fired.

stock weld: The contact of the cheek with the stock of the weapon.

supported position: Any position that uses something other than the body to steady the weapon (artificial support).

suppressive fire: Any engagement that does not have a definite or visible target. Firing in the general direction of known or suspected enemy location.

sustained rate of fire: Rate of fire that a weapon can continue to deliver for an indefinite period without overheating.

terminal ballistics: What happens to the bullet when it comes in contact with the target.

tight shot group: A shot group with all bullet holes close together.

tracer ammunition: Ammunition with a substance at the rear of the bullet that ignites soon after firing. It burns brightly so the trajectory of the bullet can be seen.

tracking: Engaging moving targets where the lead is established and maintained; moving with the target as the trigger is squeezed.

train the trainer: Describes any training that is designed to train marksmanship instructors or coaches.

trainfire: A marksmanship program using pop-up targets in a realistic environment.

trajectory: The flight path the bullet takes from the rifle to the target.

trapping: A technique for engaging moving targets. The aiming point is established forward of the target. The rifle is held stationary and fired as the target approaches the aiming point.

trigger squeeze: The fourth fundamental; squeezing the trigger so that the movement of firing is a surprise, the lay of the weapon is not disturbed, and a large target hit can be expected.

unit marksmanship: All marksmanship training that is conducted by units.

unlocking: The step in the cycle of operation that refers to the clockwise rotation of the bolt after firing, freeing the bolt from the barrel locking lugs.

unsupported position: Any position that requires the firer to hold the weapon steady using only his body (bone support).

vertical dispersion: The up-and-down displacement of bullets on a target.

Weaponeer: A training device that simulates the firing of the M16 rifle to provide performance feedback.

windage adjustment: Moving the rear sight aperture to cause the bullet to strike left or right on the target.

wind value: The effect the wind will have on the trajectory of the bullet.

wobble area: The natural movement of the weapon/sight on and around an aiming point when the weapon is being held in a steady position.

zero criterion: The standard or requirement for zeroing; 4cm or smaller group at 25 meters.

zeroing: Adjusting the rifle sights so bullets hit the aiming point at a given range.

zero target: A scaled-silhouette target with a superimposed grid for use at 25 meters.

References

SOURCES USED

These sources were used to prepare this publication.

AR 385-10, The Army Safety Program. 23 August 2007.

AR 385-63, Range Safety. 19 May 2003.

DA Pam 350-38, Standards in Weapons Training. 16 August 2004.

DA Pam 385-63, Range Safety. 10 April 2003.

FM 5-19, Composite Risk Management. 21 August 2006.

TM 9-1005-249-10, , Operator's Manual for Rifle, 5.56-mm, M16 (NSN 1005-00-856-6885), Rifle 5.56-mm, M16A1 (1005-00-073-9421). 11 May 1990.

TM 9-1005-319-10, Operator's Manual for Rifle, 5.56 MM, M16A2 W/E (NSN 1005-01-128-9936) (EIC: 4GM); Rifle, 5.56 MM, M16A3 (1005-01-357-5112); Rifle, 5.56 MM, M16A4 (1005-01-383-2872) (EIC: 4F9); Carbine, 5.56 MM, M4 W/E (1005-01-231-0973) (EIC: 4FJ); Carbine, 5.56 MM, M4A1 (1005-01-382-0953) (EIC: 4GC), 1 October 1998. 01 October 1998.

DOCUMENTS NEEDED

These documents must be available to the intended users of this publication.

ARMY PUBLICATIONS

AR 350-1, Army Training and Leader Development. 03 August 2007.

FM 7-0, Training the Force. 22 October 2002.

FM 7-1, Battle Focused Training. 15 September 2003.

TC 7-9, Infantry Live-Fire Training. 30 September 1993.

TC 25-1, Training Land. 15 March 2004.

TC 25-8, Training Ranges. 5 April 2004.

TM 9-1005-319-23&P, Unit and Direct Support Maintenance Manual (Including Repair Parts and Special Tools List) FOR RIFLE, 5.56MM, M16A2 W/E (NSN 1005-01-128-9936) (EIC: 4GM) Rifle, 5.56MM, M16A3, (1005-01-357-5112) Rifle, 5.56MM, M16A4, W/E (1005-01-383-2872) (EIC: 4F9) Carbine, 5.56MM, M4 (1005-01-231-0973) (EIC: 4FJ) and Carbine, 5.56MM, M4A1 (1005-01-382-0953) (EIC: 4GC), 1 May 1991.

TM 9-1240-413-13&P, Operator and Field Maintenance Manual Including Repair Parts and Special Tools List for M68 Sight, Reflex, w/Quick Release and Sight Mount (COMP M2: NSN 1240-01-411-1265) (COMP M4: NSN 1240-01-540-3690). 19 March 2008.

TM 9-5855-1914-13&P, Operator and Field Maintenance Manual Including Repair Parts and Special Tools List for the Advanced Target Pointer Illuminator Aiming Light (ATPIAL) AN/PEQ-15 (NSN 5855-01-534-5931). 16 February 2007.

TM 9-6920-746-12&P, Operator's and Organizational Maintenance Manual (Including Repair Parts and Special Tools List) M2 Practice Bolt Plastic Ammunition for Rifle 5.56mm, M16 Series (NSN 1005-01-184-4041). 09 September 1986.

TM 11-5855-213-10, Operator's Manual for Night Vision Sight, Individual Served Weapon, AN/PVS-4 (NSN 5855-00-629-5334) (EIC: IPJ). 01 February 1993.

TM 11-5855-261-10, Operator's Manual for Infrared Aiming Light, AN/PAQ-4 (NSN 5855-01-107-5925). 28 May 1981.

TM 11-5855-301-12&P, Operator's and Unit Maintenance Manual (Including Repair Parts and Special Tools List) Light, Aiming, Infrared, AN/PAQ-4B (NSN 5855-01-361-1362), AN/PAQ-4C (5855-01-398-4315).15 May 2000.

TM 11-5855-306-10, Operator's Manual for Monocular Night Vision Device, AN/PVS-14 (NSN 5855-01-432-0524) (EIC: IPX). 15 May 2007.

TM 11-5855-308-12&P, Operator's and Unit Maintenance Manual (Including Repair Parts and Special Tools List), Target Pointer Illuminator/Aiming Light AN/PEQ-2A (NSN 5855-01-447-8992). 15 May 2000.

TM 11-5855-312-10, Operator's Manual Sight, Thermal AN/PAS-13B(V)2 (NSN 5855-01-464-3152); AN/PAS-13B(V)3 (5855-01-464-3151). 15 February 2005.

TM 11-5855-316-10, Operator's Manual AN/PAS-13C(V)1 Sight, Thermal (NSN 5855-01-523-7707); AN/PAS-13C(V)2 Sight, Thermal (NSN 5855-01-523-7713) (MWTS); AN/PAS-13C(V)3 Sight, Thermal (NSN 5855-01-523-7715). 15 June 2006.

TM 11-5855-317-10, Operator's Manual For Sight, Thermal AN/PAS-13D(V)2 (NSN 5855-01-524-4313); AN/PAS-13D(V)3 NSN 5855-01-524-4314) (HWTS). 15 February 2007.

Army Forms

DA Form 2028, Recommended Changes to Publications and Blank Forms.

DA Form 3595-R, Record Fire Scorecard.

DA Form 3601-R, Single Target—Field Fire I Scorecard.

DA Form 5239-R, 100-, 200-, and 300-Meter Downrange Feedback Scorecard.

DA Form 5241-R, Single and Multiple Targets—Field Fire II Scorecard.

DA Form 5789-R, Record Fire Scorecard—Known-Distance Course (LRA).

DA Form 5790-R, Record Fire Scorecard—Scaled Target Alternate Course.

DA Form 7489-R, Record Night Fire Scorecard (LRA).

DA Form 7566, Composite Risk Management Worksheet.

DA Form 7649-R, Squad Designated Marksman—Record Fire I and II.

DA Form 7650-R, Squad Designated Marksman—Position Evaluation.

READINGS RECOMMENDED

These sources contain relevant supplementary information.

AR 190-11, Physical Security of Arms, Ammunition, and Explosives. 15 November 2006.

AR 350-38, Training Device Polices and Management. 15 October 1993.

DA Pam 350-9, Index and Description of Army Training Devices. 03 September 2002.

DA Pam 385-64, Ammunition and Explosive Safety Standards. 15 December 1999.

FM 1-02, Operational Terms and Graphics. 21 September 2004.

FM 3-06, Urban Operations. 26 October 2006.

FM 3-06.11, Combined Arms Operations in Urban Terrain. 28 February 2002.

FM 3-11.4, Multiservice Tactics, Techniques, and Procedures for Nuclear, Biological, and Chemical (NBC) Protection. 02 June 2003.

FM 4-25.11, First Aid. 23 December 2002.

FM 20-3, Camouflage, Concealment, and Decoys. 30 August 1999.

FM 21-10, Field Hygiene and Sanitation. 21 June 2000

FM 3-21.75, The Warrior Ethos and Soldier Combat Skills. 28 January 2008.

FM 22-6, Guard Duty. 17 September 1971.

FM 25-4, How to Conduct Training Exercises. 10 September 1984.

STP 21-1-SMCT, Soldier's Manual of Common Tasks, Skill Level 1. 14 December 2007.

STP 21-24-SMCT, Soldier's Manual of Common Tasks (SMCT), Skill Levels 2-4. 2 October 2006.

TC 25-20, A Leader's Guide to After-Action Reviews. 30 September 1993.

TC 90-1, Training for Urban Operations. 19 May 2008.

TM 9-1005-249-23&P, Unit and Direct Support Maintenance Manual (Including Repair Parts and Special Tools List) for Rifle, 5.56-mm, M16 (NSN 1005-00-856-6885) (EIC: 4F7); Rifle, 5.56-mm, M16A1 (1005-00-073-9421) (EIC: 4FC). 19 June 1991.

TM 9-6920-363-12&P, Operator's and Organizational Maintenance Manual (Including Repair Parts and Special Tools List) for Conversion Kit (Cal .22 Rimfire Adapter) M261 (NSN 1005-01-010-1561) for Rifle: 5.56-mm, M16 and M16A1. Reprinted with Changes 1 and 2. 21 August 1984.

INTERNET WEBSITES

U.S. Army Publishing Directorate, http://www.apd.army.mil/

Reimer Doctrine and Training Digital Library, http://www.train.army.mil

Index

SINGLE TARGET
FIELD FIRE I SCORECARD

For use of this form see FM 3-22.9; the proponent agency is TRADOC.

ID CODE	UNIT	DATE *(YYYYMMDD)*	EVALUATOR'S ID CODE

TABLE 1 — SUPPORTED FIRING POSITION

RD	RANGE (m)	TIME (sec)	HIT	MISS
1	75	6	☐	☐
2	175	8	☐	☐
3	300	10	☐	☐
4	175	8	☐	☐
5	75	6	☐	☐
6	300	10	☐	☐
7	300	10	☐	☐
8	75	75	☐	☐
9	175	175	☐	☐
10	175	8	☐	☐
11	300	10	☐	☐
12	175	8	☐	☐
13	75	6	☐	☐
14	300	10	☐	☐
15	175	8	☐	☐
16	75	6	☐	☐
17	300	10	☐	☐
18	75	6	☐	☐
TOTAL				

TABLE 2 — SUPPORTED FIRING POSITION

RD	RANGE (m)	TIME (sec)	HIT	MISS
1	75	6	☐	☐
2	175	8	☐	☐
3	300	10	☐	☐
4	175	8	☐	☐
5	75	6	☐	☐
6	300	10	☐	☐
7	300	10	☐	☐
8	75	6	☐	☐
9	175	8	☐	☐
10	175	8	☐	☐
11	300	10	☐	☐
12	175	8	☐	☐
13	75	6	☐	☐
14	300	10	☐	☐
15	175	8	☐	☐
16	75	6	☐	☐
17	300	10	☐	☐
18	75	8	☐	☐
TOTAL				

TABLE 3 — PRONE FIRING POSITION

RD	RANGE (m)	TIME (sec)	HIT	MISS
1	75	6	☐	☐
2	175	8	☐	☐
3	300	10	☐	☐
4	175	8	☐	☐
5	75	6	☐	☐
6	300	10	☐	☐
7	300	10	☐	☐
8	75	6	☐	☐
9	175	8	☐	☐
TOTAL				

TABLE 4 — KNEELING FIRING POSITION

RD	RANGE (m)	TIME (sec)	HIT	MISS
1	75	6	☐	☐
2	175	8	☐	☐
3	75	6	☐	☐
4	175	8	☐	☐
5	75	6	☐	☐
6	175	8	☐	☐
7	75	6	☐	☐
8	75	6	☐	☐
9	175	8	☐	☐
TOTAL				

SCORE

TABLE	HIT	MISS	NO FIRE
1			
2			
3			
4			

REMARKS

DATE INITIALED *(YYYYMMDD)*	SCORER'S INITIALS

100-, 200- AND 300-METER DOWNRANGE FEEDBACK SCORECARD

For use of this form see FM 3-22.9; the proponent agency is TRADOC.

ID CODE	UNIT	DATE (YYYYMMDD)	EVALUATOR'S ID CODE

ZERO CONFIRMATION

RANGE (m)	HIT	MISS
300		

ZERO CONFIRMATION REFIRE

RANGE (m)	HIT	MISS
300		

TABLE 1 -- DOWNRANGE SILHOUETTES, SUPPORTED FIRING POSITION

RDS	RANGE (m)	HIT	MISS
5	100		
10	200		
5	300		
TOTAL			

TABLE 2 -- DOWNRANGE SILHOUETTES, PRONE FIRING POSITION

RDS	RANGE (m)	HIT	MISS
5	100		
10	200		
5	300		
TOTAL			

ZERO-SHOT LOCATIONS

300M

WIND SPEED (mph):

WIND DIRECTION (use arrow):

SHOT LOCATIONS

SUPPORTED FIRING POSITION

TABLE 1

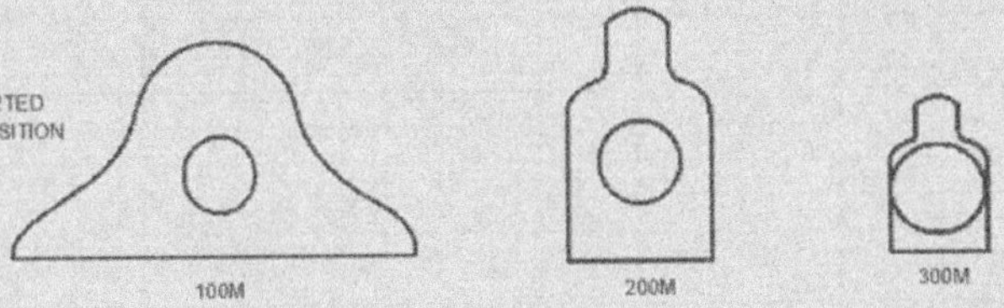

PRONE UNSUPPORTED FIRING POSITION

TABLE 2

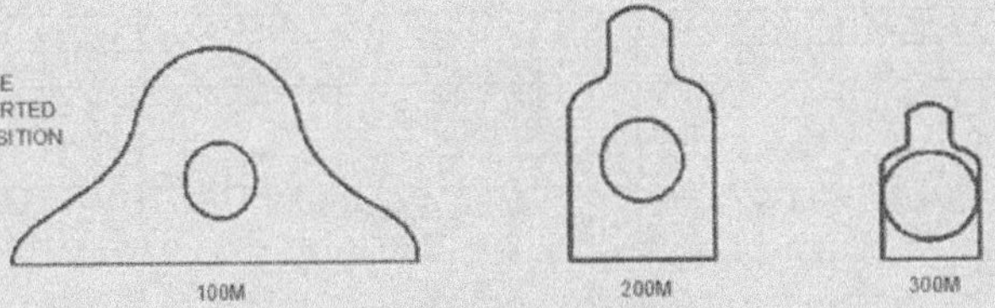

REMARKS

DATE INITIALED (YYYYMMDD)	SCORER'S INITIALS

SINGLE AND MULTIPLE TARGETS
FIELD FIRE II SCORECARD

For use of this form see FM 3-22.9; the proponent agency is TRADOC.

ID CODE	UNIT	DATE *(YYYYMMDD)*	EVALUATOR'S ID CODE

TABLE 1 SUPPORTED FIRING POSITION					TABLE 2 SUPPORTED FIRING POSITION					TABLE 3 PRONE FIRING POSITION					SCORE		
RD	RANGE (m)	TIME (sec)	HIT	MISS	RD	RANGE (m)	TIME (sec)	HIT	MISS	RD	RANGE (m)	TIME (sec)	HIT	MISS	TABLE	HIT	MISS
1	75	5			1	175	7			1	75	6			2		
2	175	7			2	75	10			2	175	8			3		
3	75	11			3	300				3	75	13					
4	300				4	75	9			4	300						
5	75	9			5	175				5	75	11					
6	175				6	300	9			6	175						
7	75	10			7	75	9			7	75	12					
8	300				8	175				8	300						
9	175	11			9	175	11			9	175	13					
10	300				10	300				10	300						
					11	75	9			11	75	11					
					12	175				12	175						
					13	175	11			13	175	8					
					14	300				14	75	6					
					15	75	5			15	75	11					
					16	175	11			16	175						
					17	300				17	75	12					
					18	75	9			18	300						
					19	175				19	75	11					
					20	75	10			20	175						
					21	300				21	175	13					
					22	175	7			22	300						
		TOTAL					TOTAL					TOTAL			TOTAL		

REMARKS

DATE INITIALED *(YYYYMMDD)*	SCORER'S INITIALS

RECORD FIRING SCORECARD -- KNOWN DISTANCE COURSE

For use of this form see FM 3-22.9; the proponent agency is TRADOC.

ID CODE	UNIT	DATE *(YYYYMMDD)*	EVALUATOR'S ID CODE

TABLE 1
PRONE SUPPORTED OR FOXHOLE SUPPORTED FIRING POSITION
(TIME: 120 SECONDS)

RD	RANGE (m)	HIT	MISS	RD	RANGE (m)	HIT	MISS
1		☐	☐	11		☐	☐
2		☐	☐	12		☐	☐
3		☐	☐	13		☐	☐
4	300	☐	☐	14	300	☐	☐
5		☐	☐	15		☐	☐
6	E-SIL	☐	☐	16	E-SIL	☐	☐
7		☐	☐	17		☐	☐
8		☐	☐	18		☐	☐
9		☐	☐	19		☐	☐
10		☐	☐	20		☐	☐
TOTAL				TOTAL			

TABLE 2
PRONE UNSUPPORTED FIRING POSITION
(TIME: 60 SECONDS)

RD	RANGE (m)	HIT	MISS
1		☐	☐
2		☐	☐
3		☐	☐
4	300	☐	☐
5		☐	☐
6	E-SIL	☐	☐
7		☐	☐
8		☐	☐
9		☐	☐
10		☐	☐
TOTAL			

TABLE 3
KNEELING FIRING POSITION
(TIME: 60 SECONDS)

RD	RANGE (m)	HIT	MISS
1		☐	☐
2		☐	☐
3		☐	☐
4	300	☐	☐
5		☐	☐
6	F-SIL	☐	☐
7		☐	☐
8		☐	☐
9		☐	☐
10		☐	☐
TOTAL			

SCORE

TABLE	HIT	MISS	NO FIRE
1			
2			
3			
FIRER'S QUALIFICATION SCORE			

QUALIFICATION SCORE RATINGS *(Check One)*

☐ 38-40 -- EXPERT ☐ 26-32 -- MARKSMAN

☐ 33-37 -- SHARPSHOOTER ☐ 25 AND BELOW -- UNQUALIFIED

LIGHT

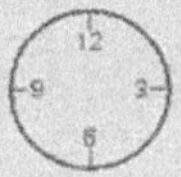

WIND

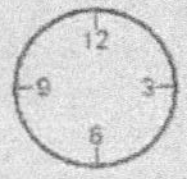

ZERO			
ELEV			
WIND			

NIGHT FIRE EXERCISE

DATE *(YYYYMMDD)*	HIT	MISS	GO	NO GO
			☐	☐

CBRN FIRE EXERCISE

DATE *(YYYYMMDD)*	HIT	MISS	GO	NO GO
			☐	☐

THE FIRER WILL BE ISSUED 40 ROUNDS. THE ROUNDS WILL BE PRELOADED IN FOUR 10-ROUND MAGAZINES - TWO FOR TABLE 1, ONE FOR EACH REMAINING TABLE.

DATE INITIALED *(YYYYMMDD)*	SCORER'S INITIALS
DATE INITIALED *(YYYYMMDD)*	OFFICER'S INITIALS

This scorecard is used to score known distance course record fire qualification when the known distance range is used. This course is used only when the standard record fire course is not available.

CONDUCT OF FIRE

The uniform for qualification ia a helmet, LBE/LBV, and interceptor body armor with front and back SAPI plates (if available). No other armor is required.

(1) Table 1 -- Supported Firing Position. The firer is given two 10-round magazines to engage an E-silhouette at 300 yards within 120 seconds.

(2) Table 2 -- Unsupported Firing Position. The firer is given a 10-round magazine to engage an E-silhouette at 200 yards within 60 seconds.

(3) Table 3 -- Unsupported Firing Position. The firer is given a 10-round magazine to engage an F-silhouette at 100 yards within 60 seconds.

SCORING

Scoring is conducted in the pits, with the results provided after each table. A hit is scored for any bullet hole that is within or touches some part of the silhouette facing.

RECORD FIRING SCORECARD -- SCALED TARGET ALTERNATE COURSE

For use of this form see FM 3-22.9; the proponent agency is TRADOC.

ID CODE	UNIT	DATE *(YYYYMMDD)*	EVALUATOR'S ID CODE

TABLE 1 — PRONE SUPPORTED OR FOXHOLE SUPPORTED FIRING POSITION (TIME: 120 SECONDS)

RD	RANGE (m)	HIT	MISS	RD	RANGE (m)	HIT	MISS
1	300	☐	☐	11	150	☐	☐
2	300	☐	☐	12	150	☐	☐
3	250	☐	☐	13	100	☐	☐
4	250	☐	☐	14	100	☐	☐
5	200	☐	☐	15	100	☐	☐
6	200	☐	☐	16	100	☐	☐
7	200	☐	☐	17	100	☐	☐
8	200	☐	☐	18	100	☐	☐
9	150	☐	☐	19	50	☐	☐
10	150	☐	☐	20	50	☐	☐
TOTAL				TOTAL			

TABLE 2 — PRONE UNSUPPORTED FIRING POSITION (TIME: 60 SECONDS)

RD	RANGE (m)	HIT	MISS
1	300	☐	☐
2	250	☐	☐
3	200	☐	☐
4	200	☐	☐
5	150	☐	☐
6	150	☐	☐
7	100	☐	☐
8	100	☐	☐
9	100	☐	☐
10	50	☐	☐
TOTAL			

TABLE 3 — KNEELING FIRING POSITION (TIME: 60 SECONDS)

RD	RANGE (m)	HIT	MISS
1	150	☐	☐
2	150	☐	☐
3	100	☐	☐
4	100	☐	☐
5	100	☐	☐
6	100	☐	☐
7	100	☐	☐
8	100	☐	☐
9	50	☐	☐
10	50	☐	☐
TOTAL			

SCORE

TABLE	HIT	MISS	NO FIRE
1			
2			
3			

FIRER'S QUALIFICATION SCORE

QUALIFICATION SCORE RATINGS *(Check One)*

- ☐ 36-40 -- EXPERT
- ☐ 23-29 -- MARKSMAN
- ☐ 30-35 -- SHARPSHOOTER
- ☐ 22 AND BELOW -- UNQUALIFIED

FIRER ISSUED 40 ROUNDS TO ENGAGE 10 TARGETS. NO MORE THAN 4 ROUNDS PER TARGET. THE ROUNDS WILL BE PRELOADED IN ONE 20-ROUND MAGAZINE FOR TABLE 1, ONE 10-ROUND MAGAZINE FOR TABLE 2, AND ONE 10-ROUND MAGAZINE FOR TABLE 3.

NIGHT FIRE EXERCISE

DATE *(YYYYMMDD)*	HIT	MISS	GO	NO GO
			☐	☐

CBRN FIRE EXERCISE

DATE *(YYYYMMDD)*	HIT	MISS	GO	NO GO
			☐	☐

REMARKS

DATE INITIALED *(YYYYMMDD)*	SCORER'S INITIALS
DATE INITIALED *(YYYYMMDD)*	OFFICER'S INITIALS

This scorecard is used to score alternate course record fire qualification when the 25-meter scaled silhouette target (NSN 6920-01-167-1396) is used. The alternate course is used only when standard record fire and known distance ranges are unavailable.

NOTE: If zeroing/grouping exercises are not performed on the day of record fire, six rounds of training/sustainment ammunition will be fired for 25-meter zero confirmation prior to conducting the qualification course.

CONDUCT OF FIRE

Alternate course qualification firers will have one 20-round magazine and two 10-round magazines. To ensure that firers do not forget which targets they engaged and shoot a given target more than the prescribed number of times, firers should adhere to the following guideline: Engage targets on the sheet from left to right and nearest to farthest. Engagement should follow this order: 50m, 100m left, 100m center, 100m right, 150m left, 150m right, 200m left, 200m right, 250m, and 300m.

Though the time between each firing position is not specified, enough time should be allotted to allow the firer to clear his weapon, quickly change firing positions, and reload before beginning the next firing table. The range RSO ensures that enough time is given between each change in firing positions to facilitate the timely flow of the record fire qualification table.

(1) Table 1-- Prone Supported Firing Position or (at the unit commander's discretion) Foxhole Supported Firing Position. The firer is given one 20-round magazine to engage 10 silhouettes on the target sheet. Table 1 includes 2 rounds for each silhouette. Firing must be completed in 120 seconds. No more than 2 hits are scored for each silhouette.

(2) Table 2 -- Prone Unsupported Firing Position. The firer is given one 10-round magazine to engage 10 silhouettes on the same target sheet. Table 2 includes 1 round for each silhouette. Firing must be completed in 60 seconds. No more than 1 hit is scored for each silhouette.

(3) Table 3 -- Kneeling Firing Position. The firer is given one 10-round magazine to engage 10 silhouettes on the same target sheet. Table 3 includes 2 rounds for each silhouette positioned 50 to 150 meters away. Firing must be completed in 60 seconds. No more than 2 hits are scored for each silhouette.

SCORING

The same target sheet is used for every 40-round qualification table that a firer completes. One hit is awarded for each round that strikes within or touches some part of the silhouette. A maximum of four hits for each silhouette on the same target sheet are scored.

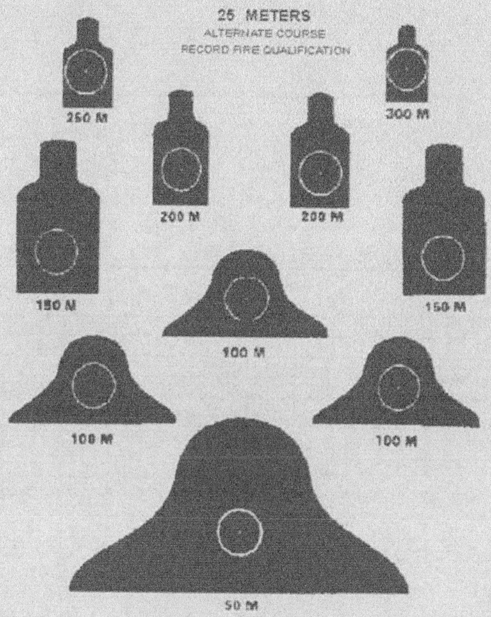

THE WHITE DOT ON EACH TARGET SHOWS THE CENTER OF MASS AIMING POINT. BULLETS SHOULD HIT WITHIN THE CIRCLE, BUT ARE SCORED AS HITS IF THEY HIT ANY PART OF THE SILHOUETTE.

RECORD NIGHT FIRE SCORECARD

For use of this form see FM 3-22.9; the proponent agency is TRADOC.

ID CODE	UNIT	DATE *(YYYYMMDD)*	EVALUATOR'S ID CODE

TABLE 1
FOXHOLE SUPPORTED FIRING POSITION

RD	RANGE (m)	TIME (sec)	HIT	MISS	NO FIRE	RD	RANGE (m)	TIME (sec)	HIT	MISS	NO FIRE
1	50L	3				11	100	8			
2	200	6				12	200				
3	100	4				13	150	8			
4	150	5				14	50R				
5	100	8				15	100	8			
6	150	5				16	150				
7	50R	3				17	200	6			
8	200	6				18	150	5			
9	150	5				19	50L	9			
10	250	7				20	250				
TOTAL						TOTAL					

TABLE 2
PRONE UNSUPPORTED FIRING POSITION

RD	RANGE (m)	TIME (sec)	HIT	MISS	NO FIRE	RD	RANGE (m)	TIME (sec)	HIT	MISS	NO FIRE
1	100	5				11	150	8			
2	200	8				12	50R	5			
3	150	6				13	100	12			
4	50L	12				14	200				
5	200					15	150	12			
6	150	12				16	50L				
7	100					17	100	12			
8	50R	12				18	150				
9	150					19	200	12			
10	100	8				20	100				
TOTAL						TOTAL					

SCORE

TABLE	HIT	MISS	NO FIRE
1			
2			
TOTAL			

FIRER'S QUALIFICATION SCORE

QUALIFICATION SCORE RATINGS *(Check One)*

- ☐ 36-40 -- EXPERT
- ☐ 17-23 -- MARKSMAN
- ☐ 24-34 -- SHARPSHOOTER
- ☐ 16 AND BELOW -- UNQUALIFIED

AIMING DEVICE USED *(Check One)*

- ☐ AN/PEQ-2A/B
- ☐ AN/PAQ-4B/C
- ☐ AN/PVS-4

REMARKS

DATE INITIALED *(YYYYMMDD)*	SCORER'S INITIALS
DATE INITIALED *(YYYYMMDD)*	OFFICER'S INITIALS

SQUAD DESIGNATED MARKSMAN
RECORD FIRE I AND II

For use of this form, see FM 3-22.9; the proponent agency is TRADOC.

ID CODE	UNIT	DATE *(YYYYMMDD)*	EVALUATOR'S ID CODE

RECORD FIRE I - MECHANICAL SIGHT ADJUSTMENT			RECORD FIRE II - ADJUSTED AIMING POINTS/OPTIC		
TARGET RANGE/POSITION	HIT	MISS	TARGET RANGE/POSITION	HIT	MISS
100m unsupported	☐	☐	100m unsupported	☐	☐
100m unsupported	☐	☐	100m unsupported	☐	☐
100m unsupported	☐	☐	100m unsupported	☐	☐
100m unsupported	☐	☐	100m unsupported	☐	☐
200m supported	☐	☐	200m supported	☐	☐
200m supported	☐	☐	200m supported	☐	☐
200m unsupported	☐	☐	200m unsupported	☐	☐
200m unsupported	☐	☐	200m unsupported	☐	☐
300m supported	☐	☐	300m supported	☐	☐
300m supported	☐	☐	300m supported	☐	☐
300m unsupported	☐	☐	300m unsupported	☐	☐
300m unsupported	☐	☐	300m unsupported	☐	☐
400m supported	☐	☐	400m supported	☐	☐
400m supported	☐	☐	400m supported	☐	☐
400m supported	☐	☐	400m supported	☐	☐
400m supported	☐	☐	400m supported	☐	☐
500m supported	☐	☐	500m supported	☐	☐
500m supported	☐	☐	500m supported	☐	☐
500m supported	☐	☐	500m supported	☐	☐
500m supported	☐	☐	500m supported	☐	☐
TOTAL			**TOTAL**		
☐ **PASS (14-20)** ☐ **FAIL (13 and below)**			☐ **PASS (14-20)** ☐ **FAIL (13 and below)**		
DATE INITIALED *(YYYYMMDD)*			SCORER'S INITIALS		
DATE INITIALED *(YYYYMMDD)*			OFFICER'S INITIALS		

SQUAD DESIGNATED MARKSMAN
POSITION EVALUATION

For use of this form, see FM 3-22.9; the proponent agency is TRADOC.

☐ FOXHOLE SUPPORTED ☐ PRONE UNSUPPORTED	ID CODE	UNIT	DATE *(YYYYMMDD)*	EVALUATOR'S ID CODE

PHASE	EYE RELIEF	TRIGGER FINGER	ELBOWS	NON-FIRING HAND	LEGS
EXAMPLE	How far is nose from charging handle?	What part of 1st joint (tip, middle, joint)?	Is platform stable?	Where is it located (Soldier's choice, stable)?	Where are they located (stability for upper body)?
Phase I -- Steady Position					
Phase II -- Borelight Dry Fire					
Phase V -- Practice Qualification					
100m					
200m					
300m					
400m					
500m					
Phase VI -- Record Qualification					
100m					
200m					
300m					
400m					
500m					

DATE INITIALED *(YYYYMMDD)*	SCORER'S INITIALS

TECHNICAL MANUAL
ARMY NO. 9-1005-319-23&P
AIR FORCE TO 11W3-5-5-42

***ARMY TM 9-1005-319-23&P**
AIR FORCE TO 11W3-5-5-42

DEPARTMENTS OF THE ARMY
AND AIR FORCE
Washington, DC, 1 May 1991

Technical Manual for 5.56MM,

M16A2, M4 and M4A1 Carbine

DISTRIBUTION STATEMENT. APPROVED FOR PUBLIC RELEASE; DISTRIBUTION IS UNLIMITED.

REPORTING ERRORS AND RECOMMENDING IMPROVEMENTS

You can help improve this manual. If you find any mistakes or if you know of a way to improve the procedures please let us know.

Army users mail your letter, DA Form 2028 (recommended Changes to Publications and Blank Forms), or DA Form 2028-2 located in the back of this manual direct to: Director, Armament and Chemical Acquisition and Logistics Activity, ATTN: AMSTA-AC-NML, Rock Island, IL 61299-7630.

Air Force users submit AFTO Form 22, Technical Order System Publication Improvement Report and Reply, to: WR-ALC/MMDET, Robins AFB, GA 31098-5609.

A reply will be furnished to you.

*This manual supersedes ARMY TM 9-1005-319-23&P dated 28 August 1987, including all changes.

WARNING

ALL WARNINGS in this technical manual pertain to both the rifle and the carbines unless otherwise specified.

Before starting an inspection, be sure to clear the rifle. Do not pull the trigger until the rifle has been cleared. Inspect the chamber to ensure that it is empty and no ammunition is in position to be chambered.

Do not keep live ammunition near work area.

To avoid injury to your eyes, use care when removing and installing spring-loaded parts.

All M1 6A2 rifles and M4 carbines must be inspected and gaged at least once annually for safety and serviceability. Initial gaging is required 1 year from receipt of the weapons. Air Force users refer to inspection requirements in Air Force Manual (AFM) 36-2227, Volume 1.

All Army Reserve and Army National Guard M16A2 rifles and M4 carbines must be inspected and gaged at least once every 2 years, after the initial inspection/gaging procedures have been accomplished. This initial gaging procedure is required 1 year from receipt of the weapons. This 2 year interval may be maintained unless preventive maintenance checks and services (PMCS) or other physical evidence indicates that an individual unit's M16A2 rifles and M4 carbines require inspection/ gaging at a more frequent interval. If it is determined that a yearly inspection is necessary for an individual unit, only that unit will be affected. This will not affect other units in regard to the interval of inspection.

It is recommended that training units inspect/gage all rifles and carbines at the end of each training cycle. Training units will inspect/gage all rifles and carbines at least once annually.

Below direct support maintenance, **DO NOT** interchange bolt assemblies from one rifle/carbine to another. Doing so may result in injury to, or death of, personnel.

Bolt cam pin must be installed or rifle/carbine will blow up while firing the first round. If the bolt cam pin is not installed, injury to or death of, personnel may result.

Dry cleaning solvent is flammable and toxic and should be used in a well ventilated area. The use of rubber gloves is necessary to protect the skin when washing rifle parts.

When using solid film lubricant or dichloromethane, be sure the area is well ventilated.

When using carbon removing compound (item 8, app D), avoid skin contact. If carbon removing compound comes In contact with the skin, wash thoroughly with running water. using a good lanolin base cream after exposure to the compound is helpful. Using gloves and protective equipment is required.

The lock plate prevents the selector lever from being placed in BURST and will be installed at the discretion of the unit commander. It is mandatory for use in civil disturbance (riot control).

Only blank cartridge M200 is to be used when the blank firing attachment is attached to the rifle/carbine.

WARNING (CONT)

Do not fire blank ammunition at a representative enemy at distances of less than 20 feet (6.10 Om). The unburned propellant grains can cause injury within this distance.

For further information on safety, care, and handling of ammunition: Army and Air Force users refer to M16A2 Rifle Operator's Manual.

For additional first aid data, see Field Manual (FM) 21-11.

HOW TO USE THIS MANUAL

Read this manual carefully before performing required maintenance. This manual will be referred to for Inspection/Maintenance and Repair procedures.

GENERAL

There are several things you need to know to use this manual efficiently.

1. All references in the manual are to pages only. Reference to maintenance procedures is to the page where the respective initial setup appears.

2. Illustrations for the maintenance procedures show only those parts affected by the operation being performed.

3. Whenever the male gender is mentioned in the manual (i.e., crewman, repairman), it also pertains to females.

4. When the term "evacuate to support maintenance" is used, the entire rifle must be evacuated.

5. When a procedure is common to M16A2 rifle, and M4/M4A1 carbine, **ONLY** the M16A2 configuration will be depicted. If a procedure is not common to both weapons, the procedure will be incorporated.

6. When the word rifle is referenced in text, it will reference the rifle and the carbines.

INDEXES

This manual is organized to help you find the information you need quickly. There are several useful indexes.

1. **Table of Contents**. Lists in order all chapters, sections, and appendixes. Gives page references.

2. **Nomenclature Cross-References List.**

3. **Chapter Overviews**. Summarize material covered in the chapter. Are located at the beginning of each chapter.

4. **Symptom Index**. Located just before the troubleshooting table in each maintenance chapter. Lists, in alphabetical order, parts of the rifle with possible malfunctions. References pages of the troubleshooting table.

5. **Alphabetical Index**. Located at the end of the manual. An extensive subject index for everything in the manual. Gives page references.

MAINTENANCE PROCEDURES

There are two maintenance chapters:

Army personnel use chapter two for unit maintenance procedures and chapter three for direct support maintenance procedures.

Air Force personnel: Only Air Force Specialty Code 753XX Combat Arms Training and Maintenance (CATM) specialists, technicians, and gunsmiths are authorized to perform maintenance procedures contained in this manual.

Each maintenance task has an initial setup containing a list of the following things you will need in order to do your maintenance task.

1. **Tools and Special Tools**. For standard and special tools, see appendixes B and C. Army users are to use the Tool Set, Gage Set, and/or Shop Set listed in the initial setup.

2. **Materials/Parts**. Lists expendable materials and 100 percent replaceable parts. Each material or part is followed by a part number or appendix reference.

3. **References**. Lists other publications containing necessary information.

4. **Equipment Condition**. Lists conditions to be met before starting the procedure. The reference on the left of the condition is a page reference to instructions for setting up the condition.

5. **General Safety Instructions**. Lists safety instructions to follow before performing maintenance procedures.

EXTERNAL VIEW OF 5.56 MM RIFLE M16A2

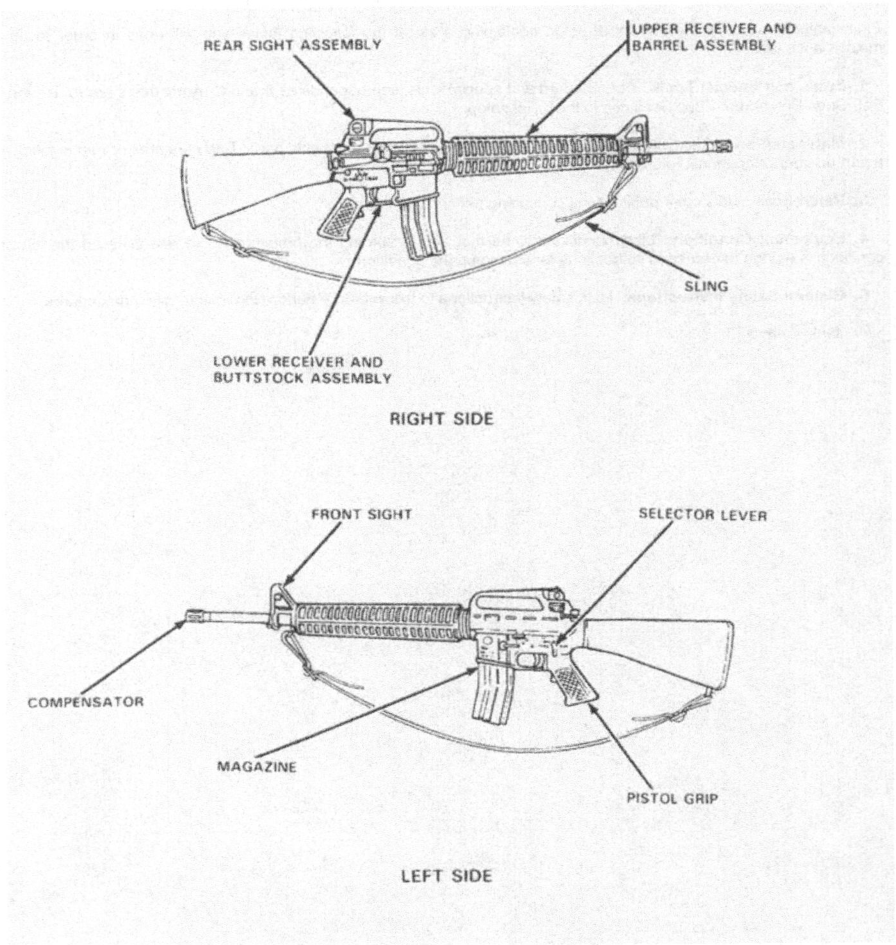

EXTERNAL VIEW OF 5.56MM CARBINE, M4/M4A1

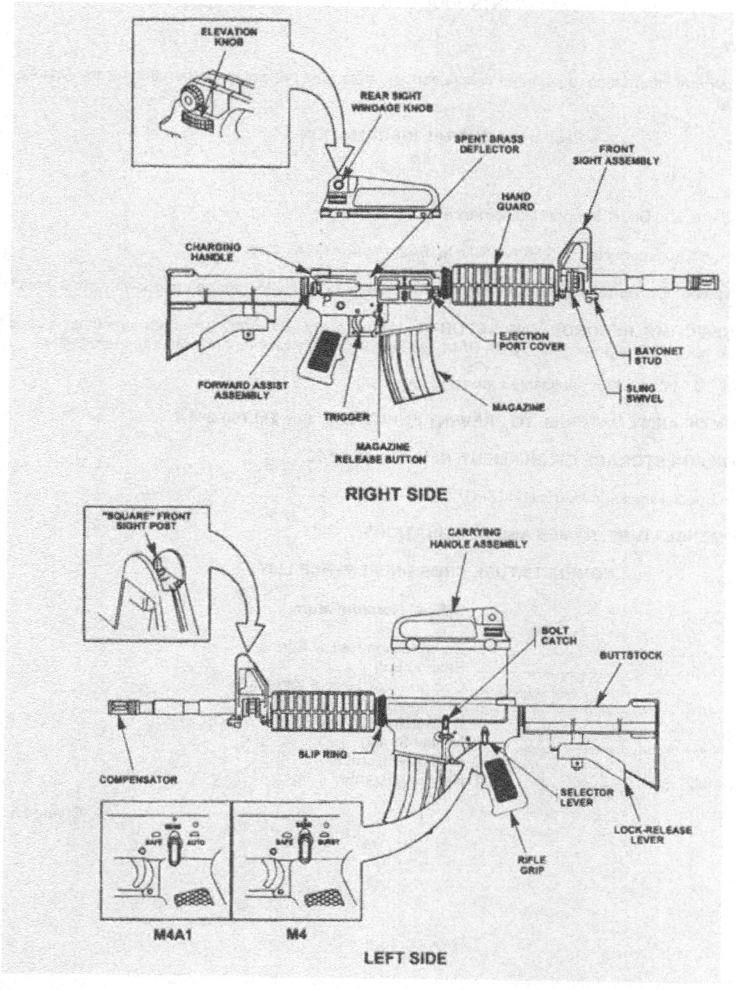

CHAPTER 1
INTRODUCTION

CHAPTER OVERVIEW

This chapter contains general information, equipment description and data, and principles of operation for the M16A2 rifle and M4/M4A1 carbines.

Section I. GENERAL INFORMATION

1-1. SCOPE.

a. Type of Manual: Unit and Direct Support Maintenance.

b. Model Number and Equipment Name: 5.56mm Rifle M16A2, M4 and M4A1 Carbines.

c. Purpose of Equipment. Provides personnel an offensive/defensive capability to engage targets with small arms fire.

1-2. MAINTENANCE FORMS, RECORDS, AND REPORTS. Department of the Army forms and procedures used for equipment maintenance will be those prescribed by DA PAM 738-750, The Army Maintenance Management System.

Air Force users refer to TO 11W-1-10 for applicable forms and records.

1-3. DESTRUCTION OF ARMY MATERIEL TO PREVENT ENEMY USE. See TM 750-244-7.

1-4. PREPARATION FOR STORAGE OR SHIPMENT. Refer to page 2-70.

Air Force users refer to Special Package Instruction (SPI) 00-856-6885.

1-5. OFFICIAL NOMENCLATURE, NAMES AND DESIGNATIONS.

NOMENCLATURE CROSS-REFERENCE LIST

Common Name	Official Nomenclature
Action Spring	Compression Helical Spring
Ball Bearing	Bearing Ball
Bolt Catch Spring	Compression Helical Spring
Bolt Carrier Key Tool	Machine Key
Burst Disconnector	Lock-Release Lever
Cam Clutch Spring	Helical Spring
Carbine	M4/M4A1 Carbine
Charging Handle Assembly	Handle Assembly

1-5. OFFICIAL NOMENCLATURE, NAMES AND DESIGNATIONS (CONT).

NOMENCLATURE CROSS-REFERENCE LIST

Common Name	Official Nomenclature
Disconnector Springs	Compression Helical Spring
Ejector Spring	Helical Spring
Extractor Spring Assembly	Spring Assembly
Hammer Spring	Torsion Helical Spring
Lower Receiver Extension	Spring Receiver Holder
Magazine	Cartridge Magazine
Magazine Catch Spring	Compression Helical Spring
Peel Washer	Shim
Pistol Grip	Rifle Grip
Pivot Pin Detent	Takedown Pin Detent
Rifle	Rifle, 5.56mm, M16A2
Rifle Barrel Assembly	Barrel Assembly
Selector Lever	Fire Control Selector
Semiautomatic Disconnector	Lock-Release Lever
Sling	Small Arms Sling
Trigger Spring	Torsion Helical Spring
Upper Receiver	Upper Cartridge Receiver

1-6. REPORTING EQUIPMENT IMPROVEMENT RECOMMENDATIONS (EIR). If your M16A2 rifle needs improvement, let us know. Send us an EIR. You, the user, are the only one who can tell us what you don't like about your equipment. Let us know why you don't like the design.

Army users submit SF 368 (Product Quality Deficiency Report) to: Commander, U.S. Army Armament Research, Development and Engineering Center, ATTN: AMSTA-AR-QAW (R)/Customer Feedback Center, Rock Island, IL 61299-7300.

Air Force users submit Materiel Deficiency Report (MDR) to: DIR MAT MGT ROBINS AFB GA//MMIBTC// and Product Quality Deficiency Report to: DIR MAT MGT ROBINS AFB GA//MMQA// IAW Technical Order 00-35D-54.

A reply will be sent to you.

1-7. CORROSION PREVENTION AND CONTROL (CPC). CPC of Army materiel is a continuing concern. It is important that any corrosion problems with this item be reported so that the problem can be corrected and improvements can be made to prevent the problem in future items,

While corrosion is typically associated with rusting of metals, it can also include deterioration of other materials such as rubber and plastic. Unusual cracking, softening, swelling, or breaking of these materials may be a corrosion problem.

If a corrosion problem is identified, it can be reported using Standard Form 368, Product Quality Deficiency Report. Use of key words such as "corrosion", "rust", " deterioration", or "cracking" will assure that the information is identified as a CPC problem.

Army users submit Product Quality Deficiency Report (SF 368) to:

Commander
U.S. Army Armament Research, Development and Engineering Center
ATTN: AMSTA-AR-QAW (R)
Rock Island, IL 61299-7300

Air Force users submit Materiel Deficiency Report (MDR) to:

DIR MAT MGT
ATTN: MMIBTC
Robins AFB, GA

and Product Quality Deficiency Report to:

DIR MAT MGT
ATTN: MMQA
Robins AFB, GA

Section II. EQUIPMENT DESCRIPTION AND DATA

1-8. EQUIPMENT CHARACTERISTICS, CAPABILITIES, AND FEATURES.

a. Characteristics.

(1) Light weight
(2) Air-cooled
(3) Gas-operated
(4) Magazine-fed
(5) Semiautomatic or burst fire

b. Capabilities. Provides personnel an offensive/defensive capability to engage targets with direct small arms fire.

c. Features.

(1) Receivers are made of light-weight aluminum alloys; however, the safety , durability, and function of the rifles are in no way reduced. The portability and logistical values are greatly increased, particularly when air transport is used.

(2) The bolt locking action is one of the mechanical features of the rifle. The bolt assembly and barrel extension contain locking lugs which engage and lock the bolt assembly firmly in the barrel extension. The initial force of the explosion of the cartridge is absorbed by the barrel, barrel extension, and bolt assembly.

1-8. EQUIPMENT CHARACTERISTICS, CAPABILITIES, AND FEATURES (CONT).

(3) The trigger guard is easily adaptable to winter operations. A spring-loaded retaining pin is depressed to allow ready access to the trigger when wearing arctic mittens.

(4) The ejection port cover prevents dirt or sand from getting into the ejection port. The ejection port cover must be closed during periods when firing is not anticipated. It opens automatically by the forward or rearward movement of the bolt carrier assembly.

1-9. LOCATION AND DESCRIPTION OF MAJOR COMPONENTS.

(A) MAGAZINE. 30 cartridge capacity.

(B) **SLING**. The sling is adjustable and provides a means to carry the weapon.

(C) **BOLT CARRIER ASSEMBLY**. Carries bolt assembly to chamber and fires the weapon. Contains the firing pin, cartridge extractor, bolt assembly, cartridge ejector, and bolt cam pin.

(D) **CHARGING HANDLE ASSEMBLY**. Provides a means of charging the weapon.

(E) **M16A2 UPPER RECEIVER AND BARREL ASSEMBLY**. Upper receiver contains rear sight assembly, ejection port, ejection port cover, and a housing for the key and bolt carrier assembly and bolt assembly. Rifle barrel assembly is air-cooled, contains compensator and front sight assembly, and holds the two handguard assemblies and the sling swivel.

(F) **LOWER RECEIVER AND BUTTSTOCK ASSEMBLY**. Lower receiver contains the trigger assembly, sear, hammer assembly, selector lever, rifle grip, bolt catch, and buttstock assembly. The buttstock assembly houses the action spring, buffer assembly, and extension assembly.

(G) **M4/M4A1 CARRYING HANDLE**. Provides a means of carrying carbine.

(H) **M4/M4A1 UPPER RECEIVER AND BARREL ASSEMBLY.** Upper receiver contains, ejection port, ejection port cover, a housing for key and bolt carrier assembly and bolt assembly, and mounting surface for the carrying handle assembly. Carbine barrel assembly is air-cooled, contains compensator and front sight assembly, and holds the two handguard assemblies and the sling swivel.

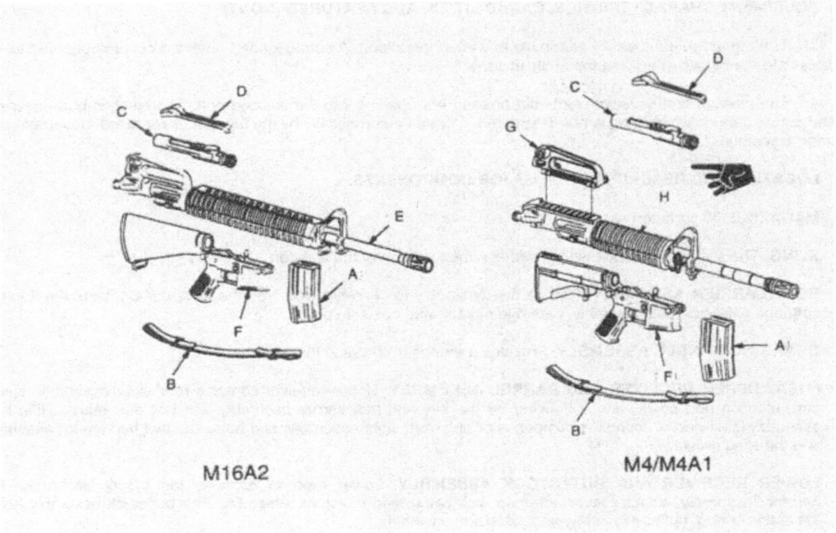
D
C
E
A
F
B
M16A2
D
C
G
H
A
F
B
M4/M4A1

1-10. EQUIPMENT DATA.

	US CUSTOMARY	METRIC
Weight:		
Carbine, M4/M4A1 without magazine and sling	6 lb 7 oz	2.91 kg
Rifle, MI 6A2 without magazine and sling	7 lb 8 oz	3.40 kg
Sling, adjustable	4 oz	0.11 kg
Empty magazine	4 oz	0.11 kg
Loaded magazine	1 lb 1 oz	0.48 kg
Carbine, M4/M4A1 w/sling and loaded magazine	7 lb 12 oz	3.51 kg
Rifle M16A2 w/sling and loaded magazine	8 lb 13 oz	4.00 kg
Bayonet-Knife M7	10.5 oz	0.30 kg
Scabbard M10 o	5 oz	0.14 kg
Length:		
Carbine with compensator, buttstock extended	33.0 in	83.82 cm
Carbine with compensator, buttstock closed	29.75 in	75.57 cm
Rifle with compensator	39.63 in	100.66 cm
Barrel (Carbine)	14.5 in	36.83 cm
Barrel (Rifle)	20 in	50.8 cm
Barrel with compensator (Carbine)	15.5 in	39.37 cm
Barrel with compensator (Rifle)	21 in	53.34 cm
Mechanical features:		
Rifling	right-hand twist 6 grooves, 1 turn in 7 inches (17.78 cm)	
Method of operation	direct gas	
Type of breech mechanism	rotating bolt	
Method of feeding	magazine	
Cooling	air	
Trigger pull (M16A2 & M4)	5.5 to 9.5 lb	2.49 to 4.31kg
Trigger pull (M4A1)	5.5 to 8.5 lb	2.49 to 3.86 kg
Ammunition:		
Caliber	223	5.56mm
Type	ball, blank, dummy, and tracer	
Firing characteristics:		
Muzzle velocity (Carbine) (approximate)	2,970 fps	905.85 mps
Muzzle velocity (Rifle) (approximate)	3,100 fps	945.5 mps
Chamber pressure	52,000 psi	358,540 kPa
Cyclic rate of fire (Carbine) (approximate)	700-970 rds/m	
Cyclic rate of fire (Rifle) (approximate)	700-900 rds/m	

1-10. EQUIPMENT DATA (CONT).

	US CUSTOMARY	METRIC
Maximum rate of fire:		
Semiautomatic	45 rds/m	
Burst	90 rds/m	
Sustained rate of fire	12/15 rds/m	
Maximum range	3,938 yards	Approximately 3,600 meters
Maximum effective range:		
Individual/point targets (Carbine)	547 yards	500 meters
Individual/point targets (Rifle)	602 yards	550 meters
Area targets (Carbine)	650 yards	600 meters
Area targets (Rifle)	875 yards	800 meters

Section III. PRINCIPLES OF OPERATION

1-11. GENERAL. The 5.56mm M16A2 and M4/M4A1 carbine:

a. Is gas-operated. It fires in either the semiautomatic or burst mode.

b. Has positive locking of the bolt. Firing pin is part of the bolt carrier assembly and cannot strike the primer until the bolt assembly is fully locked.

Section III. PRINCIPLES OF OPERATION (CONT).

1-12. PRINCIPLES OF OPERATION.

(A) **MAGAZINE**. Holds cartridges ready for feeding and provides a guide for positioning cartridges for stripping. Provides quick reload capabilities for sustained firing.

(B) **SLING**. Provides the means for carrying the weapon.

(C) **BOLT CARRIER ASSEMBLY**. Provides stripping, chambering, locking, firing, extraction, and ejection of cartridges using the drive springs and projectile propelling gases for power.

(D) **CHARGING HANDLE ASSEMBLY**. Provides initial charging of the weapon. The handle latch locks the charging handle assembly in the forward position during sustained fire to prevent injury to the operator.

(E) **M16A2 UPPER RECEIVER AND BARREL ASSEMBLY**. Provides support for the bolt carrier assembly. The barrel chambers the cartridge for firing and directs the projectile.

(F) **LOWER RECEIVER AND BUTTSTOCK ASSEMBLY**. Provides firing control for the rifle and carbine. M16A2 ONLY provides storage for basic cleaning materials.

(G) **M4/M4A1 CARRYING HANDLE ASSEMBLY**. Provides a means of carrying the carbine, contains rear sight assembly, and can be removed for mounting various optics.

(H) **M4/M4A1 UPPER RECEIVER AND BARREL ASSEMBLY**. Provides support for the bolt carrier assembly. The barrel chambers the cartridge for firing and directs the projectile.

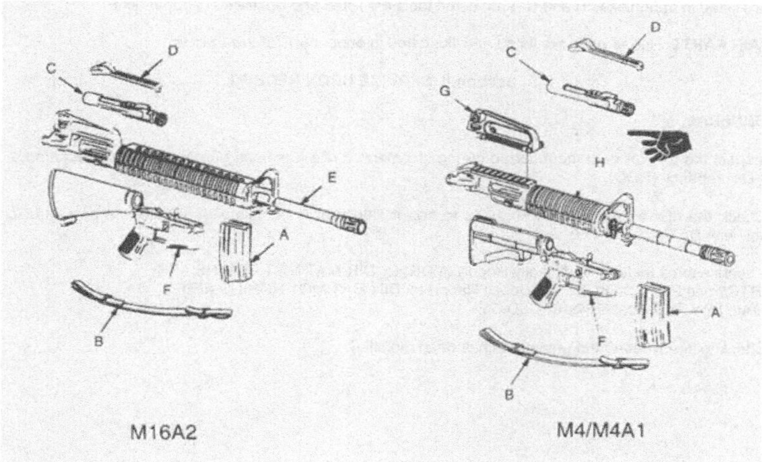

CHAPTER 2
UNIT MAINTENANCE INSTRUCTIONS

CHAPTER OVERVIEW

This chapter provides information and Instructions to help keep the rifle in good repair and contains the following sections:

a. Repair Parts, Special Tools, TMDE, and Support Equipment
b. Service Upon Receipt
c. Preventive Maintenance Checks and Services (PMCS)
d. Troubleshooting
e. Maintenance Procedures

Section I. REPAIR PARTS, SPECIAL TOOLS, TMDE, AND SUPPORT EQUIPMENT

2-1. COMMON TOOLS AND EQUIPMENT. For authorized common tools and equipment, refer to the Modified Table of Organization and Equipment (MTOE) applicable to your unit.

Air Force users must maintain the following common tools:

3-ounce soft-brass hammer
Vise
Flat tip screwdriver
Punch
Tweezers/round nose pliers
Hammer
Needle nose pliers

2-2. SPECIAL TOOLS, TMDE, AND SUPPORT EQUIPMENT. Special tools required for unit support are listed in appendixes B and C. Fabricated tools are listed and illustrated in appendix E.

2-3. REPAIR PARTS. Repair parts are listed and illustrated in appendix C of this manual.

Section II. SERVICE UPON RECEIPT

2-4. GENERAL.

a. Inspect the rifle for damage Incurred during shipment. If rifle has been damaged, report the damage on SF 364, Report of Discrepancy (ROD).

b. Check the rifle against the packing slip to see If shipment is complete. Army users report all discrepancies in accordance with DA PAM 738-750.

Air Force users submit Materiel Deficiency Report (MDR) to: DIR MAT MGT ROBINS AFB GA//MMIBTC// and Product Quality Deficiency Report to: DIR MAT MGT ROBINS AFB GA//MMQA//. IAW Technical Order 00-35D-54.

c. Check to see whether the equipment has been modified.

2-5. SERVICE UPON RECEIPT OF MATERIEL.

WARNING

Before starting an inspection, be sure to clear the rifle. Do not actuate the trigger before clearing the rifle. Inspect the chamber to make sure it is empty and free of obstructions. Check to see there are no obstructions in the barrel and no ammunition is in position to be chambered.

SERVICE UPON RECEIPT

LOCATION	ITEM	ACTION	REMARKS
1. Container	a. M16A2 rifle or M4/M4A1 carbine	a. Remove rifle from containers.	
		b. Inspect the equipment for damage incurred during shipment	If the equipment has been damaged, report the damage on SF Form 364, Report of Discrepancy (ROD).
		c. Check the equipment against the packing list to see if the shipment is complete	Report all discrepancies in accordance with the instructions of DA PAM 738-750.
	b. Basic issue items	Check for missing items	Refer to TM 9-1005-319-10 (operator's manual).
2. M16A2 rifle or M4/M4A1 carbine	a. Barrel assembly	If volatile corrosion inhibitor (VCI) is in barrel, remove and discard.	
	b. All parts	a. Field-strip rifle and inspect for missing, damaged, and rusted or corroded parts.	Refer to operator's manual.
		b. Clean and lubricate	Refer to operator's manual.
		c. Reassemble	Refer to operator's manual.
		d. Function check	Refer to page 2-69.

SERVICE UPON RECEIPT (CONT)

LOCATION	ITEM	ACTION	REMARKS
		e. Check to see whether the equipment has been modified	Refer to DA PAM 25-30.
	c. Magazine	Check for positive retention and functioning of bolt catch	Refer to operator's manual.

Section III. PREVENTIVE MAINTENANCE CHECKS AND SERVICES (PMCS)

2-6. GENERAL This section contains the procedures and instructions necessary to perform unit preventive maintenance checks and services. These services are to be performed by unit maintenance personnel with the assistance of the operator where practical.

2-7. PREVENTIVE MAINTENANCE CHECKS AND SERVICES

WARNING

Before starting an inspection, be sure to clear the rifle. Do not keep live ammunition near the work area.

a. **General**. The PMCS procedures are contained in the table following. They are arranged in logical sequence requiring a minimum amount of time and motion on the part of the persons performing them and are arranged so that there will be minimum interference between persons performing checks simultaneously on the same end item.

b. **Item No. Column**. Checks and services are numbered in disassembly sequence. This column shall be used as a source of item numbers for the "TM Number" column on DA Form 2404, Equipment Inspection and Maintenance Worksheet, in recording results of PMCS.

c. **Interval Column**. This column gives the designated interval when each check is to be performed.

d. **Item To Be Checked Or Serviced Column**. This column lists the items to be checked or serviced.

e. **Procedure Column**. This column contains a brief description of the procedure by which the check is to be performed. It contains all the information required to accomplish the checks and services. Information marked SH Indicates a specific equipment shortcoming and the procedure needed to correct the shortcoming.

NOTE

For the purpose of this technical manual, the following definition is supplied. This definition is not intended to apply to any other document

Shortcoming (SH): A fault that requires maintenance or supply action on a piece of equipment, but does not render equipment Not Mission Capable

f. **Not Fully Mission Capable If: Column**. This column contains a brief statement of the condition (e.g., malfunction, shortage) that would cause the covered equipment to be less than fully ready to perform its assigned mission.

2-7. PREVENTION MAINTENANCE CHECKS AND SERVICES (CONT).

PREVENTIVE MAINTENANCE CHECKS AND SERVICES FOR M16A2 RIFLE (CONT)

Item No.	Interval	Item to be Checked or Serviced	Procedure	Not Fully Mission Capable if:

WARNING

Before starting an inspection, be sure to clear the weapon. Do not pull the trigger until the weapon has been cleared. Inspect the chamber to ensure that it is empty and no ammunition is in position to be chambered. Do not keep live ammunition near work area.

NOTE

An inactive weapon is a weapon which has been stored in an arms room for a period of 90 days without use. The weapon may or may not have been assigned to an individual.

Inactive weapons shall receive quarterly PMCS unless inspection reveals more frequent servicing is necessary.

Normal cleaning (PMCS) of an inactive weapon will be performed every 90 days. Should the unit armorer detect corrosion on a weapon prior to the end of the 90-day period, the PMCS should be performed immediately.

Solid Film Lubricant (SFL) is the authorized touch up for the M16A2 Rifle and M4/M4A1 Carbine and may be used on up to one third of the exterior finish of the weapon.

FOR ARMY CONUS USE ONLY AND AIR FORCE TRAINING WEAPONS ONLY: Solid Film Lubricant may be used as a touch up without limitation on the upper receiver and barrel assembly. This is to say that units which **DO NOT** fall under the category of Divisional Combat Units or rapid deployment type units may have up to 100 percent of the exterior surface of the upper receiver and barrel assembly protected with SFL.. Prior to application of SFL, the surface must be thoroughly cleaned and inspected for corrosion and/or damage. If corroded or damaged, the part must be repaired or replaced prior to application of SFL. Continued use under combat conditions would result in an unprotected surface when the SFL wears off. This would result in a large light reflecting surface and accelerated deterioration of the unprotected surface. Therefore, Divisional Combat Units and units which fall under the definition of Rapid Deployment type must adhere to the limitation of **NOT** over one third of their exterior surface covered by SFL.

When determining mission capability, deadline if it is a deficiency.

PREVENTIVE MAINTENANCE CHECKS AND SERVICES FOR M16A2 RIFLE (CONT)

Item No.	Interval	Item to be Checked or Serviced	Procedure	Not Fully Mission Capable if:
1	Quarterly	Magazine (serviceability check)	Disassemble as In TM 9-1005-319-10 (operator's manual). Inspect tube (1) for bulges, dents, or damaged feeder lips (2). Inspect spring (3) and follower (4) for kinks or damage. SH-Replace the magazine if any of these conditions exist. Reassemble magazine and check for binding during operation of follower (4). SH-Replace the magazine if the follower binds. 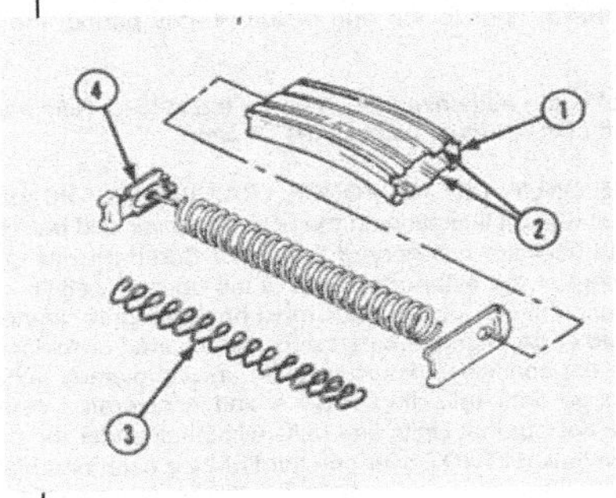	A magazine is not available fu; use with the rifle.
2	Quarterly	Charging handle assembly and selector lever	**WARNING** If the rifle falls any of the following selector lever tests, evacuate It to support maintenance. Continued use of the rifle could result In injury to, or death of, personnel.	

2-7. PREVENTIVE MAINTENANCE CHECKS AND SERVICES (CONT).

PREVENTIVE MAINTENANCE CHECKS AND SERVICES FOR M16A2 RIFLE (CONT)

Item No.	Interval	Item to be Checked or Serviced	Procedure	Not Fully Mission Capable if:
2	Quarterly (cont)	Charging handle assembly and selector lever (cont)	Pull charging handle (1) to rear. Check that chamber is clear. Let bolt carrier assembly (2) close. Leave hammer in cocked position. Do not pull trigger	Charging handle does not lock in place when in the forward position.
			Place selector lever (3) in SAFE position Pull trigger. Place selector lever (3) in SEMI position.	Hammer falls.
			NOTE For the purpose of the following test, "SLOW" is defined as 1/4 to 1/2 the normal rate of trigger release.	
			Pull trigger	Hammer does not fall.
			Hold trigger to the rear, charge weapon, and release the trigger with a slow, smooth motion, without hesi-tations or stops, until the trigger is fully forward (an audible click should be heard).	Hammer falls.
			Repeat the above SEMI position test five times	The weapon mal-functions during any of these five tests.

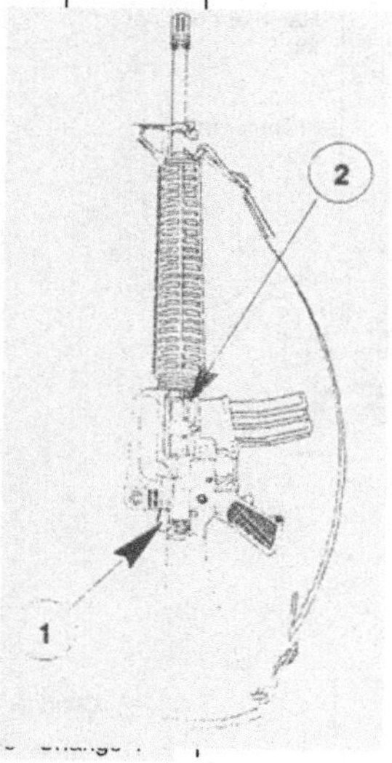

PREVENTIVE MAINTENANCE CHECKS AND SERVICES FOR M16A2 RIFLE (CONT)

Item No.	Interval	Item to be Checked or Serviced	Procedure	Not Fully Mission Capable if:
2	Quarterly (cont)	Charging handle assembly and selector lever (cont)	**M16A2 and M4 ONLY** Place selector lever (3) in BURST position. Charge weapon and squeeze trigger.	Hammer does not fall.
			Hold trigger to the rear, pull the charging handle to the rear and release it three times. Release trigger. Squeeze trigger. **NOTE** The burst disconnector should have held the hammer to the rear when it engaged the deep notch of the burst cam.	Hammer does not fall.
			M4A1 ONLY Place selector lever (3) in AUTO position Charge carbine and squeeze trigger Hammer should fall.	Hammer does not fall.
			Hold trigger to the rear, charge carbine, and release trigger. Squeeze trigger. Hammer should not fall.	Hammer falls.

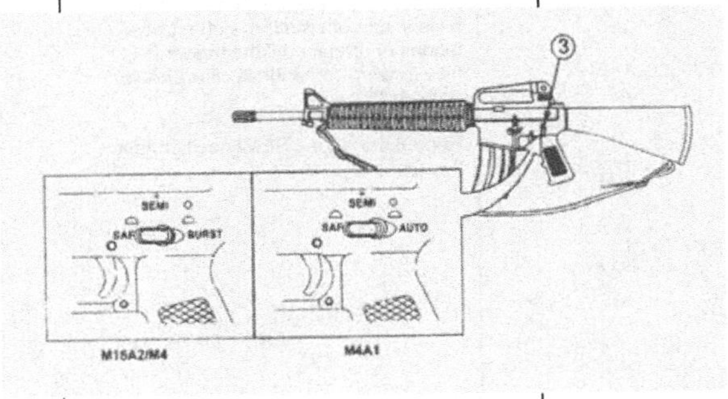

2-7. PREVENTIVE MAINTENANCE CHECKS AND SERVICES (CONT).

PREVENTIVE MAINTENANCE CHECKS AND SERVICES FOR M16A2 RIFLE (CONT)

Item No.	Interval	Item to be Checked or Serviced	Procedure	Not Fully Mission Capable if:
2	Quarterly (cont)	Charging handle assembly and selector lever (cont)	**NOTE** Automatic sear should have released hammer while holding trigger in the squeezed position before releasing and resqueezing the trigger. **All weapons** With hammer in forward position, using moderate finger/thumb pressure attempt to place the selector lever (3) in SAFE position	Moderate finger/ thumb pressure moves selector lever to SAFE position.

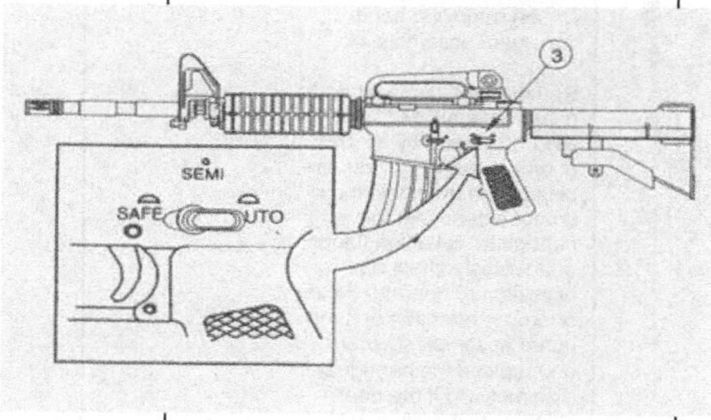

2-7. PREVENTIVE MAINTENANCE CHECKS AND SERVICES (CONT).

PREVENTIVE MAINTENANCE CHECKS AND SERVICES FOR M16A2 RIFLE (CONT)

Item No.	Interval	Item to be Checked or Serviced	Procedure	Not Fully Mission Capable if:
3	Quarterly	Upper receiver and barrel assembly (handguard assemblies) 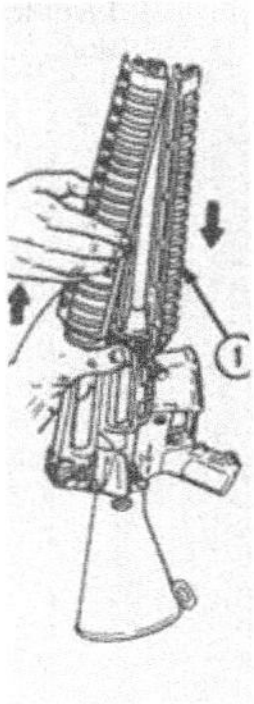	**CAUTION** Do not use screwdriver or any other tool when removing the handguard assemblies, doing so may damage the handguard assemblies and/or slip. **NOTE** Refer to operator's manual for "buddy system" procedure on removing handguard assemblies. Remove and Inspect handguard assemblies (1) internally and externally for cracks and/or damage. Cracks are acceptable providing they do not extend into the handguard retaining flange, or adversely affect rifle operation or operator safety or proper retention of handguard assembly. Discard and replace the handguard assembly (1) If the heatshield is loose enough to rattle when Installed on rifle.	Handguard missing or unserviceable.

PREVENTIVE MAINTENANCE CHECKS AND SERVICES FOR M16A2 RIFLE (CONT)

Item No.	Interval	Item to be Checked or Serviced	Procedure	Not Fully Mission Capable if:
4	Quarterly	Upper receiver and barrel assembly (serviceability check) 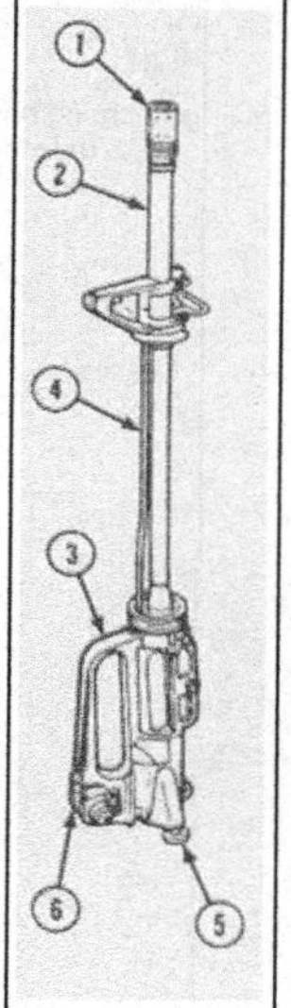	**WARNING** Dry cleaning solvent is flammable and toxic and should be used in a well ventilated area The use of rubber gloves is necessary to protect the skin when washing rifle parts **CAUTION** Damage may occur if excessive force is used to release takedown pin or pivot pin. Use hand pressure ONLY. Release takedown pins and open and separate receivers Hand check compensator (1) for looseness on barrel (2), then hand check barrel (2) for looseness on upper receiver (3). Check center slot of compensator for alignment (p 2-50). If compensator or barrel is loose, evacuate to support maintenance Check gas tube (4), forward assist assembly (5), and rear sight assembly (6) for damage, The rear sight spring should retain the rear sight assembly (6) In either position with firmness SH - If damaged, evacuate to support maintenance.	Compensator or barrel is loose

2-7. PREVENTIVE MAINTENANCE CHECKS AND SERVICES (CONT).

PREVENTIVE MAINTENANCE CHECKS AND SERVICES FOR M16A2 RIFLE (CONT)

Item No.	Interval	Item to be Checked or Serviced	Procedure	Not Fully Mission Capable if:
4	Quarterly (cont)	Upper receiver and barrel as sembly (service ability check) (cont)	**NOTE** If front or rear sight is moved, return to original position Check front sight post, de tent, and spring (7) for dam age and corrosion Clean and lubricate Check charg ing handle (8) and election port cover (9) for defects and proper function Check sling swivel (10) and rivet (11) for damage and proper function SH Other corn ponents are defective, re place as necessary.	Charging handle (8) is defective.

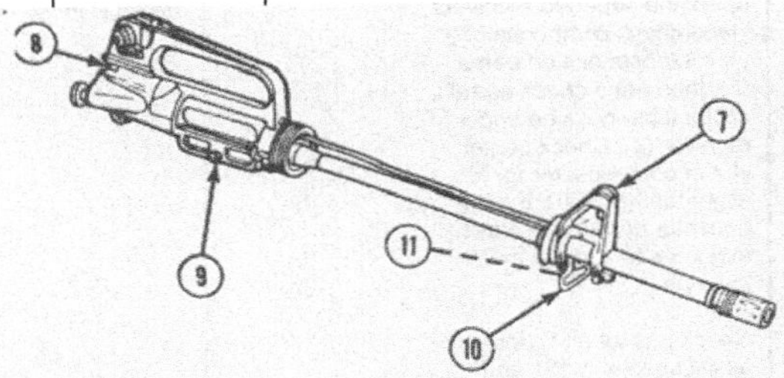

CAUTION

Do not use a wire brush to roughen surfaces Use a well ventilated area during cleaning and application of solid film lubricant. If solid film lubricant comes in contact with moving parts or functioning surfaces of the rifle, remove lubricant immediately by washing with dry cleaning solvent.

NOTE

Shiny metal exterior surfaces of the rifle should be recoated with solid film lubricant (Item 21, app D). Clean surface with dry cleaning solvent (item 16, app D); dry, roughen with abrasive cloth (Item 13, app D) and apply solid film lubricant.

PREVENTIVE MAINTENANCE CHECKS AND SERVICES FOR M16A2 RIFLE (CONT)

Item No.	Interval	Item to be Checked or Serviced	Procedure	Not Fully Mission Capable if:
4	Quarterly (cont)	Upper receiver and barrel assembly (serviceability check) (cont) 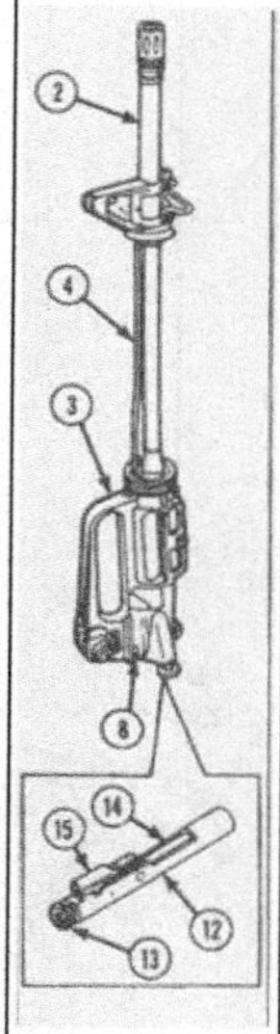	Inspect upper receiver (3) finish for scratches or worn shiny spots. If scratched or worn shiny in spots, disassemble and remove all lubricant from surface with dry cleaning solvent (item 16, app D). Wear rubber gloves (item 18, app D) and use a wash pan (item 24, app D) to apply solvent. Let parts dry thoroughly. Roughen the surface using abrasive cloth (item 1 3, app D) and apply solid film lubricant (item 16, app D). Allow 16 to 24 hours to dry before handling.	
			Hold barrel (2) at 40-degree angle (muzzle down). Pull charging handle (8) to rear. Hold bolt carrier assembly (12) to rear and push charging handle forward. Release bolt carrier assembly (12). The bolt carrier assembly should close and lock under its own weight. If It does not, remove the bolt assembly (13) from the key and bolt carrier assembly (14) and slide the key and bolt carrier assembly (without bolt) back and forth In the upper receiver and barrel assembly. If the gas tube (4) hits the carrier key (1 5), or If the gas tube binds in the carrier key, try to correct the malfunction by adjusting (slightly bending) the gas tube In the area of the handguard assemblies. If unable to adjust, evacuate to support maintenance.	Adjustment does not correct the malfunction.

2-7. PREVENTIVE MAINTENANCE CHECKS AND SERVICES (CONT).

PREVENTIVE MAINTENANCE CHECKS AND SERVICES FOR M16A2 RIFLE (CONT)

Item No.	Interval	Item to be Checked or Serviced	Procedure	Not Fully Mission Capable if:
4	Quarterly (cont)	Upper receiver and barrel assembly (serviceability check) (cont)	**M4/M4A1** Inspect carrying handle assembly (16) and mounting surface (17) of upper receiver for damage. If the carrying handle is missing or can not be correctly mounted, repair as authorized or evacuate to support maintenance. Inspect carrying handle assembly (16) to insure unit applied identification (ID) code matches unit applied ID code on carbine. If it doesn't match locate correct carrying handle assembly and match up to correct M4/ M4A1 carbine. If a match can not be found, the weapon should be re-zeroed by the operator.	

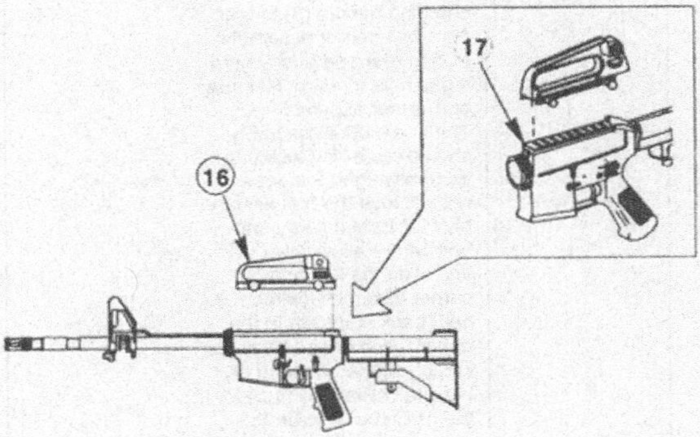

2-7. PREVENTIVE MAINTENANCE CHECKS AND SERVICES (CONT).

PREVENTIVE MAINTENANCE CHECKS AND SERVICES FOR M16A2 RIFLE (CONT)

Item No.	Interval	Item to be Checked or Serviced	Procedure	Not Fully Mission Capable if:
			WARNING Below direct support maintenance, do not interchange bolt assemblies from one rifle to another. Doing so may result in injury to, or death of, personnel.	
5	Quarterly	Key and bolt carrier assembly and bolt assembly (serviceability check)	Remove and disassemble. Visually inspect bolt assembly (1) for cracks, especially in the area of the bolt cam pin hole (2). Check for cracks on locking lugs (3), for a cluster of pits or chipped bolt face (4), and for an elongated firing pin hole (5). If cracked or broken, evacuate to support maintenance for repair.	Defects are found.
			Check for worn or missing bolt rings (6) Check for proper staggering of bolt rings Insert the bolt assembly (1) into the key and bolt carrier assembly (7). Turn key and bolt carrier assembly (7) so the bolt assembly (1) points down. The bolt assembly must not drop out. Remove bolt assembly (p 2-35). Check for broken or missing firing pin retaining pin (8) and bolt cam pin (9); replace as necessary.	The bolt assembly drops out of the key and bolt carrier assembly due to its own weight. Missing or broken firing pin retaining pin or bolt cam pin.

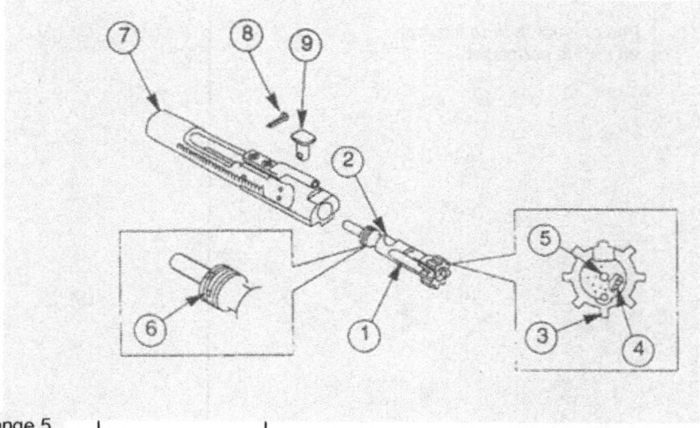

PREVENTIVE MAINTENANCE CHECKS AND SERVICES FOR M16A2 RIFLE (CONT)

Item No.	Interval	Item to be Checked or Serviced	Procedure	Not Fully Mission Capable if:
5	Quarterly (cont)	Key and bolt carrier assembly and bolt assembly (serviceability check) (cont)	Check cartridge extractor (10), extractor spring assembly (11), cartridge ejector (12), and ejector spring (13) for dirt and serviceability. If dirty, clean, lubricate and assemble. If unserviceable, replace as necessary.	Parts are missing or unserviceable
			Check key and bolt carrier assembly (7) and carrier key (14) for damage and looseness. If damaged or loose, evacuate to support maintenance	Key and bolt carrier assembly or carrier key is damaged, or carrier key is loose.
			NOTE If carrier key Is dented, evacuate to support maintenance.	
			Check firing pin 15) for chips or breaks If damaged, evacuate to support maintenance.	Firing pin is damaged.
			Pits or wear in area Illustrated (16) is permissable	

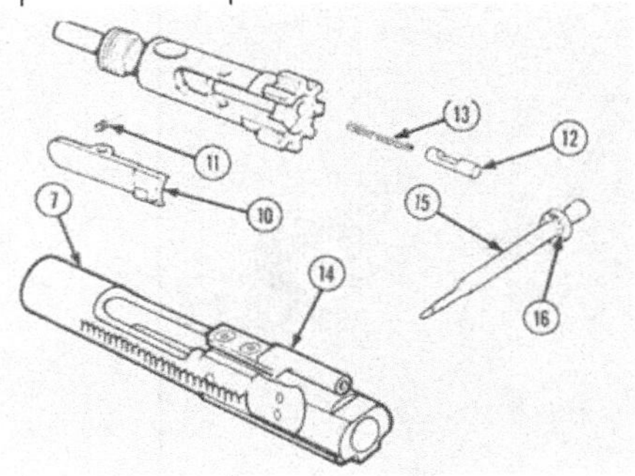

2-7. PREVENTIVE MAINTENANCE CHECKS AND SERVICES (CONT).

PREVENTIVE MAINTENANCE CHECKS AND SERVICES FOR M16A2 RIFLE (CONT)

Item No.	Interval	Item to be Checked or Serviced	Procedure	Not Fully Mission Capable if:
6	Quarterly	Lower receiver and buttstock assembly (serviceability check)	Remove buffer assembly (1) and action spring (2) Check buffer assembly for cracks SH - Buffer assembly is cracked Check action spring (2) for kinks and free length Free length should be **RIFLE**: 11 3/4 Inches (29 85 cm) minimum to 13 1./2 inches (34 29 cm) maximum SH - If action spring is kinked or does not meet free length require-ments **CARBINE**: 10 1/16 inches (25.56 cm) minimum to 11 1/4 inches (28.58 cm) maximum. Do not attempt to adjust spring length SH - If action spring is kinked or does not meet free length require-ments	

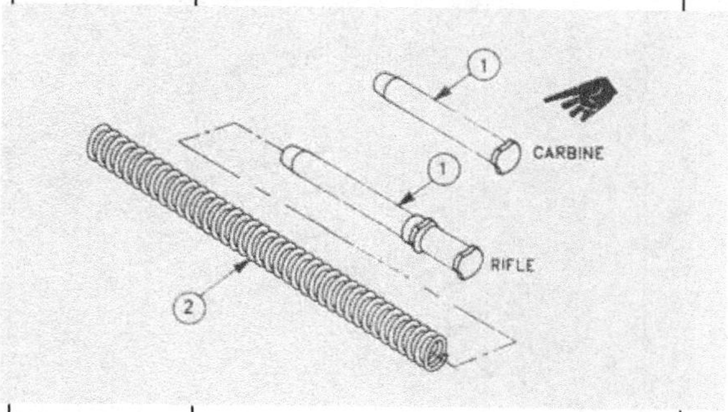

PREVENTIVE MAINTENANCE CHECKS AND SERVICES FOR M16A2 RIFLE (CONT)

Item No.	Interval	Item to be Checked or Serviced	Procedure	Not Fully Mission Capable if:
6	Quarterly (cont)	Lower receiver and buttstock assembly (serviceability check) (cont)	Remove pistol grip screw (3), lockwasher (4), pistol grip (5), helical spring (6), safety detent (7), pivot pin (8), pivot pin detent (9), and helical spring (10). Clean and lubricate metal components. Also clean and generously lubricate pivot pin holes and spring/detent holes. Replace defective/damaged components as necessary.	Components are defective/damaged.
			Disengage takedown pin (11) and pull out, push back in to re-engage takedown pin (an audible click should be heard). If an audible click is not heard, see page 2-57 for repair.	Components are defective/damaged.

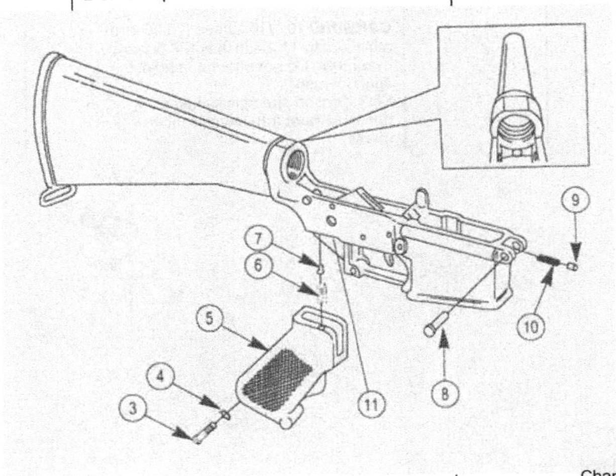

2-7. PREVENTIVE MAINTENANCE CHECKS AND SERVICES (CONT).

PREVENTIVE MAINTENANCE CHECKS AND SERVICES FOR M16A2 RIFLE (CONT)

Item No.	Interval	Item to be Checked or Serviced	Procedure	Not Fully Mission Capable if:
6	Quarterly (cont)	Lower receiver and buttstock assembly (serviceability check) (cont)	Lubricate helical compression spring and takedown pin detent (11) by placing one drop of lubricant on takedown pin detent and lowering the buttstock assembly (12) to vertical position. Allow the lubricant to work Its way around the helical compression spring and takedown pin detent (11). Check buttstock assembly(12) components for damage.	

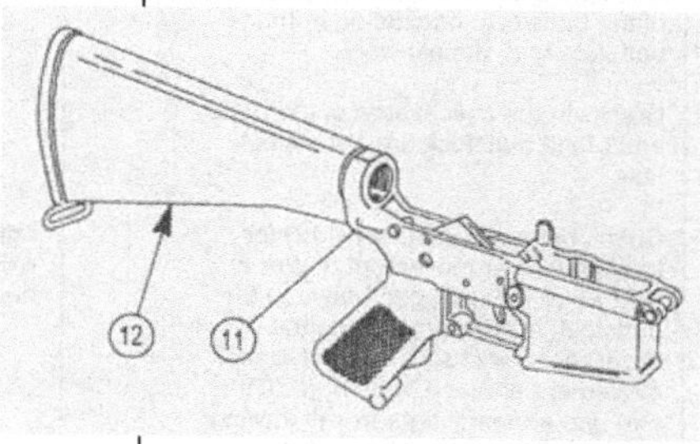

Item No.	Interval	Item to be Checked or Serviced	Procedure	Not Fully Mission Capable if:
			RIFLE ONLY Under the following conditions, hairline cracks (no chipped away material allowed) originating from the buttplate end of the buttstock are acceptable.	Components are damaged.

PREVENTIVE MAINTENANCE CHECKS AND SERVICES FOR M16A2 RIFLE (CONT)

Item No.	Interval	Item to be Checked or Serviced	Procedure	Not Fully Mission Capable if:
6	Quarterly (cont)	Lower receiver and buttstock assembly (serv-iceability check) (cont)	**RIFLE ONLY** a. One hairline crack, not to exceed 1 in. (2.54 cm) in length, per side of buttstock b. Two additional hairline cracks up to 0.25 in. (0.64 cm) in length, per side of buttstock. c. A total of three cracks per side of the buttstock, originating from the buttplate end, are allowable. Cracks in the critical area at the front end of the buttstock are not accept-able.	The buttstock is cracked in the critical area or does not meet the crack criteria.
			Check buttstock assembly (12) for forward to rear movement and/or a 1/32 in. (0.079 cm) gap between the buttstock assembly (12) and the lower receiver (13). If forward to rear movement and/or a 1/32 in. (0.079 cm) gap appears, tighten self-locking screw. If still not tight, remove butt-stock assembly and check for loose lower receiver extension. If loose evacuate to support maintenance. If not loose, replace buttplate (14).	Lower receiver extension cannot be tightened.

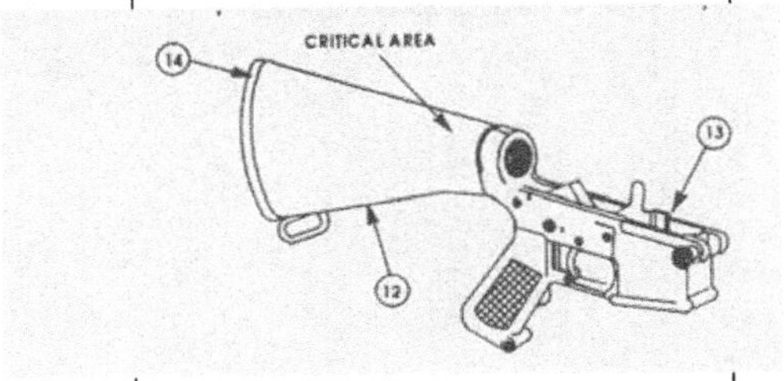

2-7. PREVENTIVE MAINTENANCE CHECKS AND SERVICES (CONS).

PREVENTIVE MAINTENANCE CHECKS AND SERVICES FOR M16A2 RIFLE (CONT)

Item No.	Interval	Item to be Checked or Serviced	Procedure	Not Fully Mission Capable if:
6	Quarterly (cont)	Lower receiver and buttstock assembly (serviceability check) (cont)	**RIFLE ONLY** Small amounts of side-to-side, up-and-down or rotational movement of the buttstock assembly is acceptable. (1) Cracks visible around the butt-plate mounting holes while screws are mounted SH - Cracks are visible around mounting holes when installed on rifle. (2) Cracks or separations around the door assembly are visible when the door assembly is closed SH - Cracks are visible when door assembly is closed. (3) If buttplate is cracked in excess of 0.25 in (O 64 cm) in length and extends through the buttplate (14). see pg. 2-64 for repair SH - If cracked in excess of 0 25 in. extends thru buttplate (4) The buttplate (14) should not be removed other than for repair or replacement of parts at which time a new self-locking screw, NSN 5305-01-147-8585, must be used.	

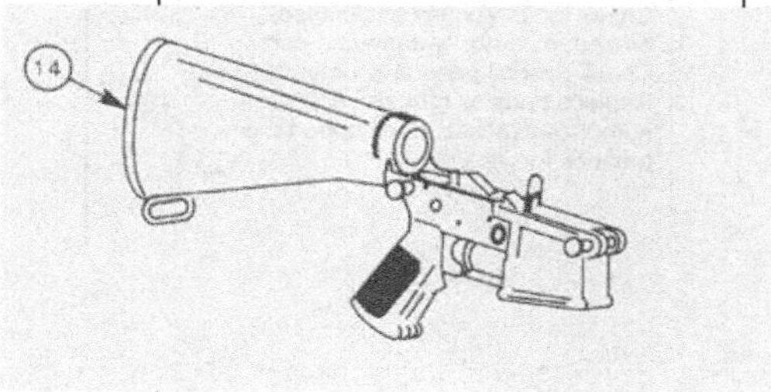

PREVENTIVE MAINTENANCE CHECKS AND SERVICES FOR M16A2 RIFLE (CONT)

Item No.	Interval	Item to be Checked or Serviced	Procedure	Not Fully Mission Capable if:
6	Quarterly (cont)	Lower receiver and buttstock assembly (serviceability check) (cont)	**NOTE** If a weapons lower receiver is missing one third or more of its exterior protective finish, resulting in an unprotected/light reflecting surface, it is candidate for overhaul. This missing finish will be considered a shortcoming. This shortcoming requires action to obtain a replacement weapon. Once a replacement has been received, evacuate the original weapon to depot for overhaul. If scratched or worn shiny in spots, repair in the same manner as outlined for upper receiver (see item 4 above).	
7	Quarterly	M16A2 Rifle	Assemble as in TM 9-1005-319-10 (operator's manual). Check sling for damage. If damaged, replace. Check for improperly assembled, broken, missing, or damaged parts. Check over all general appearance. Replace parts as required and authorize evacuation to support maintenance for repair.	

2-7. PREVENTIVE MAINTENANCE CHECKS AND SERVICES (CONT).

PREVENTIVE MAINTENANCE CHECKS AND SERVICES FOR M16A2 RIFLE (CONT)

Item No.	Interval	Item to be Checked or Serviced	Procedure	Not Fully Mission Capable if:
8	Quarterly	Annual DS safety and serviceability inspection and gaging	Check to ensure annual DS safety and serviceability inspection and gaging has been done and that the next gaging and inspection is scheduled. If annual gaging has not been performed within the last year, notify support maintenance.	Annual gaging has not been performed.

Section IV. TROUBLESHOOTING

2-8. GENERAL.

a. This section contains unit level troubleshooting information for locating and correcting most of the operating troubles which may develop in the M16A2 rifle, M4/M4A1 carbine. Each malfunction for the individual part or assembly is followed by a list of tests or inspections which will help you to determine the corrective actions in the order listed.

b. This manual cannot list all malfunctions that may occur, nor all tests or inspections and corrective actions. If a malfunction is not listed or is not corrected by listed corrective actions, see individual repair sections in the maintenance procedures on each major assembly.

2-9. TROUBLESHOOTING PROCEDURES. Refer to troubleshooting table for malfunctions, tests, and corrective actions. The symptom index is provided for a quick reference of the malfunctions covered in the table.

SYMPTOM INDEX

TROUBLESHOOTING

MALFUNCTION
TEST OR INSPECTION
CORRECTIVE ACTION

1. FAILURE OF MAGAZINE TO LOCK IN RIFLE.

Step 1. Dirty or corroded magazine catch (1).

Disassemble and clean.

Step 2 Defective magazine catch spring (2)

Evacuate to support maintenance.

Step 3 Worn or broken magazine catch (1).

Evacuate to support maintenance

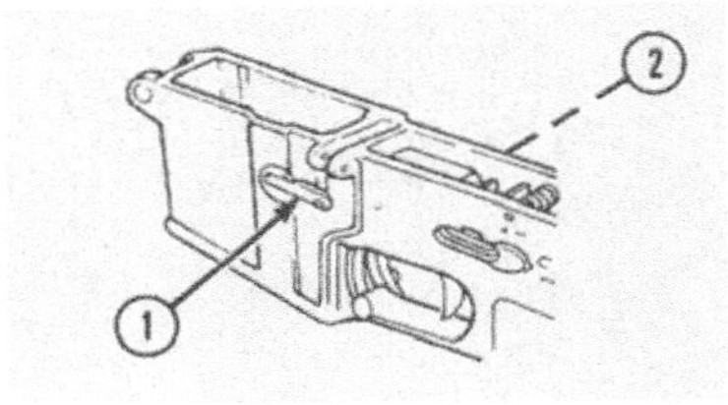

2-9. TROUBLESHOOTING PROCEDURES (CONT).

TROUBLESHOOTING (CONT)

MALFUNCTION
TEST OR INSPECTION
CORRECTIVE ACTION

2. FAILURE TO FEED

Step 1. Magazine catch spring weak or broken

Evacuate to support maintenance

Step 2. Magazine catch (1) defective.

Evacuate to support maintenance

Step 3. Magazine catch (1) out of adjustment (will not retain magazine)

Refer to operator's manual.

Step 4. Short recoil.

Refer to page 2-28

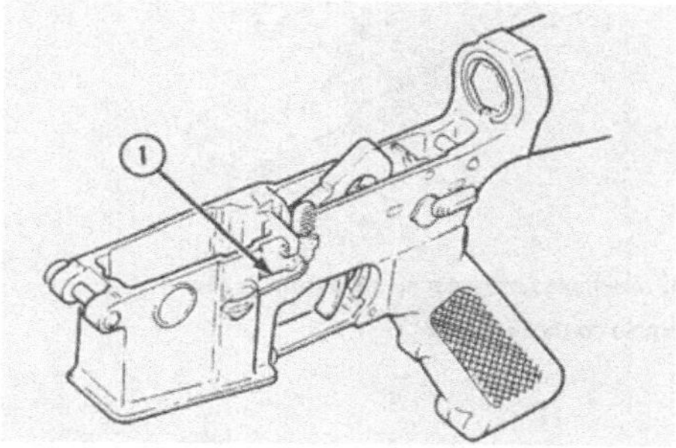

3. FAILURE TO CHAMBER.

Step 1. Weak or broken action spring (1), RIFLE ONLY (free length 11 3/4 inches (29.85 cm) minimum to 13 1.'2 Inches (34 29 cm) maximum) CARBINE ONLY (10 1 16 inches (2S 56 cm) minimum to 11 1,4 inches (28 58 cm) maximum)

Replace action spring (p 2-57)

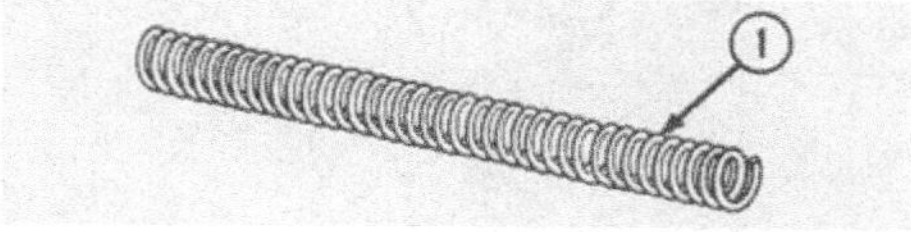

TROUBLESHOOTING (CONT)

MALFUNCTION
TEST OR INSPECTION
CORRECTIVE ACTION

Step 2. Short recoil

Refer to page 2-28

4. FAILURE TO LOCK

Step 1. Bolt cam pin (1) missing

Replace (p 2-35)

Step 2. Loose or damaged bolt carrier key (2).

a. Evacuate to support maintenance.

b. Dented bolt carrier key may be repaired (p 2 35).

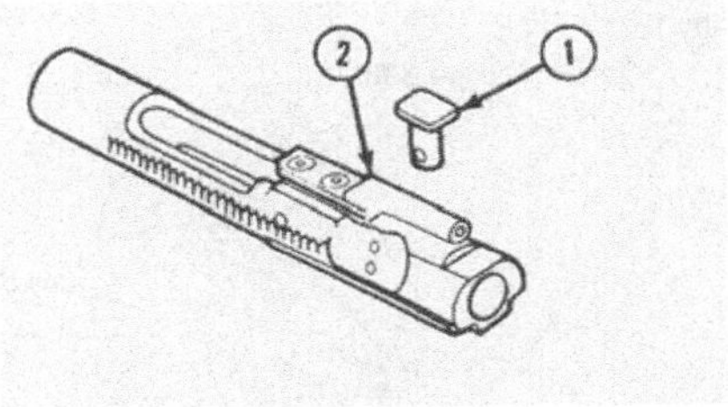

Step 3. Improperly assembled extractor spring assembly (3)

Assemble correctly (p 2-38).

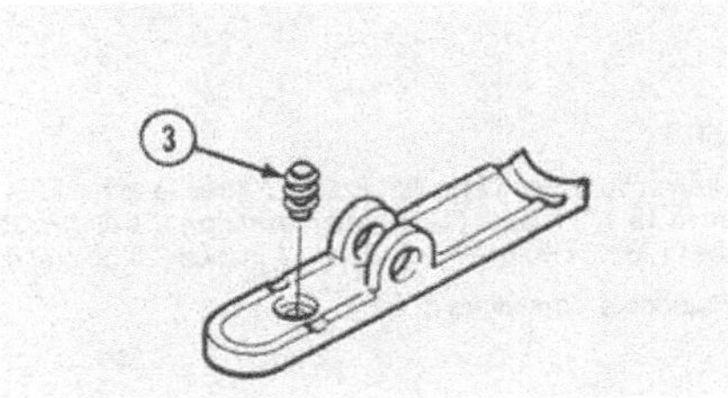

2-9. TROUBLESHOOTING PROCEDURES (CONT).

TROUBLESHOOTING (CONT)

MALFUNCTION
TEST OR INSPECTION
CORRECTIVE ACTION

4. FAILURE TO LOCK (CONT)

Step 4. Bent gas tube 14J

a Adjust to Its original configuration by bending in area of hand-guard assembly.

b If the gas tube cannot be returned to Its original configuration, evacuate the rifle to support maintenance

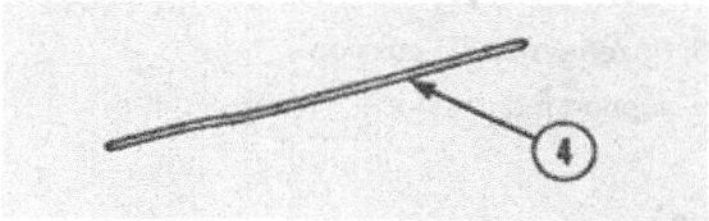

Step 5 Weak or broken action spring (5); RIFLE ONLY: (free length 11 3/4 inches (29,85 cm) minimum to 13 1,/2 Inches (34 29 cm) maximum). CARBINE ONLY (10 1/16 Inches (25 56 cm) minimum to 11 1,4 inches (28 58 cm) maximum

Replace action spring (p 2-571

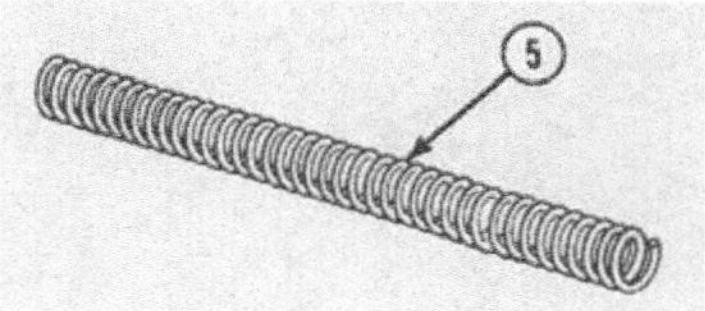

5. FAILURE TO FIRE.

Step 1 Broken or chipped firing pin (1).

Evacuate to support maintenance.

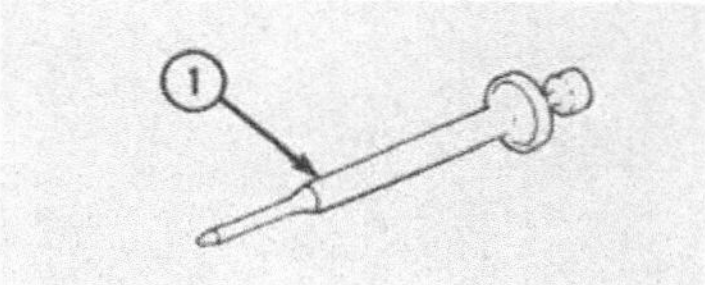

TROUBLESHOOTING (CONT)

MALFUNCTION
TEST OR INSPECTION
CORRECTIVE ACTION

Step 2 Carbon buildup in firing pin recess inside bolt assembly

Remove cartridge extractor and clean recess with pipe cleaner (Item 11, app D), refer to operator's manual.

Step 3. Firing mechanism (2) and or lower receiver assembly (31 improperly assembled or has worn, broken, or missing parts.

Evacuate to support maintenance

Step 4. Broken, defective, or missing firing pin retaining pin (4).

Replace (p 2-35)

Step 5. Selector lever (5) frozen on SAFE position.

Evacuate to support maintenance

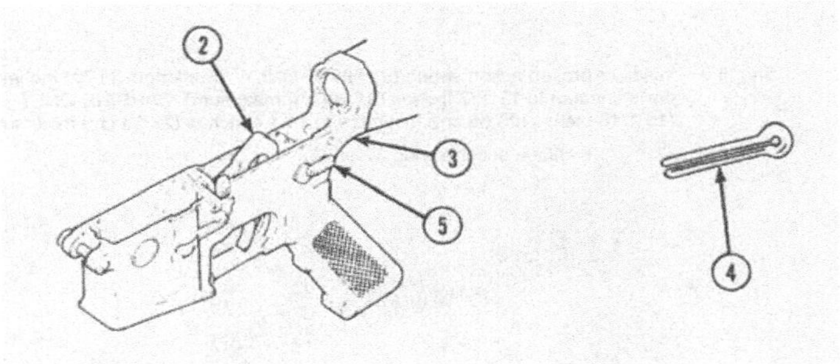

2-9. TROUBLESHOOTING PROCEDURES (CONT).

TROUBLESHOOTING (CONT)

MALFUNCTION
TEST OR INSPECTION
CORRECTIVE ACTION

6. FAILURE TO UNLOCK.

Step 1. Burred locking lugs (1) on bolt assembly.

Remove burrs.

Step 2. Burred lugs (2) on barrel extension

Remove burrs.

Step 3 Short recoil.

Refer to page 2-28.

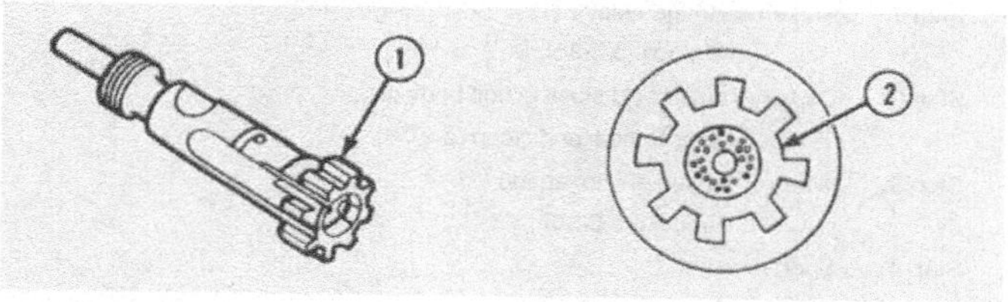

7. FAILURE TO EXTRACT.

Step 1. Defective extractor pin (1), cartridge extractor (2), and or extractor spring assembly (3).

Replace extractor pin (1), cartridge extractor (2), and or extra(:tor spring assembly (3) (p 2-381

Step 2. Short recoil

Refer to page 2-28

NOTE

Rubber insert and spring are an assembly Illustration shows Insert out of assembly for clarification only Do not remove the rubber Insert from the extractor spring assembly.

TROUBLESHOOTING (CONT)

MALFUNCTION
TEST OR INSPECTION
CORRECTIVE ACTION

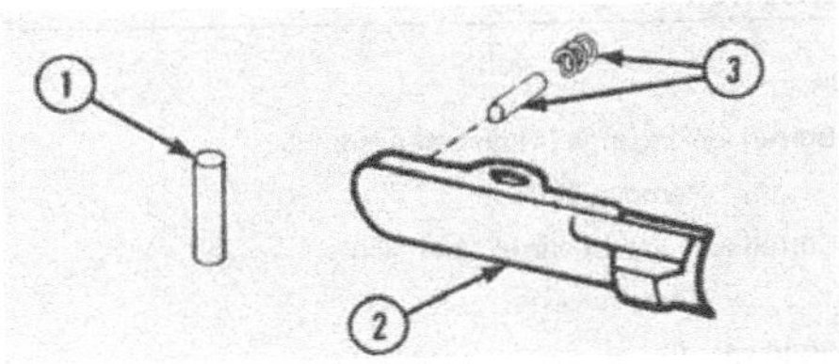

8. FAILURE TO EJECT.

Step 1. Broken cartridge ejector (1).

Replace (p 2-38).

Step 2. Cartridge ejector (1) stuck in bolt body (2).

Disassemble and clean (p 2-38).

Step 3. Weak or broken ejector spring (3).

Replace (p 2-38)

Step 4. Short recoil

Refer to page 2-28.

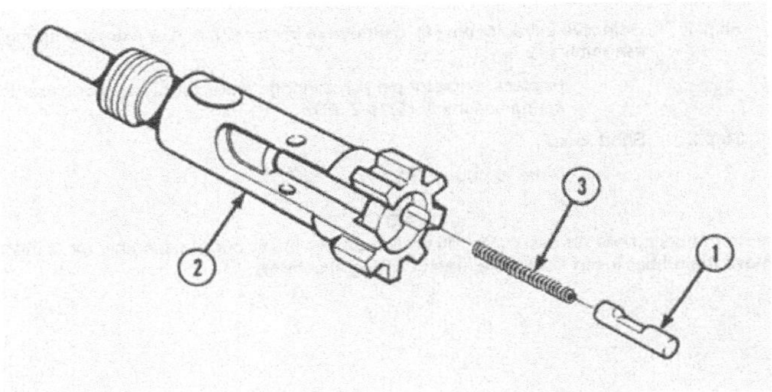

2-9. TROUBLESHOOTING PROCEDURES (CONT).

TROUBLESHOOTING (CONT)

MALFUNCTION
TEST OR INSPECTION
CORRECTIVE ACTION

9. FAILURE TO COCK

Step 1 Worn, broken, or missing parts of firing mechanism

Evacuate to support maintenance.

Step 2 Short recoil

Refer to below

10 SHORT RECOIL

Step 1 Broken or damaged action spring (1).

Replace action spring (p 2-57).

Step 2 Unlubricated or dirty action spring and receiver extension

Clean and lubricate

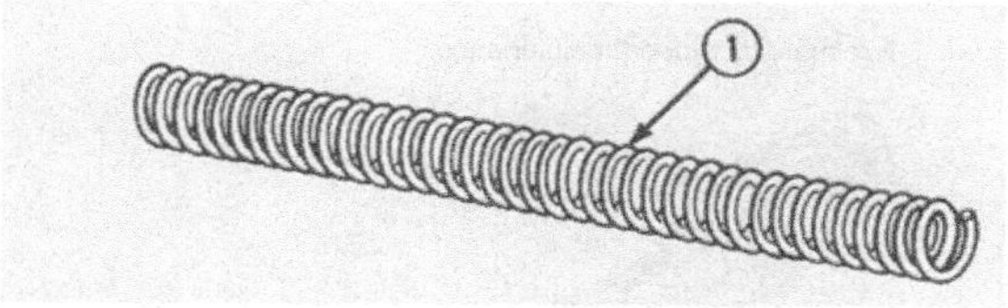

Step 3 Improper gap space or worn, missing, or broken bolt rings (2)

a. Stagger bolt ring gaps (approximately 13 turn apart}

b. Evacuate to support maintenance if bolt rings are worn, broken, or missing.

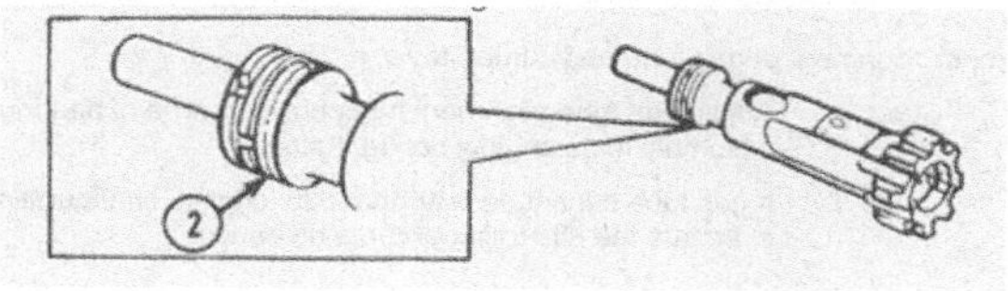

TROUBLESHOOTING (CONT)

MALFUNCTION
TEST OR INSPECTION
CORRECTIVE ACTION

Step 4. Carbon build-up or foreign matter In the narrow passage of the bolt carrier key (3).

Clean with CLP (Item 9, app D) and a pipe cleaner (Item 11, app D).

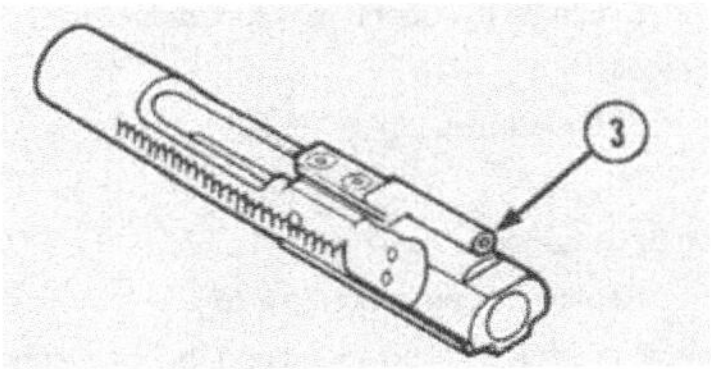

Step 5. Gas leakage caused by broken or loose gas tube (4) around front sight base.

Evacuate to support maintenance.

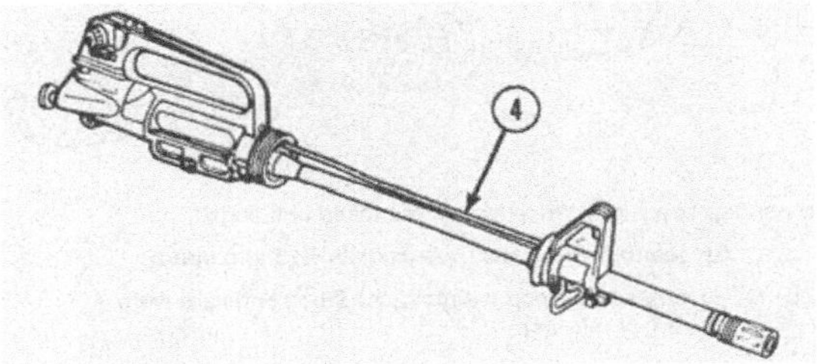

Step 6. Improper alignment of gas tube and carrier key.

a. Adjust gas tube alignment by bending in area of handguard assembly to Its original configuration

b. If gas tube cannot be returned to Its original configuration, evacuate the rifle to support maintenance.

2-9. TROUBLESHOOTING PROCEDURES (CONT).

TROUBLESHOOTING (CONT)

MALFUNCTION
TEST OR INSPECTION
CORRECTIVE ACTION

11. RIFLE CANNOT BE ZEROED.

Step 1. Defective barrel assembly (1).

Evacuate to support maintenance

Step 2. Barrel assembly out of alignment with rear sight assembly (2) on upper receiver.

Evacuate to support maintenance

Step 3. Defective front sight (3).

Remove front sight post (4), front sight detent (5), and helical spring (6). If damaged, replace.

Step 4. Defective -ear sight assembly (2).
Evacuate to support maintenance.

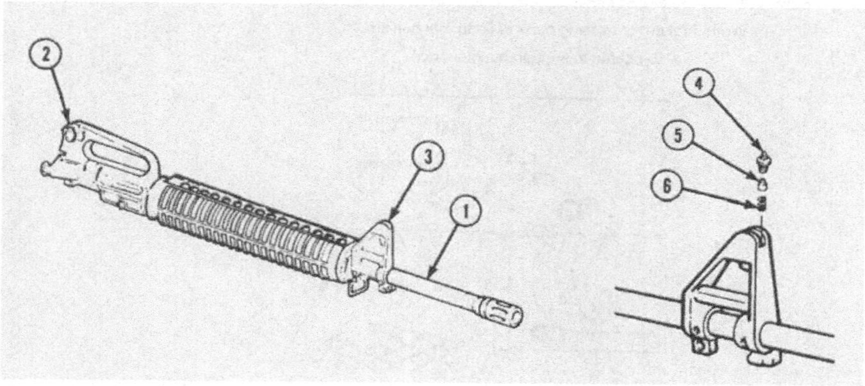

TROUBLESHOOTING (CONT)

MALFUNCTION
TEST OR INSPECTION
CORRECTIVE ACTION

12. FAILURE TO CYCLE WITH SELECTOR LEVER SET ON BURST (M16A2 AND M4).

Faulty selector lever or broken cam, cam clutch spring, or burst disconnector.

Evacuate to support maintenance.

12.1. FAILURE TO CYCLE WITH SELECTOR LEVER SET ON AUTO (M4A1).

Faulty selector lever.

Evacuate to support maintenance.

13. FIRES TWO ROUNDS WITH ONE PULL OF TRIGGER WITH SELECTOR LEVER SET ON SEMI (DOUBLE FIRING).

Perform function test.

If any part of function test (p 2-68) fails, evacuate to support maintenance.

14. FIRES WITH SELECTOR LEVER ON SAFE OR WHEN TRIGGER IS RELEASED WITH SELECTOR LEVER ON SEMI.

Worn, broken, or missing parts of firing mechanism.

Evacuate to support maintenance.

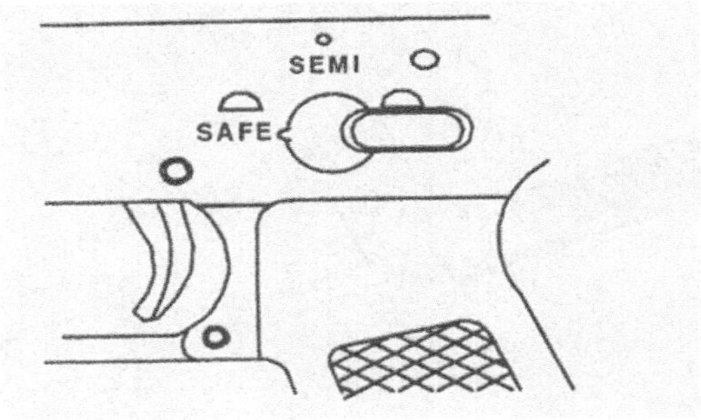

2-9. TROUBLESHOOTING PROCEDURES (CONT).

TROUBLESHOOTING (CONT)

MALFUNCTION
TEST OR INSPECTION
CORRECTIVE ACTION

15 BOLT ASSEMBLY FAILS TO LOCK TO REAR AFTER FIRING LAST ROUND.

Step 1. Magazine follower (1) worn or broken.

Replace magazine.

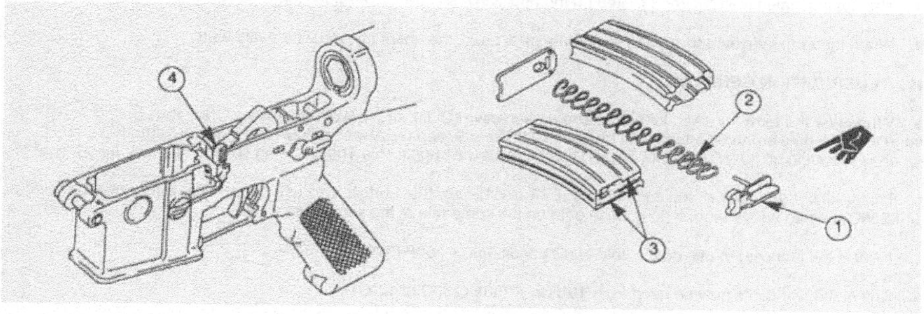

Step 2. Magazine catch spring (2) weak or broken.

Replace magazine.

Step 3. Magazine feeder lips (3) bent or broken.

Replace magazine.

Step 4. Magazine follower (1) binds during operation.

Replace magazine.

Step 5. Broken bolt catch (4) and/or spring.

Evacuate to support maintenance.

Section V. MAINTENANCE PROCEDURES

2-10. INITIAL SETUP. The following information will reduce the space required for the initial setup portion of the maintenance procedures.

a. Materials/Parts required are not listed unless they apply to the procedure

b. Personnel Required is listed only if the task requires more than one person If Personnel Required Is not listed, It means one person can do the job.

c. The normal standard equipment condition Is that the item is removed from the en(1 Item or next higher assembly and is in the assembled condition. Equipment Condition is not listed unless some other condition is required.

d. The approximate time required Is listed on the applicable Maintenance Allocation Chart (MAC).

e. When the term evacuate to support maintenance is used, the entire rifle must be evacuated.

2-11. LUBRICATION GENERAL.

a. Whenever the term "cleaner, lubricant, and preservative (CLP)" or the words "lubricant", "lube", "LSA", or "LAW" are cited In this TM, they are to be interpreted to mean CLP (Item 9, app D), weapons lubricating oil (LSA) (Item 23, app D), or weapons lubricating oil (LAW) (Item 22, app D) can be utilized as applicable The following constraints must be adhered to:

b. Under all but the coldest arctic conditions, LSA or CLP are the lubricants to use on the rifle. Either may be used at 100F (230C) and above. However, do not use both on the same rifle at the same time.

c. LAW is the lubricant to use during cold arctic conditions, + 100F (12C) and below

d. Any of the lubricants may be used from 10°F to + 100F (230C to 12°C)

e. ARMY ONLY: Do not mix lubricants on the same rifle The rifle must be thoroughly cleaned during change from one lubricant to another Dry Cleaning Solvent (SD) (Item 16, app D) is recommended for cleaning during change from one lubricant to another.

f. Rifle Bore Cleaner (RBC) (item 12, app D) may be used to remove carbon buildup In the bore and other portions of the rifle.

2-12. MAJOR COMPONENTS OF M16A2 RIFLE AND CARBINES.

This task covers disassembly

INITIAL SETUP:

References
TM 9-1005-319-10 (operator's manual)

Equipment Conditions
Rifle and Carbines assembled

General Safety Instructions
Before starting an inspection, be sure to clear the weapon. Do not keep live ammunition near the work area.

To avoid injury to your eyes, use care when removing and installing spring-loaded parts.

Below direct support maintenance, do not interchange bolt assemblies from one weapon to another. Doing so may result in injury to, or death of, personnel.

DISASSEMBLY

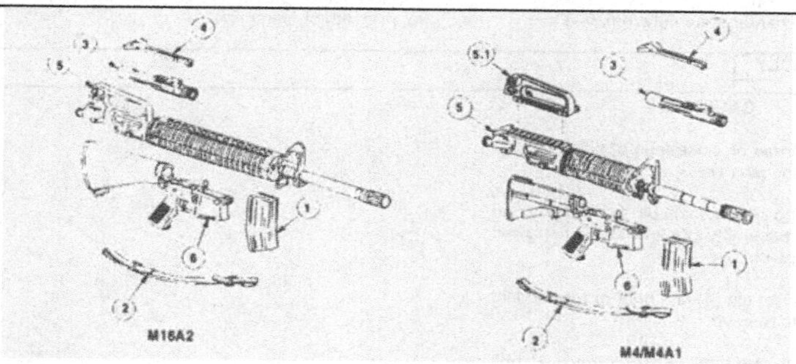

a. Refer to operator's manual.

b. Remove magazine (1), sling (2), bolt carrier assembly (3), charging handle assembly (4), and upper receiver and barrel assembly (5), carrying handle assembly (5.1), from lower receiver and buttstock assembly (6).

NOTE

Solid Film Lubricant (SFL) is the only authorized touchup for the MI 16A2 rifle and M4/M4A1 carbines and may be used on up to one third of the exterior finish of the weapon. **FOR ARMY CONUS USE ONLY AND AIR FORCE TRAINING WEAPONS ONLY**: SFL may be used as touchup without limitation on the upper receiver and barrel assembly. This is to say that units which DO NOT fall under the category of Divisional Combat Units or rapid deployment type units may have up to 100 percent of the exterior surface of the upper receiver and barrel assembly protected with SFL if necessary.

12-13. BOLT CARRIER ASSEMBLY.

This task covers:

- a. **Disassembly**
- b. **Cleaning**
- c. **Inspection/Repair**
- d. **Lubrication**
- e. **Reassembly**

INITIAL SETUP:

Tools
(ARMY) Small Arms Repairman Tool Kit (item 3, app B)
Bolt Carrier Key Tool (item 11, figure C-17)

References
TM 9-1005-319-10 (operator's manual)

Equipment Conditions
2-34 Bolt carrier assembly removed

General Safety Instructions
Bolt cam pin must be installed or rifle will blow up while firing the first round. If the bolt cam pin is not installed, injury to, or death of, personnel may result.

Do not interchange bolt assemblies from one rifle to another. Doing so may result in injury to, or death of, personnel.

a. DISASSEMBLY

CAUTION

Do not spread or close legs of firing pin retaining pin (1).

1. Remove firing pin retaining pin (1). Tilt key and bolt carrier assembly (2) and catch firing pin (3) as it drops out.
2. Rotate bolt cam pin (4) one quarter turn and lift straight up to remove.
3. Remove bolt assembly (5) from key and bolt carrier assembly (2).

NOTE

For disassembly of bolt assembly (5), see page 2-38.

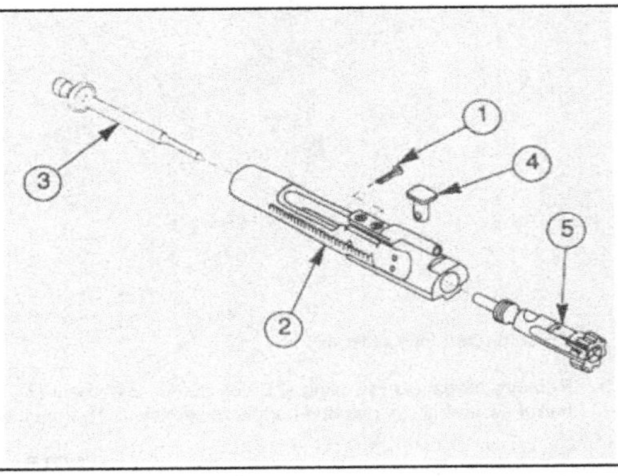

2-13. BOLT CARRIER ASSEMBLY (CONT).

b. CLEANING

Clean all items (operator's manual). Remove carbon deposits.

c. INSPECTION/REPAIR

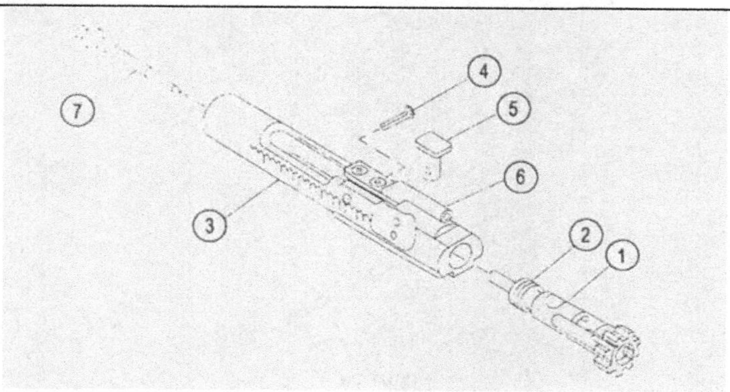

1. Inspect bolt assembly (1) as follows:

   a. Check to see that the bolt ring (2) gaps are staggered.

   b. Install bolt assembly (1) into key and bolt carrier assembly (3). Check for worn rings by holding the key and bolt carrier assembly (3) with the bolt assembly facing down. If bolt assembly falls out after firing pin retaining pin (4) and bolt cam pin (5) are removed, the rings are worn. Evacuate to support maintenance.

2.Inspect carrier key (6) for dents, distortion, or looseness. If dented or loose, evacuate to support maintenance.

3. Inspect firing pin retaining pin (4) and bolt cam pin (5) for cracks, damage, or excessive wear. Replace if unserviceable.

4. Inspect firing pin (7) for damage or If tip is chipped. Evacuate rifle to support maintenance if unserviceable

5. Inspect key and bolt carrier assembly (3) for damage or wear. If unserviceable, evacuate to support maintenance.

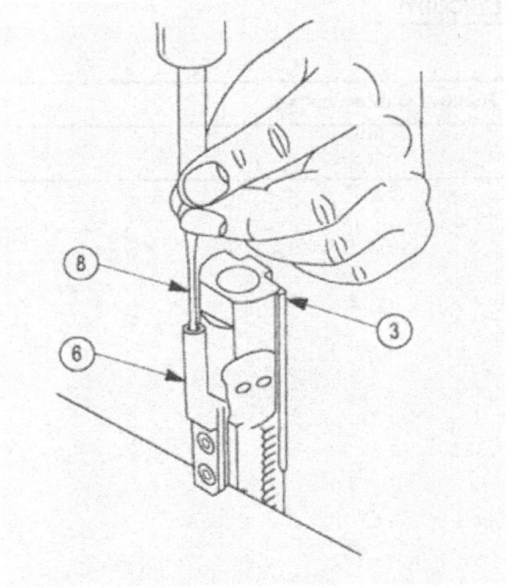

CAUTION

Extreme care must be exercised during the following procedure to assure that the striking force is not directed to the attaching screws and that the tube portion is not enlarged or flared beyond original requirement. Such enlargement would permit loss of gas pressure when the key and gas tube come together during functioning.

6. Repair small dents and/or distortions in carrier key (6) using bolt carrier key tool as follows:

a. Place the key and bolt carrier assembly (3) in a vertical position, supported In a manner that contact is made with the rear surface of the carrier key (6).

b. Insert the small end of the key tool (8) into the tube portion of the carrier key (6).

c. Strike the large end of the key tool (8) lightly with a 3 ounce, soft-brass hammer.

d. Repeat striking (gently) until carrier key (6) is reformed to original configuration.

e. If carrier key (6) cannot be reformed to original configuration, evacuate the weapon to support maintenance.

d. LUBRICATION

Lubricate all Items (p 2 33) (operator's manual)

e. REASSEMBLY

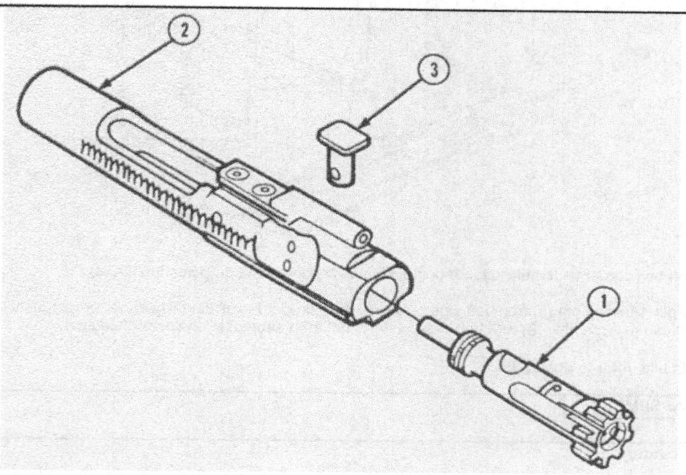

WARNING

Do not Interchange bolt assemblies from one rifle to another Doing so may result In Injury, or death of, personnel.

Bolt cam pin must be Installed or rifle will blow up while firing the first round. If the bolt cam pin is not Installed, Injury to, or death of, personnel may result

NOTE

Before installing bolt assembly, check to see that the bolt ring gaps are staggered to prevent loss of gas pressure

1. Install bolt assembly (1) In key and bolt carrier assembly (2)

2.Install bolt cam pin (3) and rotate one quarter turn

2-13. BOLT CARRIER ASSEMBLY (CONT).

e. REASSEMBLY (CONT)

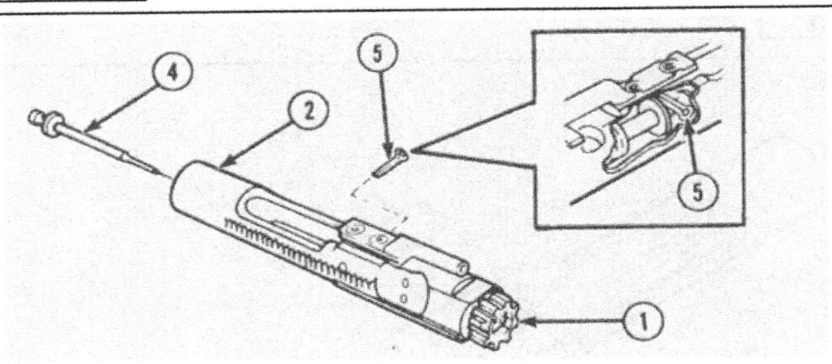

3. Hold key and bolt carrier assembly (2) with bolt assembly (1) down and drop in firing pin (4).

4. Install firing pin retaining pin (5) from left side only to ensure proper installation Check for proper installation by holding key and bolt carrier assembly (2) with bolt assembly (1) up, then attempt to shake out firing pin.

5. Reassemble rifle, refer to page 2-68.

2-14. BOLT ASSEMBLY.

This task covers:

a. **Removal**
b. **Cleaning**
c. **Inspection**
d. **Lubrication**
e. **Reassembly**

INITIAL SETUP:

Tools
(ARMY) Small Arms Repairman Tool Kit (item 3, app B)

Equipment Conditions
2-35 Bolt assembly removed

General Safety Instructions
Do not Interchange bolt assemblies from one rifle to another. Doing so may result in injury to, or death of, personnel.

To avoid injury to your eyes, use care when removing and installing spring-loaded parts.

a. DISASSEMBLY

NOTE

Do not separate cartridge extractor and extractor spring assembly unless replacement of either or both is required

Do not remove the rubber Insert from the extractor spring assembly

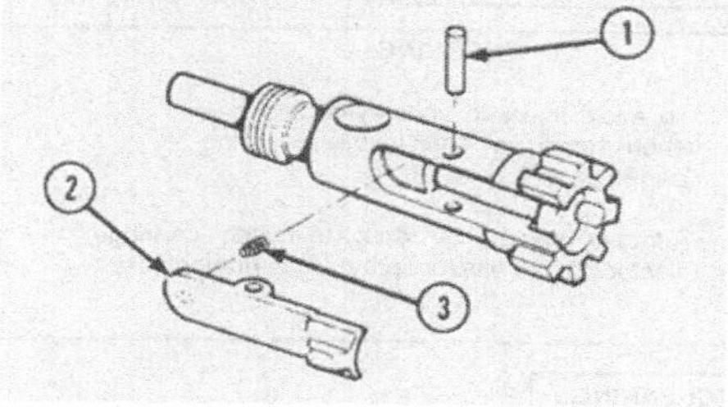

1. Push out extractor pin (1) and remove cartridge extractor (2) and extractor spring assembly (3) as a unit

2. If required, twist extractor spring assembly (3) counterclockwise to remove from cartridge extractor (2).

CAUTION

Be sure to use vise law protective caps

3. Hold bolt body In vise and remove spring pin (4) using 1 16 Inch punch and hammer

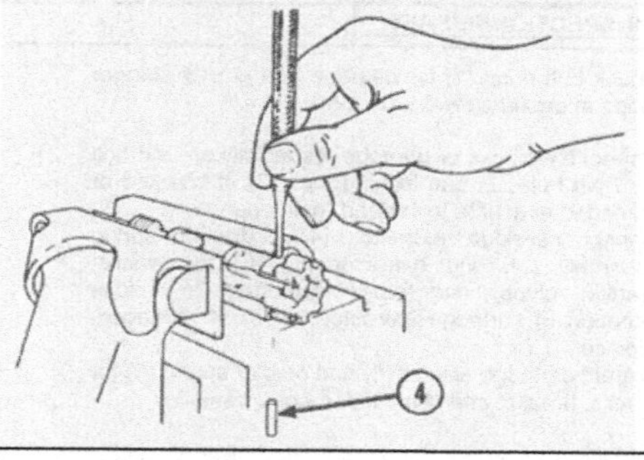

2-14. BOLT ASSEMBLY (CONT).

a. DISASSEMBLY (CONT)

WARNING

To avoid injury to your eyes, use care when removing and installing spring-loaded parts.

4. Remove punch, be sure to catch cartridge elector (5) and ejector spring (6) to prevent loss

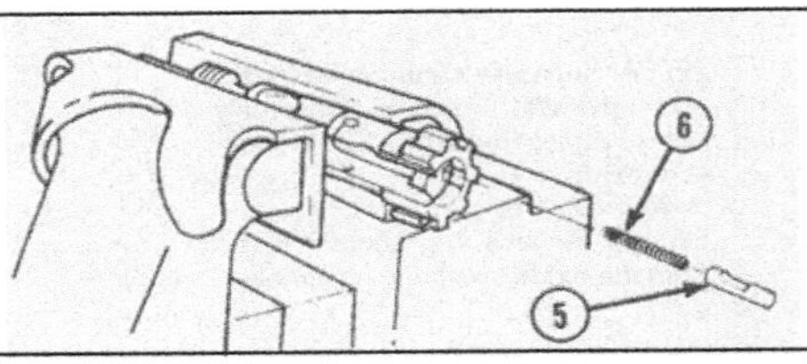

b. CLEANING

CAUTION

Do not distort extractor spring assembly during cleaning.

Clean all Items (operator's manual). Remove carbon deposits.

c. INSPECTION/REPAIR

1. Check bolt rings (1) for damage and proper stagger (gaps approximately 3 turn apart).
2. Inspect for cracks or damage, especially around bolt cam pin hole (2) and locking lugs (3). If cracked or damaged evacuate to support maintenance.
3. Inspect cartridge extractor (4), extractor spring assembly (5), and extractor pin (6) for cracks, breaks, chips, and other damage. Pay close attention to cartridge extractor lip (7). If damaged, replace.
4. Inspect cartridge ejector (8) and ejector spring (9) for cracks, breaks, and chips If damaged, replace.

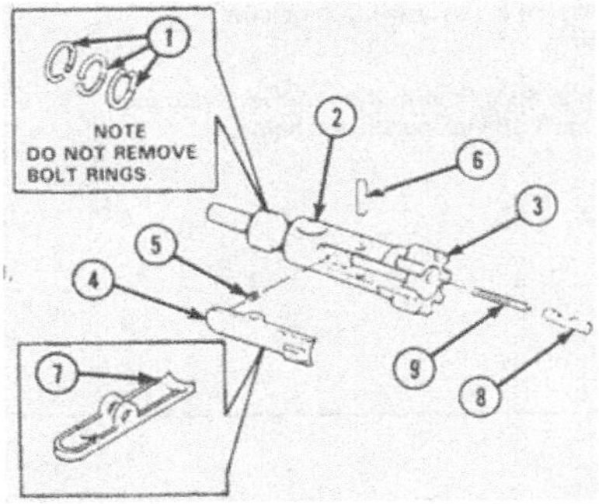

d. LUBRICATION

Lightly lubricate all items (p 2 33) (operator's manual)

e. REASSEMBLY

WARNING

To avoid Injury to your eyes, use care when removing and Installing spring loaded parts.

Do not Interchange bolt assemblies from one rifle to another Doing so may result In injury to, or death of, personnel.

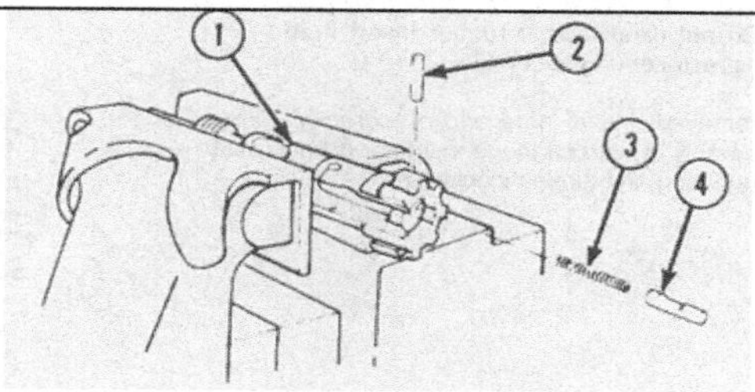

CAUTION

Be sure to use vise jaw protective caps

1. Place bolt body (1) in a vise and start spring pin (2) In hole.
2. Install ejector spring (3) and cartridge ejector (4). Align groove on cartridge ejector (4) so that spring pin 12) can be Installed
3. Compress and hold ejector spring and cartridge ejector In place with a 3 8 inch punch.
4. Using hammer and 1 16 Inch punch, complete Installation of spring pin (2) so that the ends are flush with the out side of bolt body (1).

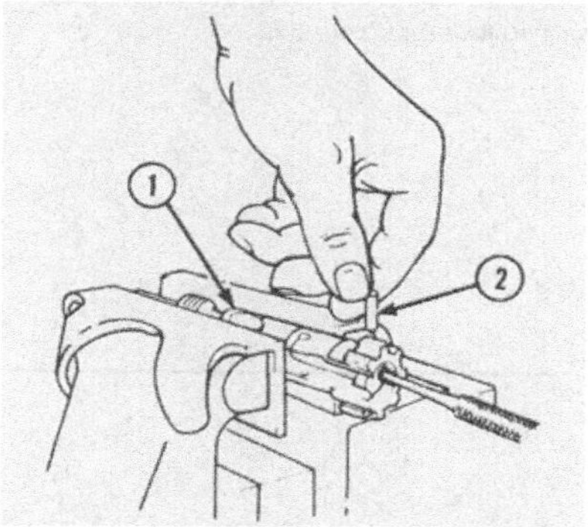

2-14. BOLT ASSEMBLY (CONT).

e. REASSEMBLY (CONT)

NOTE

Do not disassemble rubber Insert from extractor spring assembly.

5. If removed, Insert large end of extractor spring assembly (5) into cartridge extractor (6) and seat by pushing and turning clockwise.

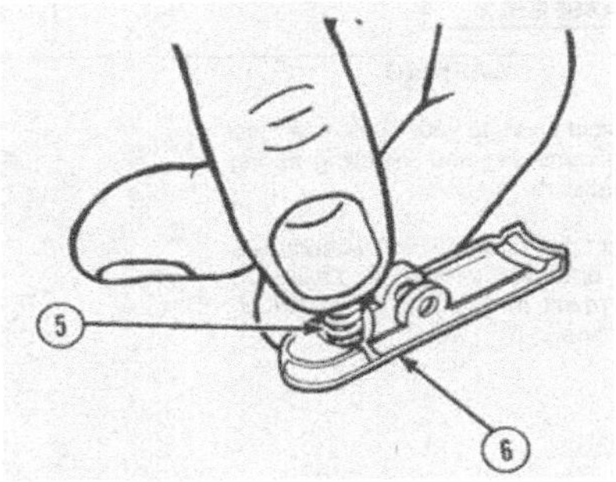

6 Position cartridge extractor (6) and ex tractor spring assembly (5) on bolt body (1).

7 Compress extractor spring assembly (5) and cartridge extractor (6) to align holes.

8. Install extractor pin (7) by hand.

9. Reassemble rifle, refer to page 2-68.

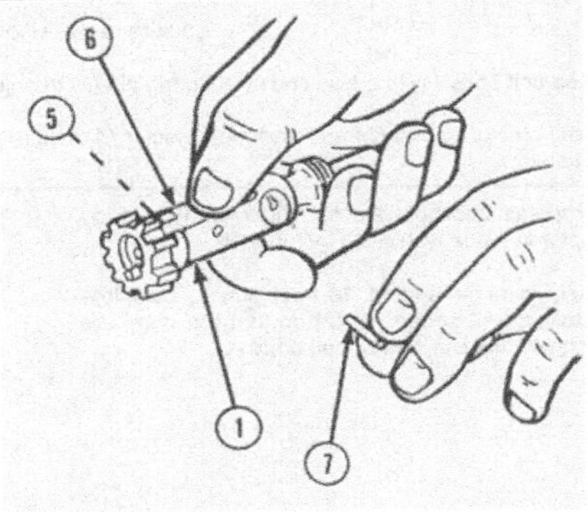

2-15. CHARGING HANDLE ASSEMBLY.

This task covers:

a. Disassembly **b. Cleaning** **c. Inspection/Repair** **d. Lubrication** **e. Reassembly**

INITIAL SETUP:

Tools
(ARMY) Small Arms Repairman Tool Kit (item 3, app B)

Equipment Conditions
2-34 Charging handle assembly removed

References
TM 9-1005-319-10 (operator's manual)

a. DISASSEMBLY

WARNING

To avoid injury to eyes, use care when removing and installing spring-loaded parts.

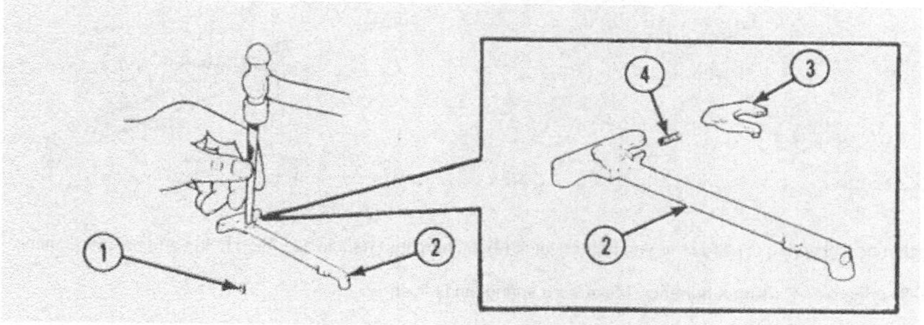

1. Remove spring pin (1) from charging handle (2) using a hammer and 1/16 inch punch.
2. As punch is withdrawn, catch charging handle latch (3) and helical spring (4) to prevent loss.

2-15. CHARGING HANDLE ASSEMBLY (CONT).

b. CLEANING

Clean all Items (operator's manual) Remove carbon deposits.

c. INSPECTION/REPAIR

Inspect all Items for breaks, cracks, or damage Replace all unserviceable Items

d. LUBRICATION

Lightly lubricate all Items (p 2-33) (operator's manual)

e. REASSEMBLY

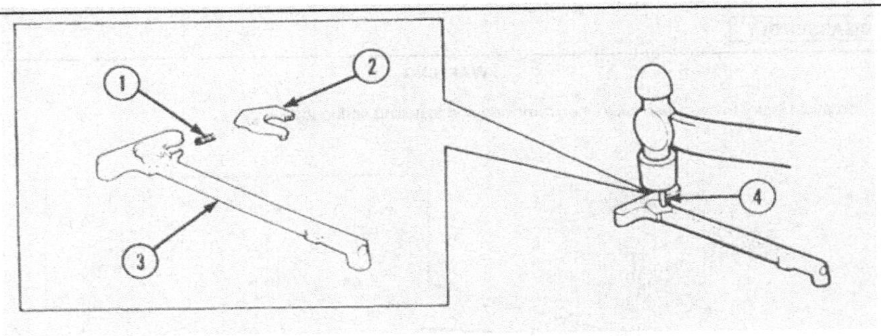

1. Position helical spring I1) and charging handle latch (2) in chdrgiin(J handle 13) Align holes and hold In position
2. Install spring pin (4) using a hammer Make sure spring pin is flush
3. Reassemble rifle, refer to page 2 68

2-16. UPPER RECEIVER AND BARREL ASSEMBLY, RIFLE BARREL ASSEMBLY, AND UPPER RECEIVER ASSEMBLY.

This task covers:

- **a.** Disassembly
- **b.** Cleaning
- **c.** Inspection
- **d.** Repair
- **e.** Lubrication
- **f.** Reassembly

INITIAL SETUP

Tools
(ARMY) Small Arms Repairman Tool Kit (item 3, app B)
Front sight post removal and installation tool (fig. E-2, app E)
Front sight detent depressor (fig. E-1, app E)

Materials/Parts
Lubricants (app D)

References
TM 9-1005-319-10 (operator's manual)

Equipment Conditions
2-34 Upper receiver and barrel assembly removed from lower receiver

General Safety Instructions
To avoid injury to your eyes, use care when removing and installing spring-loaded parts.

a. DISASSEMBLY

CAUTION

Do not use a screwdriver or any other tool when removing the handguard assemblies. Doing so may damage the handguard assemblies and/or slip ring. Do not remove heat shield for any reason. Doing so will damage the heat shield and the handguard assemblies will have to be replaced.

NOTE

Refer to operator's manual for "buddy system" procedure on removing handguard assemblies.

2-16. UPPER RECEIVER AND BARREL ASSEMBLY, RIFLE BARREL ASSEMBLY, AND UPPER RECEIVER ASSEMBLY (CONT).

a. DISASSEMBLY (CONT)

1. Push down on handguard slip ring (1) and lift upper handguard assembly (2) up and out.

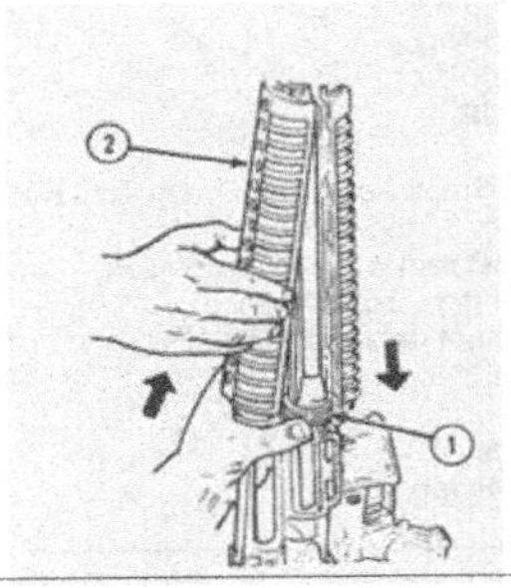

2. Push down on handguard slip ring (1) and lift the lower handguard assembly (3) up and out.

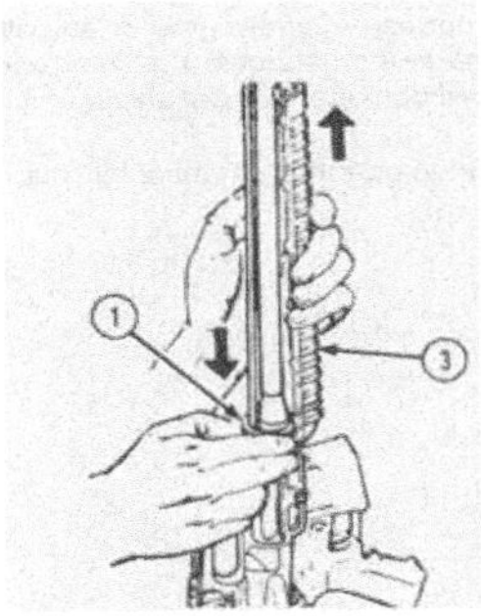

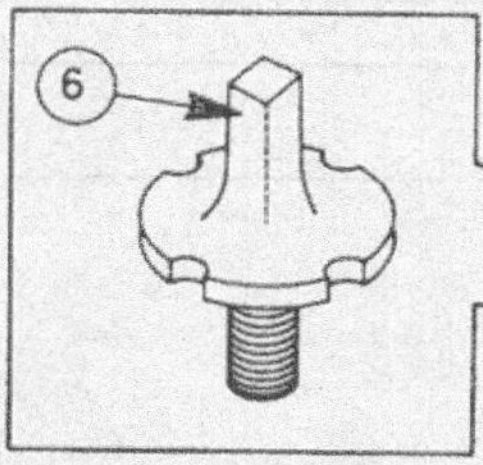

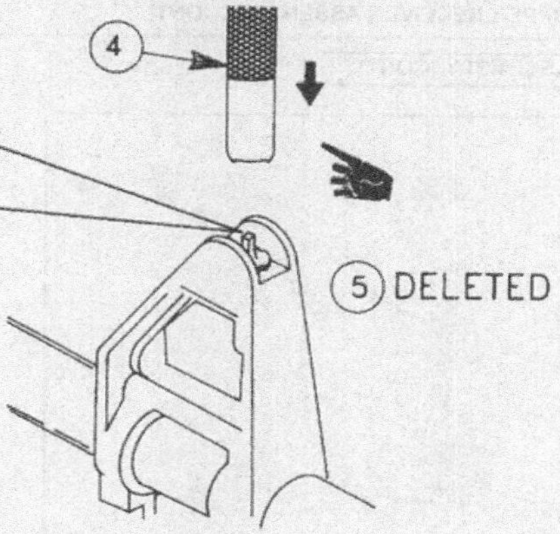

WARNING

To avoid injury to your eyes, use care when removing and installing spring-loaded parts.

NOTE

Remove front sight posts only if parts are damaged.

3. Using a dummy cartridge or a font sight post removal and installation tool (4).removed front sight post (6) by turning counter clockwise.

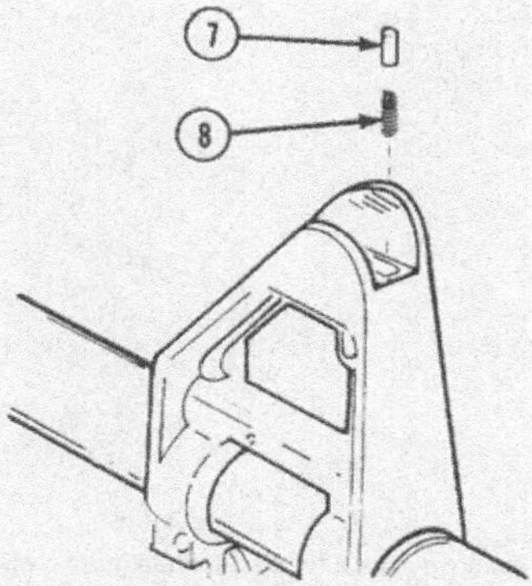

4. Catch front sight detent(7) and helical spring (8) to prevent loss.

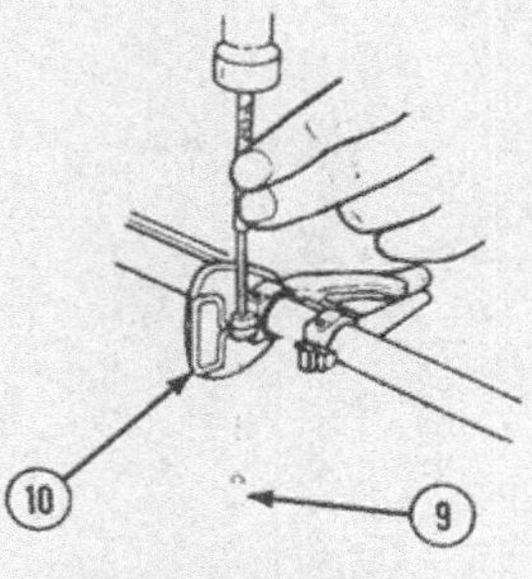

NOTE

RIFLE ONLY

Disassemble small sling swivel only if repair is necessary.

5. If necessary, knock out tubular rivet (9) with a hammer and punch and remove small sling swivel (10). Discard tubular rivet (9).

2-16. UPPER RECEIVE AND BARREL ASSEMBLY, RIFLE BARREL ASSEMBLY, AND UPPER RECEIVER ASSEMBLY (CONT).

a. DISASSEMBLY (CONT)

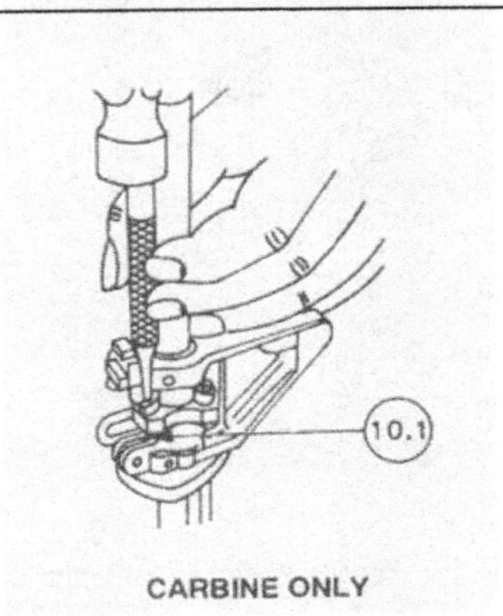

5A. Remove two spring pins (10 1)

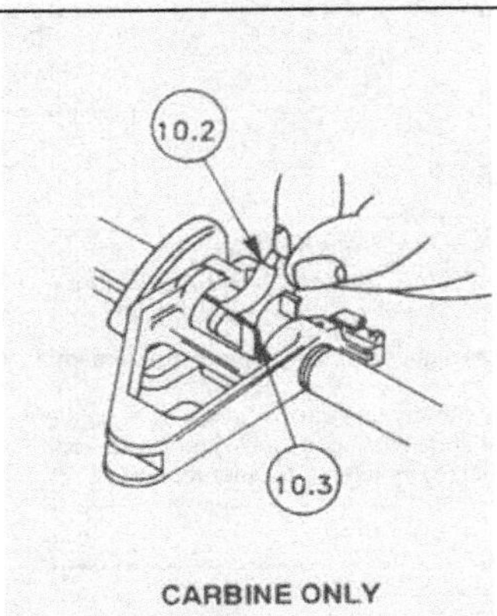

5B. Lift swivel locking bar (10 2) up and out of swivel mount (10 3)

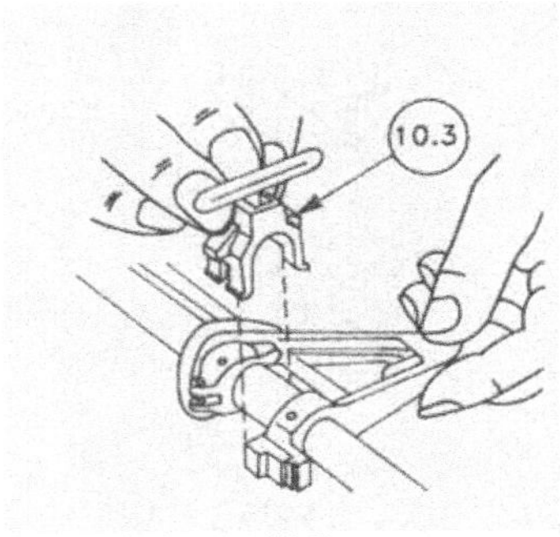

5C. Remove swivel mount (10.3).

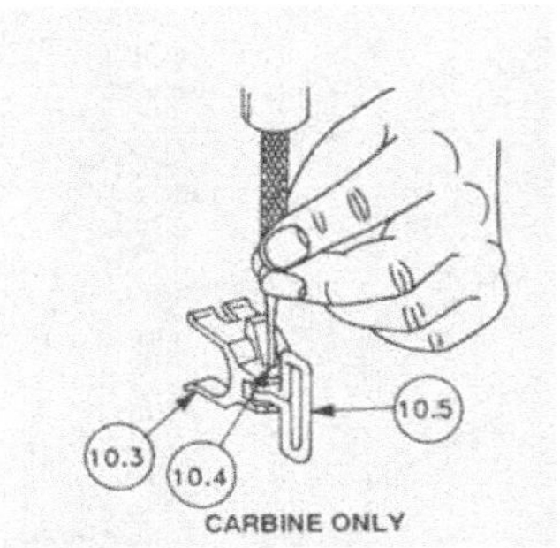

5D. Remove rivet (10 4) Separate front swivel (10 5) from swivel blank) mount (10 3)

2-16. UPPER RECEIVER AND BARREL ASSEMBLY, RIFLE BARREL ASSEMBLY, AND UPPER RECEIVER ASSEMBLY (CONT).

a. DISASSEMBLY (CONT)

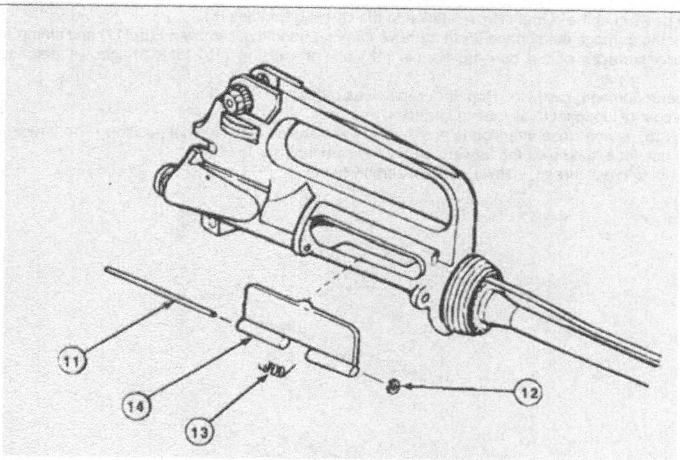

NOTE

Do not disassemble further unless repair is necessary. Headless grooved pin (11) may bind may bind against the forward assist housing and require some additional force to remove.

6. Remove retaining ring (12) and slide headles grooved pin (11) out to the rear.
7. Catch cover spring (13) and ejection port cover (14) to prevent loss as headless grooved pin is with-drawn.

M4/M4A1 ONLY

NOTE

Do not remove the carrying handle assembly nut (2 each) or clamp for normal maintenance. If carrying handle assembly nut or clamp are missing or the clamp is damaged to the point of being non-functional they may be replaced.

7.1. Disassemble as follows only if required to replace damaged or missing carrying handle assembly nut or clamp.

a. Observe the position of the clamp (15) in relation to the carrying handle (16).
b. Using care not to damage the surface finish, remove carrying handle assembly nut(s) (17) and clamp as required.
c. Inspect mating surfaces of the carrying handle (16) and the clamp (15) for damage. If damaged replace the damaged part.
d. Inspect for other damage, rust, etc. Repair or replace as required.
e. Apply a light coat of lubricant to all mating surfaces.
f. Install clamp (15) paying close attention to ensure that it is installed in its original position. The forward end of the clamp (15) must not extend past the forward end of the carrying handle (16).
g. Taking care not to cross thread, install handle assembly nut (17).

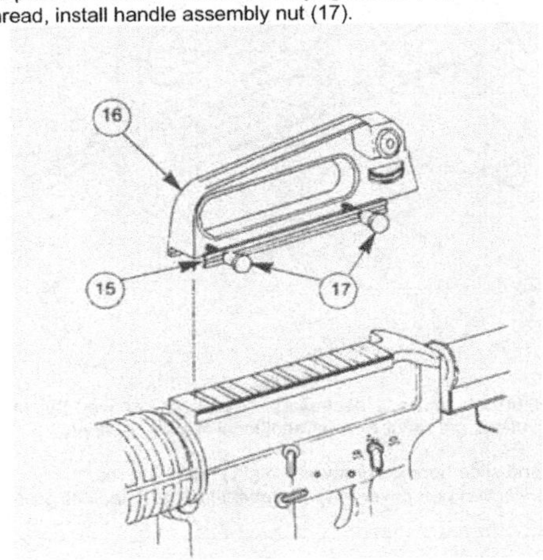

b. CLEANING

Clean all items (operator's manual).

c. INSPECTION

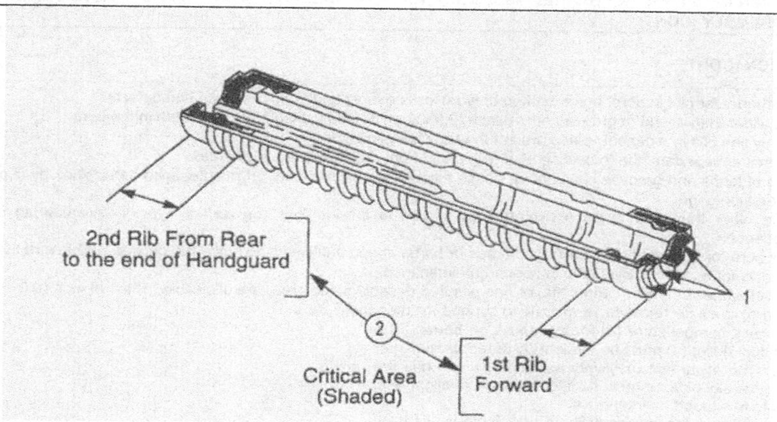

1. Inspect handguard assemblies for breaks separation, and cracks using the following guidelines:

- **a.** Breaks and separations of material which prevent proper retention or interfere with functioning of the rifle will be cause for handguard assembly rejection and replacement.
- **b.** M16A2 handguard assemblies may have up to two of the three front retaining tabs missing (1). M4/M4A1 may not have any front retaining tabs (1) missing. If all front retaining tabs for the M1 6A2 are missing, or any of the tabs for the M4/M4A1 are missing, handguard assemblies must be replaced.
- **c.** Cracks up to one inch in length are acceptable provided they do not extend into the retaining flange (CRITICAL AREA) (2).
- **d.** Replace severely cracked or damaged handguard assemblies. Handguard assemblies which have a heat shield which is loose enough to rattle when installed on the rifle must be replaced.

2. Inspect front sight assembly for chips, breaks, and cracks. Evacuate to support maintenance if broken, cracked, or bent.

3. Inspect front sight area for evidence of gas leakage around gas tube. Evacuate to support maintenance if short recoil results from gas leakage.

4. Inspect front sight post, front sight detent, and helical spring for damage. If damaged, replace.

2-16. UPPER RECEIVER AND BARREL ASSEMBLY, RIFLE BARREL ASSEMBLY, AND UPPER RECEIVER ASSEMBLY (CONT).

c. INSPECTION (CONT)

5. Inspect barrel for pits in bore, burrs, broken or worn locking lugs, and surface cracks and defects.
   a. Pits no wider than a land or groove and 3/8 inch (0.953 cm) or less in length are allowable in the bore.
   b. Uniformly fine pits in a densely pitted area of the bore are allowable.
   c. Lands that appear dark due to coating of gliding metal from projectiles are allowable.
   d. Striping of lands and grooves shall not be cause for rejection unless support maintenance determines by use of the barrel erosion gage.
   e. For pits other than mentioned above, broken or burred locking lugs, or surface cracks, evacuate to support maintenance.
6. Inspect bore for ringing. Definitely ringed bores or bores ringed sufficiently to bulge the outside surfaces of the barrel are causes for rejection. Evacuate to support maintenance.
7. Inspect chamber for pitting. Fine pits, or fine pits in a densely pitted area, are allowable. Pits 1/8 inch (0.318 cm) in length are cause for rejection. Evacuate to support maintenance.
8. Hand check compensator (3) for looseness on barrel. The third (middle) slot (4) must be straight up at top dead center (TDC). The alignment may vary as much as one half the width of the slot either direction. If loose or out of alignment, evacuate to support maintenance.
9. Inspect all items for serviceability in and tightness of latch assembly on ejection port cover. If items are damaged or nonfunctional, they are unserviceable.

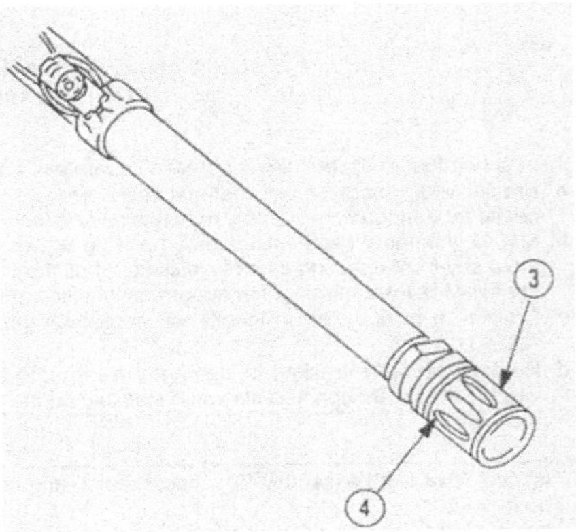

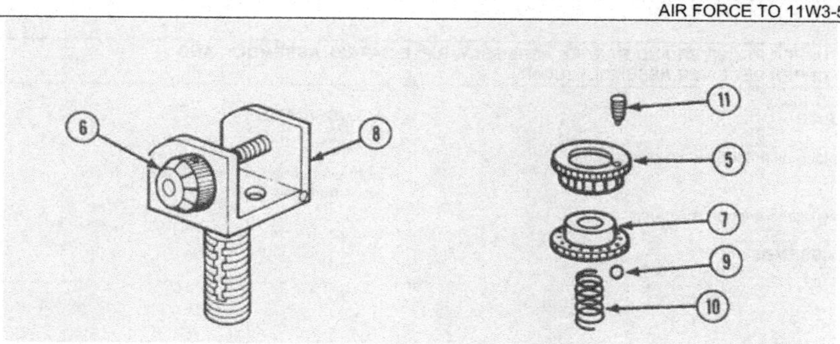

10. Rotate and test elevation index (5) and windage knob (6) for ease of functioning and legibility of markings.

11. Inspect elevation knob zero as follows:

a. Rotate elevation knob (7) counterclockwise until the rear sight assembly is all the way down. If a whole click is not felt as the rear sight assembly stops, the rear sight assembly has bottomed out and will not pivot freely.

b. Position elevation knob back slightly to Its last whole click so the rear sight assembly base (8) is under tension of the ball bearing (9) and helical spring (10). The 300 meter mark should align with the mark on the receiver.

c. If the 300 meter mark is not aligned with the mark on receiver, slip the range scale in the following manner:

(1) Position the 300 meter mark with the mark on the receiver.

(2) Insert a 1/16 inch allen wrench through the access hole of the rear sight assembly base and into the Index screw (11).

(3) Loosen the index screw three turns and leave the wrench in place.

(4) Rotate lower portion of elevation knob counterclockwise until it stops (range scale should not have moved). Elevation knob should be positioned on its last whole click.

(5) Tighten index screw and remove wrench.

(6) Check for proper setting.

2-16. UPPER RECEIVER AND BARREL ASSEMBLY, RIFLE BARREL ASSEMBLY, AND UPPER RECEIVER ASSEMBLY (CONT).

d. REPAIR

1. Replace all authorized unserviceable parts

2. Evacuate to support maintenance.

e. LUBRICATION

Lightly lubricate all Items (p 2 311 (oper ator's manual)

f. REASSEMBLY

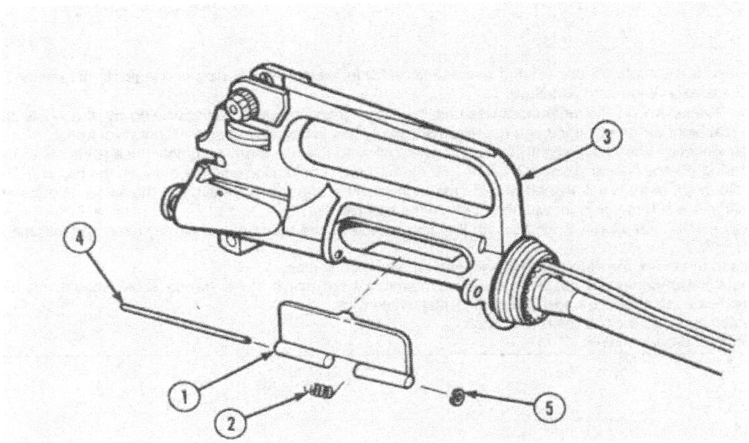

1. If previously disassembled, position election port cover 11) and helical spring (2) on Lip per receiver (3) with short leg of helical spring to the rear on Inside of ejection port cover

NOTE

Long legs of helical spring must be positioned and pretensioned before the headless grooved pin is installed

2. Hold helical spring (2) short leg In this position and turn long leg one half turn (180 degrees) with fingers of right hand

3. Position long leg of helical spring (21 against election port cover (1) Hold helical spring and election port cover in this position and Install headless grooved pin (4) Check for proper spring tension during installation of retaining ring (5)

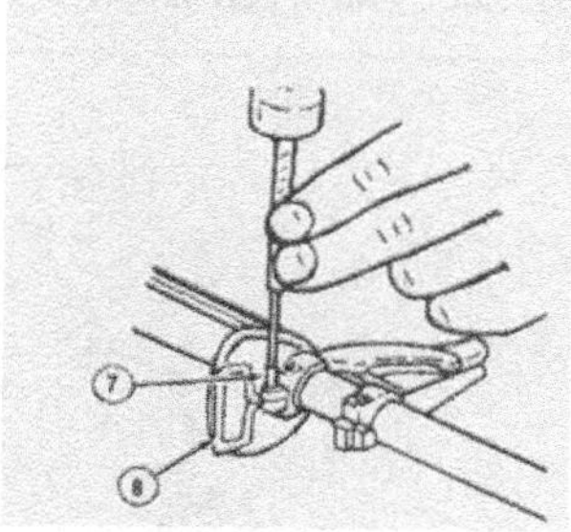

RIFLE ONLY

4. If previously disassembled, position small sling swivel (6) and install new tubular rivet (7) using center punch and hammer to spread and flare the hollow head of the tubular rivet.

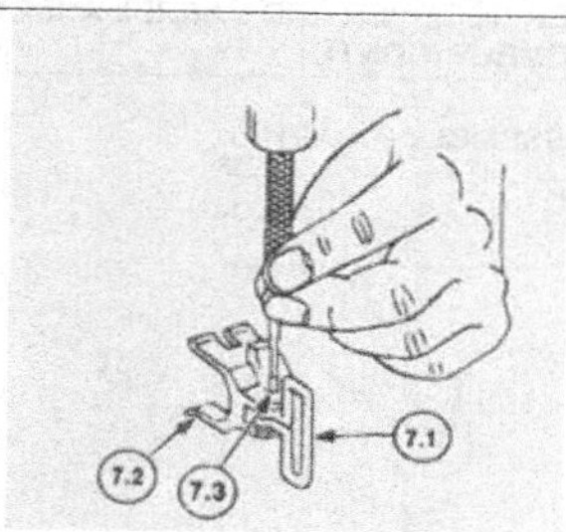

CARBINE ONLY

4A. Install front swivel (7.1) to swivel mount (7.2) with new rivet (7.3) using center punch and hammer to spread and flare the hollow head of the tubular rivet.

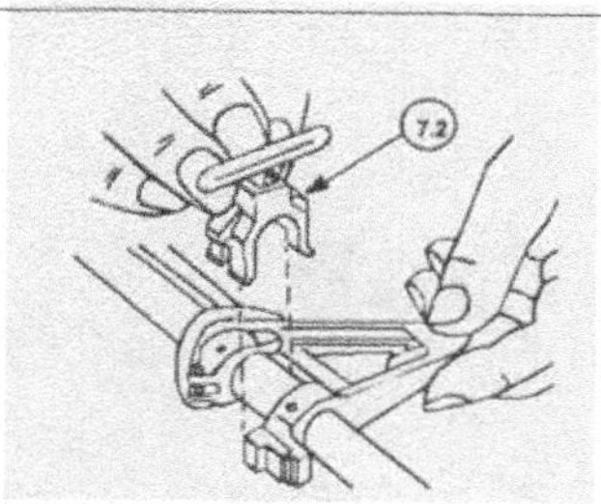

CARBINE ONLY

NOTE

It is recommended that the swivel mount be placed on the left side of the carbine for right handed operators and on the right side for left handed operators. This will keep the sling out of the way when the carbine is used.

4B. Install swivel mount (7.2).

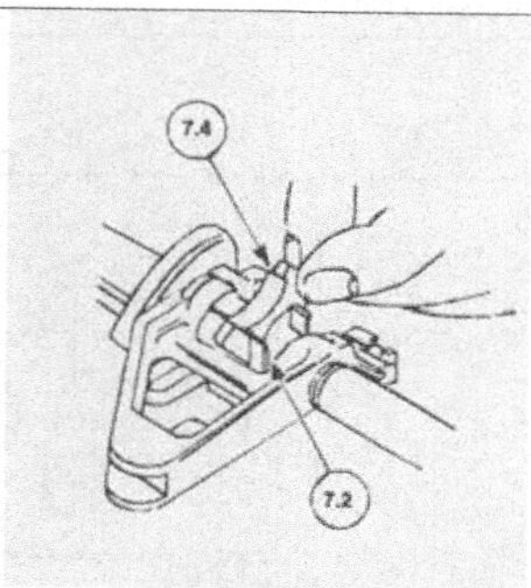

CARBINE ONLY

4C. Place swivel locking bar (7.4) in swivel mount (7.2).

2-16. UPPER RECEIVER AND BARREL ASSEMBLY, RIFLE BARREL ASSEMBLY, AND UPPER RECEIVER ASSEMBLY (CONT).

f. REASSEMBLY (CONT)

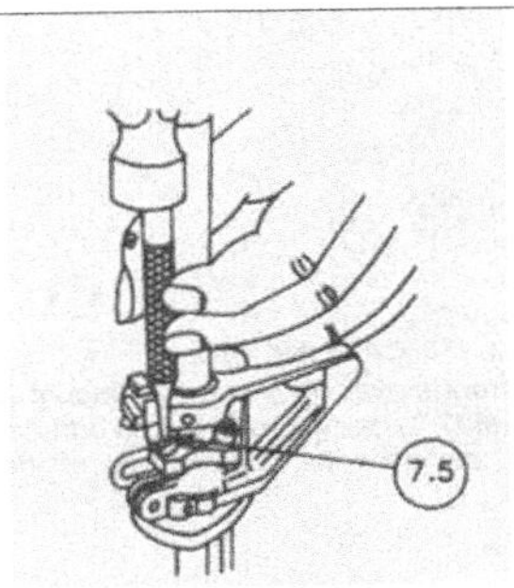

CARBINE ONLY

4D. Install two spring pins (7 5)

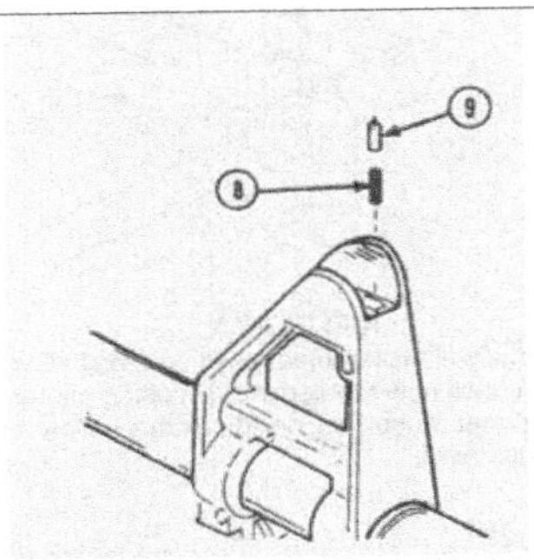

5. Position helical spring (8) and front sight detent (9).

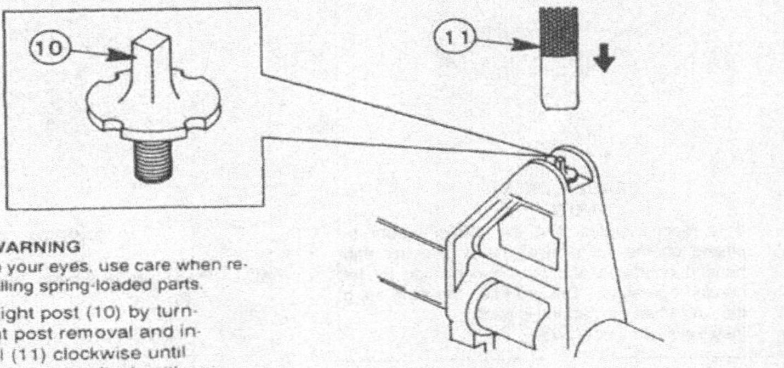

WARNING

To avoid injury to your eyes, use care when removing and installing spring-loaded parts.

6. Install front sight post (10) by turning front sight post removal and installation tool (11) clockwise until front sight post base is flush with or slightly below front sight frame.

2-16. UPPER RECEIVER AND BARREL ASSEMBLY, RIFLE BARREL ASSEMBLY, AND UPPER RECEIVER ASSEMBLY (CONT).

t. REASSEMBLY (CONT)

7. Mechanical Zero Procedures (A.F. Only).
   a. Mark a piece of plastic card stock or rigid paper with lines from 1 to 5mm in 1 mm increments Set the card on the front sight frame and check the height of the top of the front sight post.

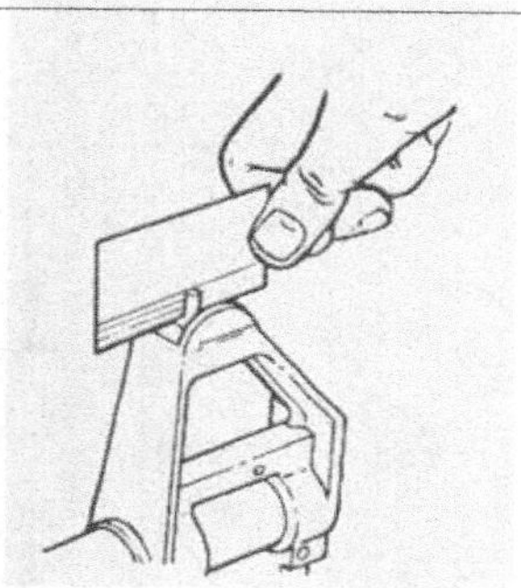

   b. Using a dummy round or front sight post tool adjust the front sight so the top of the front sight post is 5mm above the machined surfaces of the front sight frame.

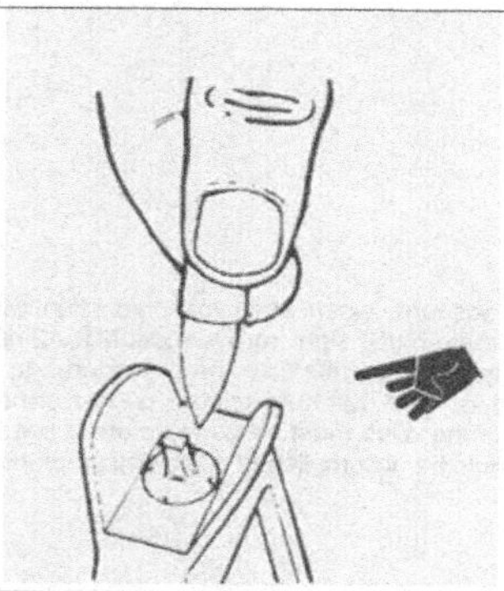

c. Visually check the front sight post top height by using the marked plastic or paper card. Card must set level on the machined surfaces of the front sight frame to obtain an accurate reading

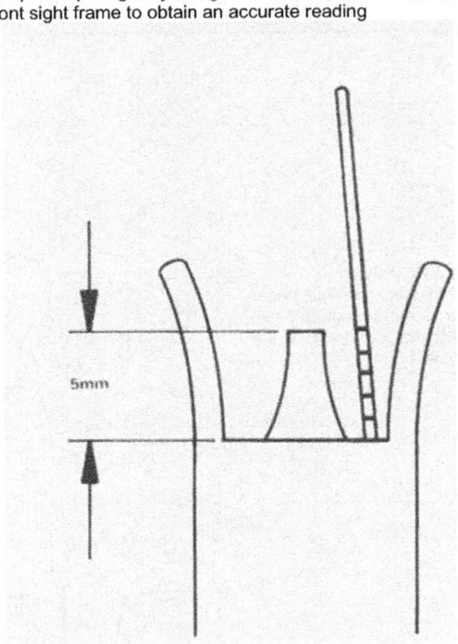

NOTE

This procedure, when used In conjunction with rear sight mechanical zero adjustment. will give an approximate battle sight zero to most M16A2 rifles The above steps can also be used before firing a new or newly assigned rifle Use the procedures to check rifles stored In preferred packaging during routine Inspections This will help ensure people armed with the rifles will stand a better chance of hitting an enemy If the rifles must be used before a live fire zero can be made Whenever possible, zeroing of the rifle should be accomplished using ball ammunition on a 25 meter zeroing target using the "L" aperture.

2-16. UPPER RECEIVER AND BARREL ASSEMBLY, RIFLE BARREL ASSEMBLY, AND UPPER RECEIVER ASSEMBLY (CONT).

f. REASSEMBLY (CONT)

NOTE
Refer to operator's manual for 'buddy system" procedure on installing hand guard assemblies

8. Install top of upper handguard assembly (12) In tube cap (13) while pushing down on handguard slip ring (14). Push bottom of upper handguard assembly (12) in place and release handguard slip ring (14) to lock handguard assembly In place

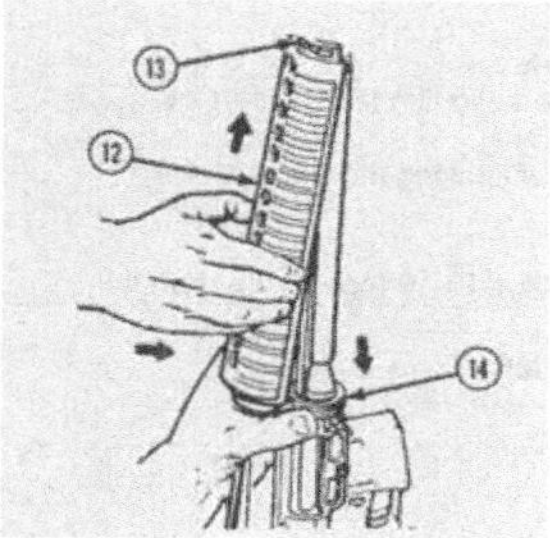

9. Install top of lower handguard assembly 1 5) In tube cap (13) while pushing down on handguard slip ring (14) Push bottom of lower handguard assembly (15) In place and release handguard slip ring {14) to lock both handguard assemblies In place

10. Reassemble rifle, refer to page 2-68

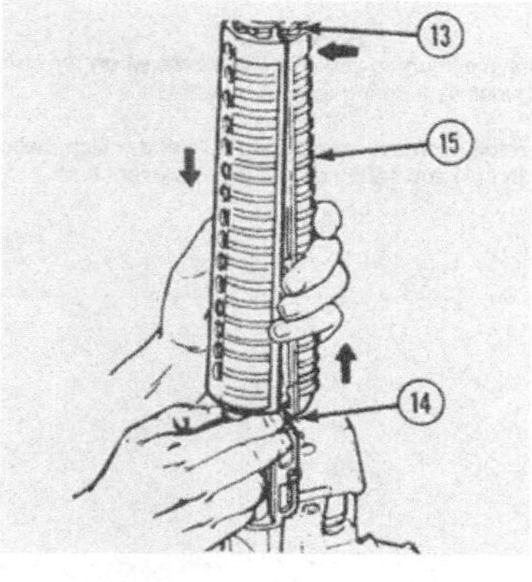

2-17. LOWER RECEIVER AND BUTTSTOCK ASSEMBLY.

This task covers:

a. Disassembly
b. Cleaning
c. Inspection
d. Repair
e. Lubrication
f. Reassembly

INITIAL SETUP

Tools
(ARMY) Small Arms Repairman Tol Kit (item 3, app B)
Pivot Pin Removal Tool (fig. E-3, app E)
Pivot Pin Installation Tool (fig. E-4, app E)

Materials/Parts
Lubricant, solid film (SFL) (item 21, app D)
Screw, self-locking (item 6, p C-1 1)

References
TM 9-1005-319-10 (operator's manual)

Equipment Conditions
2-34 Lower receiver and buttstock assembly removed

General Safety Instructions
To avoid injury to your eyes, use care when removing and installing spring-loaded parts.
When using solid film lubricant or dichloro-methane, be sure the area is well ventilated.

a. DISASSEMBLY

1. Remove screw (1) and lockwasher (2).

WARNING
To avoid injury to you eyes, use care when removing and installing spring-loaded parts.

2. Carefully remove pistol grip (3) and catch helical spring (4) and safety detent (5) to prevent loss.

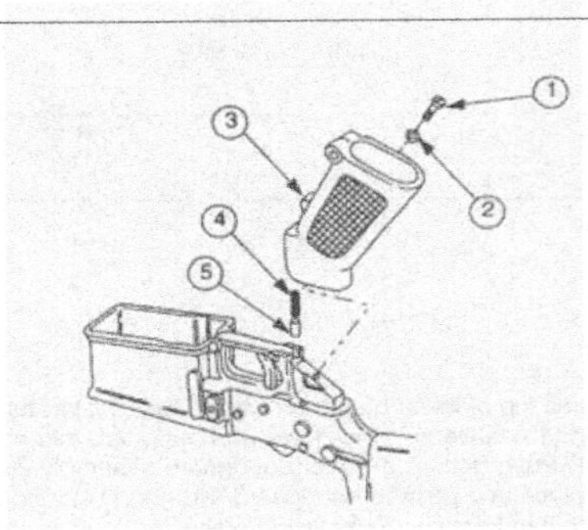

2-17. LOWER RECEIVER AND BUTTSTOCK ASSEMBLY (CONT).

a. DISASSEMBLY (CONT)

RIFLE ONLY
NOTE

If the self-locking screw is removed, it must be discarded and replaced with a new one

3. Remove self-locking screw (6).

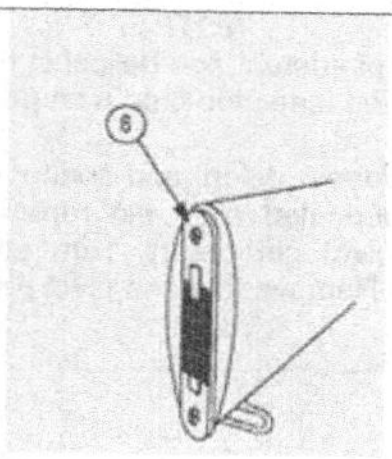

RIFLE ONLY
WARNING

To avoid injury to your eyes, use care when removing and installing spring-loaded parts.

4. Remove buttstock assembly (7) carefully and catch helical spring (8), takedown pin detent (9), takedown pin (10), and stepped spacer (11) to prevent loss.

NOTE
If detent (9) will not come out, use a wire to push it out.

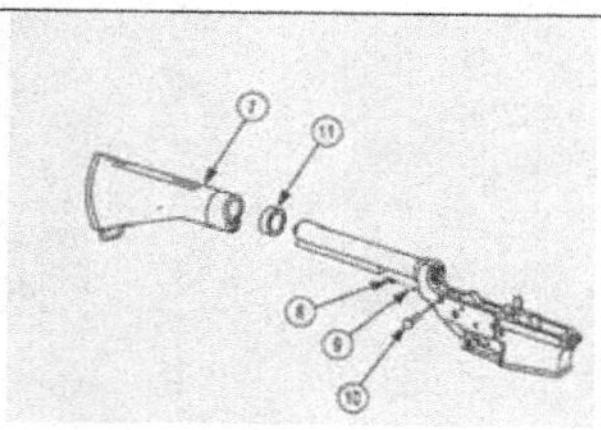

CARBINE ONLY

4A. Extend buttstock assembly (11.1).

4B Grasp the lock release lever (11.2) in the area of the retaining nut (11.3), pull down-ward, and slide buttstock to the rear to separate the buttstock assembly from the lower receiver extension

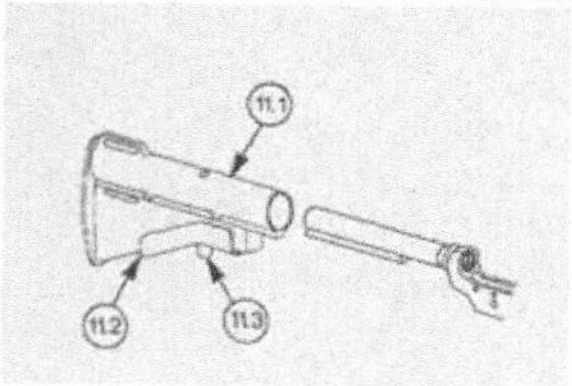

WARNING

To avoid injury to your eyes, use care when removing and installing spring-loaded parts

NOTE

Catch pivot pin detent and helical spring as pivot pin is removed (see step 6 on next page)

5. Depress pivot pin detent and spring and remove pin or Insert fabricated pivot pin removal tool (12) to compress pivot pin detent Turn pivot pin (13) a quarter turn Remove tool and pivot pin

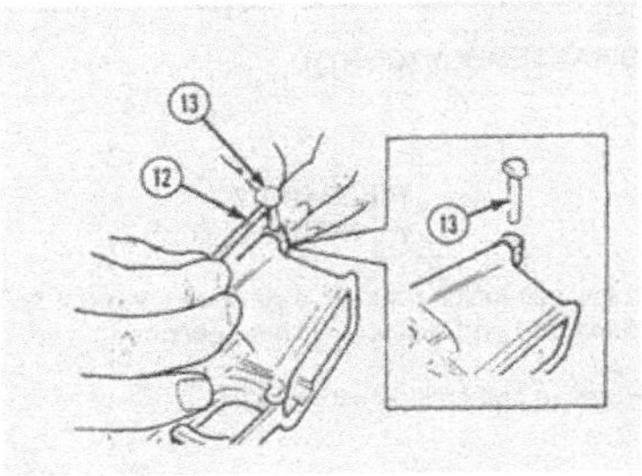

6. Be sure to hold cupped hand in front of pivot pin detent (14) and helical spring (15) to prevent loss of pivot pin detent and helical spring.

NOTE

If helical spring (15) will not come out, use a wire to pull it out.

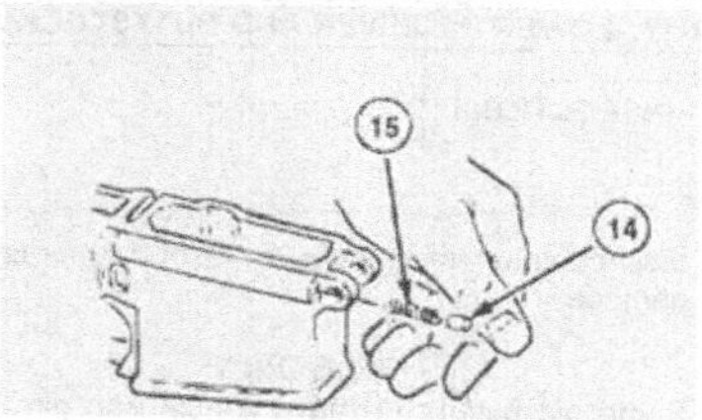

NOTE

Make sure the hammer is cocked and the selector lever is not set on BURST or AUTOMATIC before removing the buffer assembly.

7. Press buffer assembly (16) in about 1/4 inch (0.635 cm). Depress buffer retainer (17) and release buffer assembly (16) and action spring (18). Remove buffer assembly(16) and action spring (18) from receiver while depressing buffer retainer (17).

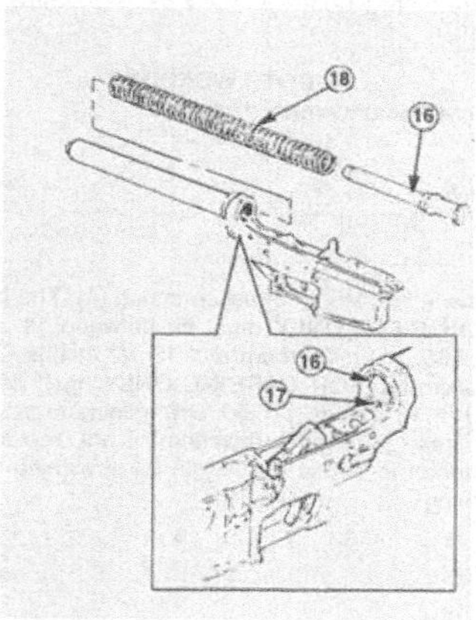

b. CLEANING

Clean all items (operator's manual). Remove carbon deposits

1 2-17. LOWER RECEIVER AND BUTTSTOCK ASSEMBLY (CONT).

c. INSPECTION I

1. Inspect buffer assembly (1, 2. 3, or 3 1) for cracks or damage

RIFLE ONLY

a. Some old buffers (1) have a hole with pin installed which protrudes equally on each side approximately 1/32 inch (O 08 cm)

b. Some buffers (2) have a hole In the housing but no pin

c. New buffers (3) do not have a hole In buffer body or a pin.

BOTH WEAPONS

2. If cracked or damaged, replace.

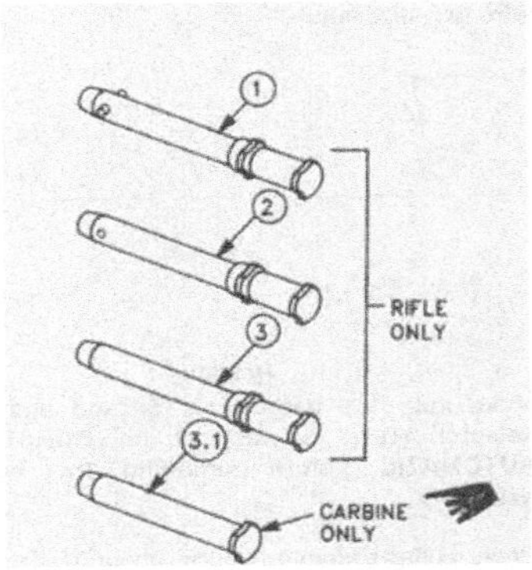

3. Check free length of action spring (4). The free length **FOR RIFLE ONLY** must be between 11 3/4 inches (29.85 cm) mini- mum and 13 1/2 inches (34 29 cm) maximum, **FOR CARBINE ONLY** must be between 10 1'/16 inches (25,56 cm) minimum and 11 114 Inches (28.58 cm) maximum, If not, replace Do not attempt to adjust the length by stretching the action spring

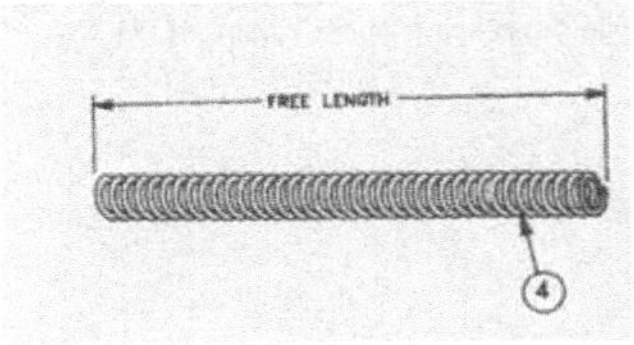

4. Inspect lower receiver (5) (without further disassembly) for legibility of serial number **ARMY ONLY**: If the serial number is hard to read, evacuate to direct support maintenance **AIR FORCE ONLY**: If the serial number is hard to read, evacuate to depot maintenance

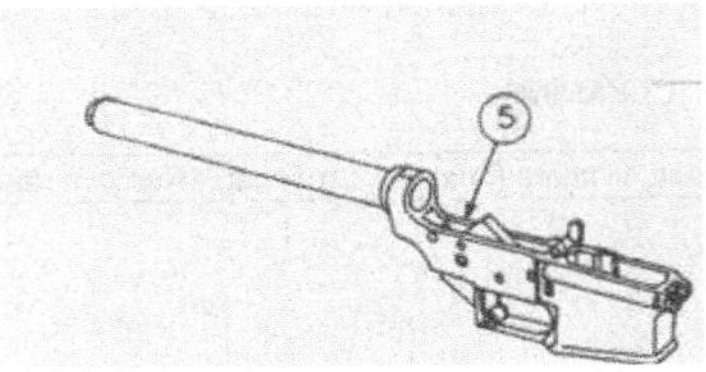

NOTE

AIR FORCE ONLY: Only depot maintenance is authorized to restamp the serial number.

5. Inspect for missing or damaged parts. Inspect finish of lower receiver for shiny spots. Touch up with solid film lubricant as required (p 2-34).

NOTE

If a M16A2 rifle or M4/M4A1 carbine lower receiver is missing one third or more of its exterior protective finish, resulting in an unprotected, light reflecting surface, it is candidate for overhaul. This missing finish will be considered a shortcoming. This shortcoming requires action to obtain a replacement weapon. Once a replacement has been received, evacuate the original weapon to depot maintenance for overhaul.

d. REPAIR

Replace all authorized unserviceable parts. If repair is not authorized at this level, evacuate to support maintenance.

e. LUBRICATION

Lightly lubricate all metal components (p 2-33) (operator's manual).

2-17. LOWER RECEIVER AND BUTTSTOCK ASSEMBLY (CONT).

f. REASSEMBLY

WARNING

To avoid Injury to your eyes, use care when removing and Installing spring loaded parts.

NOTE

Make sure the hammer Is cocked and the selector lever is not set on BURST before Installing the buffer assembly.

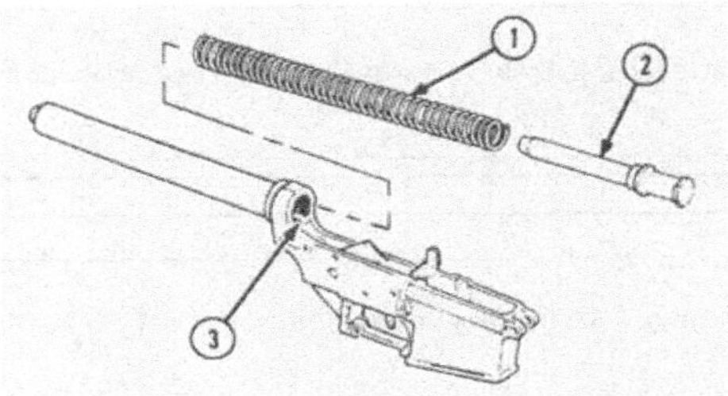

1. Press action spring (1) and buffer assembly (2) In until buffer retainer (3) snaps up and holds them In place.

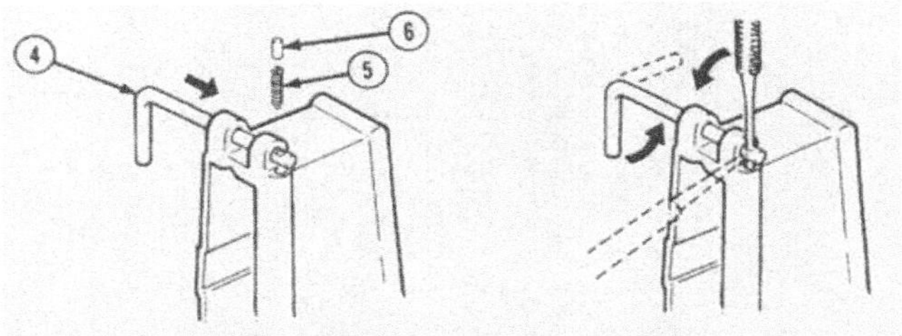

2. Install fabricated pivot pin installation tool (4) Insert helical spring 151 and pivot pin detent (6) Compress pivot pin detent in recess with 3 32 inch punch and rotate tool Remove 3'32 Inch punch

NOTE

Rounded end of pivot pin detent must be In the groove of the pivot pin (7) when assembly is complete.

3. Install pivot pin (7) while removing fabricated pivot pin Installation tool (4) Maintain pressure while sliding pivot pin (7) into hole Rotate pivot pin until pivot pin detent is inserted Into pivot pin groove

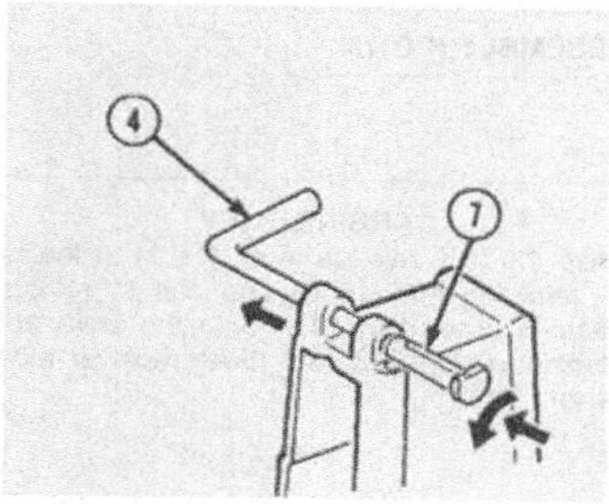

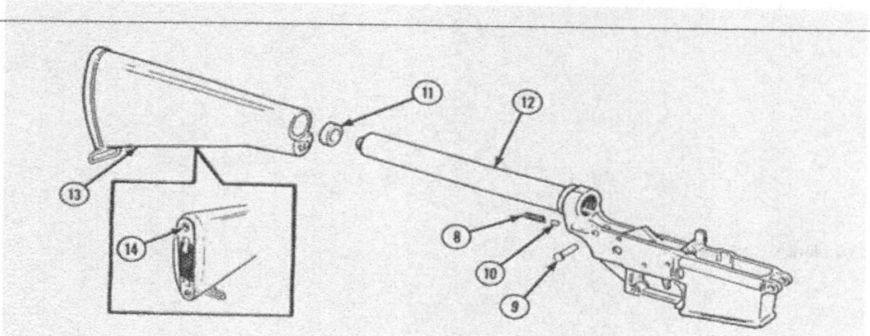

RIFLE ONLY

CAUTION

Do not kink helical spring (8) during assembly

4. Install takedown pin (9) with groove toward the rear Install takedown pin detent (10)and helical spring (8) from the rear
5. Install stepped spacer (11) on receiver extension (12) and carefully slide buttstock assembly (13) on to compress helical spring (8)

NOTE

Self-locking screw (14), If removed must be discarded and replaced with a new one 6. Install self-locking screw (14) to secure buttstock assembly (13)

6. Install self-locking screw (14) to secure buttstock assembly (13)

1 2-17. LOWER RECEIVER AND BUTTSTOCK ASSEMBLY (CONT).

f. REASSEMBLY (CONT)

CARBINE ONLY

6A. Grasp the lock release lever (14 1) In the area of the retaining nut (14 2) and pull to reinstall the buttstock assembly (14.3) onto the lower receiver extension (14 3) onto the lower receiver extension (14 4)

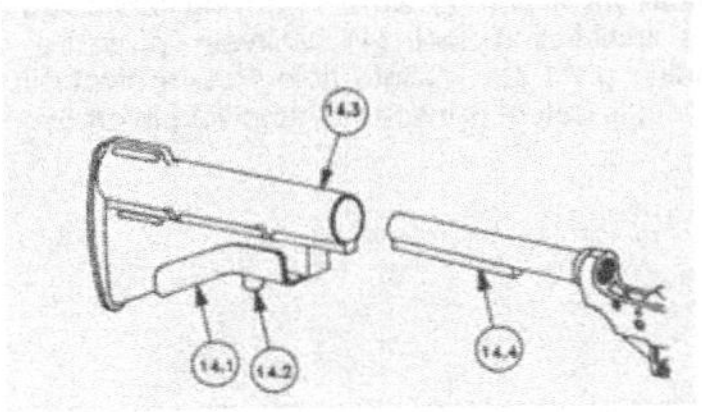

2-17. LOWER RECEIVER AND BUTTSTOCK ASSEMBLY (CONT).

f. REASSEMBLY (CONT)

WARNING

When utilizing the enhanced rifle grip (it has a bump between the second third finger for a better grip) PN 9349127, rifle grip screw. PN AN 501D416 18 (1-1/8 in.) or AN 501 D416 16 (1 in), is authorized to be(used with the enhanced grip Any screw longer than 1-1 8 In used with 9349127 could cause a hazardous situation Also, ensure the washer Is in 11place

CAUTION

Do not kink helical spring (15) during assembly

7. Install safety detent (16), pointed end first, and helical spring (1 5) into bottom of lower receiver (17)

NOTE

A portion of the helical spring will fit In a hole in the pistol grip

8. Carefully install pistol grip (18) to compress helical spring 115). Secure pistol grip 118) in place with lockwasher (19) and screw 120)

9. Reassemble rifle, refer to page 2 68

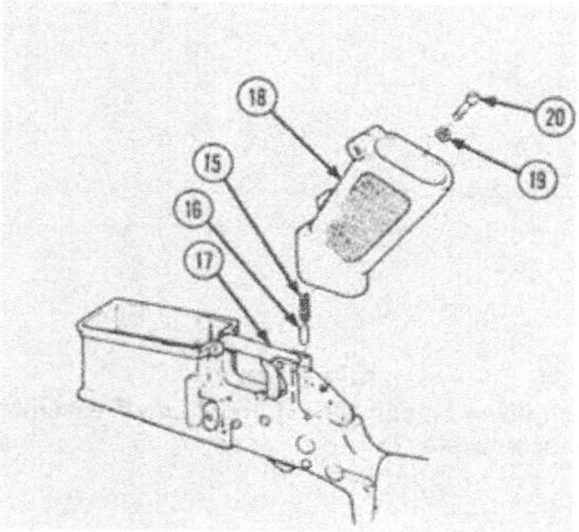

2-18. BUTTSTOCK ASSEMBLY.

This task covers:

a. Disassembly
b. Cleaning
c. Inspection
d. Repair
e. Lubrication
f. Reassembly

INITIAL SETUP

Tools

(ARMY) Small Arms Repairman Tool Kit (item 3, app B)

Equipment Conditions

2-57 Buttstock assembly removed from lower receiver and buttstock assembly

a. DISASSEMBLY

RIFLE ONLY

1. Remove self-locking screw (1), small sling swivel (2), and buttplate group (3) from buttstock (4)

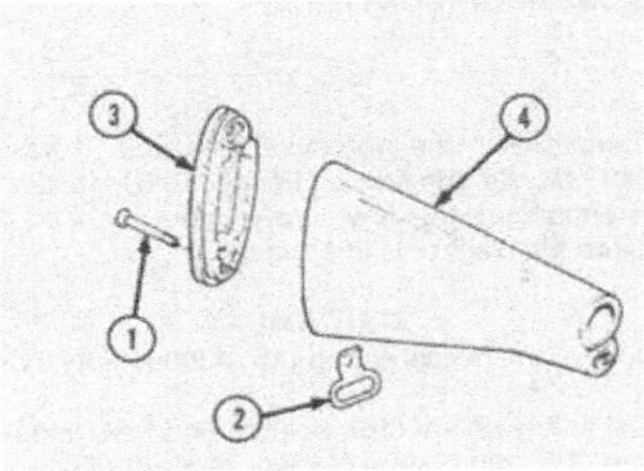

RIFLE ONLY

2. Push down on plunger (5) and lift door assembly (6) out of buttplate (7).

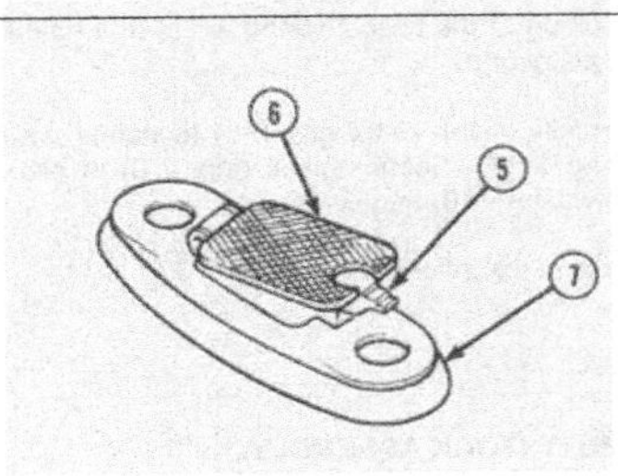

RIFLE ONLY

3. Remove straight pin 181 and separate hinge (9) and door assembly (61

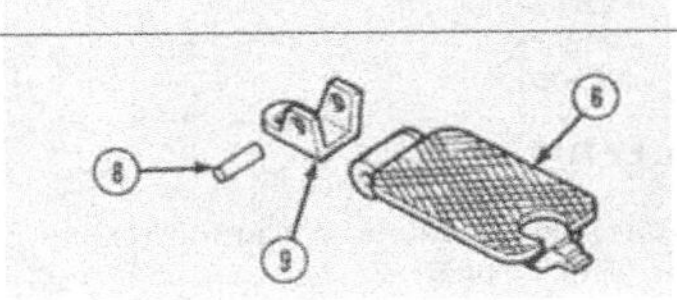

2-18. BUTTSTOCK ASSEMBLY (CONT).

a. DISASSEMBLY (CONT)

CARBINE ONLY

4. Disassemble the buttstock assembly by tapping out the spring pin (10) located in ,he oval slot of the retaining nut (1 1) This is done using a 1 16 Inch punch

5. Insert your Index finger into the forward end of the buttstock 112) and push down on the locking pin (13 Unscrew the retaining nut 11 and remove the release lever (14) locking pin (13) and spring (15)

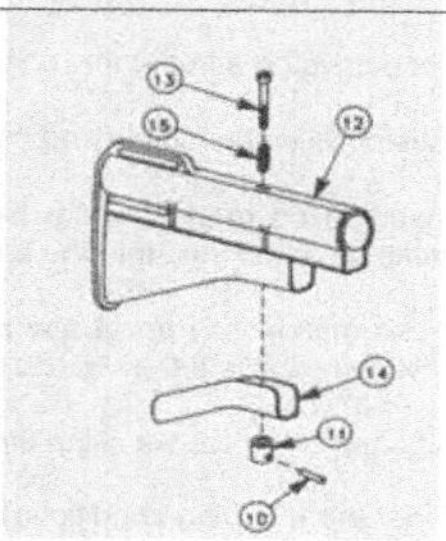

b. CLEANING (RIFLE ONLY)

Clean all parts with CLP (operator s manual) Use brush to clean knurled surface of door assembly

2-18. BUTTSTOCK ASSEMBLY (CONT).

c. INSPECTION

NOTE

M16A2 buttstocks, PN 9349121, with unauthorized markings may be used under the following conditions:

a. The only authorized markings are those which are temporary In nature, i e , paint, tape, etc

b. When marking a buttstock, only use temporary markings

c. Buttstocks with unauthorized markings that have been stamped into the surface of the buttstock will not be used

d. Unauthorized markings that have previously been scratched, etched, carved, etc may continue In use if the marks do not extend Into the fiber of the buttstock Cutting Into the fiber of the buttstock may weaken it

e. These marks may be at any location on the buttstock Unauthorized markings are not desirable. However, if previously applied, they will be allowed to continue In use due to the cost of the buttstock.

1. Inspect buttstock for cracks using the following guidelines:

   a. Under the following conditions, hairline cracks (no chipped away material allowed) originating from buttplate end of buttstock are acceptable.
      (1) One hairline crack, not to exceed 1 in. 12.54 cm) In length, per side of buttstock.
      (2) Two additional hairline cracks up to 0 25 in (0 64 cm) In length, per side of buttstock
      (3) A total of three cracks per side of buttstock, originating from buttplate end, are allowable.

   b. Cracks In the critical area at the front end of the buttstock are not acceptable and these buttstocks must be replaced.

2. While buttplate Is Installed on rifle, Inspect for cracks around the mounting holes Check for cracks In excess of 0 25 In. (O 64 cm) In length which extend through the buttplate Replace If cracked.

3. Inspect door assembly for cracks, corrosion, stuck plunger, separations on outer face, or other damage Replace If defective.

d. REPAIR

Replace all authorized unserviceable Items Unserviceable Items are those Items which are damaged.

e. LUBRICATION

Lubricate all metal components (p 2-33) (operator's manual)

f. REASSEMBLY

RIFLE ONLY

1. Position hinge (1) on door assembly (2) and Install straight pin (3)

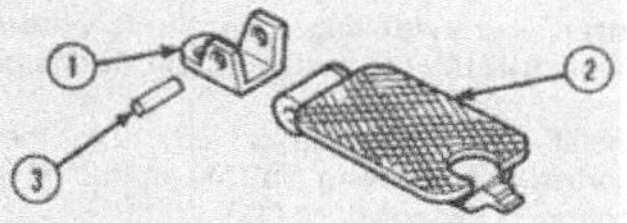

RIFLE ONLY

2. Install door assembly (2) into buttplate (4) and press plunger (5) to lock

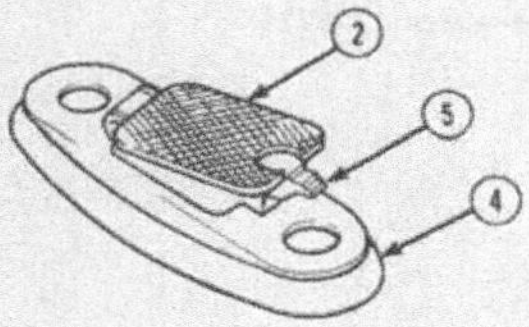

RIFLE ONLY

3. Position buttplate group (6) and small sling swivel (7) to the buttstock (8) and secure with self-locking screw (9).

NOTE

See page 2-62, reassembly, for reassembly of buttstock assembly to lower receiver

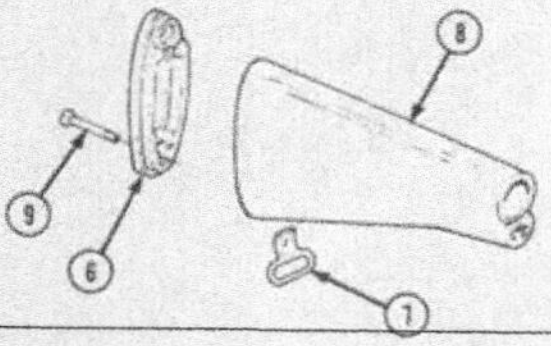

2-18. BUTTSTOCK ASSEMBLY (CONT).

f. REASSEMBLY (CONT)

CARBINE ONLY

4. Insert locking pin spring (10) onto locking pin (11)

5. Insert locking pin (11) and spring (10) Into the hole on top of the buttstock (12), threaded end first

6. Insert your index finger into the forward end of the buttstock (12) and push down on locking pin (11)

7. Install the release lever (13) onto the threaded portion of the locking pin (11) protruding through the bottom of the buttstock (12)

8. Screw on the locking nut (14) until flush with locking pin (11) Align the slot In locking nut (14) with the spring pin hole in the locking pin (11)

9. Lightly tap in spring pin (15) until flush on both sides of the retaining nut (14)

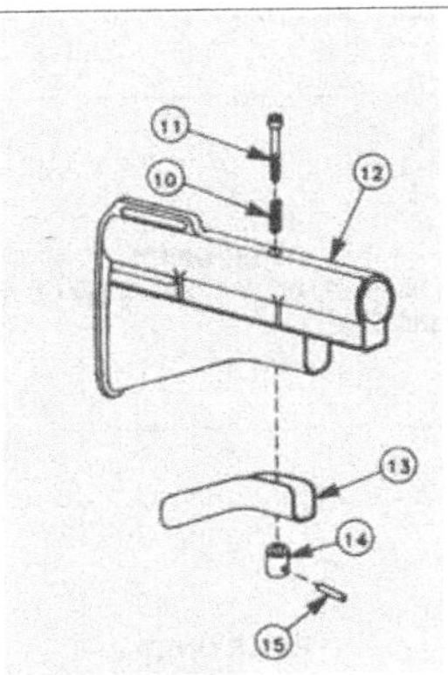

2-19. MAJOR COMPONENTS OF M16A2 RIFLE.

This task covers

a. Reassembly
b. Inspection
c. Stowage

INITIAL SETUP

References
TM 9 1005-319 10

Equipment Conditions
Rifle disassembled into major components

General Safety Instructions
To avoid injury to eyes, use care when removing and installing spring-loaded parts

Do not Interchange bolt assemblies from one rifle to another Doing so may result In Injury to, or death of. personnel

Do not keep live ammunition near the work area

a. REASSEMBLY

Refer to operator's manual and assemble lower receiver and buttstock assembly (1) upper receiver and barrel assembly (2). charging handle assembly (3), bolt carrier assembly (4), sling (5), and magazine(6).

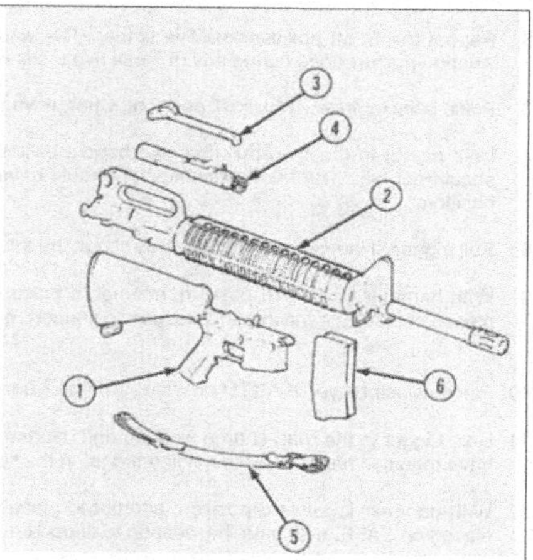

b. INSPECTION

Perform the following function checks on assembled weapon:

1. Remove magazine if installed. Pull charging handle assembly to rear. Check that chamber is clear. Let bolt and bolt carrier close. Do not pull trigger. Leave hammer in cocked position.

WARNING

If rifle fails any of the following tests, continued use of the rifle could result in injury to, or death of, personnel.

2. Place selector lever in SAFE position and pull trigger. Hammer should not fall.

3. Place selector lever in SEMI position. Pull trigger. Hammer should fall.

M16A2 and M4 ONLY
NOTE

For the purpose of the following check "SLOW" is defined as 1/4 to 1/2 the normal rate of trigger release.

4. Hold trigger to the rear, charge weapon, and release the trigger with a slow, smooth motion, without hesitations or stops, until the trigger is fully forward; an audible click should be heard. Hammer should not fall.

5. Repeat the SEMI position test five times. The weapon must not malfunction during any of these five tests. If the weapon malfunctions during any of these five tests, evacuate rifle to support maintenance for repair.

6. Place selector lever in BURST position. Charge weapon and pull trigger. Hammer should fall.

7. Hold trigger to the rear, pull charging handle assembly to rear and release three times. Release trigger. Hammer should not fall. The burst disconnector should have held the hammer to the rear while the trigger was in the pulled position.

8. Pull trigger. Hammer should fall. This should be the first round of a three round burst.

9. With hammer in forward position, attempt to place the selector lever in the SAFE position. If selector lever can be placed on SAFE, evacuate the weapon to support maintenance.

M4A1 ONLY

10. Place selector lever in AUTO position. Charge weapon and pull trigger. Hammer should fall.

11. Hold trigger to the rear, charge weapon and release trigger. Pull trigger. Hammer should not fall. AUTO sear should have released hammer while holding trigger in the squeezed position before releasing and resqueezing the trigger.

12. With hammer in forward position, attempt to place the selector lever in the SAFE position. If selector lever can be placed on SAFE, evacuate the weapon to support maintenance.

1 2-19. MAJOR COMPONENTS OF M16A2 RIFLE (CONT).

c. STOWAGE

Prior to stowing the weapon in arms room, perform the following procedures:

NOTE

Weapon passed into arms room issue window should be passed butt first with the bolt locked to the rear.

1. Clear. Refer to operator's manual.
2. Place selector lever in SEMI position.
3. Pull trigger. Hammer should fall.
4. Close ejection port (dust) cover.
5. Place weapon in rack.

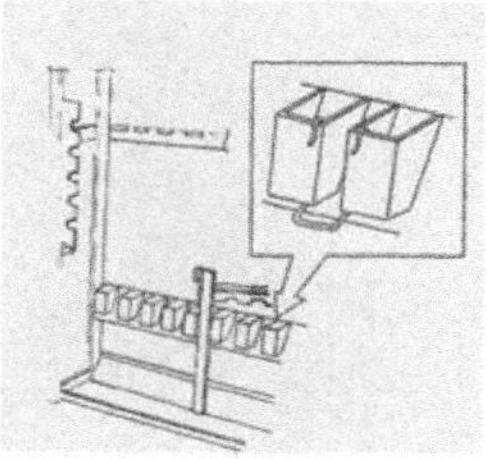

6. **M4/M4A1 Carbine Only, use M12 Arms Rack.**

a. The M12 arms rack is the correct arms rack in which to store the M4/M4A1 Carbine. The carbine must be stored with buttstock extended. When storing the M4/M4A1 Carbine in the M12 arms rack a mounting bracket, NSN 5340-01-230-3181, (app D, item 23.1) may be used for each M4/M4A1 Carbine being stored. This option is for the convenience of the person who opens and closes the arms rack to store the carbines.

b. To install the mounting bracket on the M12 arms rack, for use with the M4/M4A1 Carbine, install the bracket with the hooks of the bracket facing toward the carbine, so that the lower receiver extension will contact the bent end of the bracket. The bent end of the bracket will hold the carbine upright when the arms rack is opened. The bracket can be turned around when not in use for the carbine to allow storage of the M16A2 rifle.

c. When storing the M4/M4A1 Carbine in the M12 arms rack, an adapter bar (fig E-6, app E), MUST be used for security reasons. To install the adapter bar to the M12 arms rack:

(1) Remove all weapons from the rack and position the rack to gain access to the back.

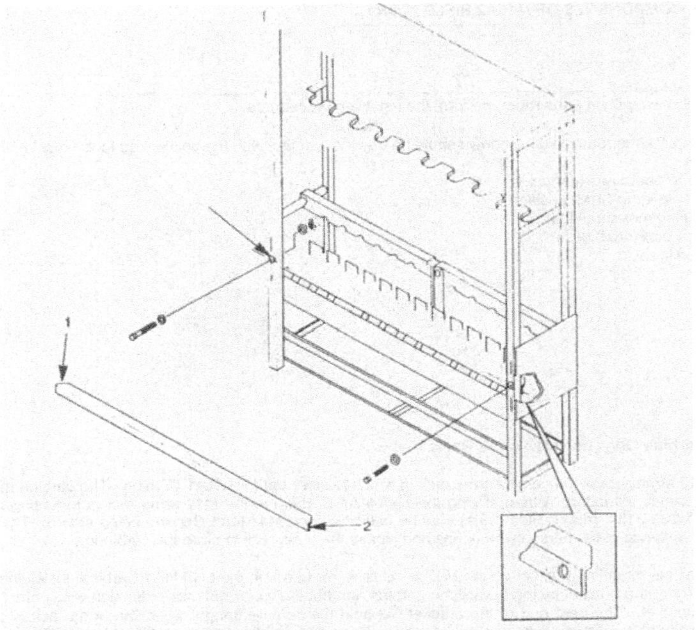

(2) The side of the adapter bar with the corners (1) cut off is the top and the side with square corners (2) is the back. The portion of the bar must be placed so the cut off corners (1) face the front of the rack.

(3) Holding the adapter bar at an angle, place one end into position inside the rear leg of the arms rack. Lower the other end of the bar into position. Allow the adapter bar to rest on the rack.

(4) Clamp both ends of the adapter bar into position. Mark the center line of the leg and adapter bar where they meet (see illustration) Using a center punch, mark the location of the holes to be drilled where the center lines cross. The holes must be centered on both the leg and the adapter bar. Drill a 1/8 inch pilot hole through both arms rack legs and the ends of the adapter bar. Drill a 3/8 inch hole through both arms rack legs and the ends of the adapter bar using the pilot hole as a guide. Remove the adapter bar. File the edges of all holes smooth Paint all bare metal surfaces with olive drab enamel paint, NSN 8010-01-350-5249, or equivalent.

2-19. MAJOR COMPONENTS OF M16A2 RIFLE (CONT).

c. STOWAGE (CONT)

(5) Reinsert the adapter bar into position on the arms rack. Using two 3/8 inch x 2 inch machine screws (MS35206-315, NSN 5305-00-984-5695) or equivalent, four washers (MS21306-1C, NSN 5310-01- 327-9713) or equivalent and two nuts (MS35649-2382, NSN 5310-00-056-3395) or equivalent assemble adapter bar to arms rack and tighten securely. The bolts can be inserted from either the back or the front to meet your requirements. If the rack is placed close to a wall or another rack It is recommended that the bolts be inserted from the back

(6) Tack weld, braze, or peen the threaded end of the bolt to the nut to prevent easy removal

(7) Place rack back into position and replace the weapons

NOTE

Adapter bar must be removed from arms rack prior to turning in the arms rack to the supply system.

d. To remove the adapter bar from the arms rack, remove the head of the bolt (or the nut), using a hand held grinder or file to avoid damage to the aluminum arms rack legs. A hammer and cold chisel may be used If no other way to remove the adapter exist

e. It is recommended that the rail protector NSN 1005-01-394-7677, (app D, item 26.1) be used during storage of the carbines when the carrying handle assembly or some other accessory is not installed on the upper receiver to prevent damage to the mounting surface on the upper receiver.

Section VI. PREPARATION FOR STORAGE OR SHIPMENT

2-20. PREPARATION FOR STORAGE OR SHIPMENT.

a. Packaging of the M16A2 Rifle and the M4/M4A1 Carbine shall be in accordance with the following

ARMY ONLY: Army users shall package the rifle and the carbine in accordance with each respective Packaging Data Sheet (PDS) for shipment or storage which may exceed 90 days The PDS is part of the Army Master Data File Retrieval Microform System (ARMS) Packaging File

AIR FORCE ONLY: Air Force users shall package the rifle in accordance with each respective Special Packaging Instruction (SPI) 00-856-6885 for shipment or storage which may exceed 90 days The SPIs are part of the Army Master Data File Retrieval Microform System (ARMS) Packaging File

b. Packaging, if required, for shipping/storage which will not exceed 90 days shall be as follows

(1) Clean in accordance with operator's manual

(2) Wrap with MIL-B-121 waterproof material.

(3) Place in barrier bag MIL-B-117, Type I, Class C, or wrap with MIL-B-121, Type I, Grade A, and seal with tape, PPP-T-76.

(4) Place one or more of item in minimum size container. Block and brace in accordance with MIL-STD-1186. Cushion the M16 and similar weight items with PPP-C-843, and use PPP-F-320 as filler, to create a tight pack.

(a) Fiber board containers shall be in accordance with PPP-B-636 and may be Class Domestic. Gross weight and size of material shall determine grade of fiberboard container. PPP-B-640 may also be used.

(b) Wood containers shall be in accordance with PPP-B-601 or PPP-B-621.

(5) Equivalent materials may be used.

c. NSNs are not assigned to all the specified material. If it is necessary to specify an NSN in the TMs, the packing materials will have to be spared and part numbers and NSNs assigned.

d. The specifications used are:

(1)	MIL-B-121	Barrier Material, Greaseproofed, Waterproofed, Flexible (NSN 8135-00-753-4661)
(2)	MIL-B-117	Bag, Sleeve and Tubing -- Interior Packaging (NSN 8135-00-543-6574)
(3)	PPP-B-636	Boxes, Shipping, Fiberboard
(4)	PPP-B-601	Boxes, Wood, Cleated Plywood
(5)	MIL-STD-129	Marking for Shipment and Storage
(6)	PPP-T-76	Tape, Packaging, Paper
(7)	MIL-STD-1186	Cushioning, Anchoring, Bracing, Blocking and Waterproofing; with Appropriated Test Methods
(8)	PPP-C-843	Cushioning Material, Cellulosic
(9)	PPP-F-320	Fiberboard, Corrugated and Solid Sheet Stock (Container Grade), and Cut Shapes
(10)	PPP-B-640	Boxes, Fiberboard, Corrugated, Triple-Wall
(11)	PPP-B-621	Boxes, Wood, Nailed and Locked -- Corner

CHAPTER 3

DIRECT SUPPORT MAINTENANCE INSTRUCTIONS

CHAPTER OVERVIEW

This chapter provides information and instructions to keep the weapon in good repair and contains the following sections:

a. Repair Parts and Special Tools
b. Direct Support Troubleshooting
c. Direct Support Maintenance Procedures for the M1 6A2 Rifle and M4/M4A1 Carbine
d. Preparation for Storage or Shipment
e. Preembarkation Inspection of Material in Units Slated for Overseas Movement

Section I. REPAIR PARTS, SPECIAL TOOLS, TMDE, AND SUPPORT EQUIPMENT

3-1. COMMON TOOLS AND EQUIPMENT. For authorized common tools and equipment, refer to the Modified Table of Organization and Equipment (MTOE) applicable to your unit. Air Force users must maintain the following common tools:

Flat tip screwdriver
Socket wrench handle and socket head screw socket wrench
Vise jaw caps
Machinist's vise
Solid center punch
Hammer
Combination wrench
Torque wrench
Retaining ring pliers
(Two) 8-inch adjustable wrenches
Flat file
Ball-peen hammer
Trigger pull test fixture rod and weights
5/64-inch drive pin punch
1/8-inch drive pin punch
1/16-inch drive pin punch
3/32-inch drive pin punch

3-2. SPECIAL TOOLS, TMDE, AD SUPPORT EQUIPMENT. Special tools required for direct support maintenance are listed in appendixes B and C. Fabricated tools are listed and illustrated in appendix E.

3-3. REPAIR PARTS. Repair parts are listed and illustrated in appendix C of this manual.

NOTE

Bolt assemblies, and/or barrel assemblies may be interchanged , at the Direct Support Maintenance level, from one rifle to another, under the provisions of the note on page C-3: If these parts are interchanged the rifle must be checked/inspected as depicted on pages 3-17, 3-21, and 3-33. While performing these checks and inspections, pay special attention to the headspace requirements on page 3-47.

Section II. TROUBLESHOOTING

3-4. GENERAL.

a. This section contains direct support level troubleshooting information for locating and correcting most of the operating troubles which may develop in the M16A2 rifle and M4/M4A1 carbine. Each malfunction for the individual part or assembly is followed by a list of tests or inspections which will help you to determine the corrective action to take. You should perform the tests/inspections and corrective actions in the order listed.

b. This manual cannot list all malfunctions that may occur, nor all tests or inspections and corrective actions. If a malfunction is not listed or is not corrected by listed corrective actions, see individual repair sections in the maintenance procedures on each major assembly.

3-5. TROUBLESHOOTING PROCEDURES. Refer to troubleshooting table for malfunctions, tests, and corrective actions. The symptom index is provided for a quick reference of the malfunctions covered in the table.

SYMPTOM INDEX

TROUBLESHOOTING

MALFUNCTION
TEST OR INSPECTION
CORRECTIVE ACTION

1. FAILURE OF MAGAZINE TO LOCK IN RIFLE.

Step 1. Dirty or corroded magazine catch (1).
Disassemble and clean.

Step 2. Defective magazine catch spring (2).
Replace magazine catch spring (2) (p 3-62).

Step 3. Worn or broken magazine catch (1).
Replace magazine catch (1) (p 3-62).

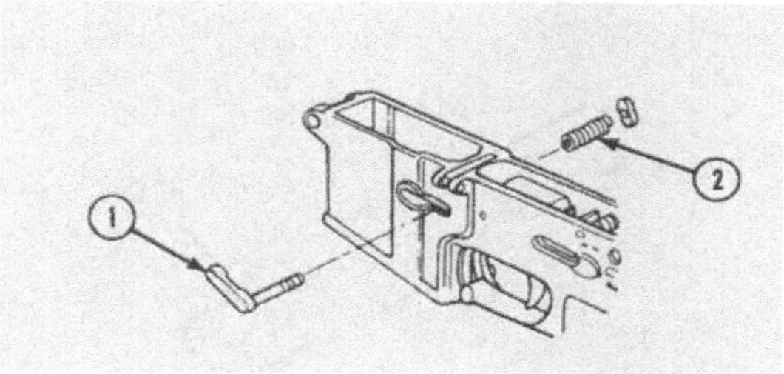

2 FAILURE TO FEED

Step 1 Magazine catch spring (1) weak or broken.
Replace magazine catch spring I1) (p 3-62)

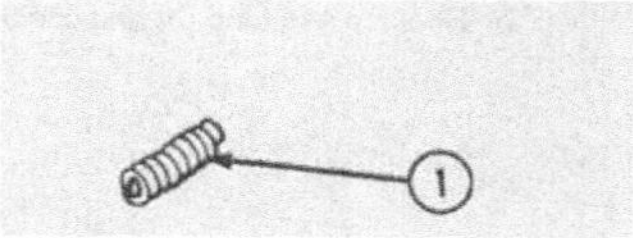

Step 2 Short recoil
Refer to page 3-9

3 FAILURE TO CHAMBER

Short recoil.
Refer to page 3-9.

3-5. TROUBLESHOOTING PROCEDURES (CONT).

TROUBLESHOOTING (CONT)

MALFUNCTION
TEST OR INSPECTION
CORRECTIVE ACTION

4. FAILURE TO LOCK.

Step 1. Damaged bolt carrier key (1).
Repair or replace bolt carrier key (1) (p 3-25)

Step 2 Loose screws (2) on bolt carrier key (1).
a. Disassemble and repair (p 3-25)
b. Reassemble using new screws.

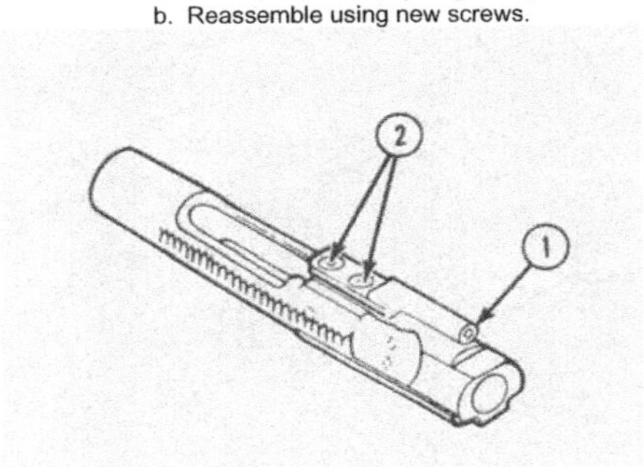

Step 3 Bent gas tube (3).
a Adjust by bending gas tube 13) In area of handguards
b Replace gas tube (3) and check alignment (p 3-29)

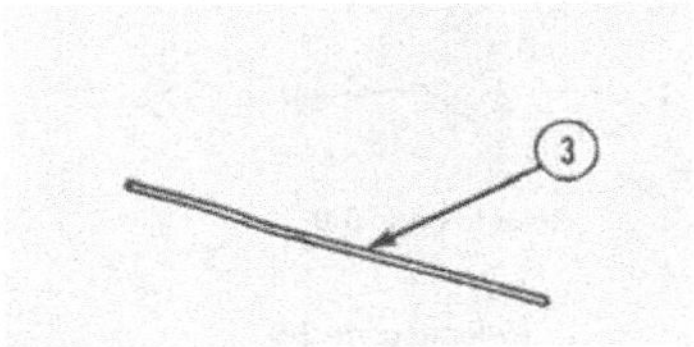

Step 4. Short recoil
Refer to page 3 9

TROUBLESHOOTING (CONT)

MALFUNCTION
TEST OR INSPECTION
CORRECTIVE ACTION

5. FAILURE TO FIRE.

Step 1. Broken hammer (1).
Replace hammer (1) (p 3-73).

Step 2. Weak or broken hammer spring (2).
Replace hammer spring (2) (p 3-73).

Step 3. Hammer spring (2) improperly assembled.
Assemble properly.

Step 4. **(M16A2 and M4 ONLY)** Burst cam (3) and/or cam spring (4) frozen or improperly assembled
Disassemble, clean, lubricate, and reassemble correctly (p 3-73).

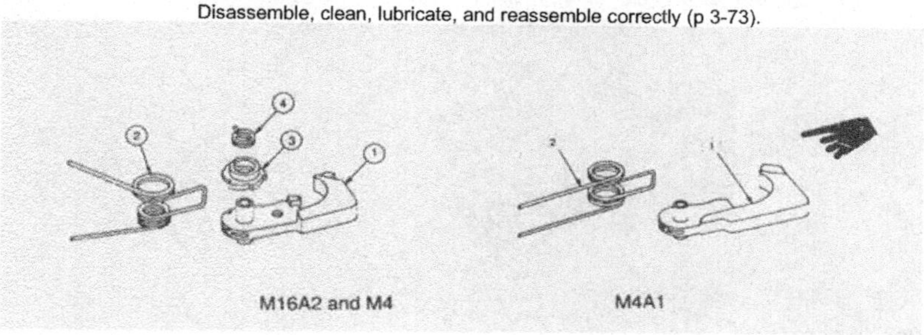

Step 5. Selector lever (5) frozen on SAFE position.
Disassemble and clean (p 3-62)

Step 6. Broken firing pin (6) or firing pin does not meet gage protrusion requirement.
Replace firing pin (6) (p 3-16).

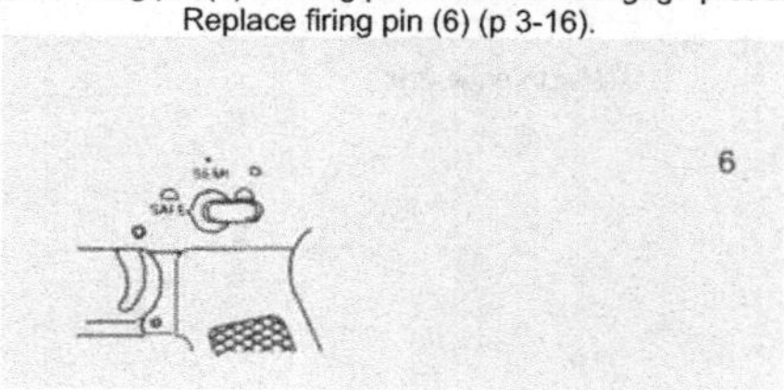

3-5. TROUBLESHOOTING PROCEDURES (CONT).

TROUBLESHOOTING (CONT)

MALFUNCTION
TEST OR INSPECTION
CORRECTIVE ACTION

6. FAILURE TO UNLOCK.

Step 1 Burred locking lugs (1) on bolt assembly.
Remove burrs.

Step 2 Burred lugs (2) on barrel extension.
Remove burrs.

Step 3. Short recoil.
Refer to page 3-9.

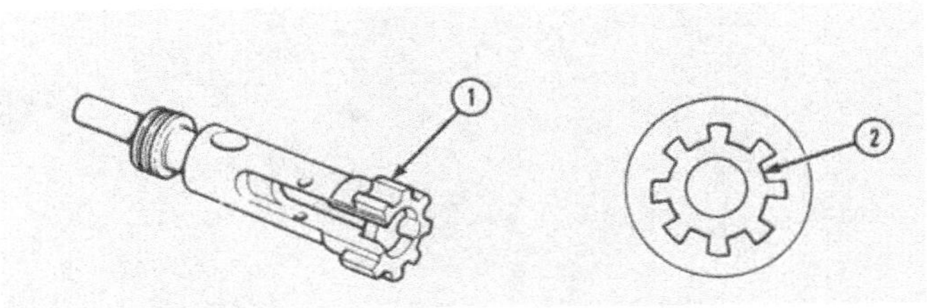

7. FAILURE TO EXTRACT

Step 1. Inspect cartridge extractor and extractor spring assembly.
Replace if cracked or broken (p 2-35).

Step 2. Inspect badly pitted chamber with reflector tool (Item 2, fig. C-1 7).
Replace barrel assembly If chamber Is badly pitted (p 3-29).

8 FAILURE TO EJECT.

Short recoil.
Refer to page 3-9.

TROUBLESHOOTING (CONT)

MALFUNCTION
TEST OR INSPECTION
CORRECTIVE ACTION

9. FAILURE TO COCK.

Step 1. Worn or broken trigger nose (1) or trigger spring (2).
Replace trigger (3) or defective trigger spring (2) (p 3-62).

Step 2. Worn or broken hammer trigger notch (4).
Replace hammer (5) (p 3-62).

Step 3. Worn or broken hammer disconnector hook (6).
Replace hammer (5) (p 3-62).

Step 4. Worn or broken hammer automatic sear hook (7).
Replace hammer (5) (p 3-62).

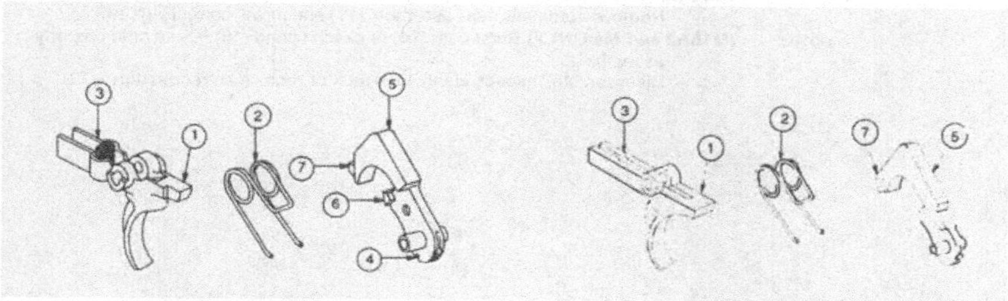

M16A2 and M4 ONLY

M4A1 ONLY

3-5. TROUBLESHOOTING PROCEDURES (CONT).

TROUBLESHOOTING (CONT)

MALFUNCTION
TEST OR INSPECTION
CORRECTIVE ACTION

Step 5 **(M4A1 has only 1 disconnector)** Worn or broken disconnector hooks (8).
Replace defective disconnectors (9) (p3-62).

Step 6 **(M4A1 has only 1 spring)** Weak, broken, or missing disconnector springs (10).
Replace disconnector springs (10) (p 3-62).

Step 7. Worn, broken, or missing automatic sear (11).
Replace automatic sear (11) (p 3-62).

Step 8 Weak or broken automatic sear spring (12).
Replace automatic sear (11) (p 3-62).

Step 9. Long leg (13) of automatic sear spring incorrectly assembled in receiver.
Remove automatic sear assembly (11) and install correctly (p 3-62).

Step 10. **(M16A2 and M4 ONLY)** Burst cam (14) or clutch spring (15) frozen or improperly assembled.
Disassemble, inspect, clean, lubricate, or replace as required (p 3-73).

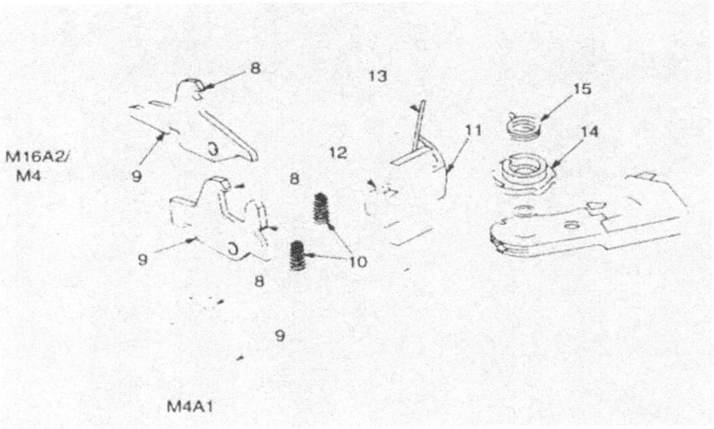

TROUBLESHOOTING (CONT)

MALFUNCTION
TEST OR INSPECTION
CORRECTIVE ACTION

10. SHORT RECOIL.

Step 1. Improper gap space or worn, missing, or broken bolt rings (1).

a. Stagger bolt ring gaps (p 3-21).

b. Replace bolt rings (1) and stagger gaps (p 3-21).

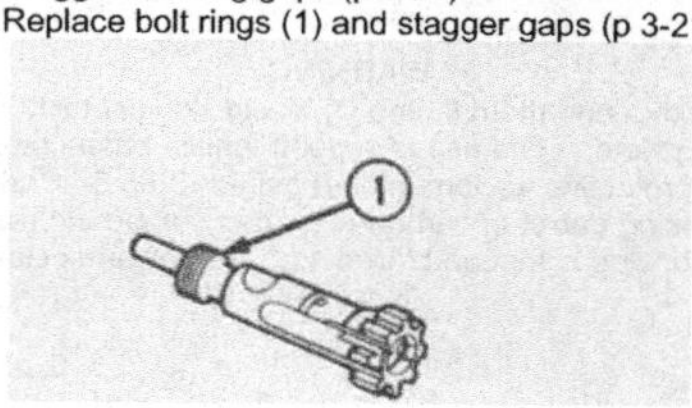

Step 2. Broken or bent gas tube (2).

Adjust by bending in area of handguards or replace gas tube (2) (p 3-29).

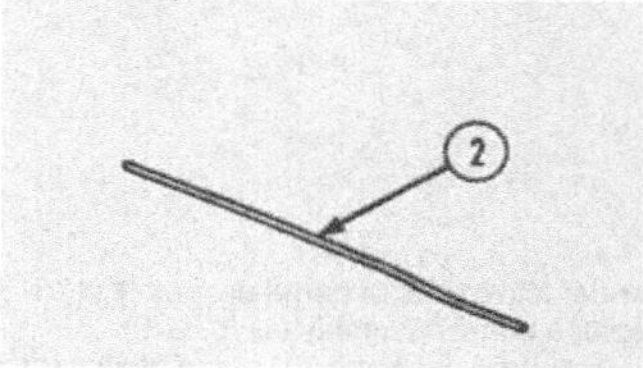

Step 3. Gas tube spring pin (3) missing from front sight (4).

Replace gas tube spring pin (3) (p 3-29).

Step 4. Partially plugged gas system because of carbon build-up in the gas tube (2).

Replace gas tube (2) (p 3-29).

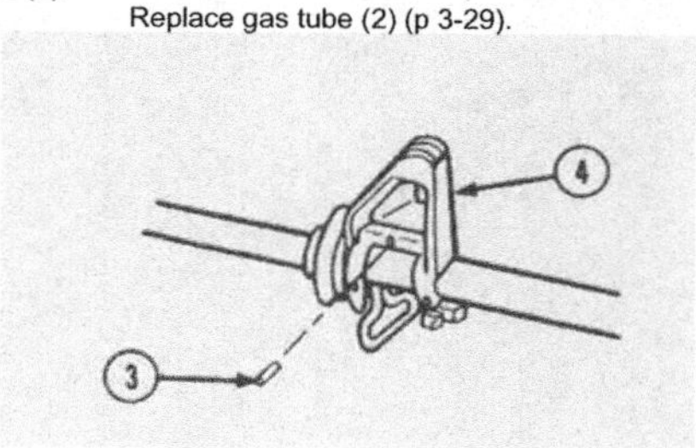

3-5. TROUBLESHOOTING PROCEDURES (CONT).

TROUBLESHOOTING (CONT)

MALFUNCTION
TEST OR INSPECTION
CORRECTIVE ACTION

10 SHORT RECOIL (CONT).

WARNING

When using carbon removing compound (Item 8, app D), avoid skin contact. If it comes in contact with the skin, wash off thoroughly with running water. The use of a good lanolin base cream after exposure to compound Is helpful. The use of gloves and protective equipment Is required. Step 5. Carbon build-up in barrel gas port (5). Remove carbon build-up by soaking barrel In carbon removing compound (item 8, app D). Use rubber gloves (Item 18, app D) with carbon removing compound. Use a bore small arms cleaning brush (Item 4, app D).

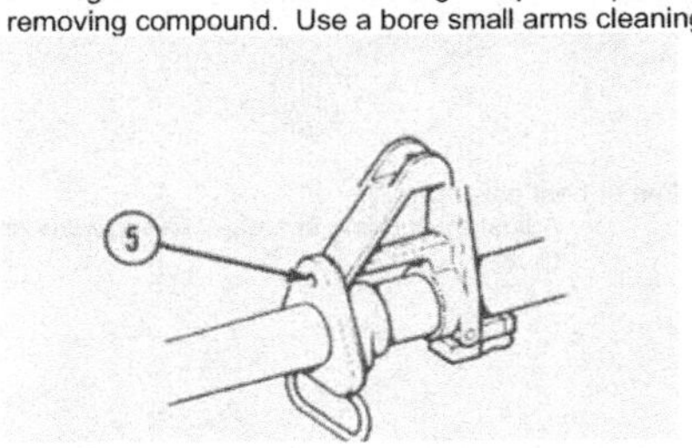

11. RIFLE CANNOT BE ZEROED.

Step 1 Inspect for defective or bent barrel assembly (1) (p 3-29).
Replace barrel assembly (1) (p 3-29).

Step 2 (For windage) barrel assembly (1) out of alignment with rear sight on up per receiver.
Align barrel assembly (1) and upper receiver (p 3-291

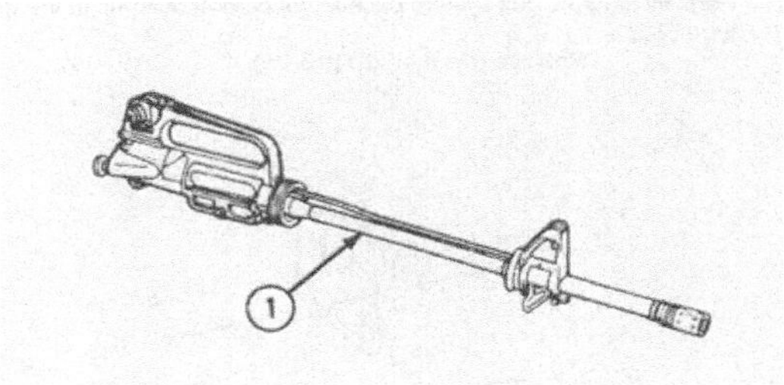

TROUBLESHOOTING (CONT)

MALFUNCTION
TEST OR INSPECTION
CORRECTIVE ACTION

Step 3. (For elevation) defective front sight (2) or rear sight (3).
Repair as required (p 3-48).

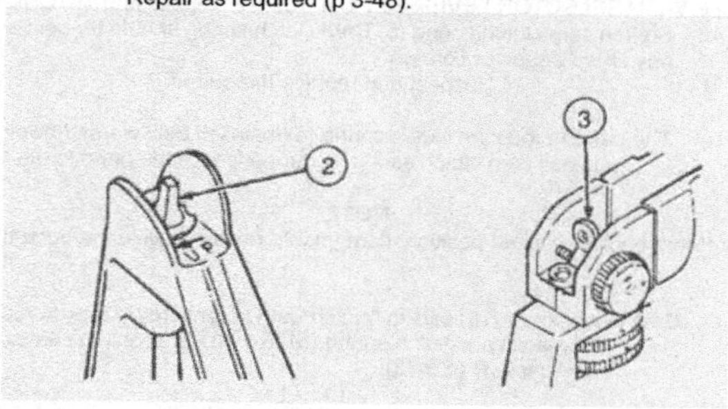

12. FAILURE TO CYCLE WITH SELECTOR LEVER SET ON BURST (M16A2 and M4 ONLY).

Step 1. Broken automatic sear (1) or spring (2).
Replace automatic sear assembly (1) (p 3-62).

Step 2. Faulty selector lever (3).
Replace selector lever (3) (p 3-62).

Step 3. Broken tooth on burst cam (4).
Replace burst cam (4) (p 3-73).

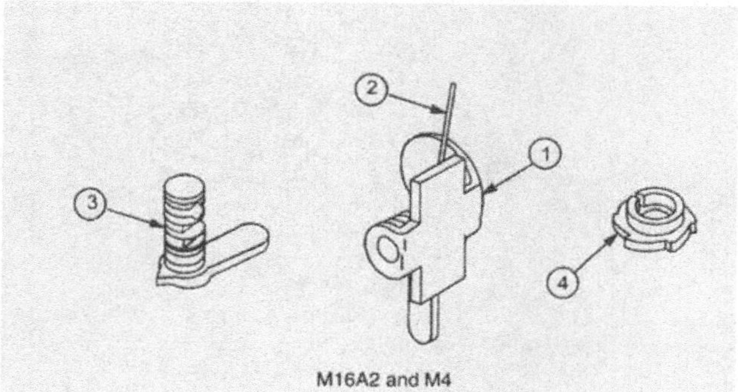

M16A2 and M4

3-5. TROUBLESHOOTING PROCEDURES (CONT).

TROUBLESHOOTING (CONT)

MALFUNCTION
TEST OR INSPECTION
CORRECTIVE ACTION

12 FAILURE TO CYCLE WITH SELECTOR LEVER SET ON BURST (M16A2 and M4 ONLY) (CONT).

Step 4 Broken cam clutch spring (5) Cam clutch spring should be bent and properly formed without any sharp edges or corners

Inspect and replace If required.

Step 5 The bend In the cam clutch spring (5) installed backwards (toward outside).

Install cam clutch spring (5) properly with the bend to the inside (p 3-73).

NOTE

When hammer Is rotated back to cocked position, cam should rotate to allow the burst disconnector to latch in the next notch.

Step 6. Cam clutch spring (5) fails to "clutch" and burst cam (4) fails to rotate back with hammer (6).

Replace cam clutch spring (5) (p 3-73). If problem continues, replace hammer (6) and cam (4) (p 3-73).

Step 7. Broken front hook (7) on burst disconnector (8).

Replace burst disconnector (8) (p 3-62).

Step 8. Short recoil.

Refer to page 3-9

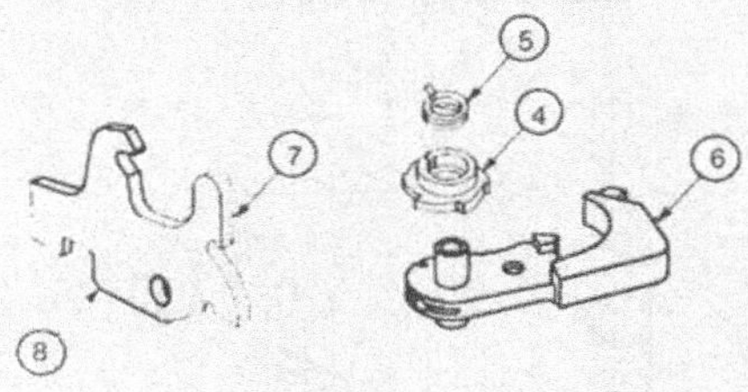

M16A2 and M4

TROUBLESHOOTING (CONT)

MALFUNCTION
TEST OR INSPECTION
CORRECTIVE ACTION

12.1. FAILURE TO CYCLE WITH SELECTOR LEVER SET ON AUTO (M4A1 ONLY)

Step 1. Broken automatic sear (1) or spring (2).
Replace automatic sear assembly (1).

Step 2. Faulty selector lever (3).
Replace selector lever (3).

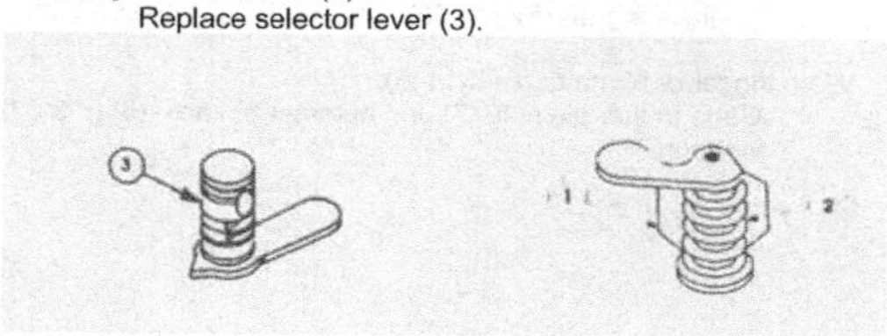

Step 3. Short recoil.
Refer to page 3-9.

13. FIRES TWO ROUNDS WITH ONE PULL OF TRIGGER WITH SELECTOR LEVER SET ON SEMI (DOUBLE FIRING).

Step 1. Defective semiautomatic disconnector (1). Replace semiautomatic disconnector (1) (p 3-62).

Step 2 Worn or broken trigger notch (2) of hammer (3) (searing portion). Replace hammer (3) (p 3-62).

Step 3. Worn or broken disconnector notch (4) of hammer (3). Replace hammer (3) (p 3-62).

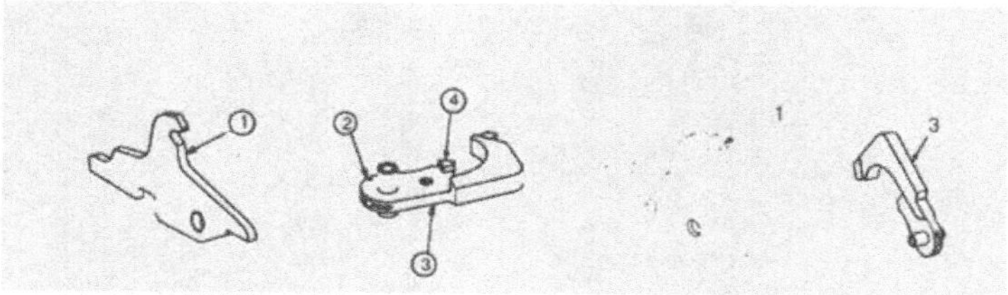

M16A2 and M4 M4A1

3-5. TROUBLESHOOTING PROCEDURES (CONT).

TROUBLESHOOTING (CONT)

MALFUNCTION
TEST OR INSPECTION
CORRECTIVE ACTION

13 FIRES TWO ROUNDS WITH ONE PULL OF TRIGGER WITH SELECTOR LEVER SET ON SEMI (DOUBLE FIRING) (CONT).

Step 4 Worn or broken trigger (5) (searing portion).
Replace trigger (5) (p 3-62).

Step 5 Worn trigger or hammer pin hole (6).
Gage trigger pin hole (7) and hammer pin hole (6) (p 3-62). If test falls, replace weapon.

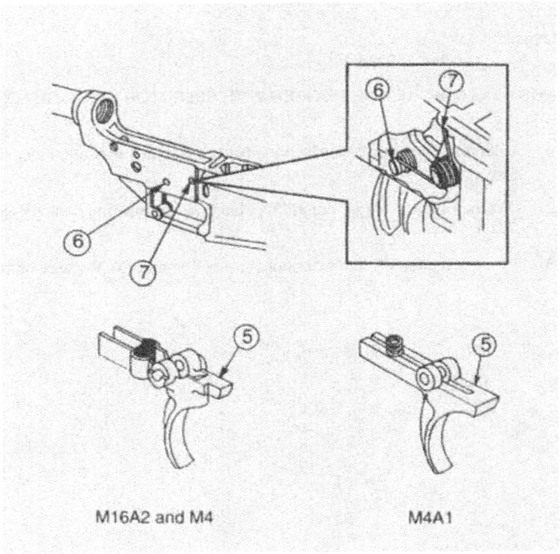

TROUBLESHOOTING (CONT.)

MALFUNCTION

TEST OR INSPECTION

CORRECTIVE ACTION

14. FIRES WITH SELECTOR LEVER ON SAFE OR WHEN TRIGGER IS RELEASED WITH SELECTOR LEVER ON SEMI.

Step 1. Defective selector lever (1).

Replace selector lever (1) (p 3-62).

Step 2. Worn or broken trigger (rear portion) (2).

Replace trigger (3) (p 3-62).

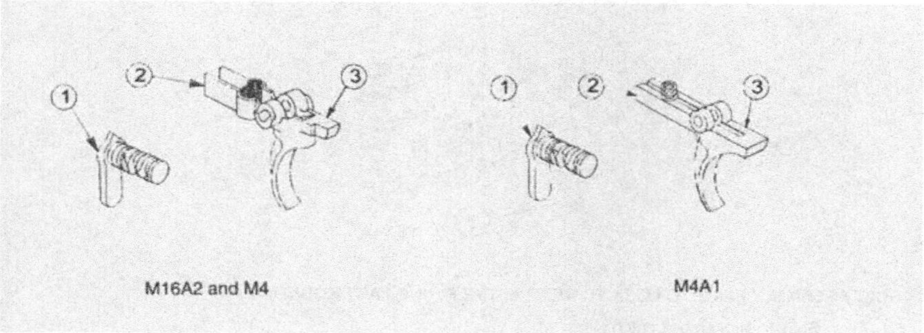

3-5. TROUBLESHOOTING PROCEDURES (CONT.).

TROUBLESHOOTING (CONT.)

MALFUNCTION

TEST OR INSPECTION

CORRECTIVE ACTION

15. HAMMER PIN "WALKS".

Hammer pin (1) "walks" or works loose during firing or hammer pin Is very easy to push out of receiver when hammer Is Installed.

Replace hammer assembly (p 3-62).

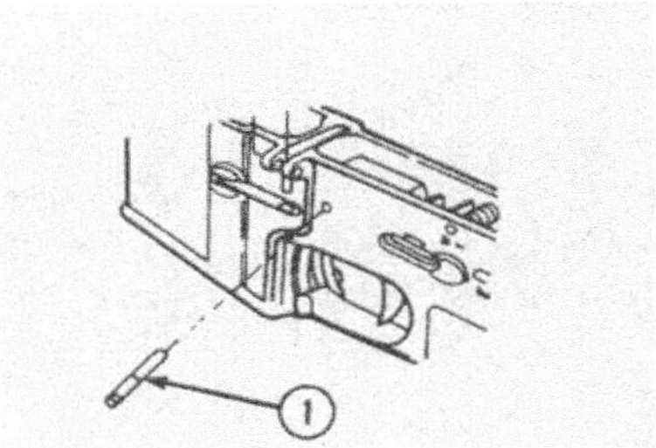

16. BOLT ASSEMBLY FAILS TO LOCK TO REAR AFTER FIRING LAST ROUND.

Step 1. Broken bolt catch (1).

Replace bolt catch (1) (p 3 62,.

Step 2. Weak or broken bolt catch spring (2).

Replace bolt catch spring (2) (p 3-62)

Step 3 Restricted movement of bolt catch (1).

Disassemble and clean.

Section III. MAINTENANCE PROCEDURES FOR THE M16A2 RIFLE AND M4/M4A1 CARBINES

3-6. MAJOR COMPONENTS OF M16A2 RIFLE AND M4/M4A1 CARBINES.

This task covers disassembly.

INITIAL SETUP

Tools
(ARMY) Small Arms Repairman Tool Kit (item 3, app B)

References
TM 9-1005-319-10 (operator's manual)

Equipment Conditions
Weapon assembled

General Safety Instructions
Before starting an inspection, be sure to clear the weapon Do not pull the trigger until the weapon has been cleared.
Inspect the chamber and receiver to ensure no ammunition is present Do not keep live ammunition near work area.

DISASSEMBLY

a. Refer to operator's manual.

b. Remove magazine (1), sling (2), bolt carrier assembly (3), charging handle assembly (4), and upper receiver and barrel assembly (5), carrying handle assembly (5.1), from lower receiver and buttstock assembly (6).

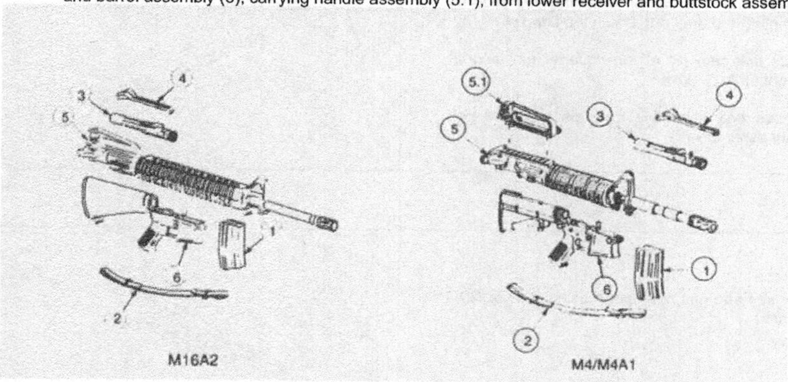

NOTE

Solid Film Lubricant (SFL) is the only authorized touchup for the M16A2 rifle and M4/M4A1 carbines and may be used on up to one third of the exterior finish of the weapon. **FOR ARMY CONUS USE ONLY AND AIR FORCE TRAINING WEAPONS ONLY**: SFL may be used as touchup without limitation on the upper receiver and barrel assembly. This is to say that units which DO NOT fall under the category of Divisional Combat Units or rapid deployment type units may have up to 100 percent of the exterior surface of the upper receiver and barrel assembly protected with SFL if necessary.

3-7. BOLT CARRIER ASSEMBLY.

This task covers:

- **a** Disassembly
- **b** Cleaning
- **c** Inspection
- **d** Test
- **e** Repair
- **f** Reassembly

INITIAL SETUP

Test Equipment
  Tool and Gage Set (item 2, app B)

Tools
  (ARMY) Small Arms Repairman Tool Kit (item 3, app B)

Materials/Parts
  Cleaner, lubricant, and preservative (CLP) (item 9, app D)
  Cleaning compound, rifle bore (RBC) (item 12, app D)
  Pipe cleaner (item 11, app D)

Equipment Conditions
  3-1 5 Bolt carrier assembly removed

a. DISASSEMBLY

1. Remove firing pin retaining pin (1).
2. Tip key and bolt carrier assembly (2) allowing firing pin (3) to drop out. Catch the firing pin.
3. Rotate bolt cam pin (4) one quarter turn and lift straight up to remove.
4. Remove bolt assembly (5) from key and bolt carrier assembly (2)

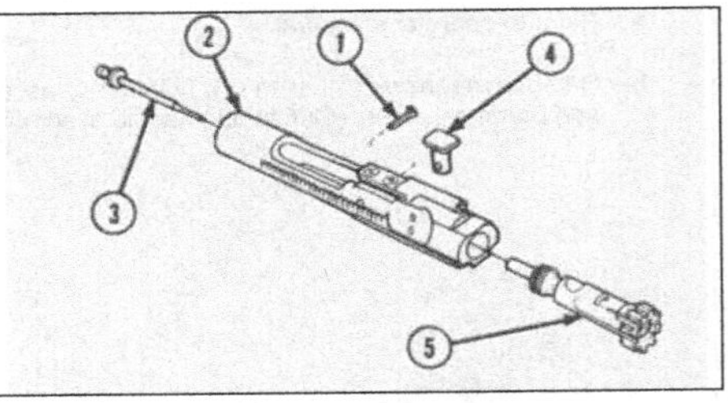

b. CLEANING

Clean carrier key and gas relief ports using CLP or RBC and pipe cleaner

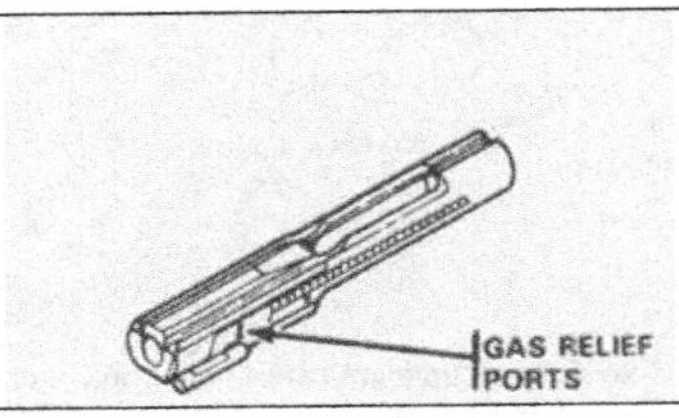

c. INSPECTION

1. Inspect bolt carrier assembly using the following guidelines.

(a) Inspect bolt carrier assembly (1) for burrs, cracks, wear, and evidence of gas loss.

(b) Visually inspect the carrier and key screws (2) for looseness and proper staking as shown below.

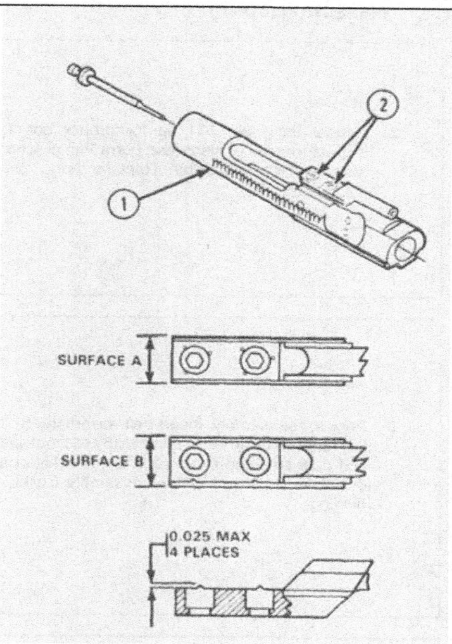

NOTE

Do not attempt to retorque if there is no loosening of the screws indicated by the staking marks.

Surface "A" must not Indicate distortion or damage which Impairs parallelism.

NOTE

A maximum of 0.025 in. (0.064 cm) protrusion In an upward direction is permissible.

NOTE

There are bolts and bolt carriers on fielded rifles, some with chrome-plated exterior surface finishes and some with phosphate coating Both finishes are acceptable under certain operational requirements and or restrictions Phosphate-coated bolt carriers are required for divisional combat units Chrome plated bolt carriers are acceptable for divisional noncombat units and training center units. Chrome-plated and phosphate-coated bolt assemblies, bolt carrier assemblies, and repair parts for these assemblies may be intermixed In any combination, with the following exception:

Phosphate-coated bolt carriers are required for all deployable and deploying units Chrome-plated bolt carriers are acceptable for nondeployable and training center units.

3-7. BOLT CARRIER ASSEMBLY (CONT.).

c. INSPECTION (CONT.)

2. Inspect firing pin 131 tip for proper con tour Inspect for pitting, wear, and burrs Pits or wear in area (4) Is permissible Replace firing pin If defective

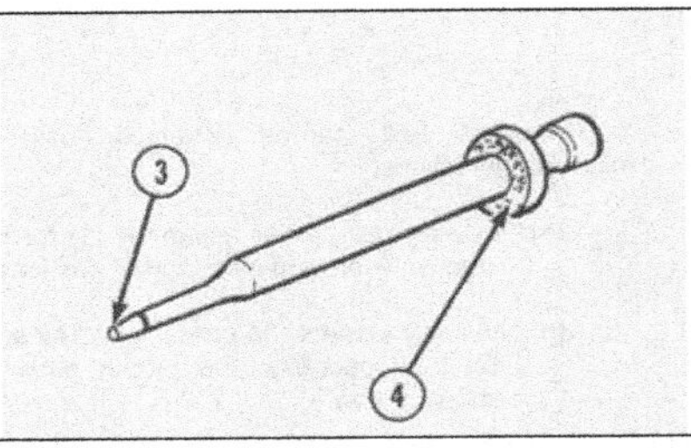

3. Prior to reassembly, insert bolt assembly (5) Into key and bolt carrier assembly (6) (do not insert bolt cam pin) and exercise bolt assembly in and out of key and bolt carrier assembly Check for binding

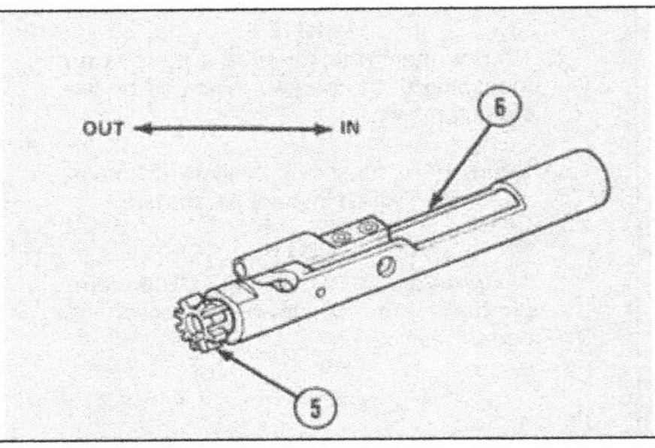

4. Check bolt assembly (5) for proper fit with bolt cam pin removed Turn key and bolt carrier assembly (61 and suspend so the bolt assembly is pointed down

NOTE

The bolt assembly must not drop out If weight of bolt assembly allows It to drop out of key and bolt carrier assembly, replace bolt rings (p 3-21)

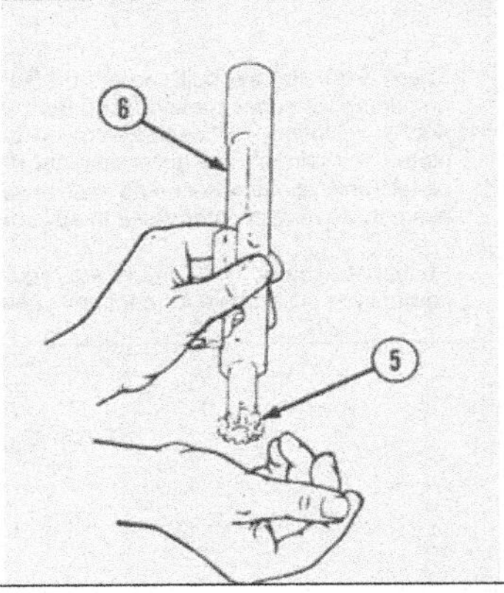

This page is blank. Procedures and figure were moved to page 2-36.

3-7. BOLT CARRIER ASSEMBLY (CONT.).

d. TEST

1. Insert firing pin (1) through bolt (2).
2. Position firing pin protrusion gage (3) PN 7799735 to check for proper firing pin (1) protrusion (minimum 0.028 in. (0 07 cm)-maximum 0 036 In. (0.09 cm))

NOTE

Firing pin should touch the gage on minimum but should not touch on maximum.

3. Replace a defective firing pin.

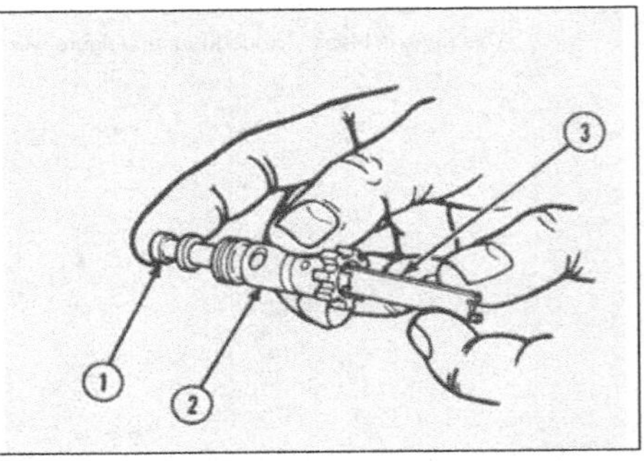

e. REPAIR

Replace all authorized unserviceable items. Retest all replaced parts.

f. REASSEMBLY

1. Install bolt assembly (1) into key and bolt carrier assembly (2).
2. Install bolt cam pin (3) and rotate one quarter turn to secure bolt assembly (1)
3. Hold key and bolt carrier assembly (2) with bolt assembly (1) down and drop In firing(pin (4)
4. Install firing pin retaining pin (5) from left side only to ensure proper Installation Check Installation by attempting to shake out firing pin
5. Reassemble rifle, refer to page 3-77

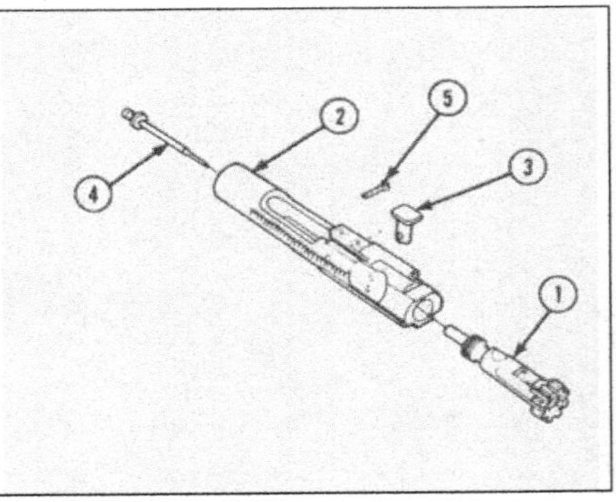

3-8. BOLT ASSEMBLY.

This task covers:
a. Disassembly
b. Inspection/Repair
c. Test
d. Reassembly

INITIAL SETUP

Test Equipment
Tool and Gage Set (Item 2, app B)

Tools
(ARMY) Small Arms Repairman Tool Kit (Item 3, app B)

Materials/Parts
Penetrant kit (item 25, app D)
Rag, wiping (item 26, app D)

Equipment Conditions
3-16 Bolt assembly removed

a. DISASSEMBLY

NOTE
Do not remove bolt rings unless they require replacement and three new replacement bolt rings are on hand.

Using small flat tip jeweler's screwdriver, remove the three bolt rings (1) from the bolt (2).

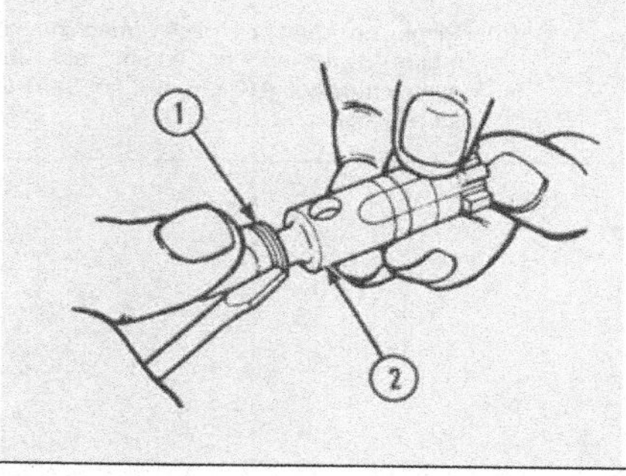

b. INSPECTION/REPAIR

1. Visually Inspect bolt rings for cracks, kinks, and bends. Replace all three bolt rings if one or more bolt rings is damaged. See page 3-16 for bolt ring wear check.

3-8. BOLT ASSEMBLY (CONT.).

b. INSPECTION/REPAIR (CONT.)

2. Inspect bolt for pits, burrs, and wear as follows.

(a) Bolt faces with a cluster of pits which are touching or tightly grouped, covering an area measuring approximately 1,8 Inch across, will be rejected and replaced.

(b) Bolts which contain individual pits or a scattered pattern will not be cause for rejection.

(c) Bolts that contain pits extending Into the firing pin hole will not be rejected unless firing pin hole gaging check determines excess wear.

(d) Rings on the bolt face (machine tool marks), grooves, or ridges less than approximately 0.010 inch will not be cause for rejection.

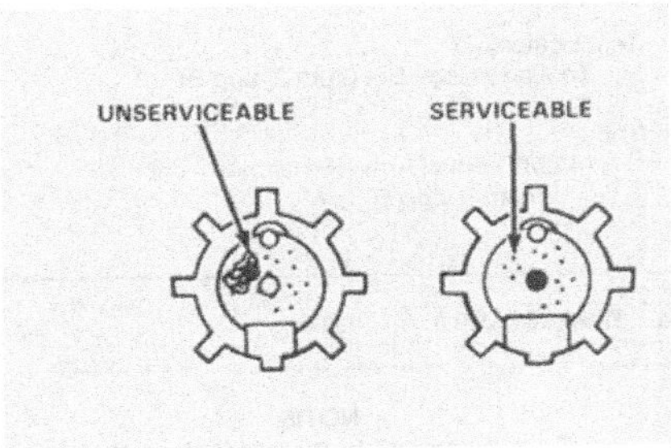

3. Inspect bolt for cracks In the locking lugs and the bolt cam pin hole area. Use black light If available; otherwise, use a glass of no more than 3X magnification or use a penetrant kit (item 25, app D). Pay close attention to the area where the locking lugs meet the body. Replace bolt assembly if bolt is defective.

WARNING

Dry cleaning solvent is flammable and toxic and should be used in a well-ventilated area. The use of rubber gloves is necessary to protect the skin when washing rifle parts.

4. Use penetrant kit (Item 25, app D) to check for cracks in bolt as follows:

(a) The area to be inspected must be clean, free of oil, etc. Spray a small amount of remover on the area to be inspected, let dry and wipe off with a wiping rag.

(b) Spray penetrant (only enough to wet the area) on the area of the bolt (1) to be inspected.

(c) Spray developer over the penetrant and let the developer work. Cracks will be indicated by a change in color where there is a crack. If there are cracks, the component 1' unserviceable.

(d) Pay close attention to the area where the locking lugs (2) meet the body.

(e) If there are no cracks, spray remover on the area; let dry and wipe off with a wiping rag. Oil the area to prevent corrosion.

(f) Replace bolt assembly If bolt (1) is defective.

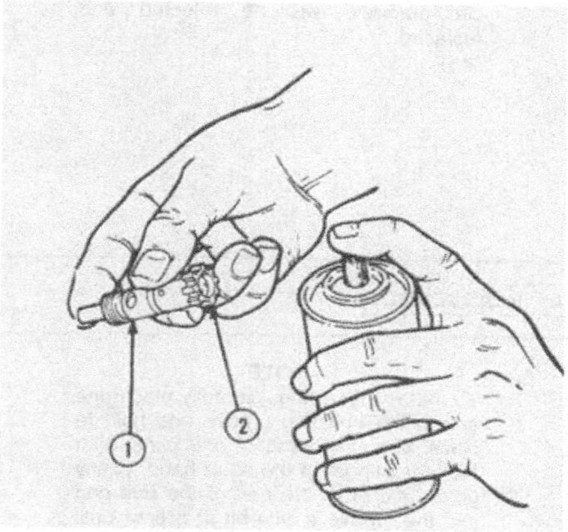

NOTE

Replacement of the bolt assembly will require that the headspace be tested (p 3-45, TEST).

3-8. BOLT ASSEMBLY (CONT.).

c. TEST

Test bolt (1) for elongated or oversized firing pin hole using special no-go plug gage (2) PN 12620101.

NOTE

Bolts with firing pin holes which permit the special no-go plug gage to fully penetrate at any position on the circumference will be rejected and replaced.

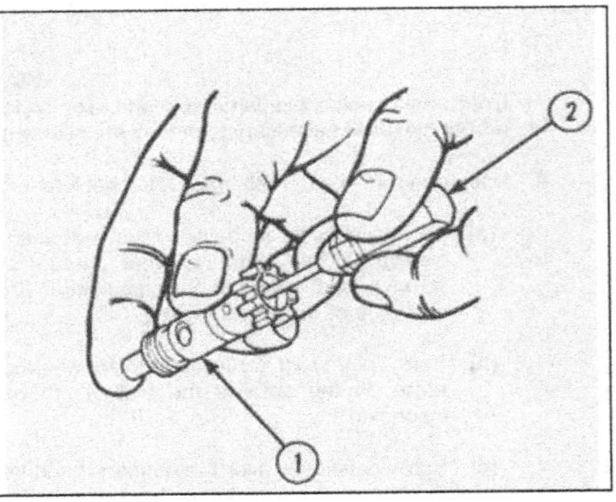

d. REASSEMBLY

NOTE

To install a bolt ring, carefully place one end in the bolt ring groove and hold In place with the thumb of one hand With the index finger of the other hand, gently guide and push the rest of the bolt ring into the groove a little bit at a time until the entire bolt ring is in place.

1. Install the three bolt rings (1) one at a time onto the bolt (2) using care not to bend or "spring" new bolt rings Stagger the bolt ring gaps (approximately ' turn apart).

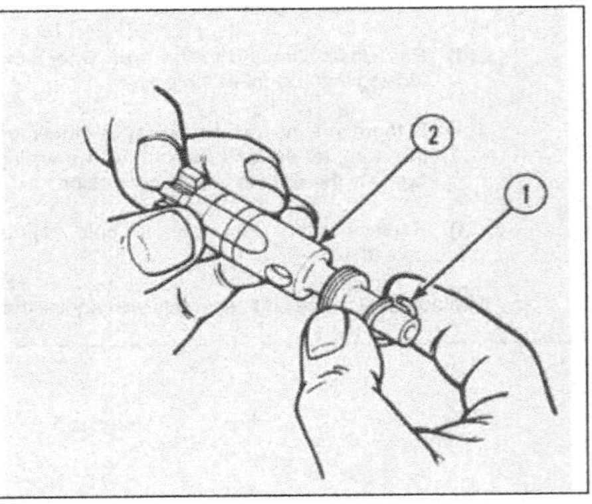

NOTE

Make certain bolt ring gaps are staggered to prevent loss of gas pressure. New bolt rings will make installing the bolt assembly difficult.

Lubricate inside key and bolt carrier assembly and use gentle pressure when installing.

2. Reassemble rifle, refer to page 3-77.

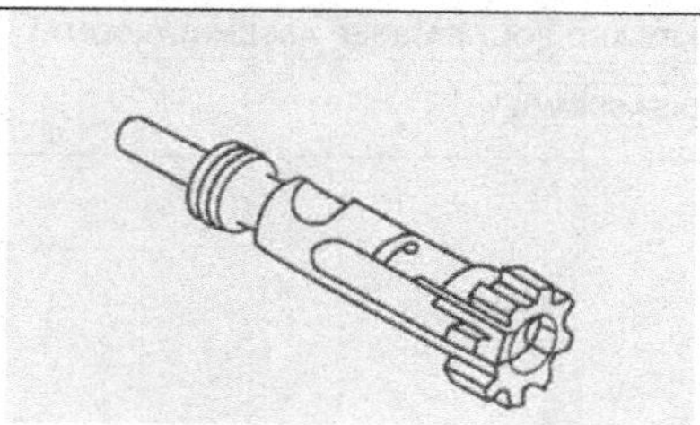

3-9. KEY AND BOLT CARRIER ASSEMBLY.

This task covers:

a Disassembly
b Repair
c Reassembly

INITIAL SETUP

Tools
Field Maintenance Basic Less Power Small Arms Shop Set (item 1, app B)
(ARMY) Small Arms Repairman Tool Kit (item 3, app B)

Materials/Parts
Carrier and key screws (2) (8448508)

Equipment Conditions
3-16 Key and bolt carrier assembly removed

3-9. KEY AND BOLT CARRIER ASSEMBLY (CONT.).

a. DISASSEMBLY

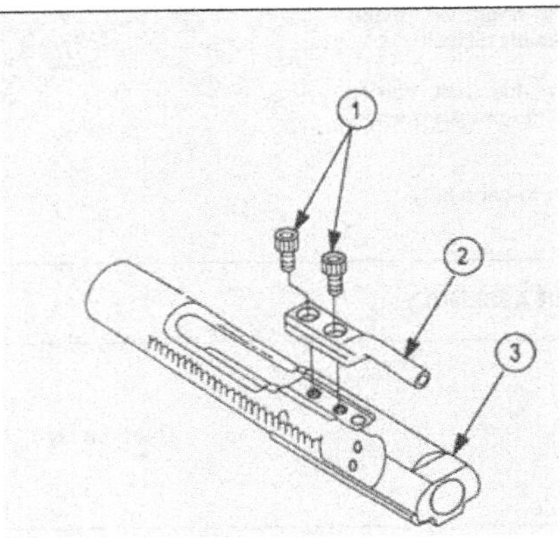

NOTE

Do not disassemble the key and bolt carrier assembly unless the bolt carrier key is defective as determined by inspection procedures described in section III, preceding.

1. Using socket wrench handle and tight fitting 1/8 inch socket head screw socket wrench attachment, remove two carrier and key screws (1).

NOTE

The heads and part of the bolt carrier key may be ground off in order to remove bolt carrier key from bolt carrier if carrier and key screws cannot otherwise be removed.

2. Remove bolt carrier key (2) from bolt carrier (3).

b. REPAIR

NOTE

Do not retorque carrier and key screws if staking marks do not indicate loosening screws.

Repair by replacing, torquing, and restaking carrier and key screws. Refer to the following reassembly procedures.

c. REASSEMBLY

NOTE

Do not reuse old carrier and key screws New carrier and key screws must be used at assembly.

1. If disassembled, place bolt carrier (1) In vise using vise law caps Install and position bolt carrier key (2) on bolt carrier (1).

2. Install two carrier and key screws (3). Always use new screws.

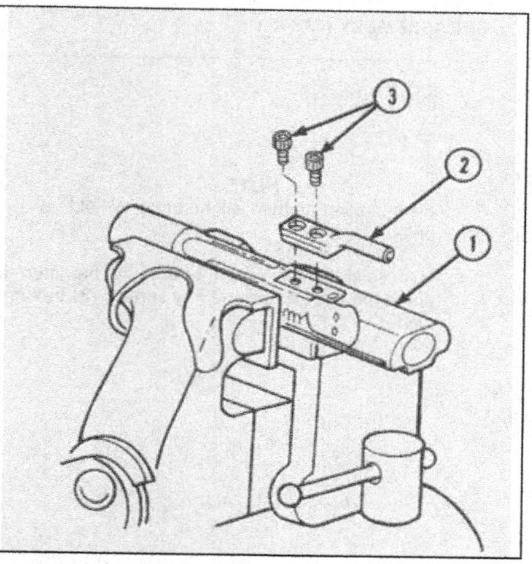

3. Use a tight-fitting 1 8 inch socket head screw wrench attachment and an Inch-pound torque wrench to torque the carrier and key screws (3) to 35 to 40 Inch-pounds (3 95 to 4 52 N m)

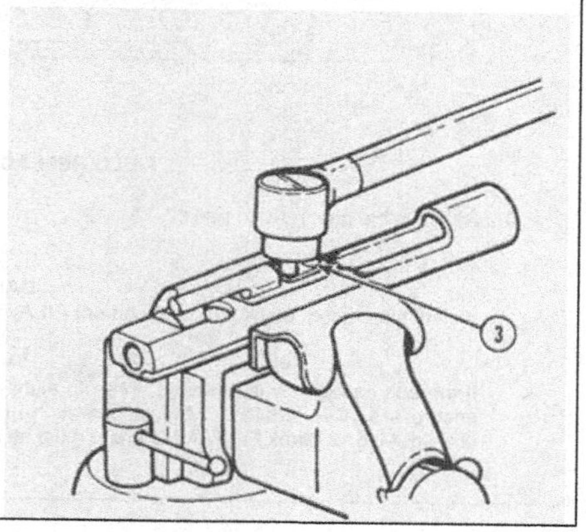

3-9. KEY AND BOLT CARRIER ASSEMBLY (CONT.).

c. REASSEMBLY (CONT.)

NOTE
Field staking method will be used by field units.

4. Use solid center punch and hand hammer to stake the two carrier and key screws (3) In three places.

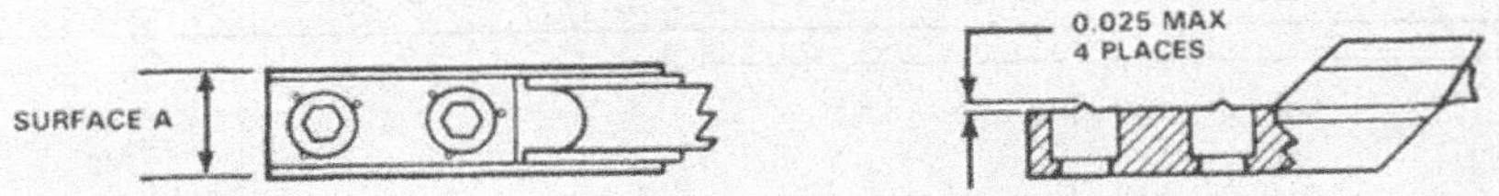

FIELD REPLACEMENT STAKING

5. Reassemble rifle, refer to page 3-77

CAUTION
If blanks are used, blank firing attachment (BFA) must be attached

NOTE
If the bolt carrier key is replaced, three to eight rounds of blank or ball ammunition must be fired to ensure a seal is created. Manual operation of the rifle may be required. If blank ammunition is utilized, M16A2 Blank Firing Attachment must be adapted

3-10. UPPER RECEIVER AND BARREL ASSEMBLY.

This task covers:

a Disassembly
b Inspection/Cleaning
c Repair
d Reassembly
e Test

INITIAL SETUP

Test Equipment
Tool and Gage Set (item 2, app B)

Tools
(ARMY) Small Arms Repairman Tool Kit (item 3, app B)
Field Maintenance Basic Less Power Small Arms Shop Set (item 1, app B)
Modified Needle Nose Pliers (fig E-7, app E)

Materials/Parts
Brush, cleaning, small (item 3, app D)
Carbon removing compound (item 8, app D)
Cloth, abrasive (item 13, app D)
Dichloromethane, technical (item 15, app D)
Dry cleaning solvent (item 16, app D)
Gloves, chemical and oil protective (item 18, app D)
Grease, molybdenum disulfide (item 19, app D)
Lubricant, solid film (item 21, app D)
Pan, wash (item 24, app D)
Polyethylene (item 32, app D)
Sealing compound (item 28, app D)
Target (item 31, app D)

References
FM 23-9
TM 9-1005-319-10

Equipment Conditions
3-15 Upper receiver and barrel assembly removed
2-43 Handguard assemblies removed

General Safety Instructions
To avoid injury to your eyes, use care when removing and installing spring-loaded parts.
When using solid film lubricant or dichloromethane, be sure the area is well ventilated.
When using carbon removing compound, avoid skin contact If carbon removing compound comes in contact with the skin, wash thoroughly with running water Using a good lanolin base cream after exposure to the compound is helpful Using gloves and protective equipment is required.

3-10. UPPER RECEIVER AND BARREL ASSEMBLY (CONT.).

a. DISASSEMBLY

1. Using ball-peen hammer and 5/64 inch diameter drive pin punch, drive spring pin (1) (which retains the gas tube) out of front sight assembly (2).

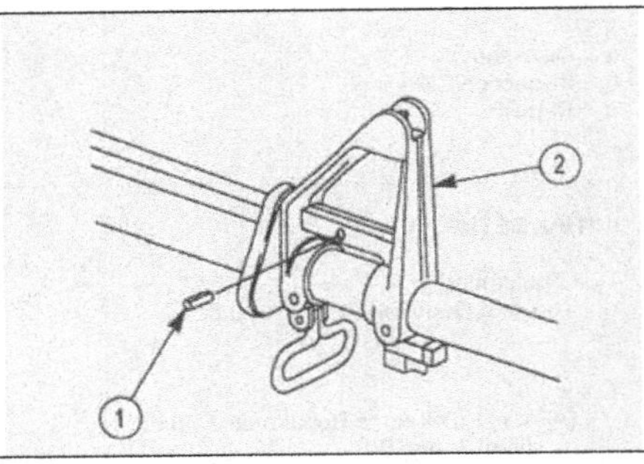

2. Slide gas tube (3) back into upper receiver assembly (4) to clear front sight assembly (2). Then lift slightly, pull forward, and remove gas tube (3).

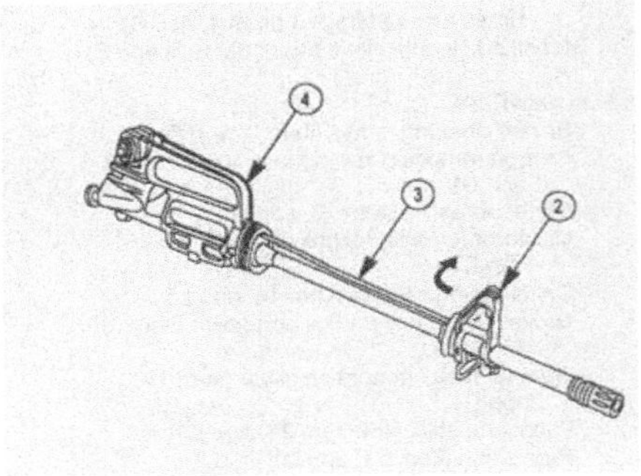

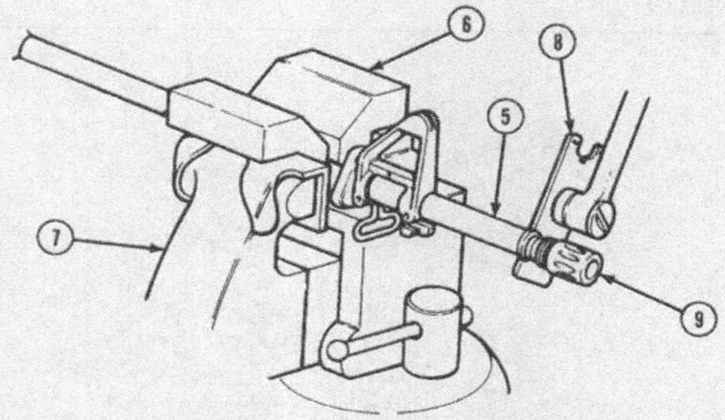

3. Position upper receiver and barrel assembly (5) in barrel removal fixture (6) and secure both In machinist's vise (7).

4. Using combination wrench (8) and 12 inch drive handle, remove compensator (9).

5. Remove peel washer (10) being careful not to lose or bend thin sections.

6. Remove upper receiver and barrel assembly (5) from barrel removal fixture and machinist's vise.

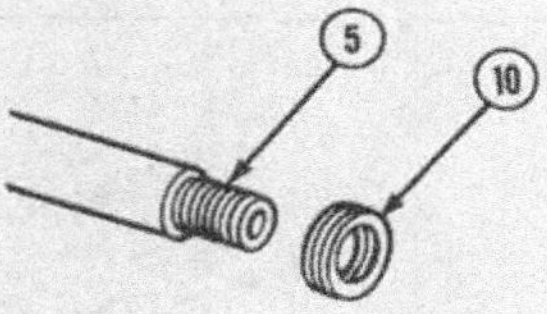

3-10. UPPER RECEIVER AND BARREL ASSEMBLY (CONT.).

a. DISASSEMBLY (CONT.)

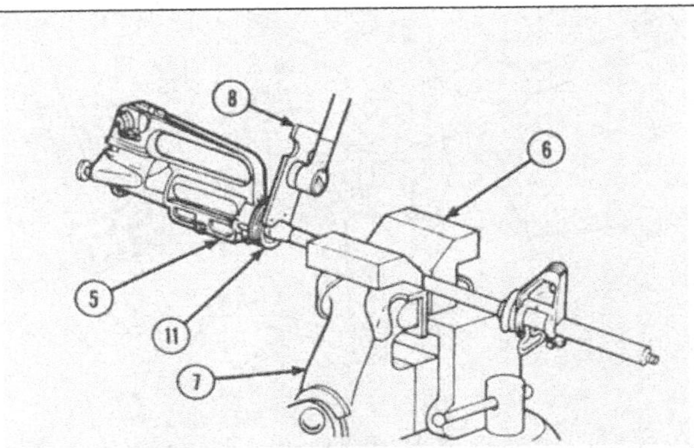

7. Place upper receiver and barrel assembly (5) Into barrel removal fixture (6) and clamp Into machinist's vise (7).

NOTE

Be sure all three drive pins on combination wrench are fully engaged with barrel nut assembly. Wrench must be pushed toward upper receiver assembly to compress the slip ring spring In barrel nut assembly. Do not use a torque wrench to loosen the barrel nut assembly.

8. Using '2 Inch drive handle and combination wrench (8), loosen barrel nut assembly (11)

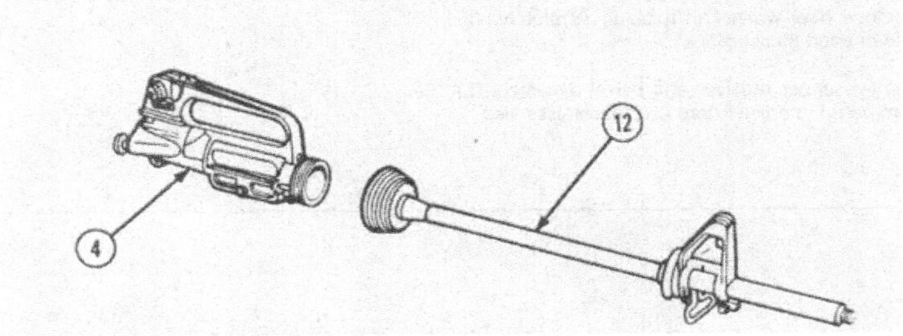

9. Separate upper receiver assembly (4) from barrel assembly (12)

10. Remove barrel assembly (12) from machinist's vise and barrel removal fixture

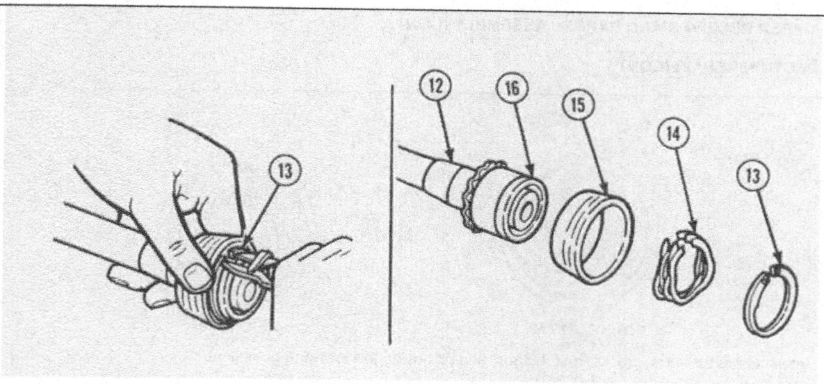

11. Remove retaining ring (13) using retaining ring pliers.

12. Remove slip ring spring (14) and handguard slip ring (15).

13. Do not remove barrel nut (16) from barrel assembly (12).

b. INSPECTION/CLEANING

WARNING

When using carbon removing compound, avoid skin contact. If carbon removing compound comes In contact with the skin, wash thoroughly with running water. Using a good lanolin base cream after exposure to compound is help ful. Using gloves and protective equipment is required.

1. Inspect gas tube for cracks. Replace if defective

2. Use carbon removing compound to remove carbon deposits from interior and exterior of gas tube. If a large amount of carbon is found and cannot be removed, replace the gas tube

NOTE

A small arms cleaning brush (bore) (item 4, app D) may be used to clean Interior of front sight assembly where gas tube is secured.

3-10. UPPER RECEIVER AND BARREL ASSEMBLY (CONT.).

b. INSPECTION/REPAIR (CONT.)

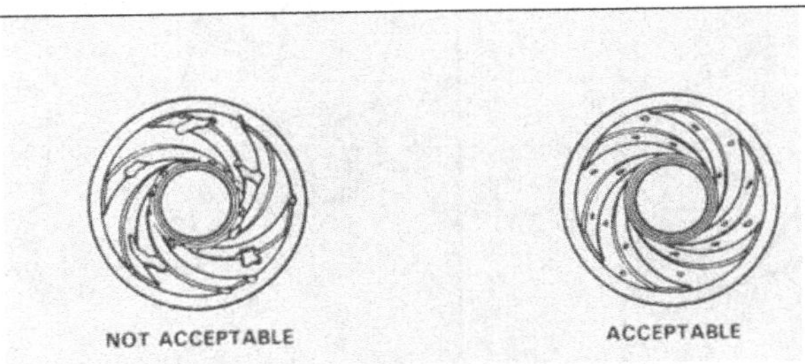

3. Inspect bore for burrs, cracks, rust, bulges, and pits using the following guidelines

(a) Pits no wider than a land or groove and no longer than 3'8 inch (0.95 cm) are al lowed in the bore.

(b) Lands that appear dark blue due to coating of gliding metal from projectiles are allowable.

(c) Definitely ringed bores or bores ringed sufficiently to bulge the outside surface of the barrel are cause for rejection. Replace barrel assembly if defective

4. If the upper receiver is separated from the barrel assembly, Inspect chamber for pits utilizing a flashlight. Pits 1'8 inch 10 32 cm) In length are cause for rejection Replace rifle barrel assembly If defective

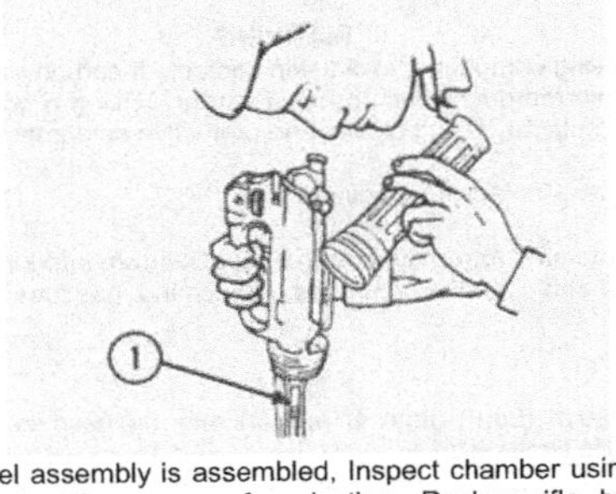

5. If upper receiver and barrel assembly is assembled, Inspect chamber using reflector tool (1) and flashlight. Pits 1 '8 inch (O 32 cm) In length are cause for rejection. Replace rifle barrel assembly If defective. If rifle barrel assembly Is replaced, Inspect headspace (p 3-45).

6. Inspect upper receiver assembly for cracks, corrosion, wear, or damage

   (a) Small dents or gouges that do not affect functioning will not be cause for rejection.

   (b) If upper receiver assembly contains cracks or holes, the upper receiver assembly will be replaced

7. Inspect all parts for damage and wear. Replace all defective parts.

NOTE

Damaged or missing teeth of the barrel nut is not cause for rejection provided the proper torque value can be obtained during installation using the tools depicted. If removal of the barrel is not possible with the combination tool, a pipe wrench or other such tool may be used during removal.

8. Inspect front sight guards for bends, if bent see page 3-37 for repair procedures.

c. REPAIR

1. Repair corroded upper receiver assembly surfaces as follows:

   (a) Sand corroded area with abrasive cloth and make sure all corrosion has been removed.

WARNING

When using solid film lubricant or dichloromethane, be sure the area is well ventilated

   (b) Wash area with technical dichloromethane (methylenechloride) to re move all dirt, grease, and foreign material

   (c) Apply sealing compound, mixed In accordance with manufacturer's directions, to areas to be filled

   (d) Spread sealing compound as smoothly as possible Into defective area using a putty knife or similar tool

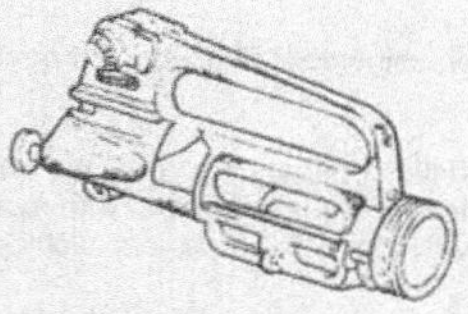

CORRODED
(REPARABLE)

CORRODED
(NONREPARABLE)

3-10. UPPER RECEIVER AND BARREL ASSEMBLY (CONT.).

c. REPAIR (CONT.)

NOTE

Do not feather edges

(e) Place a sheet of polyethylene (Item 32, app D), cut to size, over filled area Rub by hand to smooth.

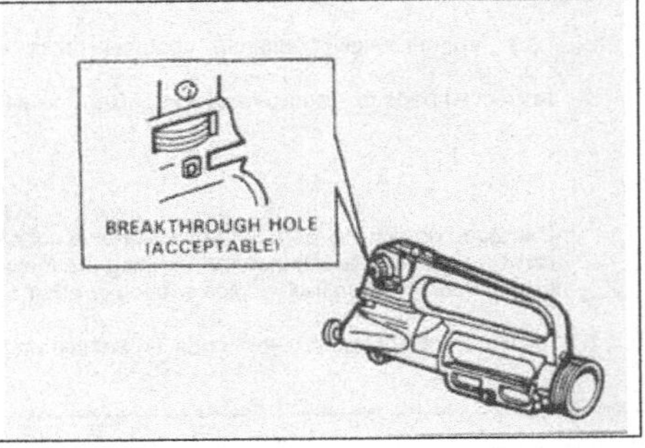

2. After curing, remove polyethylene sheet In accordance with Instructions by the manufacturer.

WARNING

When using solid film lubricant or dichloromethane, be sure the area is well ventilated.

CAUTION

Solid film lubricant is to be used only as an exterior surface protective finish and touchup If solid film lubricant comes in contact with recoiling parts or functional surfaces of the rifle, remove immediately by washing with technical dichloromethane.

3. Wash area with technical dichloromethane (methylenechloride) ; to remove dirt and grease, and foreign material.

4. Roughen area to be refinished with abrasive cloth and clean surface again Do not touch the area with fingers.

5. Repair shiny surfaces by spraying a coat of solid film lubricant in accordance with Instructions supplied by the manufacturer Dry 24 hours before handling.

SHINY SURFACES (REPARABLE).

6. Straighten bent front sight guards (1) as follows:

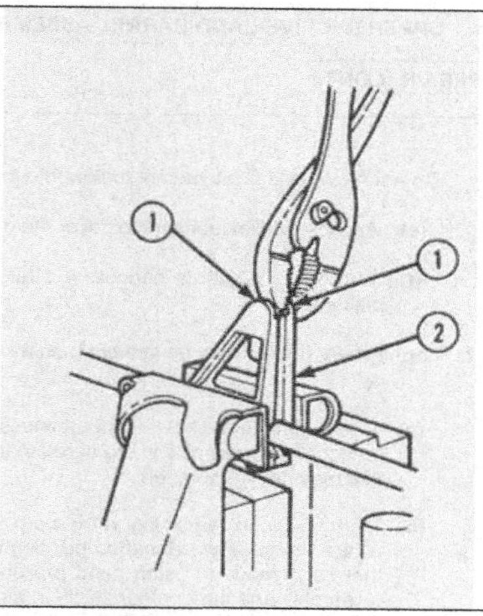

NOTE

Remove spring before heating. (Heat will damage spring.) The sight post and plunger may be reused unless damaged.

(a) Remove front sight post, detent, and helical spring (see p 2-43).

NOTE

Use copper or brass caps (jaw Inserts on bench vise to prevent damage to front sight base (2) during clamping.

(b) Place front sight base (2) in a bench vise.

(c) Heat front sight guards (1) and bend with pliers. The front sight guards (1) should be put back as nearly as possible to the original position. Allow front sight housing to air cool.

WARNING

Dry cleaning solvent is flammable and toxic and should be used in a well ventilated area. The use of rubber gloves is necessary to protect skin when washing rifle parts.

(d) Roughen any damaged surface of front sight guards with abrasive cloth and clean with dry cleaning solvent. Always wear rubber gloves when using dry cleaning solvent.

3-10. UPPER RECEIVER AND BARREL ASSEMBLY (CONT.).

c. REPAIR (CONT.)

CAUTION

Do not allow solid film lubricant to flow into front sight post threaded well.

(e) Apply solid film lubricant to cover the damaged finish.

(f) If front sight guards cannot be straightened utilizing the above procedures, reduce the rifle barrel assembly.

7. Slightly bent barrels may be straightened as follows'.

(a) Check straightness using straightness gage 8448202 (p 3 45) If the barrel falls the straightness test, and the gage remains in the barrel In the area of the front sight assembly, perform step (b) to determine If it may be straightened.

(b) With the gage remaining in the bore, hold the rifle in a vertical position with the end of the barrel into which the gage was inserted pointing up Ensure that if when the gage passes through the barrel it will not be damaged Using hand pressure' ONLY, flex the portion of the barrel between the front sight assembly and the compensator in all four directions (left, right, forward, and back) If the barrel is only slightly bent, the gage will drop through when the barrel is flexed in one of these directions Note the direction which allowed the gage to drop through the' barrel.

CAUTION

Remove the gage from the barrel before continuing

NOTE

If the gage does not pass through the barrel when it is flexed, replace the rifle barrel assembly.

(c) Place the barrel in a vise using appropriate protective laws . Clamp the barrel between the front sight assembly and the compensator approximately 1 inch (2.54 cm) from the front sight assembly The rifle barrel assembly should be in a horizontal position with the side noted in step (b) toward you.

CAUTION
Do not apply pressure to the receiver.

(d) Grasp the BARREL near the receiver so that when force is applied the barrel will flex In the same direction as noted In step (b).

(e) Give the barrel a sharp jerk of approximately 20 to 40 pounds of force (f) Remove the barrel from the vise and recheck straightness (step (al).

(g) If gage still will not pass through the barrel, perform step (b) to determine direction of bend If the barrel is still bent in the same direction as before, perform steps (c) through (f) using slightly more force. If the barrel Is now bent In the opposite direction, replace the rifle barrel assembly.

(h) If the gage passes freely through the barrel, the barrel shall be considered straight and continue In service.

(i) If the barrel has been straightened, the rifle must be targeted (p 3-45).

d. REASSEMBLY

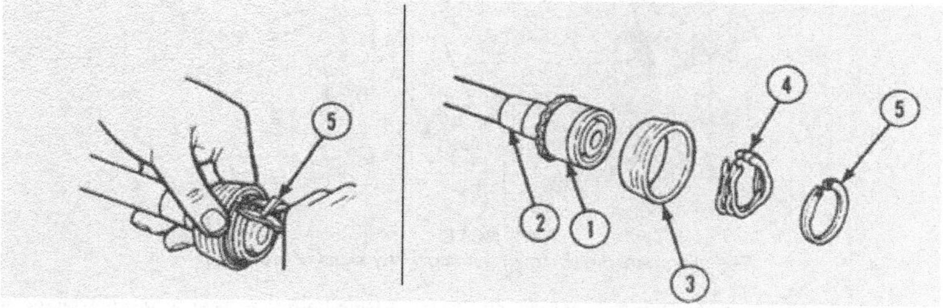

1. Position barrel nut (1) by sliding it to the rear of rifle barrel assembly (2) as far as possible
2. Slide handguard slip ring (3) over barrel nut (1).
3. Press slip ring spring (4) from both sides and insert It Into handguard slip ring (3)
4. Install retaining ring (5) against slip ring spring (4) using retaining ring pliers Snap retaining ring (5) to barrel nut (1).

NOTE
After cleaning, apply molybdenum disulfide grease to threads of barrel nut assembly before Installation.

3-10. UPPER RECEIVER AND BARREL ASSEMBLY (CONT).

d. REASSEMBLY (CONT)

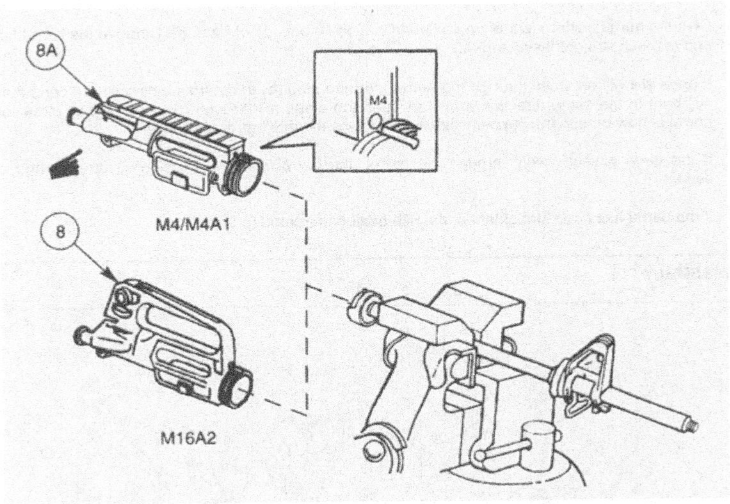

NOTE

The alignment pin (6) must not show any signs of looseness.

5. Position rifle barrel assembly (2) with alignment pin (6) up. Using barrel remover fixture (7), clamp rifle barrel assembly (2) in vise.

NOTE

The slot should fit the alignment pin perfectly with very little or no rotational play present. Note the play in a new rifle barrel assembly and new upper receiver assembly and use this as a guide.

6. Align upper receiver assembly (8) or (8A) using alignment pin (6) and the slot in upper receiver assembly (8) or (8A). Install over end of rifle barrel assembly (2).

7. Wipe upper receiver threads clean and ensure there are no burrs. Apply molybdenum disulfide grease to the threads before installation.

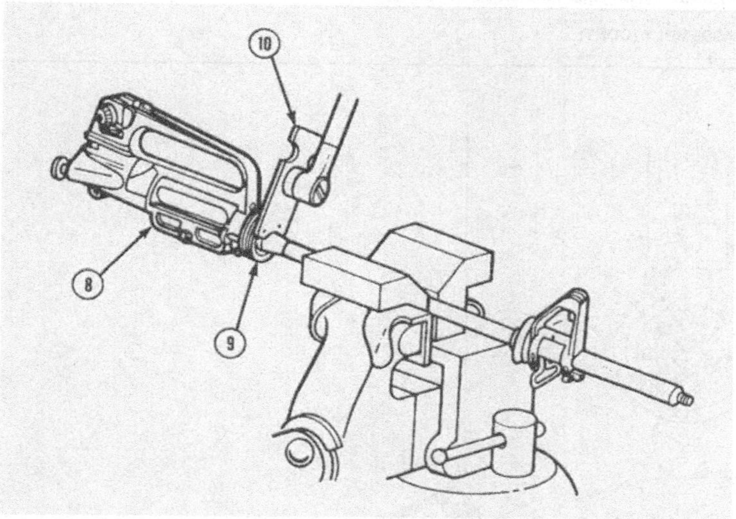

8. Engage threads of barrel nut assembly (9) with upper receiver assembly (8).

9. Using combination wrench (10) and torque wrench, torque barrel nut assembly (9) to 30 ft-lb (40.5 N-m). Torque is measured when both wrenches are used together.

NOTE

Three times torquing procedure provides for a better thread fit and prevents barrel nuts from becoming loose. Do not use the torque wrench for loosening.

10. Make certain all three drive pins on combination wrench are engaged with barrel nut assembly (91 Loosen and repeat torque operation. Then loosen the barrel nut again.

3-10. UPPER RECEIVER AND BARREL ASSEMBLY (CONT).

d. REASSEMBLY (CONT)

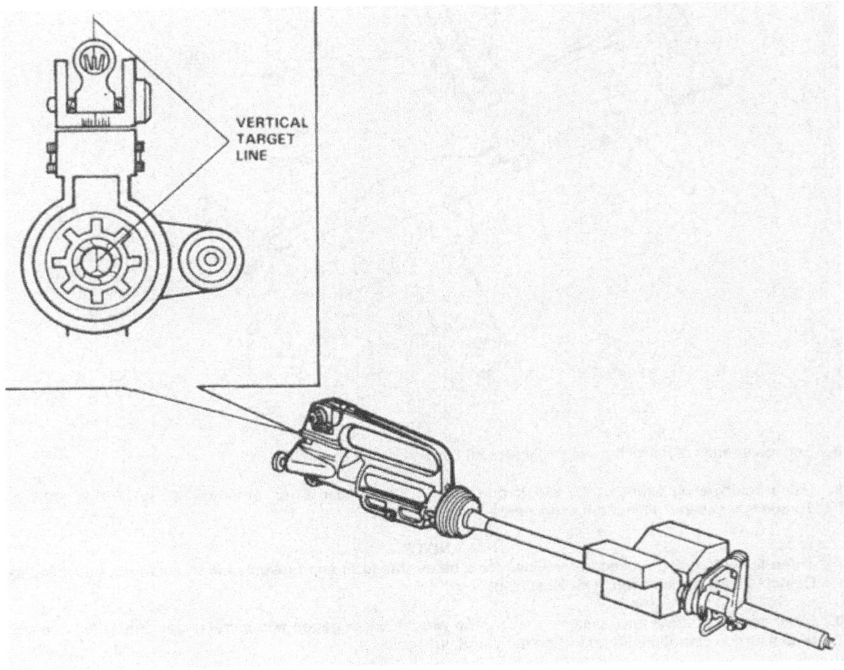

11. Front sight post must be Installed. Loosen the vise and align the bore on a distant vertical target Center the target In the bore from 12 o'clock through 6 o'clock The, front sight post should be on line and vertical with the target. Tighten vise Adjust the rear sight windage until a proper sight picture is obtained on the vertical target The rear sight aperture will be approximately In the center of the rear sight base If the rifle barrel assembly is properly aligned in the upper receiver assembly

NOTE

If rifle barrel assembly (usually new) is not properly aligned In the upper receiver assembly (usually an old part), excessive windage will be present and the upper receiver assembly will require replacement to obtain the proper fit between the alignment pin and slot.

CAUTION

Do not torque over 80 ft-lb (108 N-m) while tightening the barrel nut assembly to the next hole, to allow for proper alignment of gas tube.

NOTE

Do not attempt to hold the upper receiver assembly with a pry bar, however, If the rifle barrel assembly turns In the holding fixture, a pry bar may be used through the front sight assembly base to help prevent the rifle barrel assembly from turning In the holding fixture. Use care not to distort or bend front sight assembly or retaining pins. Use "buddy system" to hold pry bar.

12. Torque the barrel nut assembly again to 30 ft-lb (40.5 N-m) while maintaining sight alignment. The barrel nut assembly may be tightened beyond 30 ft-lb (40.5 N-m) to align the barrel nut assembly serrations for proper gas tube clearance Never loosen the barrel nut assembly to align for gas tube clearance.

13. Check alignment of barrel nut assembly (9) with upper receiver assembly (8). The front 8 inches (20.32 cm) of a gas tube may be used as an alignment tool (see Illustration). This Is Inserted Into the bolt carrier key and then Inserted into the rear of the receiver. If the parts of the barrel nut assembly are properly aligned, the tool will pass freely and lay top dead center along the top of the barrel. A number 15 twist drill (0.180 inch) may also be needed as an alignment tool. If necessary, tighten barrel nut assembly to next hole to allow proper alignment.

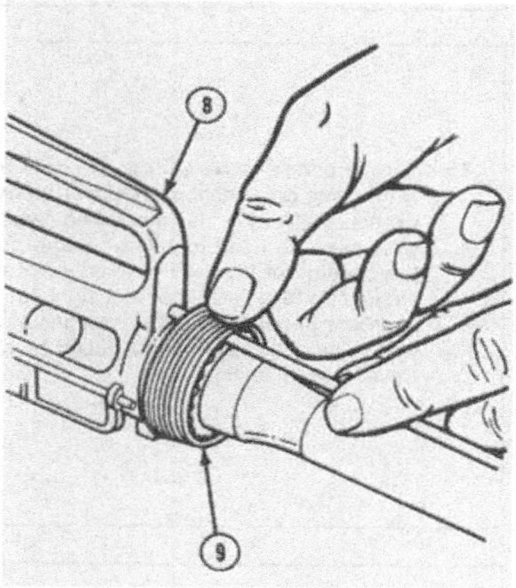

3-10. UPPER RECEIVER AND BARREL ASSEMBLY (CONT).

d. REASSEMBLY (CONT)

NOTE

The peel washer may be heated to remove thin sections. Always place thin section to rear.

14. Install peel washer (11) and compensator (12) on rifle barrel assembly (2).

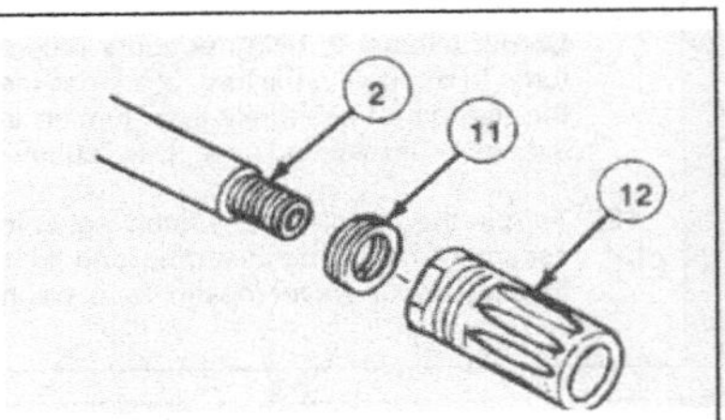

15. Torque compensator (12) to 15-20 ft-lb (20-27 N-m) using combination wrench (10) and torque wrench. Torque is measured when both wrenches are used together. Some portion of the openIng of the third or middle slot must be straight up (align with the front sight post) (12.1) at proper torque level. Thin sections of the peel washer may be removed or added as required. Save unused sections.

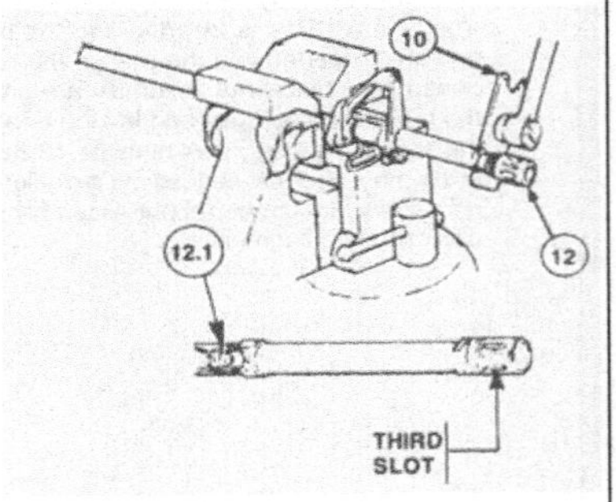

16. Slide gas tube (13) through the barrel nut assembly (9) and then slide forward, Inserting gas tube into hole In the front sight assembly base (14).

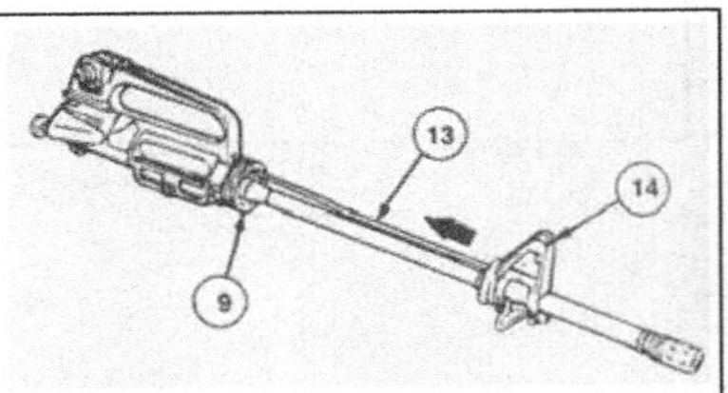

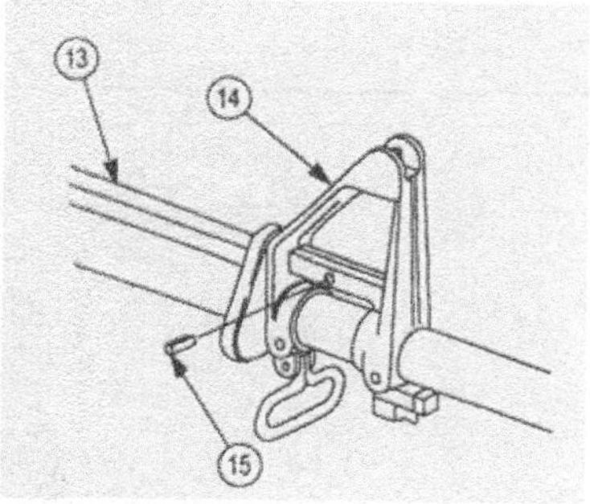

17. Align the holes in gas tube (13) and front sight assembly base (14).

NOTE
To assist in installing spring pin (1) modified nose pliers may be used.

18. Using ball-peen hammer and 5/64-inch diameter drive pin punch, drive spring pin (15) into front sight assembly base (14) to secure gas tube (13).

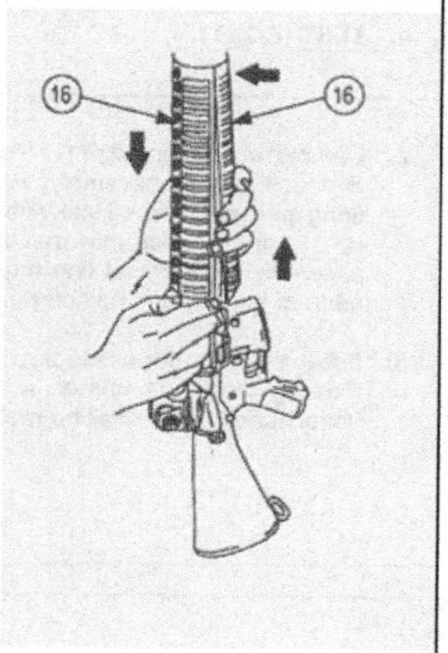

19. Install upper and lower handguard assemblies (16).

e. TEST

1. The following information pertains to the use of breech, bore, and other gages:

(a) All M1 6A2 barrels and chambers are chromed.

(b) Barrel erosion gage, PN 8448496 (normally used on M16A1 fully-chromed barrels), can be used to gage M16A2 barrels.

(c) The bore straightness gage, PN 8448202, is required for use on all barrels. The gage must pass through the barrel without being forced.

3-10. UPPER RECEIVER AND BARREL ASSEMBLY (CONT).

e. TEST (CONT)

2. Use barrel erosion gage (1) PN 8448496. Install key and bolt carrier assembly with bolt assembly and firing pin removed. Hold rifle vertical with receiver up. Insert gage into rear of key and bolt carrier assembly. The reject line must be read at the rear edge of the key and bolt carrier assembly.

3. If the reject mark passes beyond the rear surface of the key and bolt carrier assembly, the barrel is unserviceable and shall be replaced.

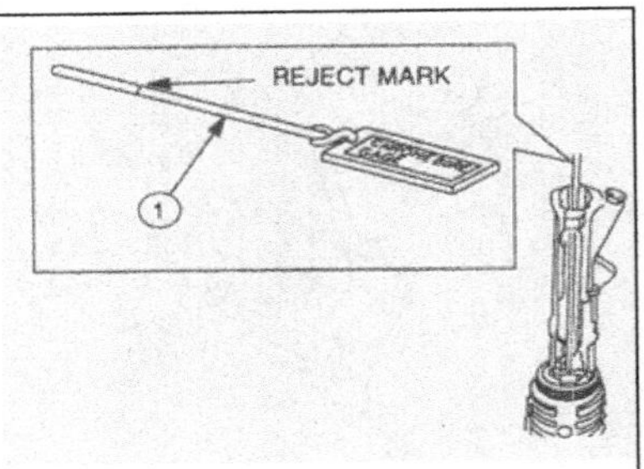

NOTE
Ensure barrel is clean prior to performing the following test.

4. Check straightness of bore using straightness gage (2) PN 8448202. Put gage in barrel. Tilt barrel and allow gage to fall through. Catch gage.

5. Gage must pass freely through barrel. If gage does not pass through barrel, recheck as follows. Hold upper receiver and barrel assembly (3) in vertical position with muzzle pointed down; insert gage into chamber end of barrel. Release gage and catch it as it exits muzzle end. If gage passes freely through the barrel, barrel is acceptable. If it does not, the barrel must be straightened or replaced. (See page 3-38 for straightening instructions/ procedures.)

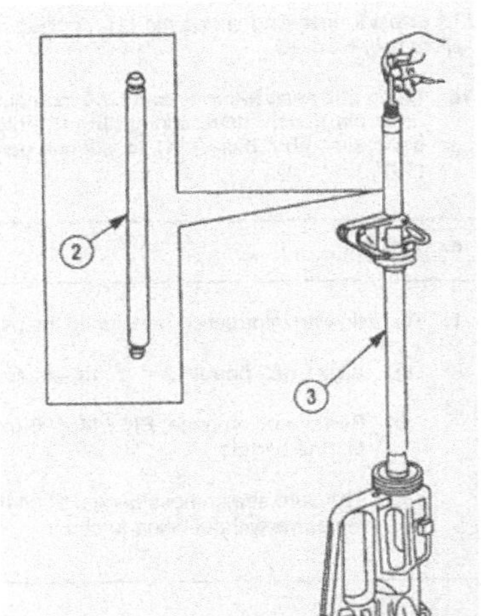

6. Assemble charging handle assembly (4), bolt assembly, and key and bolt carrier assembly (5) into upper receiver assembly (6).

7. Insert headspace gage (7) PN 7799734 in chamber.

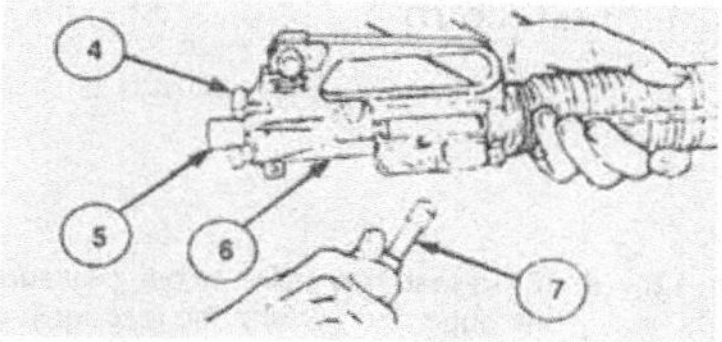

NOTE
For the purpose of this test "light finger pressure" is defined as 8 1/2 to 8 3/4 pounds.

8. Check headspace by pressing key and bolt carrier assembly (5) and charging handle assembly (4) forward using light finger pressure.

9. Bolt should not rotate to locked position. Key and bolt carrier assembly (5) must protrude from rear of upper receiver assembly (6) for proper headspace. If excessive headspace, first replace old bolt assembly with an unused bolt assembly and then recheck. If headspace is not corrected, replace rifle barrel assembly; then recheck with the original bolt assembly to determine if the bolt assembly is still good or if the bolt assembly should be replaced also.

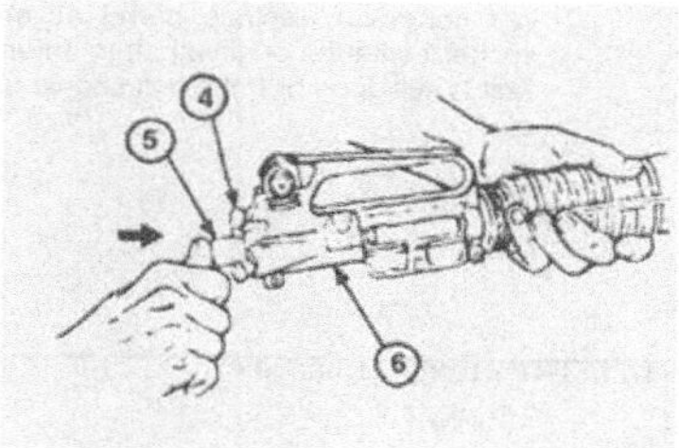

10. Remove key and bolt carrier assembly, bolt assembly, charging handle assembly, and headspace gage.

11. Reassemble rifle, refer to page 3-77.

NOTE
Rifles which have been rebarreled must be function-fired with seven rounds of 5.56mm ball ammunition. After rebarreling, the rifle must be targeted with three rounds of 5.56mm ball ammunition at 25 meter range using target. Refer to TM 9-1005-319-10 and FM 23-9

3-10. UPPER RECEIVER AND BARREL ASSEMBLY (CONT).

e. TEST (CONT)

12. (a) If the test fails using finger pressure, remove the gage and perform the test again as follows: With the muzzle down, stack 8 1/2 to 8 3/4 pounds of trigger weights (8) on a locally fabricated spacer/weight (9) on the bolt carrier assembly (10) Insert headspace gage (11) and test per above instructions.

(b) Remove trigger weights (8), spacer/weight (9), bolt carrier assembly (10), charging handle (12), and headspace gage (11).

(c) If excessive headspace, first replace bolt assembly and then recheck If headspace is not corrected, replace barrel assembly; then recheck with the original bolt to determine if the bolt is still good or If the bolt should be replaced also.

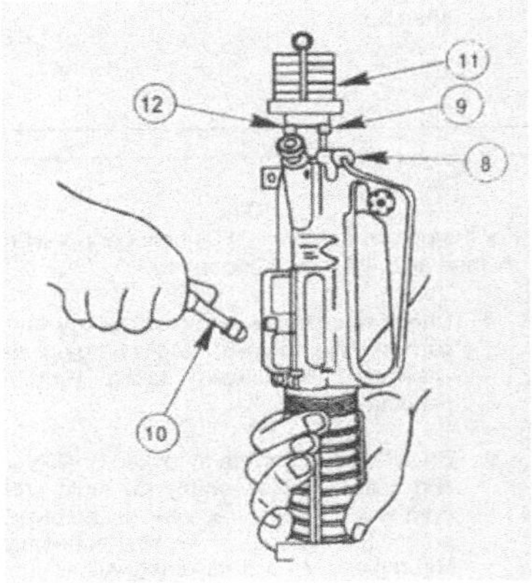

3-11. UPPER RECEIVER ASSEMBLY AND REAR SIGHT ASSEMBLY.

This task covers:

a. Disassembly
b. Inspection
c. Repair
d. Lubrication
e. Reassembly
f. Mechanical Zero Procedures (A.F. Only)

INITIAL SETUP:

Tools
(ARMY) Small Arms Repairman Tool Kit (item 3, app B)
Field Maintenance Basic Less Power Small Arms Shop Set (item 1, app B)

Materials/Parts
Cleaner, lubricant, and preservative (CLP) (item 9, app D)
Lubricant, solid film (item 21, app D)
Index screw (9349065)

Equipment Conditions
3-29 Upper receiver assembly removed.

General Safety Instructions
To avoid injury to your eyes, use care when removing and installing spring-loaded parts.
When using solid film lubricant or dichloromethane, be sure the area is well ventilated.

a. DISASSEMBLY

WARNING

To avoid injury to your eyes, use care when removing and installing springloaded parts.

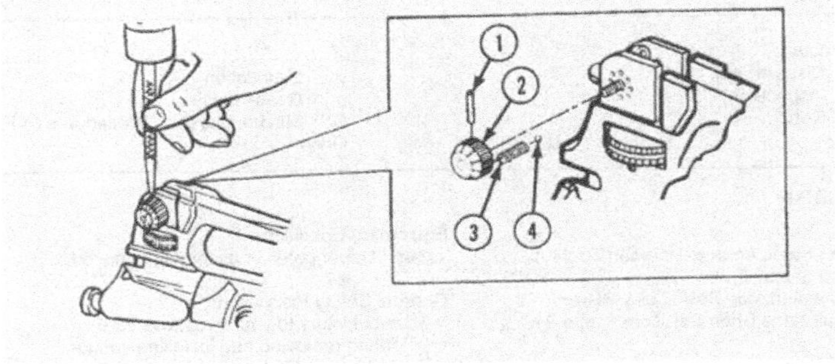

CAUTION

Be sure to catch small parts (1, 2, 3, and 4). AIR FORCE ONLY: Use magnet to keep from losing small parts (1, 3, and 4).

1. Drive out spring pin (1) using a hammer and 1/16 inch punch.
2. Catch rear sight assembly windage knob (2), helical spring (3), and ball bearing (4).

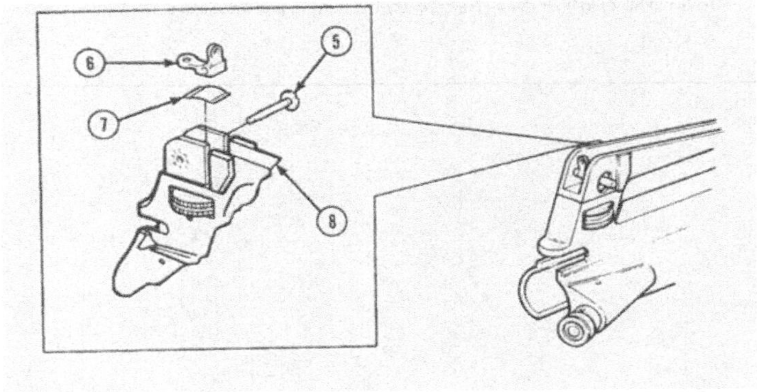

3. Using a flat-bladed screwdriver, remove rear sight assembly windage screw (5)
4. Remove sight aperture (6) and flat spring (7) from upper receiver (8).

3-11. UPPER RECEIVER ASSEMBLY AND REAR SIGHT ASSEMBLY (CONT).

a. DISASSEMBLY (CONT)

WARNING
To avoid injury to your eyes, use care when removing and installing spring-loaded parts.

5. Drive out spring pin (9) using a 3/32 Inch punch Catch helical spring (10) when punch is withdrawn.

6. Rotate elevation index (11) until rear sight assembly base (12) clears upper receiver (8). Catch ball bearing (13) and helical spring (14) as rear sight base clears.

7. Push elevation index (11) out with thumb using slight rotation motion. Catch ball bearing (15) and helical spring (16).

8. Use 1 16 Inch allen wrench to remove index screw (17). Discard Index screw (1 7) Separate elevation index (11) from elevation knob (18) by hand.

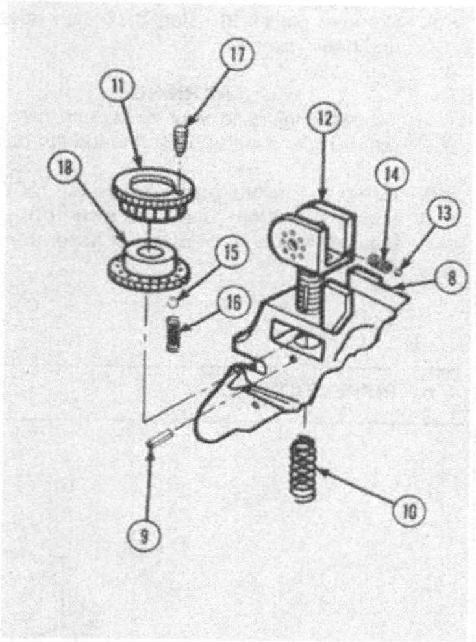

9. Remove pin (1 9) using 3,32 Inch drive pin punch and hand hammer .

WARNING

To avoid Injury to your eyes, use care when removing and installing spring-loaded parts.

10. Remove forward assist assembly (20) and helical spring (21) from upper receiver (8). For further disassembly of forward assist assembly see page 3-59.

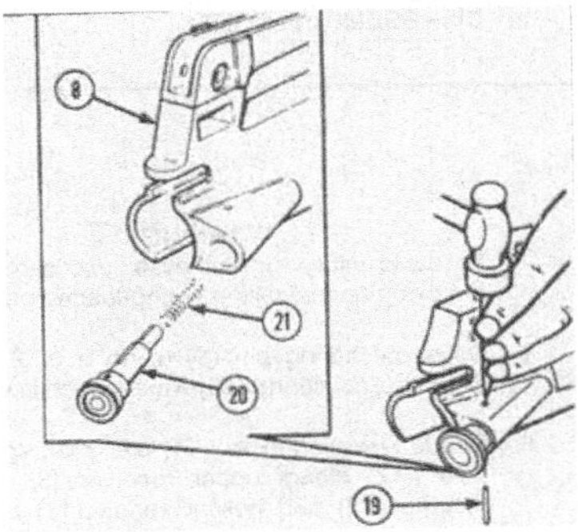

b. INSPECTION

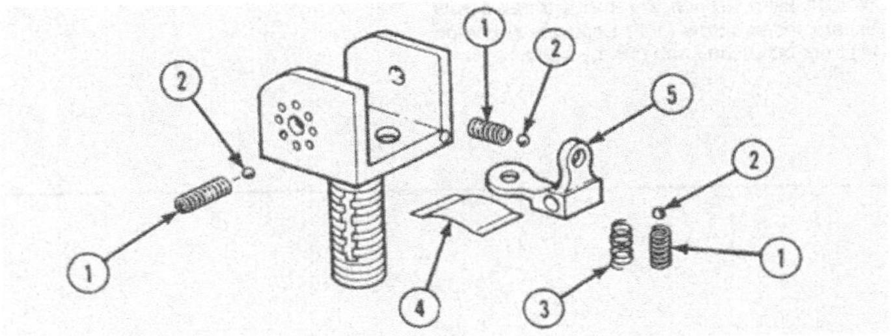

1. Check rear sight parts for serviceability. Inside of apertures should be round and distinct. Replace If defective
2. Visually inspect rear sight assembly helical springs (1) ball bearings (2), and helical spring (3) for breaks, bends, and missing parts Ball bearings should be smooth and round. Replace If defective
3. Check upper receiver for cracks, corrosion, and damage Clear drain hole with a piece of wire. Repair (p 3-29) or replace If defective.
4. Check that flat spring (4) retains sight aperture (5) firmly In either position. Replace flat spring (4) If sight aperture is not firm.

3-11. UPPER RECEIVER ASSEMBLY AND REAR SIGHT ASSEMBLY (CONT).

b. INSPECTION (CONT)

5. Check elevation Index 161 and windage knob (71 for legibility of markings Check underside of windage knobs for cracks. Detent indexing surfaces should be well formed.

6. Check rear sight assembly base (81 for serviceability Clear drain holes for springs Threaded portion of rear sight assembly base and elevation knob should be well formed.

7. Inspect rear sight guards for bends, if bent refer to the following page for repair procedures.

8. Inspect all parts for damage and wear Replace all defective parts.

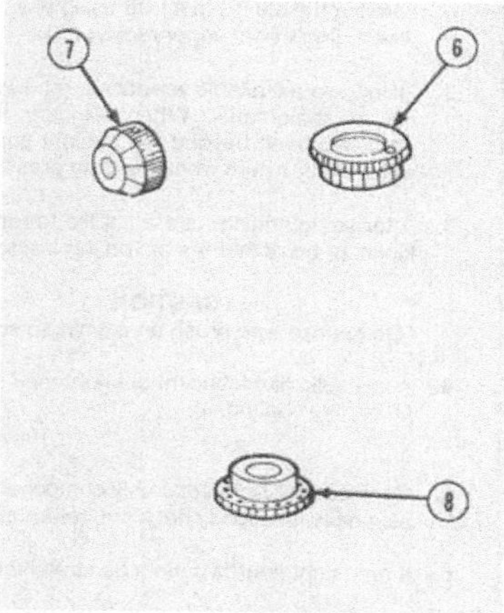

c. REPAIR

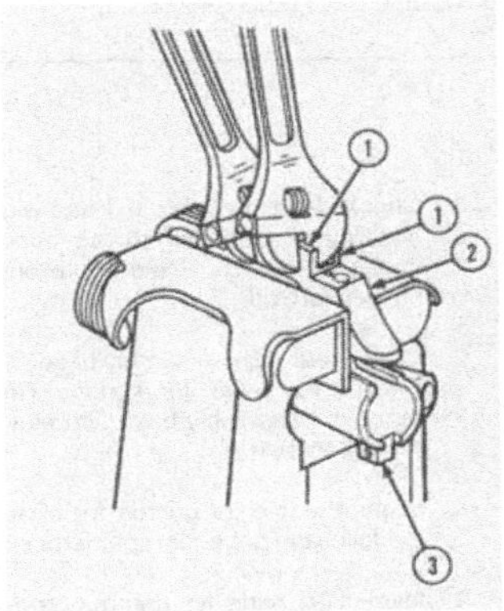

1. To straighten bent rear sight assembly guards (1), remove rear sight assembly components. Place carrying handle (2) In a vise using jaw clamps. Tighten vise to firmly hold upper receiver (3).

2. Using two adjustable wrenches, gradually bend guards (1) to straighten. When bending the guards (1), gradually bend beyond the straight point as the guard will partially return when bending pressure is stopped.

3. After straightening, use a flat file to remove any nicks, kinks, or burrs that remain on the inside of guards (1).

CAUTION

Do not use wire brush on aluminum surfaces

4. Apply solid film lubricant to brightened area for final protective coating.

5. Replace rear sight assembly components and check that rear sight assembly functions properly. If rear sight assembly functions check out, return upper receiver assembly to service.

6. If rear sight guards cannot be straightened utilizing the above procedures, replace the upper receiver

d. LUBRICATION

Lubricate upper receiver assembly and rear sight assembly. Apply CLP to helical springs and ball bearings (three each) and threaded portion of screws before installation. Lubricate helical springs and ball bearings through their respective drain holes.

3-11. UPPER RECEIVER ASSEMBLY AND REAR SIGHT ASSEMBLY (CONT).

e. REASSEMBLY

WARNING
To avoid injury to your eyes, use care when removing and Installing spring-loaded parts.

1. Apply CLP to helical spring (1) and forward assist assembly (2) and Install them into upper receiver (3).

2. Install spring pin (4) using 3, 32 Inch drive pin punch and hand hammer

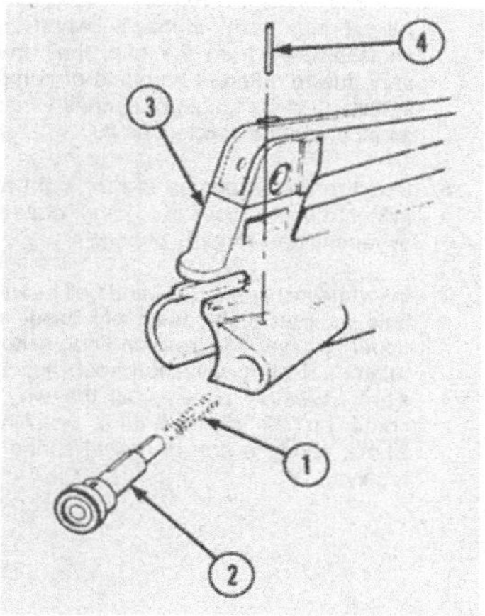

3. Assemble elevation knob (5), elevation index (6), and new index screw (7) using 1 16 Inch alien wrench. Do not overtighten Index screw as scale will require adjustment

4. Install ball bearing (8) and helical spring (9) using needle nose pliers or tweezers

5. Depress ball bearing (8) with a punch Inserted through access hole, and slide elevation knob assembly In upper receiver (3) from the side Center elevation knob assembly

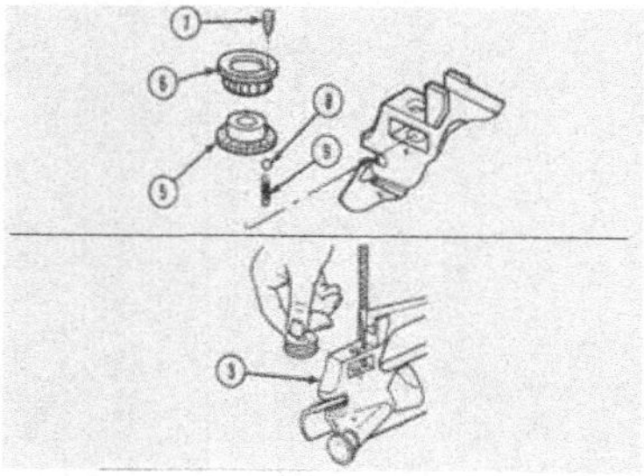

WARNING

To avoid injury to your eyes, use care when removing and installing spring-loaded parts.

NOTE

All springs are identical when new. Once disassembled from the rifle, their free length may vary due to different amounts of compression when installed. If the length of springs varies, use longest as #22, and the shortest as #9

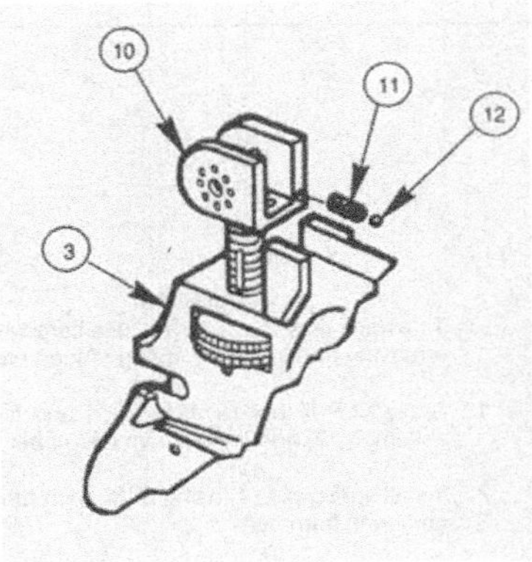

6. Insert threaded portion of rear sight assembly base (10) into upper receiver (3) and rotate elevation knob assembly until threads engage.

7. Insert helical spring (11) and ball bearing (12) in their hole as rear sight assembly base is lowered into upper receiver as elevation knob assembly is further rotated. Rotate elevation knob assembly until rear sight assembly base is all the way down. Then come up 22 clicks before installing spring pin. Check spring action of helical spring (11) on upper receiver.

3-11. UPPER RECEIVER ASSEMBLY AND REAR SIGHT ASSEMBLY (CONT).

e. REASSEMBLY (CONT)

CAUTION

Ensure that spring pin (1 3) passes over helical spring (14), not through its coils

8. Insert helical spring (14) through underside of upper receiver (15). Compress helical spring with a small tip screwdriver (16) to install spring pin (1 3) Spring pin must pass over helical spring, not through its coils Rotate elevation knob (17) until rear sight base (18) Is all the way down.

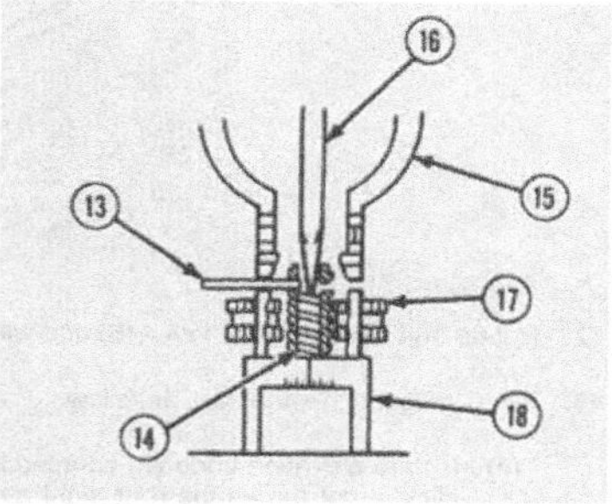

9. Install flat spring (19) and sight aperture (20) In rear sight assembly base (10). Install windage screw (21) with screwdriver.

10. Insert helical spring (22) and ball bearing 123) In windage knob (24)

NOTE

Tilt upper receiver toward wind age knob during positioning to prevent loss of ball bearing.

11. Position windage knob on shaft of windage screw (211 Align holes In windage knob with hole of shaft In windage screw and hold In alignment with a small screwdriver or punch Install spring pin (25).

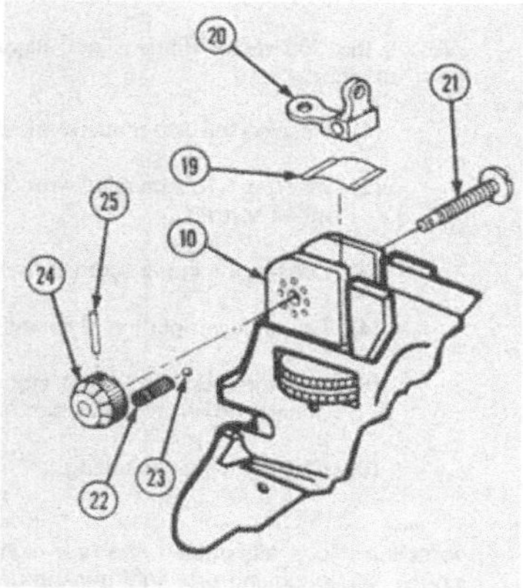

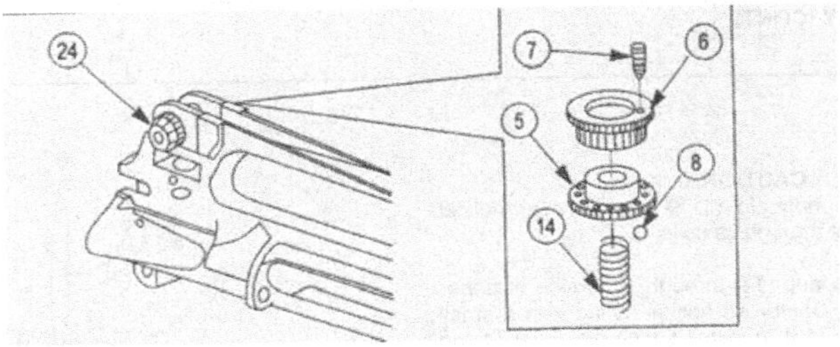

12. Rotate and test elevation index (6) and windage knob (24) for ease of functioning.

13. Inspect elevation knob zero as follows:

(a) Rotate elevation knob (5) counterclockwise until the rear sight assembly is all the way down. If a whole click is not felt as the rear sight assembly stops, the rear sight assembly has bottomed out and will not pivot freely.

(b) Position elevation knob back slightly to its last whole click so the rear sight assembly base is under tension of the ball bearing (8) and helical spring (14). The 300 meter mark should align with the mark on to receiver.

(c) If the 300 meter mark is not aligned with the mark on receiver, slip the range scale in the following manner:

(1) Position the 300 meter mark with the mark on the receiver.

(2) Insert a 1/16 inch allen wrench through the access hole of the rear sight assembly base and into the index screw (7).

(3) Loosen the index screw three turns and leave the wrench in place.

(4) Rotate lower portion of elevation knob

(5) counterclockwise until it stops (range scale should not have moved). Elevation knob should be positioned on its last whole click. (5) Tighten index screw (7) and remove wrench.

(6) Check for proper setting.

NOTE

After the rifle is assembled, the rear sight is centered, and placed at the 300 meter mark, perform the owing check: While looking at a light background, obtain good sight alignment. If the hole in the rear sight aperture appears oval Instead of round, the rear sight base or upper receiver should be replaced. To determine which part requires replacement, replace the rear sight base first. If this does not resolve the problem, replace the upper receiver.

14. Assemble rifle, refer to page 3-77

3-11. UPPER RECEIVER ASSEMBLY AND REAR SIGHT ASSEMBLY (CONT).

f. MEACHANICAL ZERO PROCEDURES (A.F. ONLY)

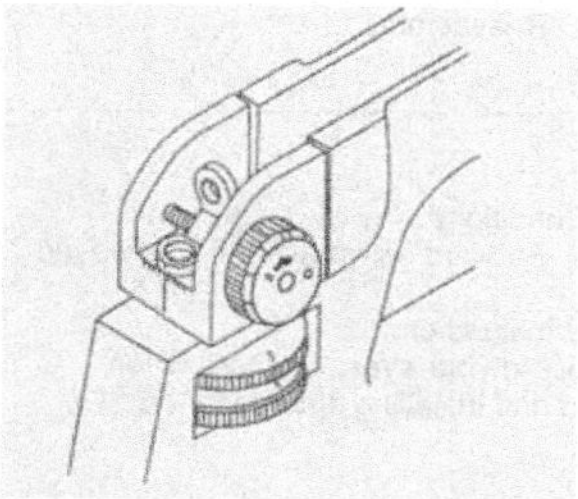

1. Center rear sight by moving windage knob in the appropriate direction.

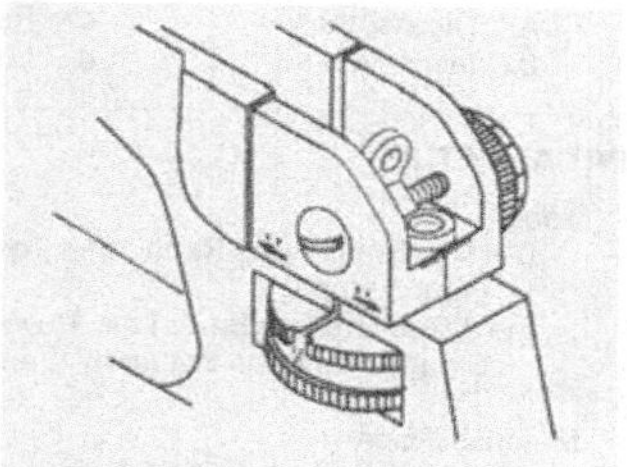

2. Always push in on windage screw head after making rear sight adjustments.

3. Visually check rear sight to ensure it is centered after making adjustments. Also, ensure the rear sight is set in the shortrange (unmarked aperture) position.

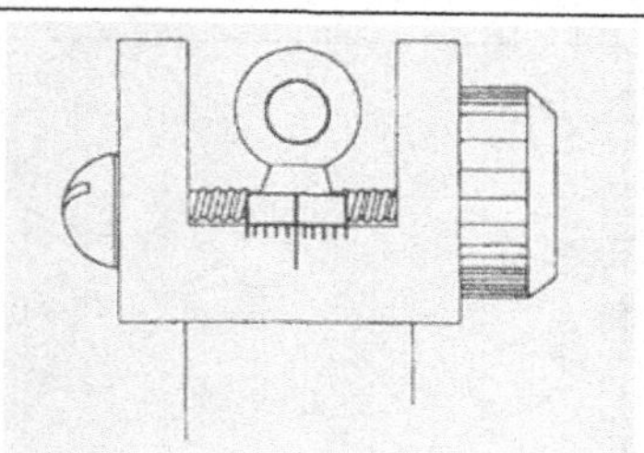

NOTE

This procedure, when used in conjunction with front sight mechanical zero adjustment (p 2-54), will give an approximate battle site zero to most M16A2 rifles. The above steps can also be used before firing a new or newly assigned rifle. Use the procedure to check rifles stored in preferred packaging during routine inspections. This will help ensure people armed with the rifles will stand a better chance of hitting an enemy if the rifles must be used before a live fire zero can be made. Whenever possible, zeroing of the rifle should be accomplished using ball ammunition on a 25 meter zeroing target using the "L" aperture.

After the rifle is assembled, the rear sight is centered, and placed at the 300 meter mark, perform the following check: While looking at a light background sight through the rear sight and obtain good sight alignment with the front sight placed in the center of the rear sight aperture. If while performing this check the hole in the rear sight aperture appears oval instead of round, the rear sight base or upper receiver should be replaced. To determine which part requires replacement, replace the rear sight base first. If this does not resolve the problem, replace the upper receiver.

3-12. FORWARD ASSIST ASSEMBLY.

This task covers:

- **a.** Disassembly
- **b.** Inspection
- **c.** Repair
- **d.** Lubrication
- **e.** Reassembly

INITIAL SETUP:

Tools
(ARMY) Small Arms Repairman Tool Kit (item 3, app B)
Field Maintenance Basic Less Power Small Arms Shop Set (item 1, app B)

Materials/Parts
Cleaner, lubricant, and preservative (CLP) (item 9, app D) | a

Equipment Conditions
3-48 Forward assist assembly removed

General Safety Instructions
To avoid injury to your eyes, use care when removing and installing spring-loaded parts

a. DISASSEMBLY

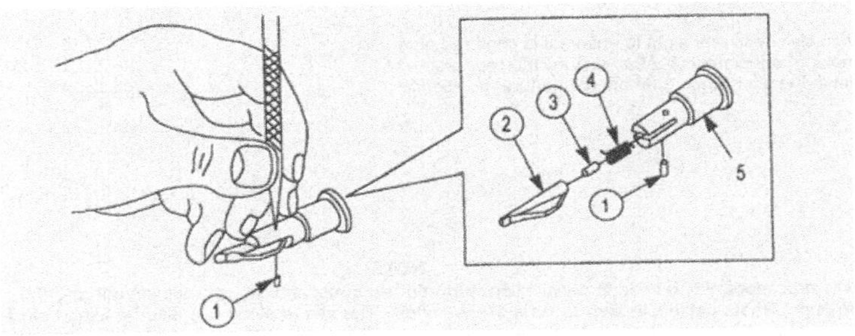

WARNING
To avoid injury to your eyes, use care when removing and installing spring-loaded parts.

1. Drive out spring pin (1) using 1/16 inch drive pin punch and hand hammer
2. Remove forward assist pawl (2), pawl detent (3), and helical spring (4) from plunger assembly (5)

3-12. FORWARD ASSIST ASSEMBLY (CONT)

b. INSPECTION

1. Inspect forward assist pawl (1) for burrs, chips, and cracks. Minor burrs may be removed using fine files or stones, as required. Replace forward assist pawl if defective.

2. Inspect pawl detent (2) for burrs and cracks. Minor burrs may be removed using fine files or stones, as required. Replace pawl detent if defective.

3. Inspect helical spring (3) for kinks, breaks, and wear. Replace helical spring if defective.

4. Inspect plunger assembly (4) for wear, burrs, chips, and breaks. Minor burrs may be removed using fine files or stones, as required. Replace forward assist assembly if defective.

5. Inspect spring pin (5) for wear. Replace if defective.

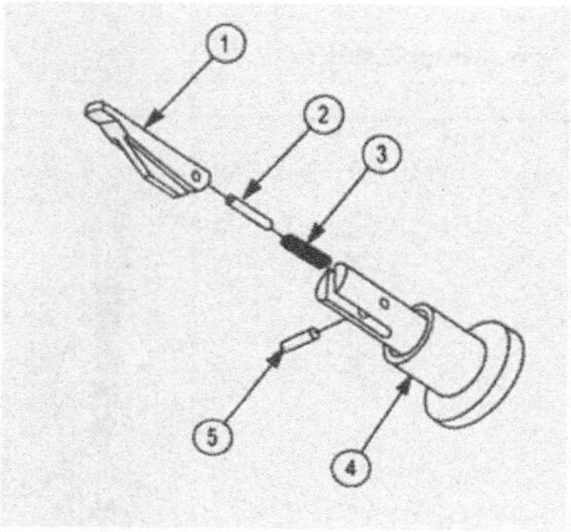

c. REPAIR

1. Repair forward assist pawl using fine files or stones, as required, to smooth burrs. Do not deform forward assist pawl.

2. Repair pawl detent using fine files or stones, as required, to smooth burrs. Do not deform pawl detent.

3. Repair plunger assembly using fine files or stones, as required, to smooth burrs. Do not deform plunger assembly.

d. LUBRICATION

Lubricate helical spring, pawl detent, and forward assist pawl with CLP (p 2-33) before installation.

e. REASSEMBLY

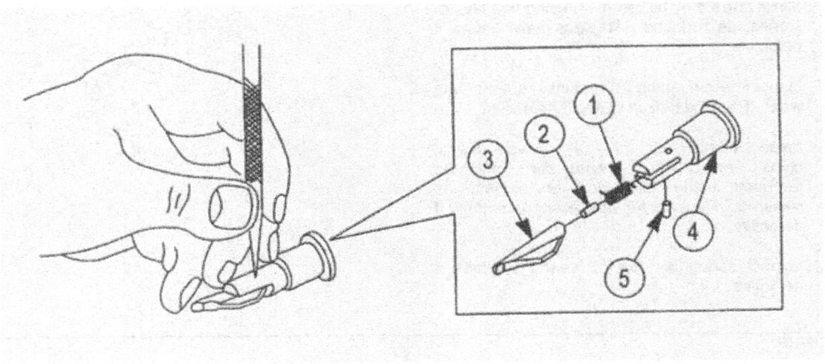

WARNING

To avoid injury to your eyes, use care when removing and installing spring-loaded parts.

1. Install helical spring (1), pawl detent (2), and forward assist pawl (3) into plunger assembly (4).
2. Align holes and install spring pin (5) using 1/16 inch drive pin punch and hand hammer. Spring pin must be flush or slightly below flush after reassembly.
3. Assemble rifle, refer to page 3-77.

3-13. LOWER RECEIVER AND BUTTSTOCK ASSEMBLY.

This task covers:

a. Disassembly
b. Inspection
c. Repair
d. Test
e. Reassembly

INITIAL SETUP

Test Equipment
Tool and Gage Set (item 2, app B)

Tools
(ARMY) Small Arms Repairman Tool Kit (item 3, app B)
Field Maintenance Basic Less Power Small Arms Shop Set (item 1, app B)
Pivot pin removal tool (fig E-3, app E)
Slave pin (fig E-5, app E)

Materials/Parts
Cleaner, lubricant, and preservative (CLP) (item 9, app D)
Dichloromehtane, technical (item 15, app D)
Lubricant, solid film (item 21, app D)
Screw, self-locking (item 6, p C-11)

Equipment Conditions
3-15 Lower receiver and buttstock assembly removed
2-57 Buttstock assembly and pistol grip removed

General Safety Instructions
To avoid injury to your eyes, use care when removing and installing spring-loaded parts.
When using solid film lubricant or dichloro methane, be sure the area is well ventilated.

a. DISASSEMBLY

WARNING
To avoid injury to your eyes, use care when removing and installing spring-loaded parts.

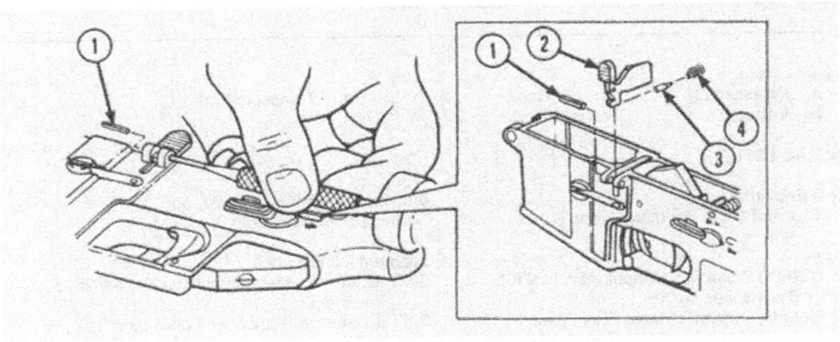

1. Remove spring pin (1) using 3/ 32 inch drive pin punch and hand hammer.

2. Remove bolt catch (2), bolt catch plunger (3), and bolt catch spring (4).

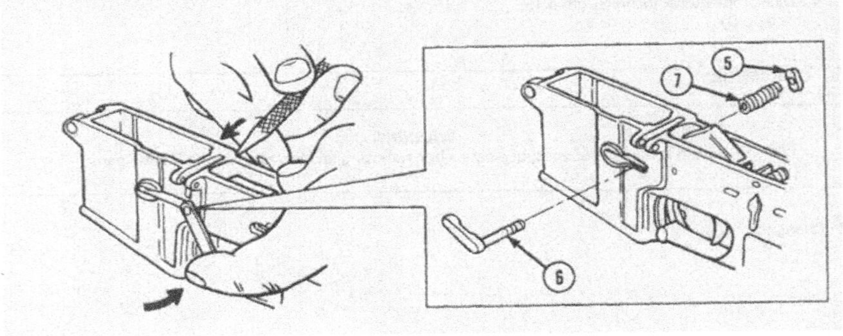

WARNING

To avoid injury to your eyes, use care when removing and installing spring-loaded parts.

3. Using drive pin punch, press in on magazine button (5) and turn magazine catch 16) counterclockwise to unscrew and remove.

4. Remove magazine button (5) and magazine catch spring (7).

3-13. LOWER RECEIVER AND BUTTSTOCK ASSEMBLY (CONT).

a. DISASSEMBLY (CONT)

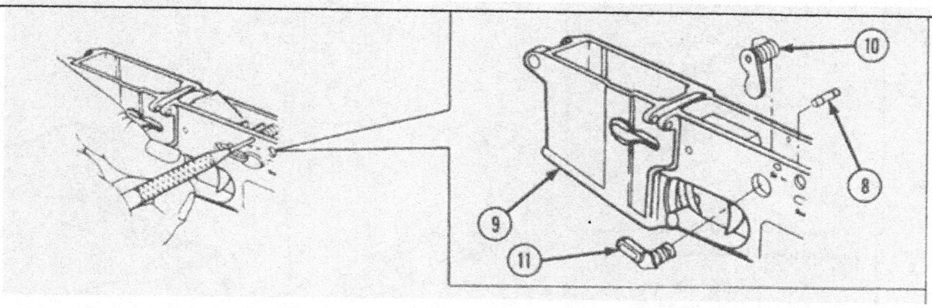

5. Use a rounded tip drive pin punch to push automatic sear pin (8) out of lower receiver (9).

NOTE
To remove automatic sear, selector lever must be positioned to BURST (if installed).

6. Remove automatic sear (10) and selector lever (11).

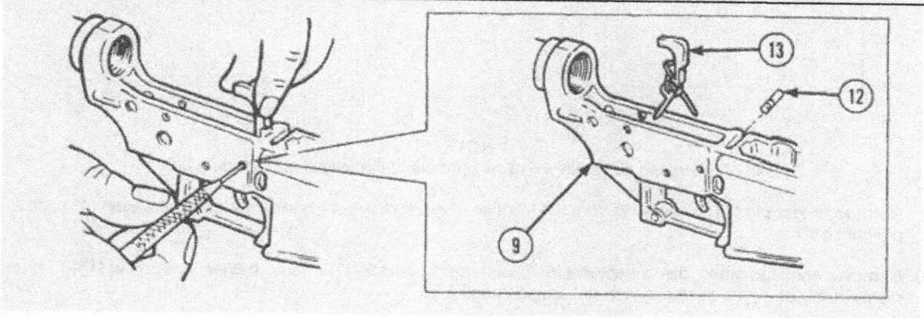

NOTE
To remove (hammer assembly should be forward), place selector lever (if installed) to SEMI position.

7. Use drive pin punch to push hammer pin (12) from lower receiver (9).

8. Remove hammer assembly (13). If further disassembly is required, see page 3-73.

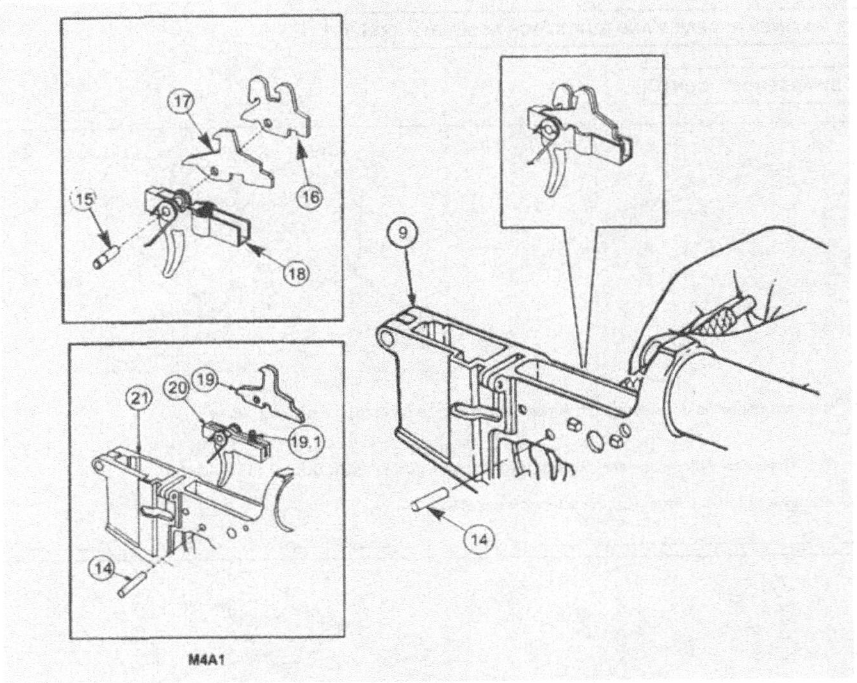

NOTE

Use of fabricated slave pin will allow removal of the following parts as a unit.

9. Remove trigger pin (14) by pushing from the left side of lower receiver (9) with fabricated slave pin (15) and a drive pin punch.

10. Remove semiautomatic disconnector (16), burst disconnector (17), and trigger assembly (18). If further disassembly of trigger assembly is required, see page 3-75.

M4A1 ONLY

11. Remove disconnector (19), disconnector spring (19.1), and trigger assembly (20) from lower receiver and extension subassembly (21).

3-13. LOWER RECEIVER AND BUTTSTOCK ASSEMBLY (CONT).

b. INSPECTION

NOTE
Refer to page 3-35 for repair of corroded surfaces.

Inspect lower receiver and buttstock assembly for corrosion in the lower receiver lobes of the pivot area or hinge pin area. If extensive corrosion appears in these areas, the receiver will not be repaired; turn rifle in for replacement.

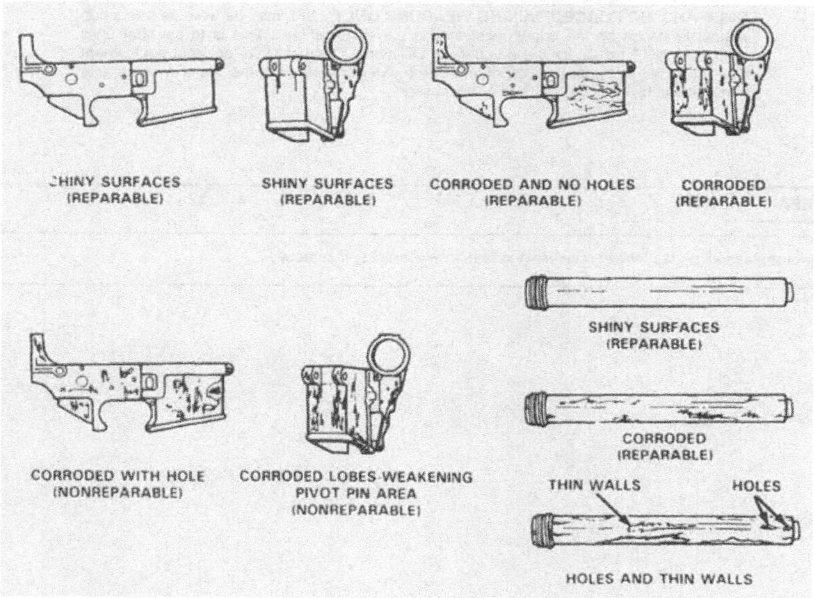

WARNING

When using solid film lubricant or dichloromethane, be sure the area is well ventilated.

CAUTION

Do not use a wire brush on aluminum surfaces.

NOTE

If a weapons lower receiver is missing one third or more of its exterior protective finish, resulting in an unprotected, light reflecting surface, it is candidate for overhaul. This missing finish will be considered a shortcoming. This shortcoming requires action to obtain a replacement weapon. Once a replacement has been received, evacuate the original weapon to depot for overhaul.

Solid Film Lubricant (SFL) is the only authorized touchup for the weapon and may be used on up to one third of the exterior finish of the weapon. **FOR ARMY CONUS USE ONLY AND AIR FORCE TRAINING WEAPONS ONLY:** SFL may be used as a touchup without limitation on the upper receiver and barrel assembly. This is to say that units which DO NOT fall under the category of Divisional Combat Units or rapid deployment type units may have up to 100 percent of the exterior surface of the upper receiver and barrel assembly protected with SFL if necessary.

c. REPAIR

Repair or replace all parts of lower receiver and buttstock assembly if defective.

3-13. LOWER RECEIVER AND BUTTSTOCK ASSEMBLY (CONT).

d. TEST

1. With the upper receiver attached to the lower receiver, and the pivot pin and takedown pins in place, perform the following test:

   (a) Apply hand pressure to push the upper receiver as far to one side as possible.

   (b) Attempt to insert a 0.020 inch thickness gage between the pivot pin lugs of the upper and lower receivers.

   (c) If the thickness gage penetrates to the pivot pin at all accessible locations, repair by replacement of the upper receiver (see (b) below) or replacement of rifle is required.

2. If the rifle fails the above test, remove the upper receiver and install a "NEW" upper receiver and perform the test again.

3. If the rifle now passes the above test, it shall be considered serviceable and continue in use.

4. If the rifle fails the test with a new upper receiver, this failure shall be considered a shortcoming. This shortcoming requires action to obtain a replacement rifle. Once a replacement has been received, evacuate the original rifle to depot for overhaul.

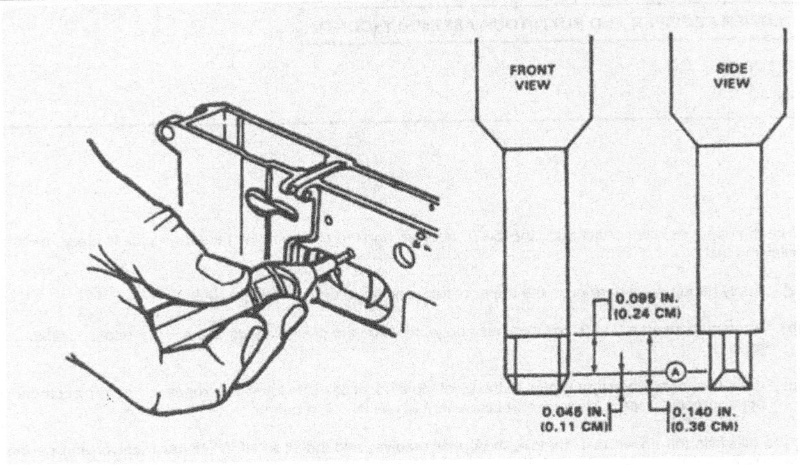

NOTE

If the lower receiver is not disassembled, visually inspect for broken or damaged parts, and to ensure that the hammer and trigger springs are correctly installed before beginning this test. It is not necessary to disassemble the lower receiver for the sole purpose of this visual Inspection. If broken or damaged parts are found, disassemble (p 3-62) and repair as authorized.

5. Test two hammer pin holes and two trigger pin holes using not-go plug gage PN 12006472. This test may be conducted by disassembly of the lower receiver (p 3-62) or by pushing the pin far enough to disengage the end of the pin from the side of the receiver which is being tested. If the lower receiver is not disassembled and the not-go plug gage enters any hole to first shoulder (A), the lower receiver must be disassembled and all four holes must be tested again.

6. Gently insert the not-go plug gage and rotate it 180 degrees. If the not-go plug gage passes through any one of the four pin holes, the rifle is unserviceable and will be turned in for replacement. The gage must extend through the wall thickness to be unserviceable.

7. After completion of gaging operation, visually inspect hammer and trigger springs to ensure proper location of spring legs.

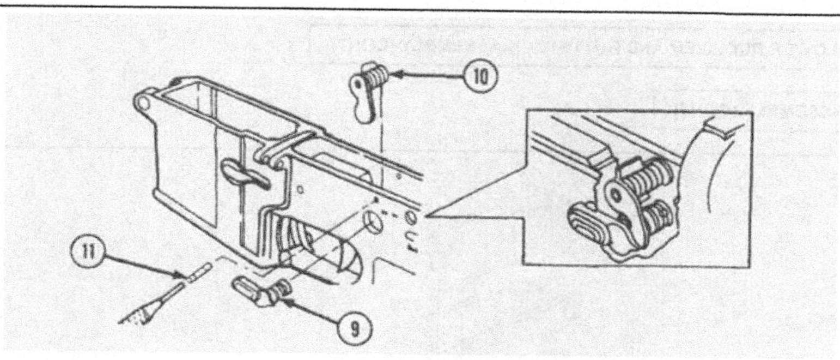

NOTE

Hammer assembly should be cocked prior to installing the selector lever.

Selector lever (9), if installed, must be positioned to BURST. Long leg of automatic sear (10) spring must rest on top of selector lever.

5. Install selector lever (9) and automatic sear (10).

6. Install automatic sear pin (11) (install on the right side) into receiver using drive pin punch. Push in until flush.

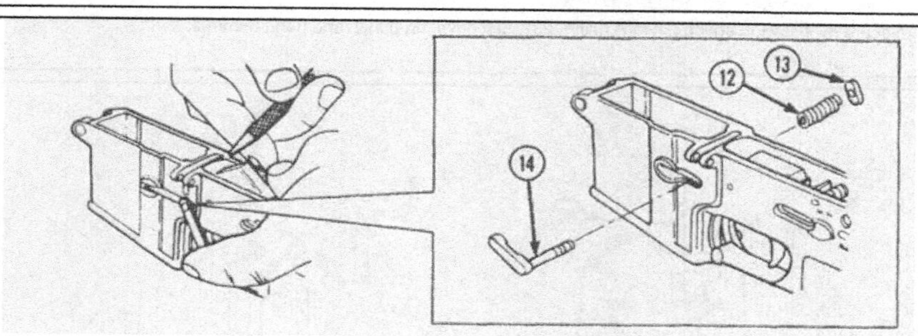

7. Install magazine catch spring (12) and magazine button (13).

NOTE

Drive pin punch should be larger than hole in magazine button.

8. Install magazine catch (14). Push in on magazine button (13) using a drive pin punch and turn magazine catch (14) clockwise until threaded end of magazine catch is flush with magazine button head.

3-13. LOWER RECEIVER AND BUTTSTOCK ASSEMBLY (CONT).

e. REASSEMBLY (CONT)

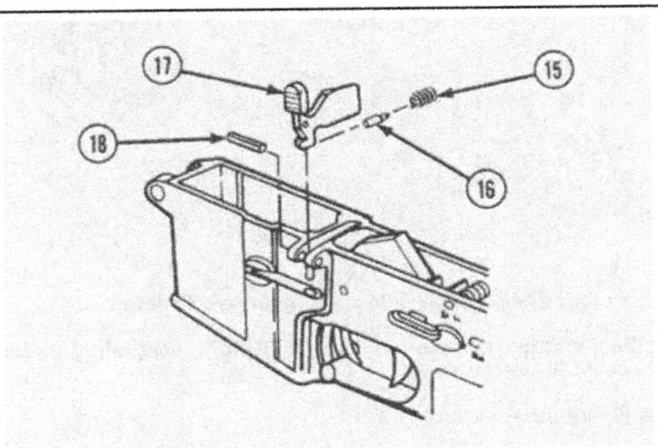

9. Install bolt catch spring (15), bolt catch plunger (16), and bolt catch (17).

10. Secure by installing spring pin (18) using 3/32 inch drive pin punch and hand hammer.

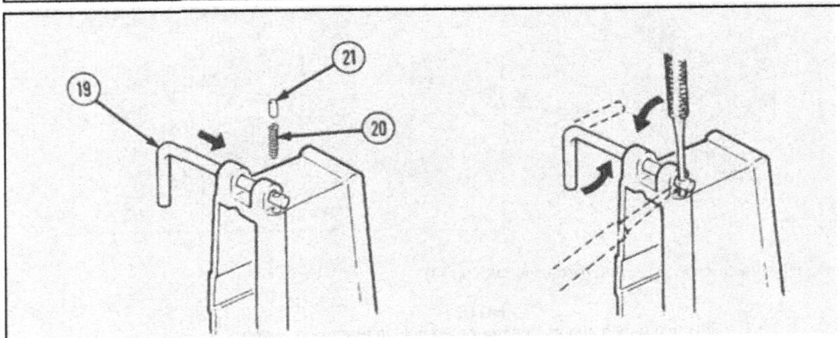

11. Install fabricated pivot pin installation tool (19). Insert helical spring (20) and pivot pin detent 21). Compress pivot pin detent in recess with punch and rotate tool. Remove punch.

WARNING
To avoid injury to your eyes, use care when removing or installing spring-loaded parts.

NOTE
Rounded end of pivot pin detent must be in the groove of the pivot pin (22) when assembly is complete.

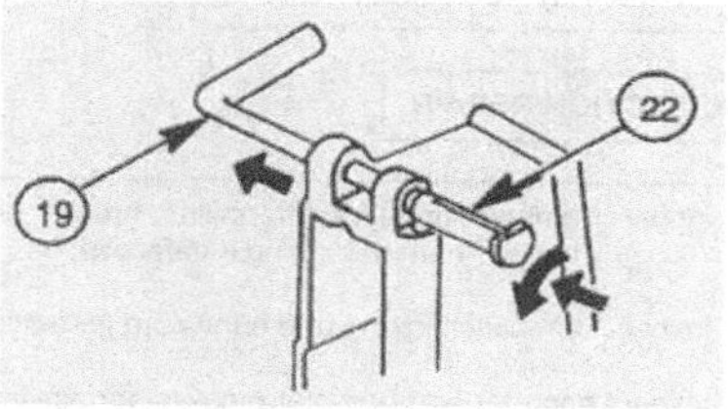

12. Position pivot pin (22) and removing fabricated pivot pin installation tool (19) while maintaining pressure, slide pivot pin (22) into hole. Rotate pivot pin to receive pivot pin detent.

13. Assemble rifle, refer to page 3-82.

3-14. HAMMER ASSEMBLY

This task covers:
- a. Disassembly
- b. Inspection/Repair
- c. Reassembly

INITIAL SETUP

Equipment Conditions
3-62 Hammer assembly removed

a. DISASSEMBLY

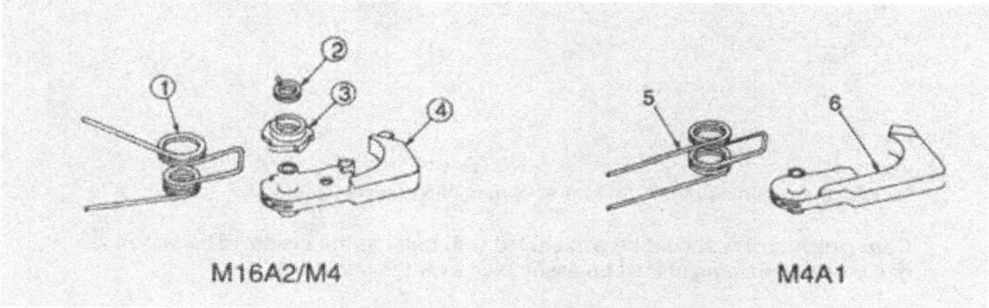

NOTE
M16A2 burst cam is black. M4 burst cam is nickel colored (shiny).

(M16A2/M4) remove hammer spring (1), cam clutch spring (2), and burst cam (3) from hammer and hammer pin retainer assembly (4). **(M4A1)** remove hammer spring (5) from hammer (6).

3-14. HAMMER ASSEMBLY (CONT).

b. INSPECTION/REPAIR

1. Inspect hammer spring for deformities, breaks, and bends. Pay special attention to the large coil. Replace hammer spring if defective.

2. Inspect cam clutch spring and burst cam for deformities, breaks, and bends; replace if defective.

3. Inspect hammer and hammer pin retainer assembly for chips and breaks. Hammer pin should click home under strong finger pressure. Install hammer pin into hole in hammer to check spring retention of the hammer pin. Replace hammer and hammer pin retainer assembly if defective.

c. REASSEMBLY

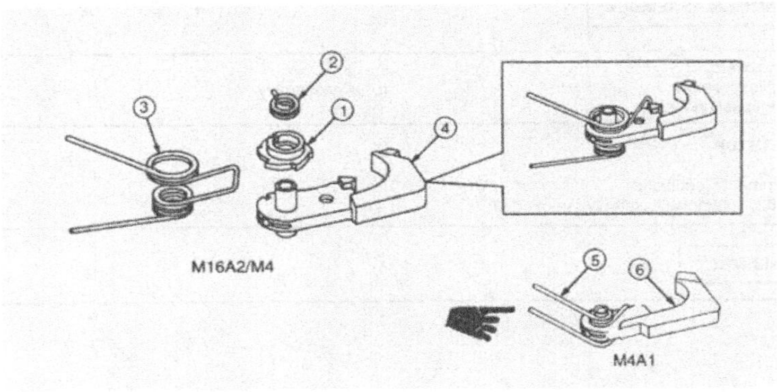

NOTE
M16A2 burst cam is black. M4 burst cam is nickel colored (shiny).

Cam clutch spring should be assembled with bend to the inside. The large loop of the hammer spring should be assembled over the burst cam.

(M16A2/M4) install burst cam (1), cam clutch spring (2), and hammer spring (3) on hammer and hammer pin retainer assembly (4). **(M4A1)** install hammer spring (5) on hammer (6).

3-15. TRIGGER ASSEMBLY AND TRIGGER SUBASSEMBLY (CONT).

b. INSPECTION/REPAIR

1. Inspect trigger spring for kinks, deformities, and weakness. Replace if defective.

2. Inspect disconnector springs for deformities, bends, breaks, and weakness. Replace if defective.

3. Inspect trigger for chips, wear, and cracks. Inspect for damaged searing surface on the trigger nose. Replace if defective.

c. REASSEMBLY

M16A2 AND M4 ONLY

NOTE

Use bolt carrier key tool to install disconnector springs.

1. Use the following procedures to install two disconnector springs (1) using the bolt carrier key tool:

(a) Secure the trigger (2) in a soft vise jaws or similar device.

(b) Place one spring (1) firmly on the tool with large diameter coils outward.

(c) Press spring (1) into recess to solid height.

(d) Hold spring at solid height and slide spring into one of the holes until the punch is flush and perpendicular with the recess wall.

(e) Turn the spring one to two turns opposite of coil winding of the spring.

(f) Discontinue winding when an audible click or snap is heard or felt. This indicates that the spring is seated.

(g) Hold the spring (1) in place when removing the tool to avoid unseating or damaging the spring.

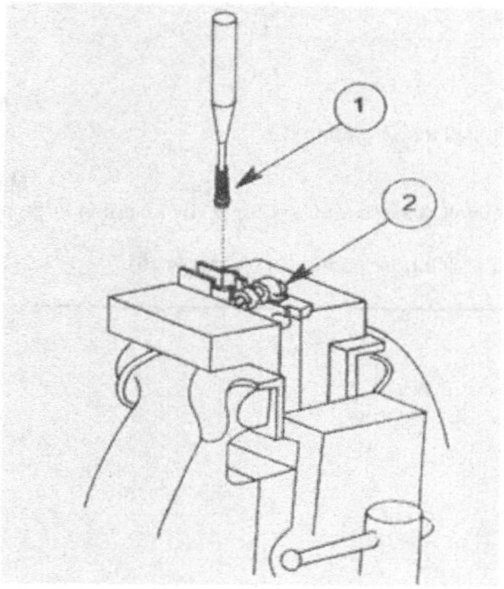

M4 CARBINE ONLY

NOTE

The semi and burst disconnector springs are not the same. The Semi disconnector spring (left side) (figure C-15, item 1B) is black while the burst disconnector spring (right side) (figure C-15, item 1A) is nickel (shiny). Ensure that the correct spring is installed on each side for proper functioning.

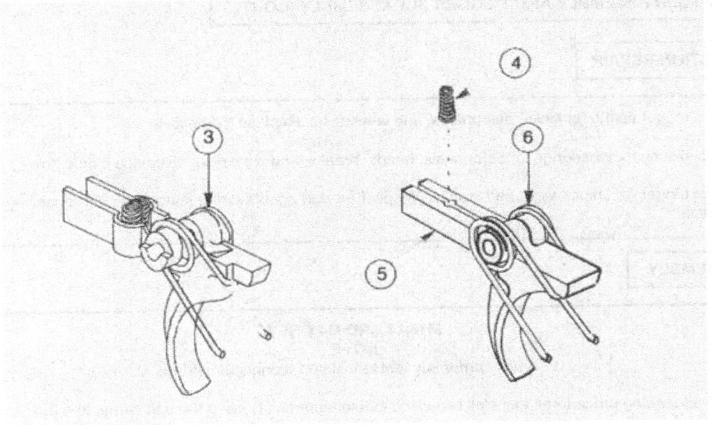

M16A2/M4 M4A1

M16A2 and M4

2. Install trigger spring (3).

M4A1 ONLY

3. Install disconnector spring (4) by inserting large end of spring into trigger (5).

4. Install trigger spring (6) on trigger (5).

3-16. LOWER RECEIVER AND RECEIVER EXTENSION ASSEMBLY.

This task covers:

a. Disassembly
b. Inspection
c. Repair/Modify
d. Reassembly

INITIAL SETUP

Tool Equipment
Tool and Gage Set (item 2, app B)

Tools
(ARMY) Small Arms Repairman Tool Kit (item 3, app B)
Field Maintenance Basic Less Power Small Arms Shop Set (item 1, app B)

Materials/Parts
Cloth, abrasive (item 13, app D)
Grease, molybdenum disulfide (item 19, app D)
Lubricant, solid film (item 21, app D)

Equipment Conditions
3-62 Lower receiver and receiver extension assembly removed

General Safety Instructions
To avoid injury to your eyes, use care when removing and installing spring-loaded parts.
When using solid film lubricant or dichloromethane, be sure the area is well ventilated.

a. DISASSEMBLY

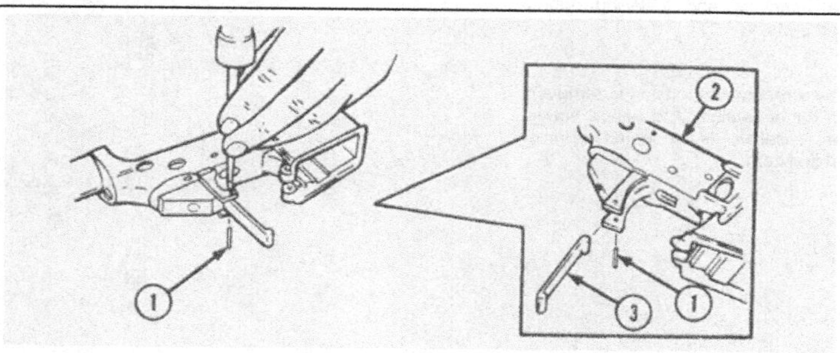

1. Remove spring pin (1) from lower receiver (2) using 1/8 inch drive pin punch and hand hammer.

2. Remove trigger guard (3).

3-16. LOWER RECEIVER AND RECEIVER EXTENSION ASSEMBLY (CONT).

a. DISASSEMBLY (CONT)

3. Use padding between lower receiver and brass vise laws. Grip the solid portion of the lower receiver with brass vise jaws which conform to the shape of the lower receiver in this area.

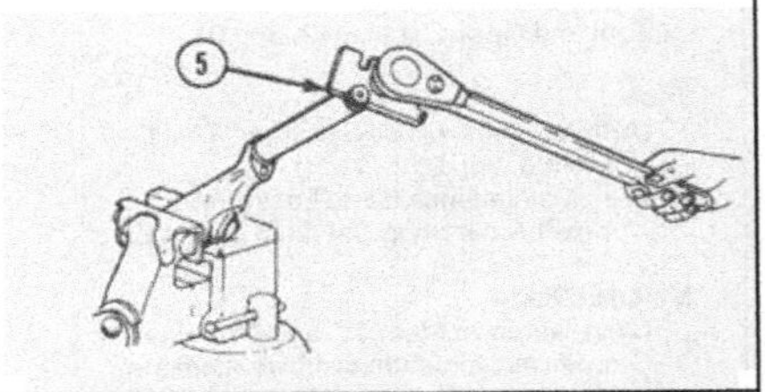

RIFLE ONLY

4. Clamp lower receiver (2) in a machinist's vise using vise jaw caps and tighten on solid portion just tight enough to hold.

WARNING

To avoid injury to your eyes, use care when removing and installing spring-loaded parts.

NOTE

As lower receiver extension is removed, catch buffer retainer and helical spring. Lower receiver is a serial number controlled item.

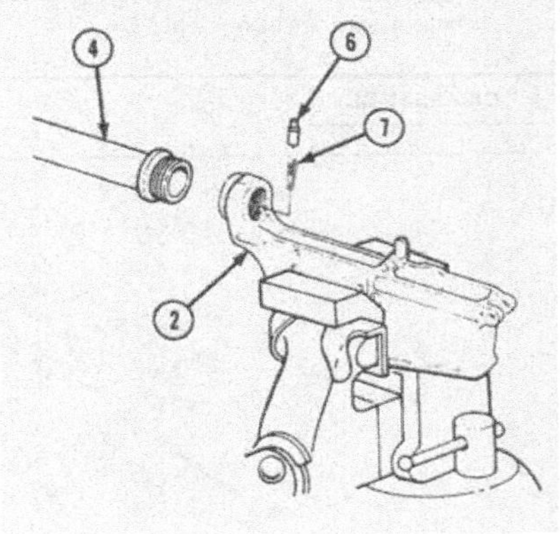

5. Remove lower receiver extension (4) from lower receiver (2) using combination wrench (5) and socket wrench handle. Catch buffer retainer (6) and helical spring (7).

CARBINE ONLY

6. Clamp lower receiver (8) in vise and tighten on solid portion just tight enough to hold.

NOTE
Use wooden vise jaws in place of brass vise jaw caps.

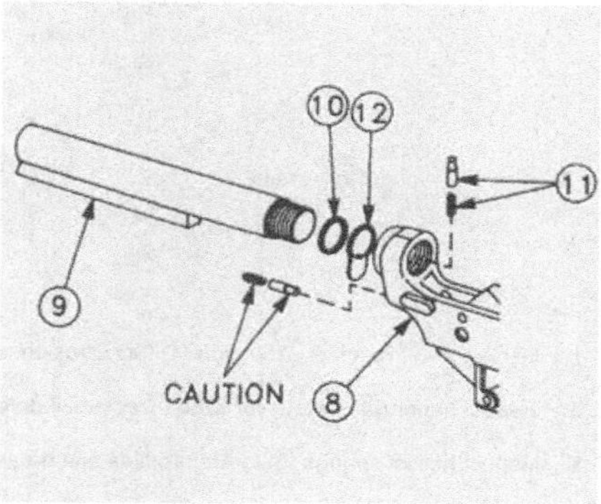

7. Remove the lower receiver extension (9), by loosening the extension locking nut (10) using the special tool (item 12, app C). Catch buffer retainer and spring (11).

CAUTION
While performing the following step, care should be taken to restrain the pivot pin spring and detent.

8. Loosen the locking nut (10) to allow the receiver end plate (12) to disengage from the lower receiver. Hold the buffer retainer and spring (11) in position with your index finger and unscrew the lower receiver extension (9) from the lower receiver (8).

b. INSPECTION

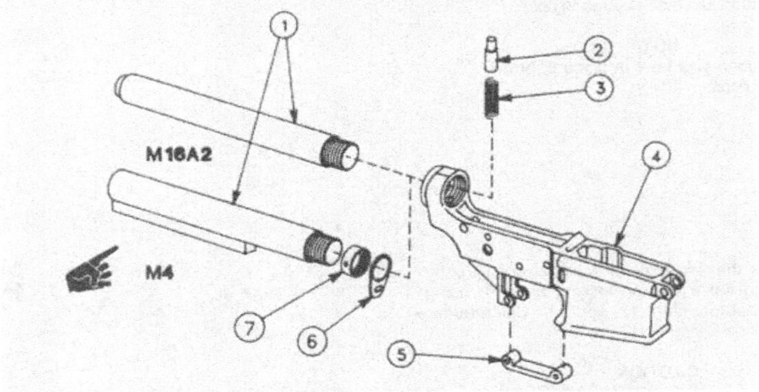

1. Inspect lower receiver extension (1) for corrosion, dents, and wear. Repair or replace if defective.
2. Inspect buffer retainer (2) for wear. Replace if defective.
3. Inspect helical springs (3) for deformities and breaks. Replace if defective.
4. Inspect lower receiver (4). See page 3-66, INSPECTION.
5. Inspect trigger guard (5) for deformities and check operation of plunger and spring. Replace trigger guard if defective.
6. **CARBINE ONLY:** Inspect receiver end plate (6) and locking nut (7) for damage. Replace if damaged.

c. REPAIR/MODIFY

1. Repair lower receiver extension by using abrasive cloth to remove light corrosion. Retouch using solid film lubricant.

NOTE

AIR FORCE ONLY: Only depot maintenance is authorized to restamp the serial number on rifle.

ARMY ONLY: Only direct support level is authorized to restamp serial number.

3-16. LOWER RECEIVER AND RECEIVER EXTENSION ASSEMBLY (CONT).

c. REPAIR/MODIFY (CONT)

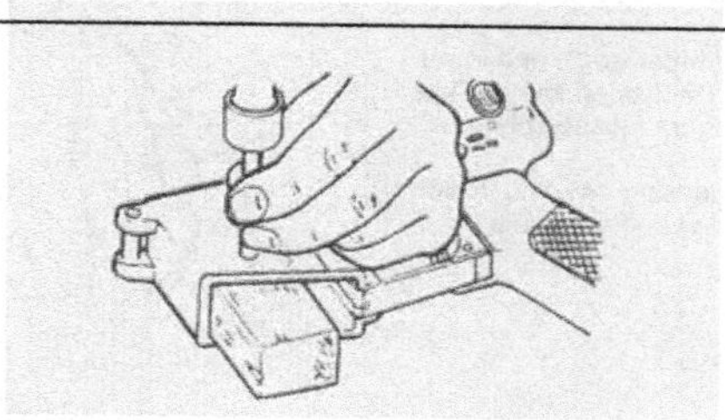

2. **ARMY ONLY:** If serial number is hard to read on rifle, restamp as follows:

   (a) Support the receiver In the stamping area to prevent bending and distortion of the receiver.

   (b) Exercise extreme care to restamp the same serial number as the original.

   (c) Restamp the serial number the same size as the original serial number.

NOTE

Most rifle serial numbers are 1/8 inch (0.31 cm) in height, or close enough that this size is acceptable for such restamping. In the event that a rifle has a serial number that cannot be reproduced by the use of the die sets contained In the Set D Field Maintenance Post, Camp, and Station Small Arms Shop Set, local purchase of an appropriate size die set is authorized.

d. REASSEMBLY

RIFLE ONLY

WARNING

To avoid injury to your eyes, use care when removing and installing spring-loaded parts.

1. Install helical spring (1) and buffer retainer (2) into lower receiver (3).

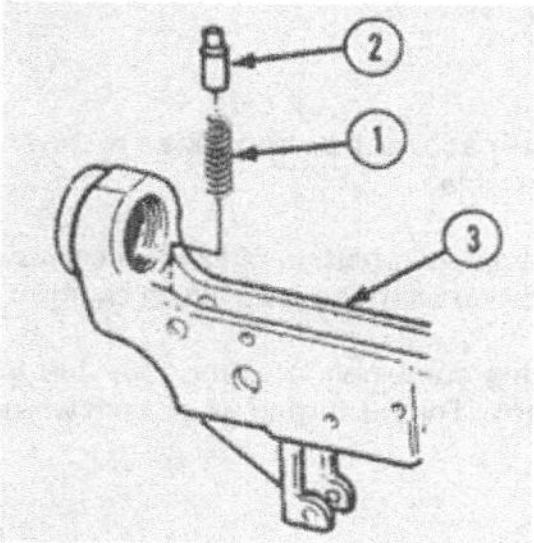

RIFLE ONLY

2. Lubricate threads of lower receiver (3) and lower receiver extension (4) with molybdenum disulfide grease (item 19, app D) before reassembly.

3. Install lower receiver extension (4) into lower receiver (3) while depressing buffer retainer.

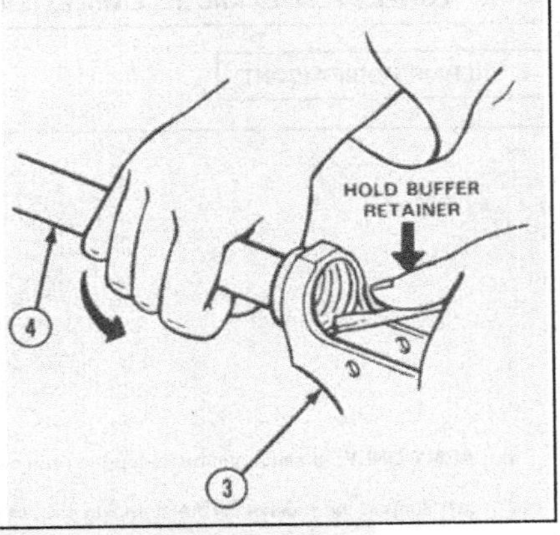

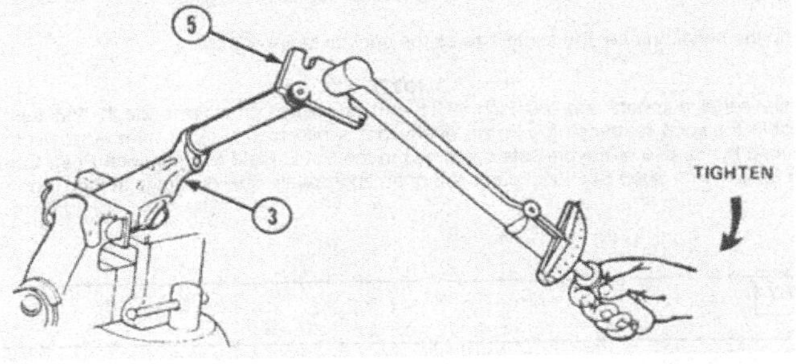

RIFLE ONLY

4. Use padding between lower receiver and brass vise jaws. Use vise jaws in vise and brass vise jaw caps, i available.

5. Clamp solid portion of lower receiver (3) in a machinist's vise using vise jaws. Grip the solid portion of the lowe receiver with vise jaws which conform to the shape of the lower receiver in this area.

6. Using combination wrench (5) and torque wrench, torque lower receiver extension to 35-39 ft-lb (47 25 52 65 N•m). Torque is read when both wrenches are used to gether.

3-16. LOWER RECEIVER AND RECEIVER EXTENSION ASSEMBLY (CONT).

d. REASSEMBLY (CONT)

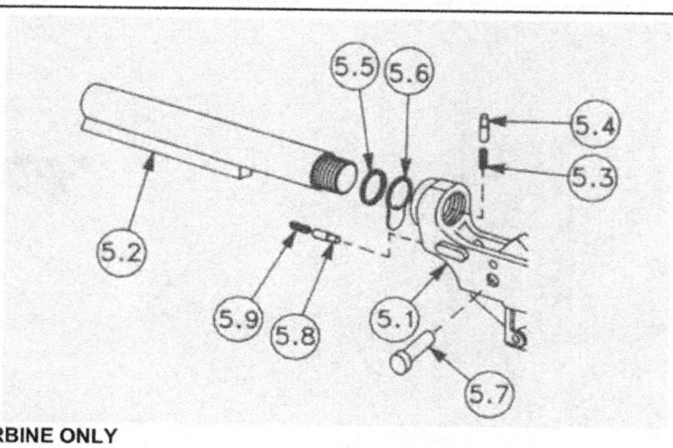

CARBINE ONLY

6A. Lubricate threads of lower receiver (5.1) and lower receiver extension (5.2) with molybdenum disulfide grease (item 19, app D) before reassembly.

6B. Pre-position the spring (5.3) and buffer retainer (5.4) into the retaining hole of the lower receiver (5.1). Screw the locking nut (5.5) onto the lower receiver extension (5.1) with the three notches on the locking ring (5.5) facing forward.

6C. Align the receiver end plate (5.6) onto the lower receiver extension (5.2) with the lug of the receiver end plate (5.6) facing forward.

6D. Pre-position the takedown pin (5.7), detent (5 8), and spring (5.9) in lower receiver assembly (5.1).

6E. Push down on the buffer retainer (5.4) and spring (5.3) and at the same time, screw the lower receiver extension (5.2) in until it retains the buffer retainer (5 4) in position.

6F. Align the lug of the receiver end plate (5.6) into the rear of the lower receiver (5.1). Screw the locking nut (5.5) forward until it contacts the receiver end plate (5.6).

6G. Using the special tool (Item 12, app C) tighten the locking nut (5.5) until snug.

6H. Using the special tool (item 12, app C), and torque wrench, torque locking nut (5.5) to 40 ±2 inch pounds.

6J. Stake the receiver end plate (5.6) in 2 places across from the notches in the locking nut (5.5).

3-16. LOWER RECEIVER AND RECEIVER EXTENSION ASSEMBLY (CONT).

d. REASSEMBLY (CONT)

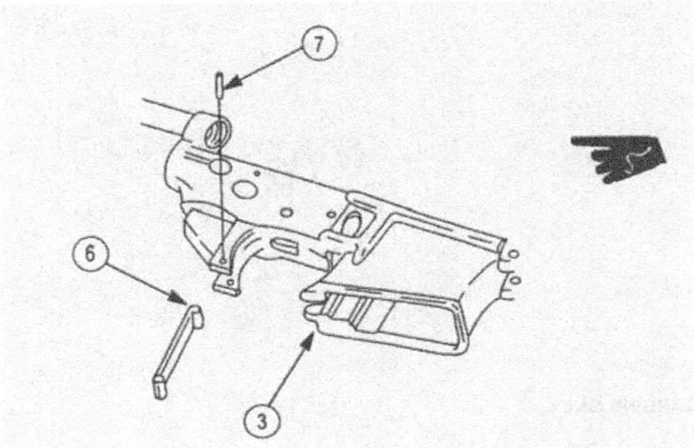

7. Install trigger guard (6) into lower receiver (3).

8. Install spring pin (7) using 1/8 inch drive pin punch and hand hammer.

9. **Deleted.**

3-17. MAJOR COMPONENTS OF M16A2 RIFLE

This task covers:

a. Reassembly b. Test c. Inspection

INITIAL SETUP

Test Equipment
Tool and Gage Set (item 2, app B)

Tools
(ARMY) Small Arms Repairman Tool Kit (item 3, app B)

Reference
TM 9-1005-319-10

General Safety Instructions
To avoid injury to your eyes, use care when removing and installing spring-loaded parts.
Live ammunition must not be near the work area.

a. REASSEMBLY

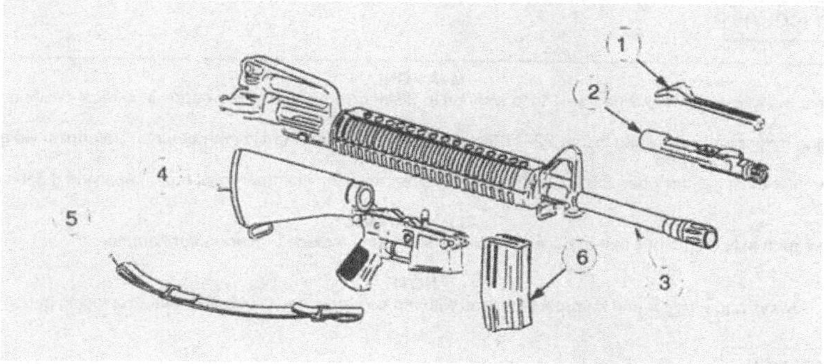

1. Refer to TM 9-1005-319-10.
2. Install charging handle assembly (1) and bolt carrier assembly (2) into upper receiver and barrel assembly (3). Join upper receiver and barrel assembly (3) and lower receiver and buttstock assembly (4).
3. Snap on sling (5) and install magazine (6).

b. TEST

Test trigger pull as follows:

M16A2 and M4 ONLY

(a) Clear the weapon. Place selector to BURST. Pull the trigger and hold it to the rear. Pull the charging handle to the rear and return to the bolt closed position three times. (This will place the BURST disconnector in the deep notch of the BURST cam.)

(b) Release the trigger. Place the selector to SEMI. Hold the rifle in the vertical position. Using trigger pull measuring fixture, 7274758, add weights until hammer trips. Determine weight applied.

(c) Hammer must not trip when 5.5 lbs (2.49 kg) have been applied, hammer must trip on applying 9.5 lbs (4.31 kg).

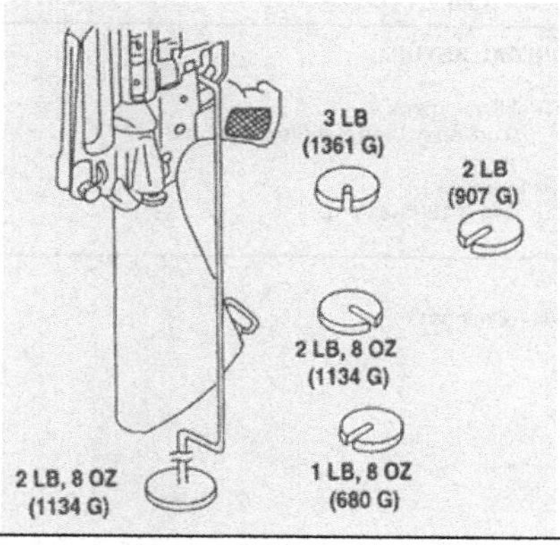

3-17. MAJOR COMPONENTS OF M16A2 RIFLE (CONT)

b. TEST (CONT)

M4A1 ONLY

(d) Clear the weapon. Cock weapon. Place weapon in SEMI position and hold weapon in vertical position.

(e) Using trigger pull measuring fixture 7274758 as shown, add weights until hammer trips. Determine weight applied.

(f) Hammer must not trip when 5.5 lbs (2.49 kg) have been applied. Hammer must trip on applying 8.5 lbs (3.86 kg).

ALL WEAPONS

(g) If weapon fails trigger pull test or excessive creep is present, replace trigger and/or hammer.

NOTE

Always gage trigger and hammer pin holes with not go plug gage 12006472 before replacing parts.

c. INSPECTION

Perform final inspection procedures below.

3-18. M16A2 RIFLE FINAL INSPECTION.

This task covers:

a. Final inspection
b. Test
c. Functional theory of three-round burst control

INITIAL SETUP

Test Equipment
Tool and Gage Set (item 2, app B)

References
TM 9-1005-319-10

General Safety Instructions
Live ammunition must not be near the work area.

Air Force rifles will be inspected in accordance with AFM 36-2227, Vol. 1.

a. FINAL INSPECTION I

1 Visually inspect general appearance of rifle should look almost new. All metal surfaces are to have a dull, rustor corrosion-resistant finish with no burrs or deep scratches.

2 Visually inspect barrel for serviceability. Barrels must be straight, clean, free of rust, powder fouling, and free of bulges and rings. Fine pitting is allowable.

Using moderate hand pressure, check for rotational movement of the front sight in relation to the barrel. Also, using moderate hand pressure, check for rotational movement of the barrel in relation to the upper receiver. If movement between the front sight and the barrel exist, the barrel assembly must be replaced. If movement between the barrel and the upper receiver exist, the barrel assembly must be aligned and tightened (page 3-29).

3 Visually inspect rifle for missing parts. All parts must be attached and all modifications must be applies. Steel parts must be rust free. Spring pins must be secure and screws must be tight.

4 Functionally inspect key and bolt carrier assembly and gas tube alignment. Refer to TM 9-1005-319-10 and use the following procedures:

(a) Disengage the takedown pin and open the receiver.

(b) Remove bolt carrier assembly.

(c) Remove bolt assembly from bolt carrier assembly.

(d) Insert key and bolt carrier assembly into upper receiver and barrel assembly. The bolt assembly must not be installed while performing test.

(e) Slide key and bolt carrier assembly forward to detect binding between key and bolt carrier assembly and gas tube by feel. Badly bent gas tube could cause damage to both the key and bolt carrier assembly and the gas tube. A slightly bent gas tube will cause unnecessary wear of the key and bolt carrier assembly and gas tube.

(f) Correct slight binding by removing handguard assemblies and slightly bending gas tube in the handguard area while repeating step (e) above until no binding is detected. Badly bent gas tubes will be replaced and realigned.

(g) Remove key and bolt carrier assembly from upper receiver and barrel assembly.

(h) Reassemble bolt assembly into key and bolt carrier assembly.

(i) Reinstall key and bolt carrier assembly into upper receiver and barrel assembly.

5 Make a functional check on an assembled weapon with selector lever in SAFE, SEMI, and BURST/AUTO positions Any portion of this check may be used alone to determine the operational condition of any specific firing position selected

3-18 M16A2 RIFLE FINAL INSPECTION (CONT)

a. FINAL INSPECTION (CONT)l

6 Check rear sight assembly as follows:

(a) Rotate elevation knob counterclockwise until the rear sight assembly is all the way down. If a whole click Is not felt as the rear sight assembly stops, the rear sight assembly has bottomed out and will not pivot freely.

(b) Position elevation knob back slightly to its last whole click as the rear sight assembly base is under tension of the ball bearing and helical spring. The 300 meter mark should align with the mark on the receiver (c) If the 300 meter mark is not aligned with the mark on receiver, slip the range scale in the following manner:

(1) Position the 300 meter mark with the mark on the receiver.

(2) Insert a 1/16 inch alien wrench through the access hole of the rear sight assembly base and into the index screw.

(3) Loosen the index screw three turns and leave the wrench in place.

(4) Rotate lower portion of elevation knob counterclockwise until it stops (range scale should not have moved). Elevation knob should be positioned on its last whole click.

(5) Tighten index screw and remove wrench.

(6) Check for proper setting.

7 Pull charging handle assembly to rear. Check that chamber is clear. Leave hammer in cocked position.

8. Place selector lever is SAFE position and pull trigger. Hammer should not fall.

WARNING

If rifle falls any of the following inspections, continued use of the rifle could result In injury to, or death of, personnel.

NOTE

For the purpose of the following tests, "SLOW" is defined as 1/4 to 1/2 the normal rate of trigger release

9. Place selector lever in SEMI position.

10 Pull trigger. Hammer should fall.

11 Hold trigger to the rear, charge rifle and release the trigger with a slow, smooth motion without hesitations or stops, until the trigger is fully forward; an audible click should be heard. Hammer should not fall.

12 Repeat the above selector lever SEMI position test (steps 10 and 11) five times. The rifle must not malfunction during any of these five repetitions. If the rifle malfunctions during any of these five tests, refer to page 3-7.

M16A2 and M4 ONLY

13 Place selector lever in BURST position. Charge rifle and pull trigger. Hammer should fall.

14 While holding the trigger to the rear, pull charging handle assembly to the rear and release it three times. Hammer should not fall the third time the charging handle assembly is released. It may or may not fall the first and second time the charging handle assembly is released. When the burst disconnector reached the deep notch of the burst cam, the burst disconnector should have held the hammer to the rear while the trigger was in the pulled position. Release the trigger with a slow, smooth motion without hesitations or stops, until the trigger is fully forward. The hammer should not fall.

15 Pull trigger. Hammer should fall. This would be the first round of a three-round burst.

NOTE

A detailed explanation of the three-round burst control can be found in the following pages.

M4A1 ONLY

15.1 Place selector lever in AUTO position. Charge weapon and pull trigger. Hammer should fall.

15.2 Hold trigger to the rear, charge weapon, and release trigger. Pull trigger. Hammer should not fall. Automatic assembly sear should have released the hammer as the bolt closed.

ALL WEAPONS

16 With the hammer in the forward position, using moderate finger/thumb pressure, attempt to place the selector lever in the SAFE position. Selector lever should not go in the SAFE position.

17 Perform the following additional functional checks and adjustments on assembled weapon:

(a) Press magazine catch button. Make sure it functions properly.

(b) Press bolt catch. Make certain it operates smoothly and holds bolt in open position.

(c) Inspect front sight and rear sight assembly. Make certain they can be adjusted properly.

(d) Actuate forward assist assembly It must work freely.

3-18 M16A2 RIFLE FINAL INSPECTION (CONT)

a. FINAL INSPECTION (CONT)

(e) Inspect upper receiver and barrel assembly. Barrel assembly should not rotate within upper receiver assembly.

(f) Check that third or middle slot of compensator is straight up (TDC).

b. TEST

1 Check headspace using headspace gage PN 7799734. See page 3-45, TEST.

2 Check firing pin protrusion using firing pin protrusion gage PN 7799735. See page 3-45, TEST.

3 Check extent of barrel erosion using barrel erosion gage PN 8448496. See page 3-45, TEST.

4 Check barrel straightness using barrel straightness gage PN 8448202. See page 3-45.

c. FUNCTIONAL THEORY OF THREE-ROUND BURST CONTROL (M16A2 AND M4 ONLY).

NOTE

First become familiar with the functioning of the firing mechanism especially when in the SAFE and SEMI positions. You should also understand the role that the automatic sear plays when firing in the BURST position. Functioning of the mechanism is explained below In a step by step manner. This actually will seem to complicate something that is really very simple and happens in less than 1 second. The diagrams below and on the following pages do not show the associated springs for the sake of simplicity The positioning of the burst cam is shown in detail

Functional check of three-round burst is as follows:

NOTE

Assume the rifle is fully loaded with a live round in the chamber and the selector lever on BURST.

(a) Hammer is cocked

(b) Front hook of burst lever is in stop notch

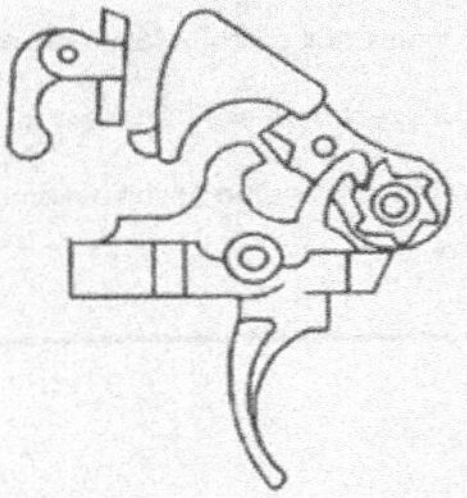

(c) Trigger Is pulled.

(d) Trigger nose drops and hammer falls firing the FIRST ROUND.

(e) Front hook of burst disconnector holds burst cam In place as hammer falls.

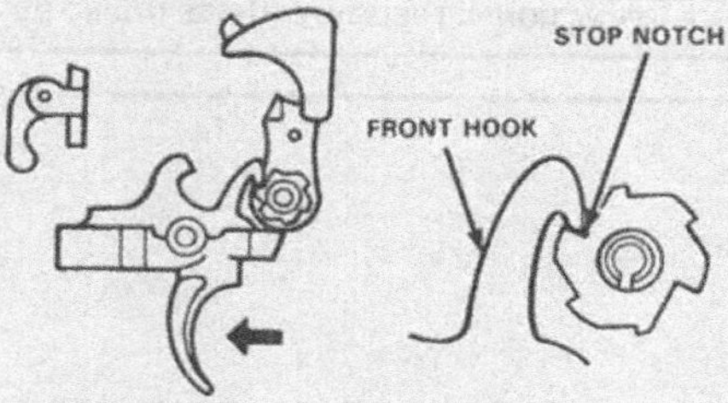

NOTE

Anytime the hammer falls forward, the clutch spring releases the burst cam and allows the front hook of the burst disconnector to keep It In place.

(f) As the key and bolt carrier assembly moves to the rear, the hammer is forced to the rear.

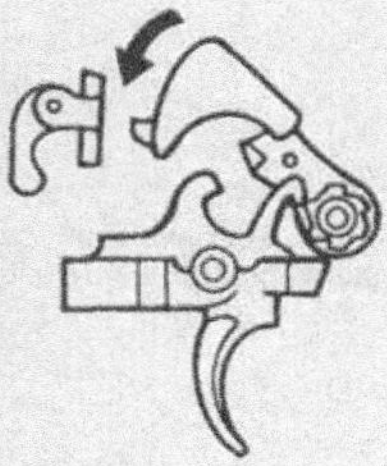

(g) The clutch spring of the burst cam clutches the burst cam and causes It to rotate one notch as the hammer is forced back.

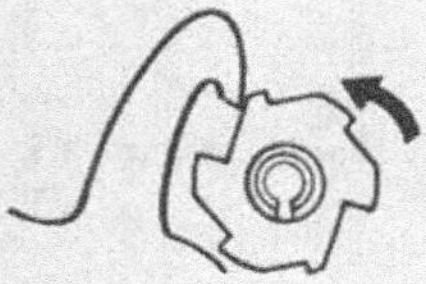

(h) When hammer is fully to the rear, the automatic sear catches It.

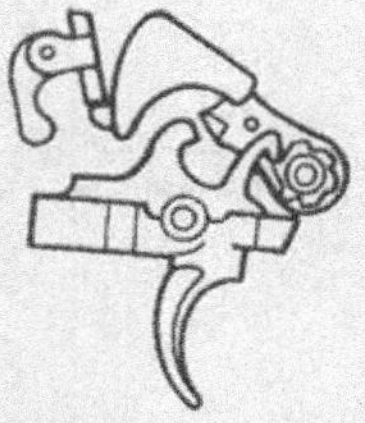

3-18 M16A2 RIFLE FINAL INSPECTION (CONT).

c. FUNCTIONAL THEORY OF THREE-ROUND BURST CONTROL (CONT)

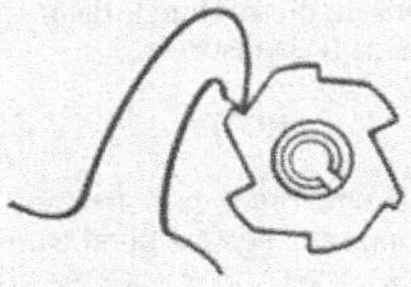

(i) The front hook of the burst disconnector is now fully in the second notch.

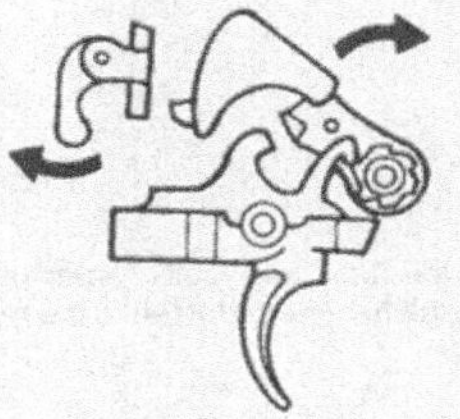

(j) As the key and bolt carrier assembly travels forward, the automatic sear releases the hammer and the hammer falls.

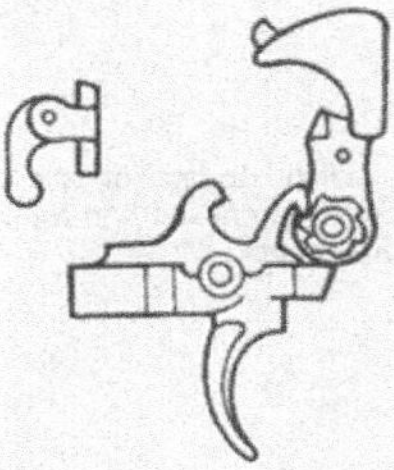

(k) When the hammer falls, the SECOND ROUND is fired.

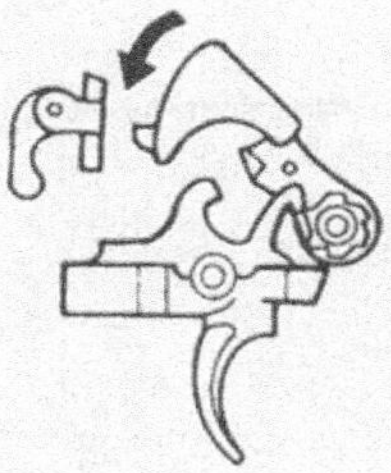

(l) As the key and bolt carrier assembly moves to the rear, the hammer is forced back to the rear.

(m) The clutch spring of the burst cam clutches against the burst cam and causes It to rotate one notch as the haiimmer is forced back

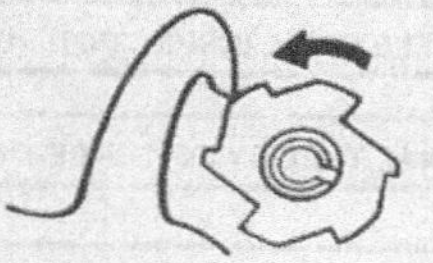

(n) When the hammer is fully to the rear, the automatic sear catches It

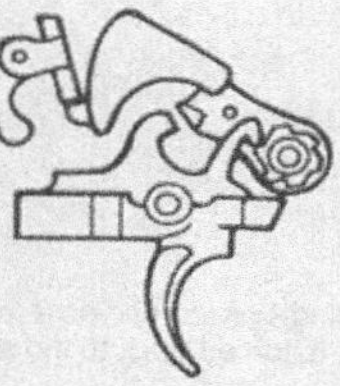

(o) The front hook of the burst disconnector IS now fully In the third notch

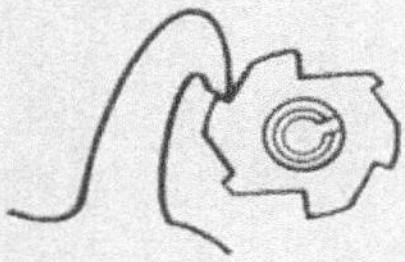

(p) As the key and bolt carrier assembly) travels forward, the automatic sear releases the hammer and the hammer falls

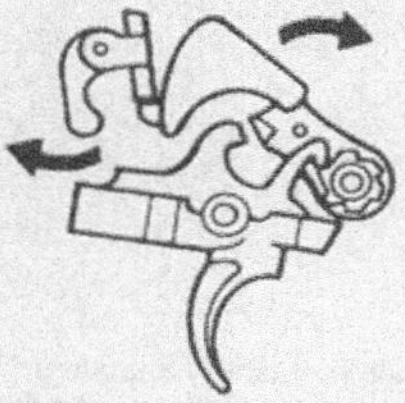

3-18. M16A2 RIFLE FINAL INSPECTION (CONT).

c. FUNCTIONAL THEORY OF THREE-ROUND BURST CONTROL (CONT)

(q) When the hammer falls, the THIRD ROUND Is fired

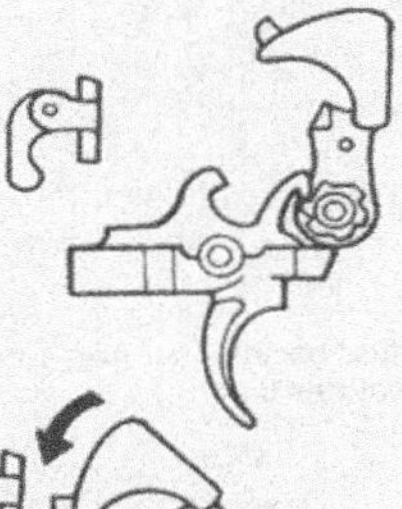

(r) As the key and bolt carrier assembly moves to the rear, the hammer is forced back to the rear

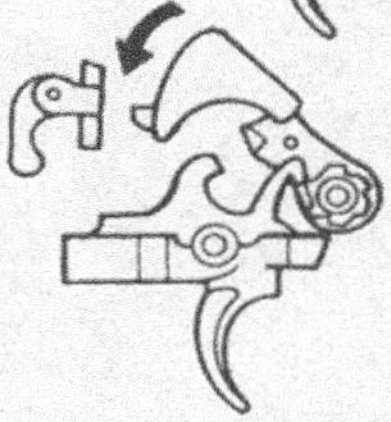

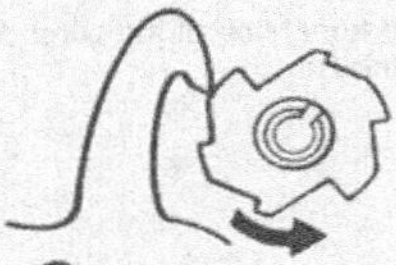

(s) The clutch spring of the burst cam clutches against the burst cam and causes it to rotate one notch as the hammer is forced back.

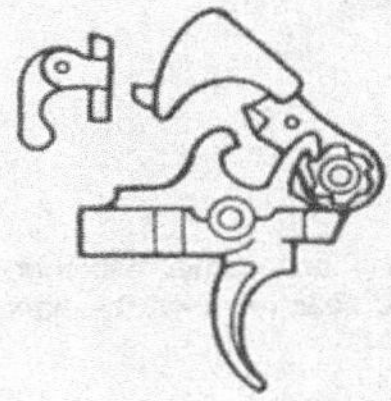

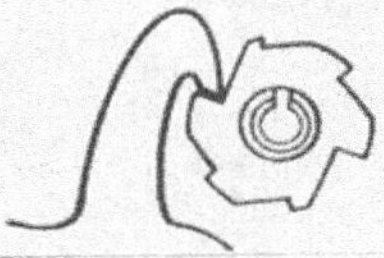

(t) When the hammer Is fully to the rear, It Is initially caught the, ;automatic sear However, the front hook of the first disconnector Is now fully in the next stop notch which is deeper than the others

(u) Because a stop notch is deeper than the others, It allows the front hook of the burst disconnector further forward than before. This allows the rear hook of the burst disconnector to latch on the rear hammer notch. This holds the hammer fully to the rear even though the trigger is still to the rear. This happens when the burst Is over and the firing Is stopped.

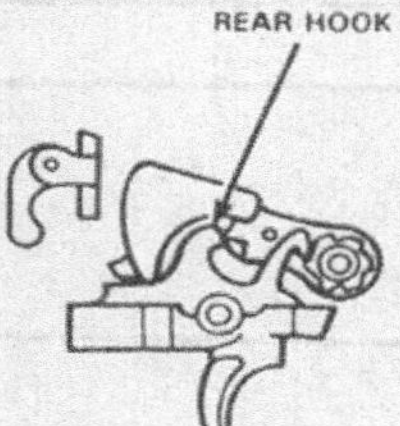

(v) Once the trigger is released, the trigger nose comes up and holds the hammer back.

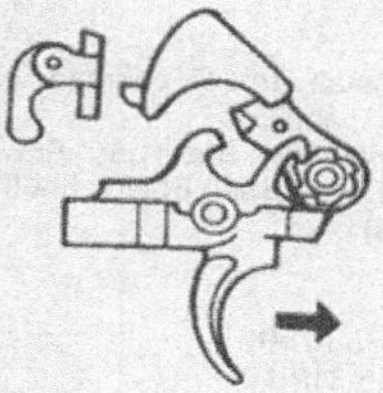

NOTE

Pulling the trigger to the rear and holding it back again will fire another three round burst This will continue until the magazine is empty However the trigger must be released between each burst

3-19 MI 6A2 RIFLE AND M4/M4A1 CARBINE ANNUAL GAGING REQUIREMENTS.

This task covers.

a Inspection

b Gaging

INITIAL SETUP

Test Equipment
Tool and Gage Set (item 2, app B)

Tools
(ARMY) Small Arms Repairman Tool Kit (item 3, app B)
Field Maintenance Basic Less Power Small Arms Shop Equipment (item 1, app B)

References
DA PAM 738-750
TM 9-1005-319-10
AFR 50-36
AFTO Form 105

General Safety Instructions
- To avoid injury to your eyes, use care when removing and installing spring-loaded parts
- Initial gaging is required 1 year from receipt of the weapons.
- All rifles and carbines must be gaged at least once annually for safety.
- All Army Reserve and Army National Guard M16A2 rifles must be inspected and gaged at least once every 2 years, after the initial inspection/gaging procedures have been accomplished. This 2 year interval may be maintained unless preventive maintenance checks and services (PMCS), or other physical evidence, indicates that an individual unit's rifles require inspection/ gaging at a more frequent interval. If it is determined that a yearly inspection is necessary for an individual unit, only that unit will be affected. This will not affect other units in regard to the interval of inspection.
- It is recommended that training units inspect/ gage all rifles at the end of each training cycle. Training units will inspect/gage all rifles at least once annually.
- Air Force M16 rifles will be inspected in accordance with AFR 50-36, Vol 1, Chapter 5.

a. INSPECTION

1 Visually inspect general appearance of weapon Overall appearance will be approximately that of a new weapon. For inspection criteria refer to final inspection, page 3-84 All visual and functional Inspection requirements must be met.

NOTE

To perform the following tests, disassemble weapon only as far as allowed in chapter 2 (Unit Maintenance Instructions). unless a deficiency is uncovered

2 Perform a general inspection of weapon per section III of this chapter Repair as required and authorized

b. GAGING

1	Gage bolt carrier assembly for firing pin protrusion using firing pin protrusion gage PN 7799735. See page 3-20, TEST.
2.	Gage bolt carrier assembly for firing pin hole wear using not-go plug gage PN 12620101. See page 3-24, TEST.
3	Inspect chamber in upper receiver and barrel assembly using chamber reflector tool PN 8448201 See page 3-33, INSPECTION/CLEANING.
4	Gage barrel in upper receiver and barrel assembly using barrel erosion gage PN 8448496 and bore straightness gage PN 8448202. See page 3-45, TEST.
5	Check headspace in upper receiver and barrel assembly by inserting headspace gage PN 7799734 in chamber. See page 3-45, TEST.
6	Gage pivot pin lug area clearance in lower receiver assembly using 0.020 thickness gage. See page 3-68.
7	Gage hammer and trigger pin holes in lower receiver assembly using taper plug gage PN 12006472. See page 3-68, TEST.
8	Gage trigger pull using trigger pull measuring fixture PN 7274758. See page 3-83, TEST.
9	Document inspection with DA PAM 738-750, AFTO Form 105, or NAVMC 11003 when completed.

Section IV. PREEMBARKATION INSPECTION OF MATERIEL IN UNITS ALERTED FOR OVERSEAS MOVEMENT

NOTE

This section refers to all the weapons (MI 16A2 rifle, M4, and M4A1) unless otherwise stated.

3-20 PURPOSE: This section establishes standards for overseas shipment (pre-embarkation in section criteria) for all weapons. These standards are provided to ensure that the user is furnished equipment which will perform its mission without early failure or major maintenance problems.

3-21 SCOPE.

a. The standards prescribed provide for a high percentage of remaining life in affected rifles; therefore, rifles designated for overseas shipment must qualify under the standards contained in the following paragraph, table, and in referenced DA publications, before they can be approved for shipping action.

b. Provisions of this standard apply to all US Army agencies/activities selecting or preparing rifles for shipment to US troops overseas. It also applies to CONUS troops preparing rifles for shipment overseas. Provisions do not apply either to rifles being prepared for shipment to MAP/MAS recipients unless specifically prescribed by MAP/MAS transaction for the materiel or to rifle being returned to CONUS from overseas. The maintenance instructions and standards contained herein do not apply to rifles once the material has arrived at the overseas destination. At that time, maintenance instructions contained in the applicable TM's will be used.

c. This applies to rifles which are the logistic responsibility of the US Army Armament, Munitions and Chemical Command.

3-22. GENERAL.

a. Only rifles which have been classified as serviceable condition code A, B, or C under AR 725-50 will be considered for overseas shipment. All items of equipment for which equipment serviceability criteria have been published must, as a minimum, be rated green under the ESC as a prerequisite to overseas shipment. In addition to the condition code standard, as enumerated above, and the required ESC rating prescribed herein, the rifle being considered for overseas shipment must meet the requirements of this section. The ESC will be discontinued as new operator manuals are revised which will be used to determine serviceability condition of rifle.

b. Waivers to provisions can only be granted by the gaining command of any particular end Item being considered for issue, deployment, or shipment. The issuing services may recommend issue or shipment of rifles not meeting the provisions when all the following conditions exist:

(1) Repair parts in required quantities cannot be obtained from the supply system prior to delivery of the end item.

(2) The gaining command concurs in the receipt of the end item for storage until required repair parts become available. The gaining command must also state that capability, facilities, and funds are available to perform the necessary work when parts become available.

(3) Department of the Army approval is obtained on a case-by-case basis.

(4) Required repair parts are requisitioned by the issuing command for delivery to the Framing command.

c. All Department of the Army MWO's applicable to the specific rifle being considered for shipment overseas must have been applied.

d. Refer to SB 746-1 for pertinent publications relating to equipment processing and marking information

e. Refer to AMC-P 310-9 for publications containing applicable overhaul standards.

3-23. SHIPMENT OR ISSUE.

a. **Organizational Repair Parts, Tools, and Equipment**. Rifles must be complete with all Items required by applicable Department of the Army publications, including those In the basic Issue items list of the appropriate operator's manual.

b. **Publications**. Operator publications applicable to the equipment log book must accompany the equipment. All log book entries must be complete and up-to-date including those covering any repairs, replacements, or adjustments made to the rifle In complying with this section.

c. **Documentation**. Prepare DA Form 2408-9 (Equipment Control Record) at time of overseas shipment or Issue to another stock record or property book account, in accordance with the provisions of DA PAM 738-750.

d. **Preparation**. Process rifles for shipment as required by shipping documents and pertinent regulations.

3-24. DISPOSITION. Disqualified rifles which do not qualify for shipment will either be redistributed within the camp, post, or station, be repaired, or become candidates for overhaul, cannibalization, or other disposition as required by existing regulations.

3-25. GENERAL INSPECTION CRITERIA.

WARNING

Before starting an Inspection, be sure to clear the rifle Do not actuate the trigger until the rifle has been cleared. Inspect the bore and chamber to ensure that It is empty and free from obstructions, and check to see that no ammunition Is In position to be Introduced.

a. Before Inspection, the materiel must be thoroughly cleaned of all grease, dirt, or other foreign matter that might Interfere with Its proper function or the use of gages and tools during inspection.

b. Materiel must be free of burrs, rust, or corrosion on functional surfaces.

3-25. GENERAL INSPECTION CRITERIA (CONT).

c. Parts must not be cracked, bent, distorted, or damaged and must be free of detrimental wear or looseness

d. Minor defects In metal components do not normally affect their acceptability For example, scratches and tool marks are ordinarily of no Importance

e. Inspect finish of metal surface

(1) General. Satisfactory metal surfaces for rifles range from black to light gray. A worn shiny metal surface is objectionable only when It is capable of reflecting light No rifle will be reelected unless exterior parts have a shine All rear sights must have a dull gray or black finish of all surfaces that would cause a glare

(2) M16A2 Rifle. Minor loss of finish (shiny spots, nicks, scratches) on exterior surfaces of the barrel and flash suppressor shall not be cause for rejection of M16A2 rifles located In hands of troops at training centers Large shiny surfaces, nicks, scratches, etc., can be restored by the use of solid film lubricant (Item 21, app D) Rifles (small arms) missing in excess of one-third or more of the exterior finish resulting In an unprotected, light reflecting surface, are considered candidates for overhaul The only authorized level of maintenance to phosphate finish small arms Is depot.

f. Plastic components must not be cracked or damaged In such a way as to Interfere with their structural strength Surface cracks, bruises, or dents that do not affect their strength will not be cause for rejection. Cracks will be cause for rejection. Criteria for determining which cracks are repairable are on page 2-1 4.

g. Barrels must be clean and free of corrosion such as that caused by moisture and powder fouling Standards of serviceability are indicated In (1) through (9).

(1) Pits In the chamber are allowable If they do not cause extraction difficulties

(2) Pits as wide as a land and 3'8 Inch (O 95 cm) or less in length are allowable for 5 56-mm barrels Pits not greater than the width of a land and less than 3 8-inch (O 95 cm) long are permissible

(3) Scattered or uniformly fine pits, or fine pits In a densely pitted area are allowable

(4) Tool marks are acceptable regardless of length They will appear as lines running laterally In the grooves, or may run spirally across the top of lands

(5) Ringed bores or bores ringed sufficiently to bulge the outside surface of the barrel are cause for rejection. However, faint rings or shadowy depressions do not indicate an unserviceable barrel and will not be cause for rejection. Gap in lined barrels will not be classified as a ringed bore.

(6) Lands that appear dark due to coating of gilding metal from projectiles will not be cause for rejection.

(7) Breech bore diameter will be checked on unlined barrels using the appropriate breech bore gage.

(8) Barrel erosion gages are provided for lined barrels. Bore wear will be checked using barrel erosion gage for the M16A2. For detailed instructions in the use of the above gage and for serviceability limits, refer to page 3-46.

(9) Flaking or checking (fine cracks) of chromium plate in barrels or chambers will not be cause for rejection, unless accompanied by pitting to the degree that extraction difficulties are encountered or accuracy is unacceptable.

h. Springs must be free of distortion and broken coils. Springs must have sufficient tension to perform their intended function.

i. Screw heads must be in serviceable condition and threads must not be stripped. Internal threads must not be stripped.

j. The sear, hammer, and/or cocking notches must be in good condition. Chipped engaging corners will be cause for rejection. Slight wear on functional surfaces, including engaging corners, shall be acceptable, providing the minimum trigger pull requirements and selector lever checks are met in accordance with instructions on page 3-83.

k. Chips, flat spots, or bent striker points on firing pins will be cause for rejection.

l. The cartridge engaging surfaces on extractors must not be chipped or deformed.

m. Evidence of any damage to sights will be cause for a sight alignment check. Rear sight bases should have no movement.

n. Rear sight elevating and windage mechanisms must operate with distinct clicks, without binding. Sights must have sufficient tension to retain their setting during firing. Graduations and numerals must be legible. Graduation filler Is not required.

o. Safeties must positively position in both the ON and OFF position. When in the ON or safe position, the rifle must not fire when the trigger is squeezed; when in the OFF or fire position, the rifle must fire when the trigger is squeezed.

3-25. GENERAL INSPECTION CRITERIA (CONT).

p. All locking devices such as latches, magazine latches, or detents must be positive in action and must not become disengaged due to normal handling and firing. Retaining pins and similar devices must not be subject to accidental loss during use or transportation.

q. Each rifle must be hand functioned to check for unusual binding, positive cocking action, and general operation. Dummy ammunition must be used to assure positive feeding, chambering, extraction, and ejection action.

Table 3-1. 5.56MM Rifle M16A2

Item	Standard
RIFLE:	
General	Clear rifle of any ammunition and Inspect In accordance with paragraph 3-25.
Barrel and barrel extension	Check barrel erosion. Use barrel erosion gage 8448496 for chrome lined barrels. Stripping of lands and grooves shall not be cause for rejection unless so determined by barrel erosion gage. Visually inspect, using chamber reflector tool 8448201. Pits 1/8 inch (0.31 cm) In length and those pits large enough to extend from the body of the chamber into the shoulder stop area and forcing cone area are cause for rejection. Large pits are defined as those 1/8 inch (0.31 cm) or more in diameter as determined by visual Inspection. Only closed flash suppressors are acceptable. Check barrel for straightness using bore straightness gage 8448202. Gage must pass freely through the bore to be acceptable, either dropped from the muzzle or chamber end.
Front sight and gas tube	Inspect gas tube for proper alignment with carrier key. Gas tube must not bind when mating with the key. Evidence of gas leaks around the front sight connection of the gas tube shall be cause for rejection until rifle has been function fired to determine if the loss of gas is sufficient to cause malfunction. If function firing malfunctions occur, repairs are necessary. Inspect front sight for damage.

Table 3-1. 5.56MM Rifle M16A2, M4/M4A1 Carbine - Cont

Item	Standard
Bolt carrier group	Inspect bolt for elongated or oversized firing pin hole using plain cylinder gage 12620101. Firing pinholes which permit the plain cylinder plug gage to fully penetrate at any position on the circumference will be rejected. Bolt face with a cluster of pits which are touching or tightly grouped, covering an area measuring approximately 1/8 inch (0.31 cm) across will be rejected. Bolts which contain pits extending into the firing pinhole will not be rejected unless firing pin hole gaging check determines rejection. Bolts which contain individual pits or scattered pits will not be cause for rejection. Only phosphated bolt carriers are acceptable. Both phosphated and chrome plated bolts are acceptable.
Bolt locking lugs and bolt cam pin hole	Inspect for cracks in the locking lugs and cam pin hole area. Use a black light, if available; otherwise, use a glass of no more than 3X magnification or use inspection penetrant (item 25, app D). Use instructions contained in kit for application. If cracks are detected, the bolts will be replaced. **NOTE** Particular attention must be given to the area where the lugs meet the bolt body and around the side walls of the cam pin hole. Bolt rings must not be broken. Ring gaps must be properly spaced approximately 1/3 turn apart and not in line. Firing pin protrusion must be not less than 0.028 inch (0.071 cm) or more than 0.036 inch (0.091 cm). (Use firing pin protrusion gage 7799735.) Socket head capscrews must be staked. Carrier key must not be dented where end mates with gas tube. Repair or replace damaged carrier keys.
Headspace	Inspect headspace using headspace gage 7799734. Excessive headspace will be cause for rejection.
Trigger pull	Inspect trigger pull using trigger measuring fixture 7274758. Trigger pull must be minimum 5.5 pounds (2.49 kg), maximum for M16A2 and M4 is 9.5 pounds (4.31 kg) and maximum for M4A1 is 8.5 pounds (3.86 kg). Test trigger pull, refer to page 3-83, steps a, b, c, and d.

3-25. GENERAL INSPECTION CRITERIA (CONT).

Table 3-1. 5.56MM Rifle M16A2, M4/M4A1 Carbine - Cont

Item	Standard
Lower receiver group	Inspect hammer and trigger pin holes using plain cylinder plug gage 12006472 Penetration of the gage in any one or more of the four holes will be cause for rejection. Inspect for cracks, corrosion, or mutilation which would affect functioning. Small dents or gouges will not be cause for rejection. Inspect receiver for corrosion in the lobes of the pivot or hinge pin area. Width between lobes shall not exceed 0 515 inch (1.30 cm). Inspect receiver for break through of metal. Inspect receiver and receiver extension for initial loss of protective coating.
Action spring	Free length of spring shall be between 11-3/4 and 13-1/2 inches (29.84 and 34.29 cm) **M16A2 ONLY** and 10-1/16 and 11-1/4 inches (25.56 and 28.58 cm) **M4/M4A1 ONLY**.
Handguard	Inspect handguard assembly for breaks, separations, and cracks. Breaks and separations of material which prevent proper retention or interfere with functioning of the weapon will be cause for handguard rejection and replacement. Cracks up to 1 inch (2.54 cm) in length are acceptable provided they do not extend into the handguard retaining flange (1) (critical area). **M16A2 ONLY** Each handguard assembly may have up to two of the three front retaining tabs (2) missing. If all three front tabs are missing, the handguard assembly must be replaced. **M4/M4A1 ONLY** No tabs can be missing. **ALL WEAPONS** Replace severely cracked handguards. Handguards that have a heat shield loose enough to rattle when installed on the weapon must be replaced.
Stock assembly	Inspect buttstock assembly for dents, cracks, and chips. Check for breaks and separation of material which could prevent proper functioning of weapon. M4/M4A1 inspect for proper functioning. Repair as required

Table 3-1. 5.56MM Rifle M16A2, M4/M4A1 Carbine - Cont

Item	Standard

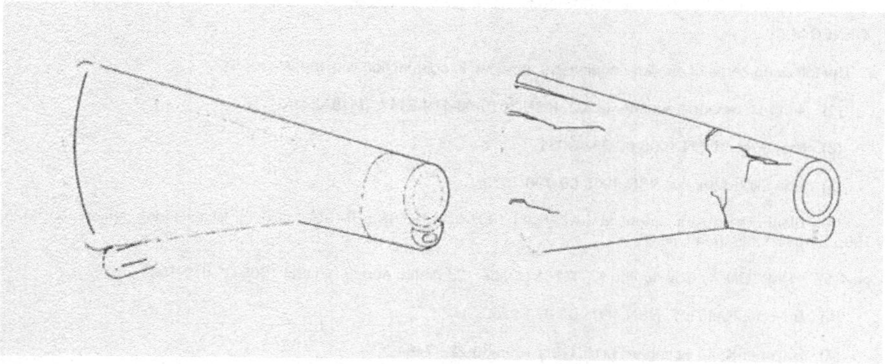

M16A2 ONLY

Under the following conditions, hairline cracks originating from buttplate end of buttstock are acceptable. No chipped away material is allowed.

a. One hairline crack, not to exceed 1 inch (2.54 cm) in length, per side of buttstock.

b. Two additional hairline cracks up to 0.22 inch (0.55 cm) in length, per side of buttstock.

Buttstocks with unauthorized markings stamped into their surfaces will be replaced. Unauthorized markings, scratched, etched, carved, etc., are acceptable if they do not extend into the fiber of the buttstock which may weaken it. These marks may lie at any location on the buttstock.

Cracks in the critical area at the front end of the buttstock are not acceptable and these buttstocks must be replaced.

CHAPTER 4

MAINTENANCE OF AUXILIARY EQUIPMENT

CHAPTER OVERVIEW

This chapter contains information and instructions to keep auxiliary equipment used with your weapon in good repair.

Section I. AUXILIARY EQUIPMENT REPAIR

4-1 GENERAL.

a The following items of auxiliary equipment are used in conjunction with the Weapons:

(1) 40mm Grenade Launcher M203, NSN 1010-00-179-6447, (M16A2 ONLY)

(2) Lock Plate, NSN 1005-00-233-9031.

(3) Top Sling Adapter, NSN 1005-00-406-1570.

(4) Blank Firing Attachment M15A2, NSN 1005-00-118-6192, (M16A2 ONLY), Blank Firing Attachment M23, NSN 1005-01-361-8208 (M4 ONLY)

Change 4**(5)** ARMY ONLY: Conversion Kit, M261 (caliber .22 rimfire adapter), NSN 1005-01-010-1561.

(6) Bayonet-Knife M7, NSN 1005-00-073-9238.

(7) Bayonet-Knife Scabbard M10, NSN 1095-00-223-7164.

(8) M9 Multi-Purpose Bayonet System, NSN 1005-01-227-1739.

(9) Night Vision Sight, Individual Served Weapon, AN/PVS-4, NSN 5855-00-629-5334 (M16A2 ONLY)

(10) M2 Practice Bolt, NSN 1005-01-184-4041 (M16A2 ONLY)

b Refer to TM 9-1010-221-23&P for unit maintenance for the Grenade Launcher M203.

c ARMY ONLY: Refer to TM 9-6920-363-12&P for unit maintenance for the M261 Conversion Kit (caliber rimfire adapter) M16, M16A1, and M16A2 Rifles.

d Refer to TM 9-1005-237-23&P for repair instructions and repair parts for Bayonet-Knife M7 and Bayonet Knife Scabbard M10 and M9 Multi-Purpose Bayonet System.

e Refer to TM 11-5855-213-23&P for maintenance for the Night Vision Sight, Individual Served Weapon, AN/PVS-4.

f Refer to TM 9-6920-746-12&P for unit maintenance for the M2 Practice Bolt.

4-2. LOCK PLATE

This task covers:

a Installation
b Removal
c Inspection

INITIAL SETUP

Tools
(ARMY) Small Arms Repairman Tool Kit (item 3, app B)

WARNING

The lock plate prevents the selector lever from being placed in BURST and will be installed at the discretion of the unit commander. It is mandatory for use in civil disturbance (riot control). Do not keep live ammunition near work area.

a. INSTALLATION

1 Using a screwdriver, reach inside pistol grip (1) and remove screw (2) and lockwasher (3).

WARNING

To avoid injury to your eyes, use care when removing and installing spring-loaded parts.

2 Carefully remove pistol grip (1). Hold detent helical spring (4) in place.

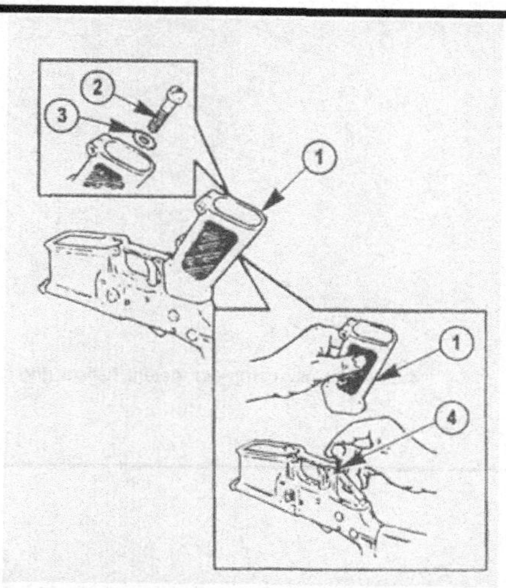

3 Install lock plate (5) with the detent helical spring (4) passing through the hole In the right side of the lock plate and the arm on the outside of the receiver The selector lever just point to the SAFE position

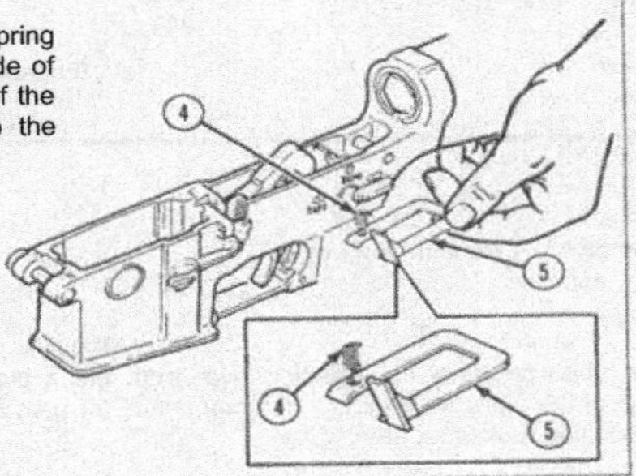

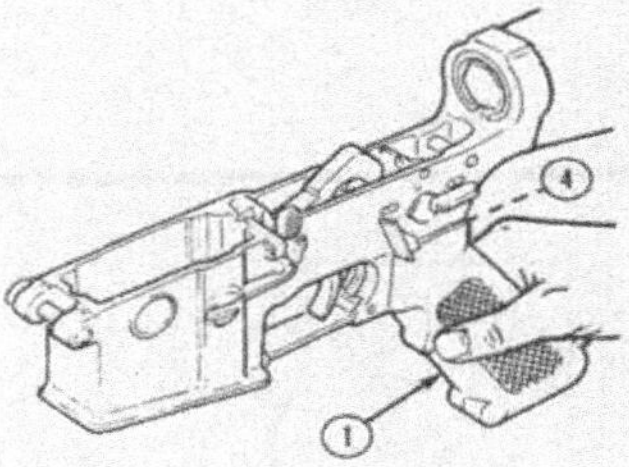

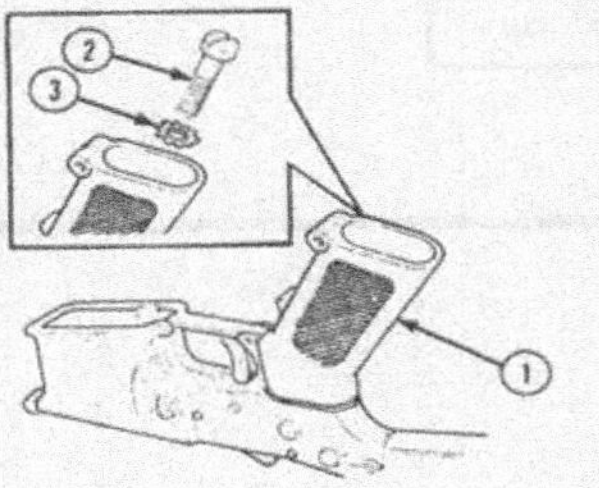

4 Carefully compress detent helical grip (1)

5. Using a screwdriver secure pistol by Installing lockwasher 131 spring (4) and position pistol grip (1) and screw 12)

4-2. LOCK PLATE (CONT).

b REMOVAL

1. Using a screwdriver, reach inside pistol grip (l) and remove screw (2) and lockwasher (3)

WARNING

To avoid Injury to your eyes, use care when removing and installing spring loaded parts

2 Carefully remove pistol grip (1) Hold detent helical spring (4) In place

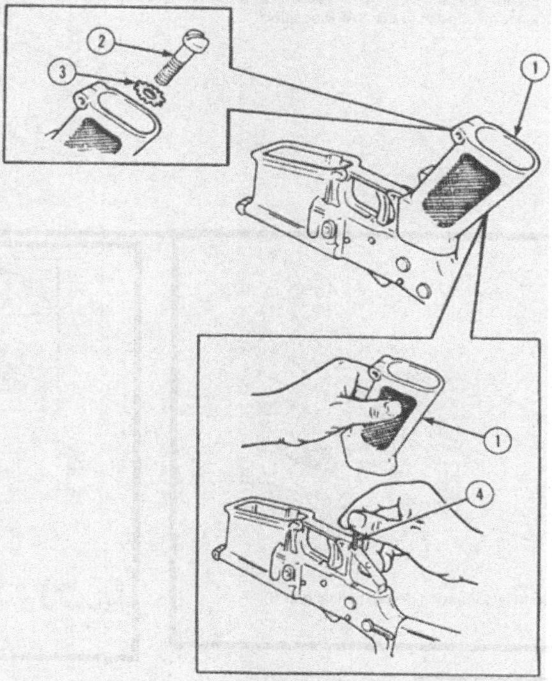

3 Remove lock plate (51 while continuing to hold detent helical spring (4) In place The selector lever must point to the SAFE position

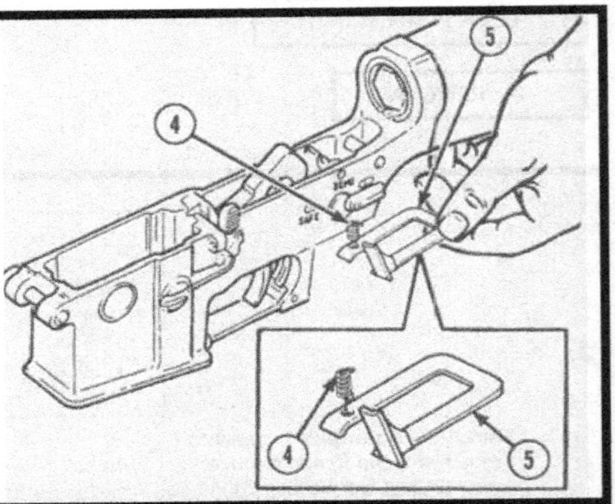

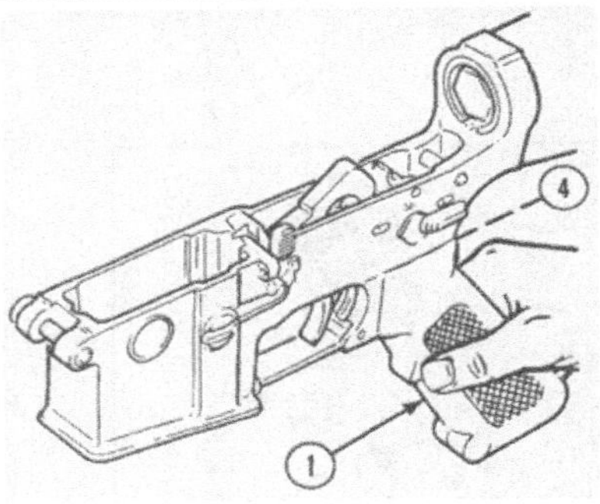

4 Carefully compress detent helical grip (I)

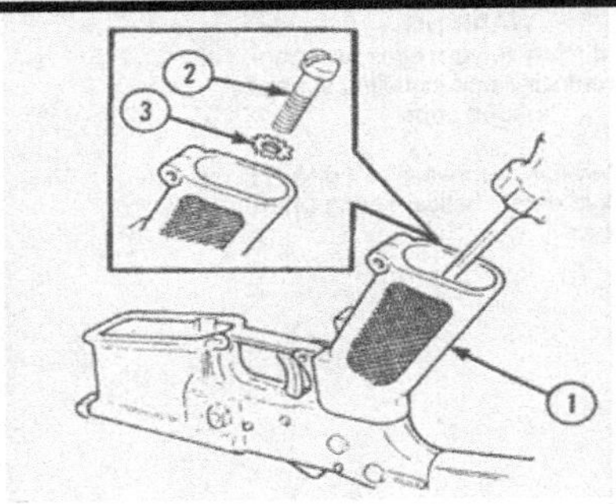

5. Using a screwdriver, secure pistol by installing lockwasher (3) spring (4) and position pistcl grip (1) and screw (2)

c. INSPECTION

Inspect lock plate for serviceability and broken arm Replace If unserviceable or if arm is broken off

4-3. TOP SLING ADAPTER

This task covers:

a Installation
b Removal
c Inspection

INITIAL SETUP

Materials/Parts
Top sling adapter kit PN 8448471

References
TM 9-1005-319-10

a. INSTALLATION I

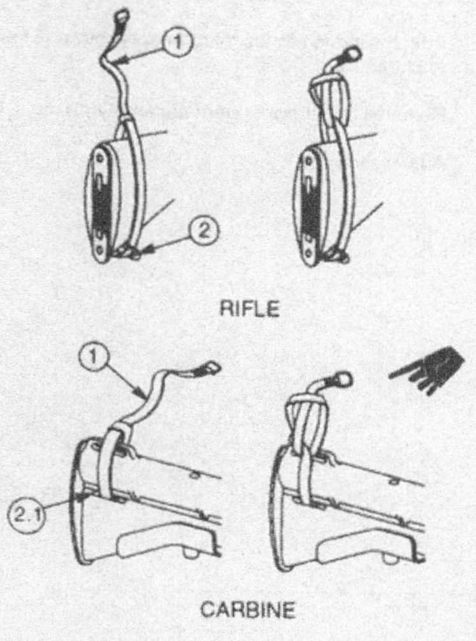

1 Refer to TM 9-1005-319-10 to remove sling.

RIFLE ONLY

2 Install top sling adapter strap (1) through sling swivel (2) and tie.

CARBINE ONLY

2.1. Install top sling adapter strap (1) through sling opening (2.1) and tie.

3 Remove upper handguard assembly.

4 Use pliers to install clamp (3) on front sight base (4).

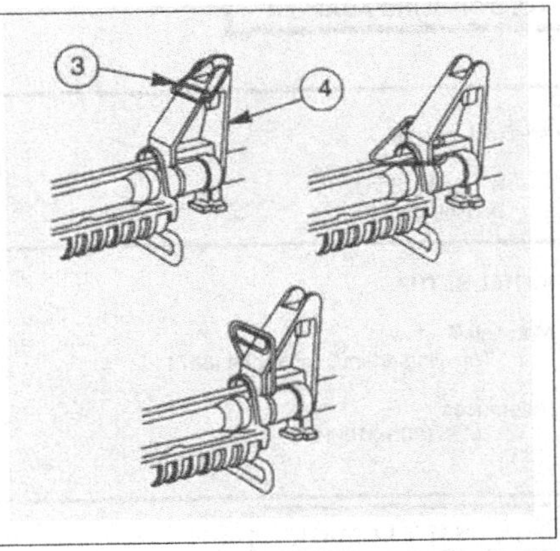

5 Attach sling (5) to top sling adapter strap (1) and to clamp (3).

6 Reinstall upper handguard assembly (6).

7 Adjust sling (5).

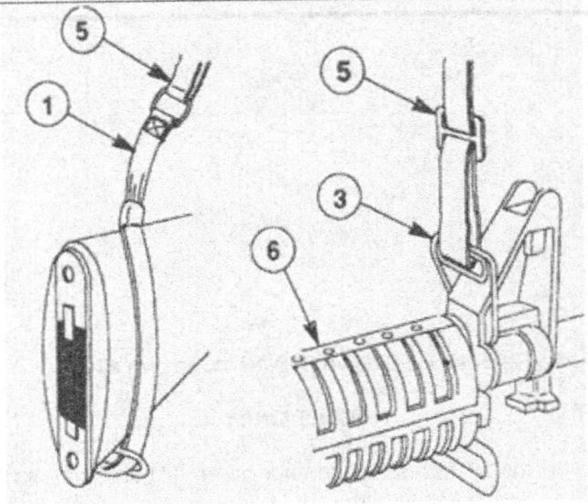

4-3. TOP SLING ADAPTER (CONT).

b. REMOVAL

1 Remove upper handguard assembly.

2 Remove sling (1) from clamp (2) and top sling adapter strap (3).

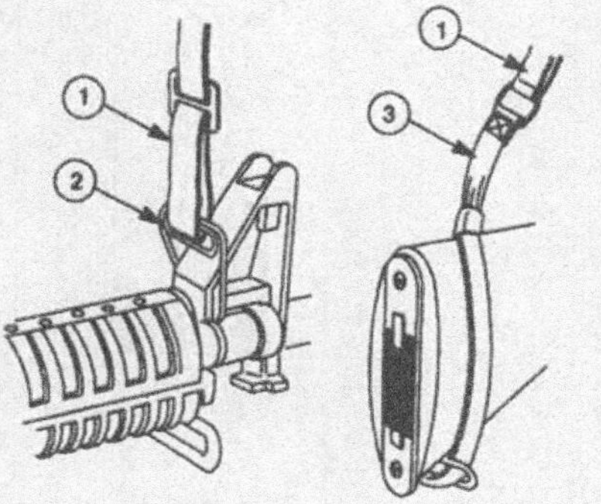

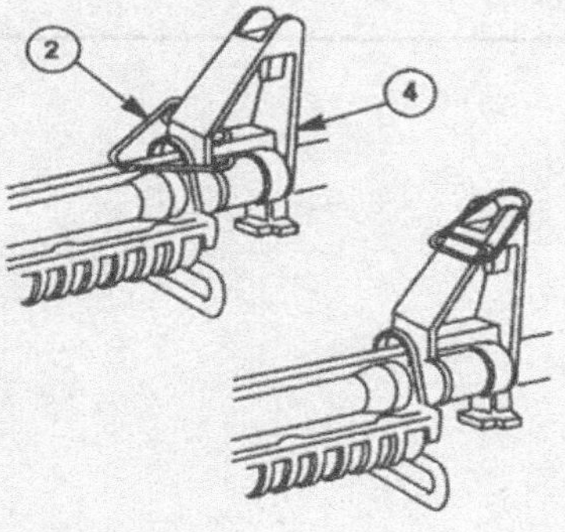

3 Use pliers to remove clamp (2) from front sight base (4). (5).

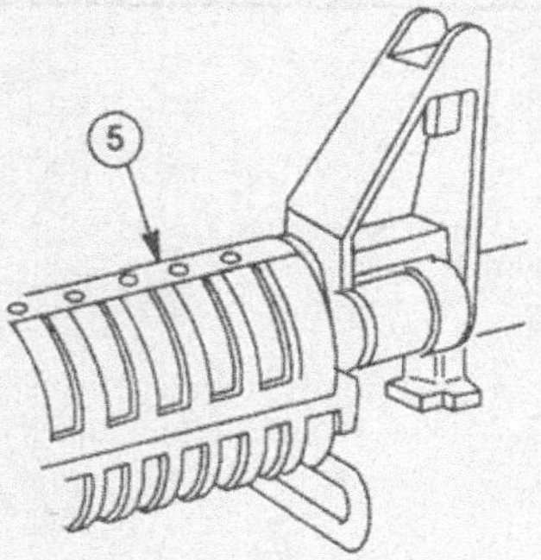

4 Reinstall upper handguard assembly

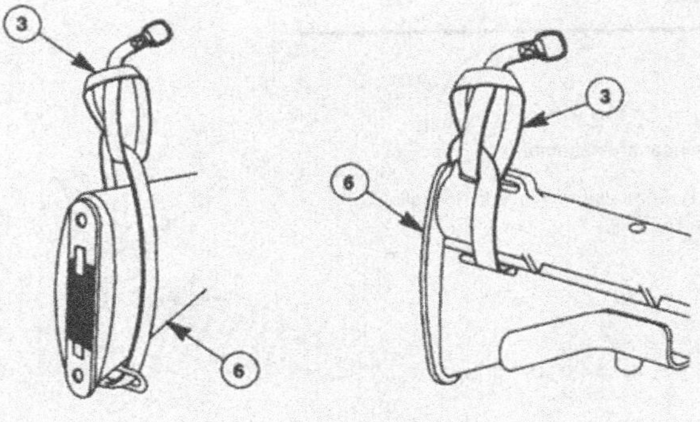

RIFLE

CARBINE

5 Untie top sling adapter strap (3) and remove from buttstock (6).

6 Refer to TM 9-1005-319-10 and install sling.

4-3. TOP SLING ADAPTER (CONT)

c. INSPECTION

Visually Inspect top sling adapter strap and replace If it is badly worn or damaged

4-4. BLANK FIRING ATTACHMENTS M15A2 (M16A2) AND M23 (M4/M4A1).

This task covers:

a Installation
b Removal
c Cleaning
d Inspection
e Repainting
f Replacement

INITIAL SETUP

Materials/Parts
Cleaner, lubricant, and preservative (CLP) (item 9, app D)
Coating compound, enamel (Red - M16A2) (item 23.2, app D)
Coating compound, enamel (Yellow - M4/M4A1) (item 23.3, app D)

General Safety Instructions
Do not keep live ammunition near the work area.

Only blank cartridge M200 is to be used when the blank firing attachment is attached to the weapon.
Do not fire blank ammunition at a representative enemy at distances of less than 20 feet (6.10m).
The unburned propellant grains can cause injury within this distance.

a. INSTALLATION

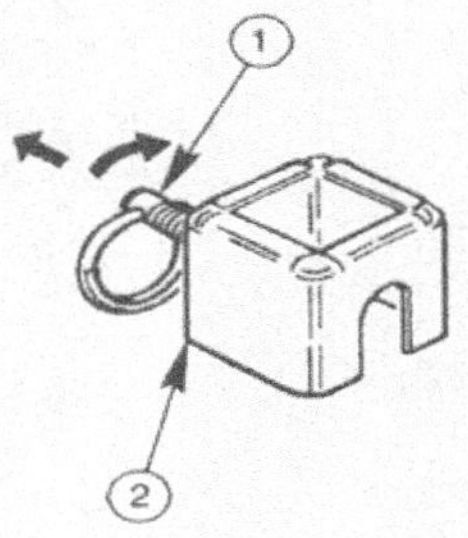

NOTE
M23 blank firing attachment is stamped "M4 Carbine Only" painted yellow and may be used on the M4 and M4A1 carbines. M15A2 BFA is painted red and is used on the M16A2 rifle.

1 Unscrew and pull slide (1) all the way out on blank firing attachment (2)

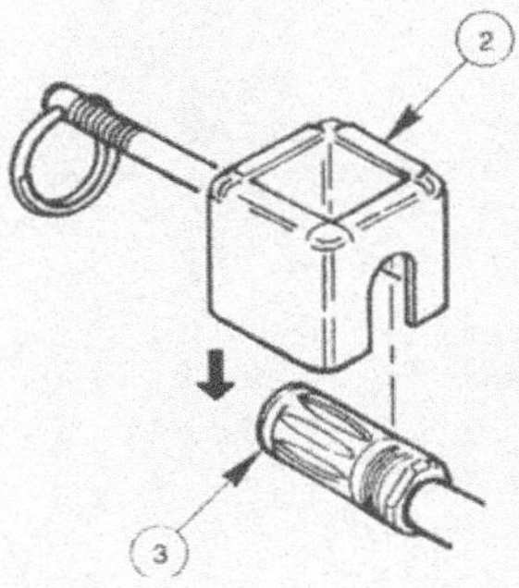

2 Hook blank firing attachment (2) behind the first groove of the compensator (3).

4-4 BLANK FIRING ATTACHMENTS M15A2 (M16A2) AND M23 (M4/M4A1)(CONT).

a. INSTALLATION (CONT)

CAUTION

Do not use tools to tighten the blank firing attachment. **USE HANDS ONLY.**

3 Push slide (2) into compensator (3) and hand tighten.

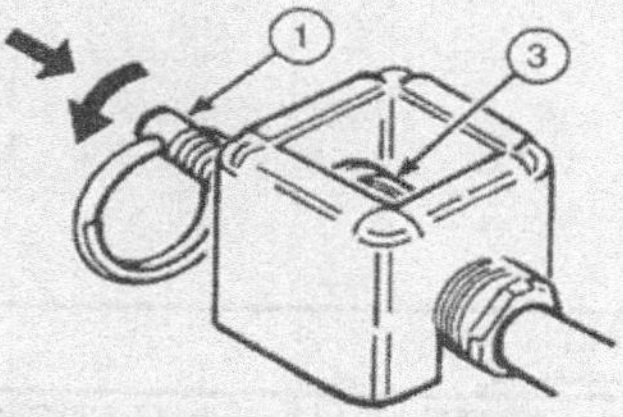

NOTE

Check for tightness after firing approximately 50 blank rounds.

b. REMOVAL

CAUTION

Do not use tools to tighten the blank firing attachment. USE HANDS ONLY.

1 Unscrew slide (1) to remove from compensator (2).

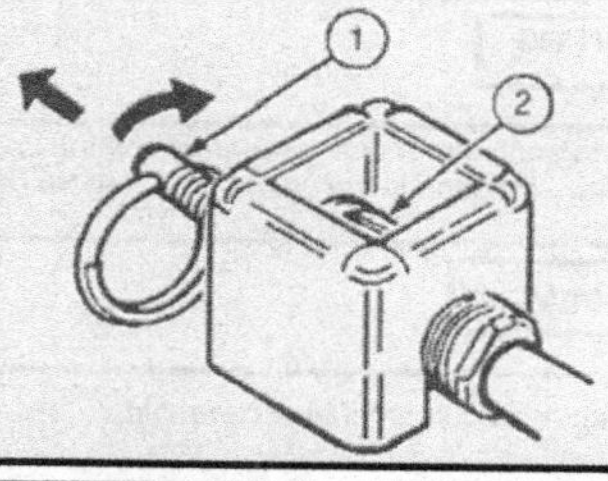

2. Unhook blank firing attachment (3) from behind the first groove of the compensator (2)

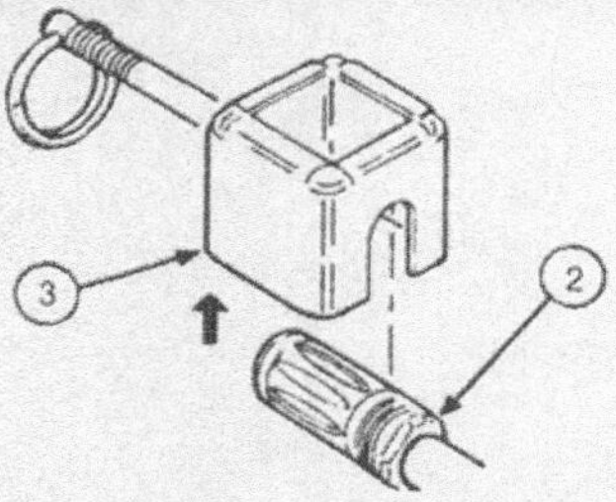

3 Screw slide (1) all the way in on blank firing attachment (3).

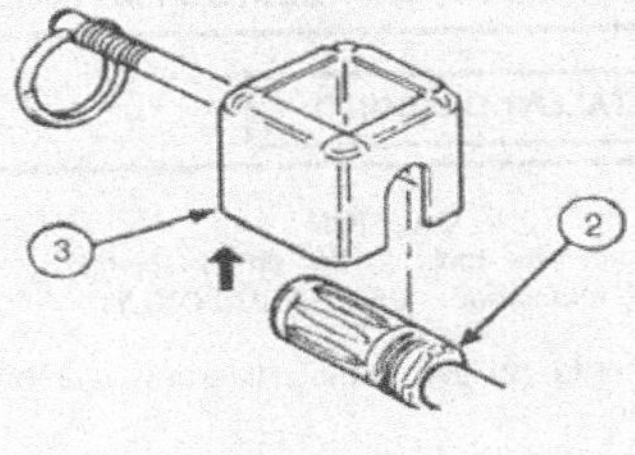

c. CLEANING

Clean blank firing attachment with CLP, wipe dry, and coat with CLP.

d. INSPECTION

Inspect blank firing attachment for cracks or distortion. Be sure the parts in the slide are clear and clean. If blank firing attachment is cracked or distorted, it is unserviceable.

e. REPAINTING

Repaint blank firing attachment using enamel coating compound (Red for M16A2 rifle or Yellow for M4/M4A1 carbine). Painting is the only repair authorized.

f. REPLACEMENT

Replace blank firing attachment if unserviceable.

APPENDIX A
REFERENCES

A-1. TECHNICAL BULLETINSINSTRUCTIONS/MANUALS/ORDERS.

TB 9-1000-247-34	Standards for Overseas Shipment or Domestic Issue of Small Arms, Aircraft Armament, Toward Howitzers, Mortars, Recoilless Rifles, Rocker Launchers and Associated Fire Control Equipment
TB 43-0002-73	Maintenance Expenditures Limits for FSC Group 10; FSC Classes 1005, 1010, 1015, 1030, 1055, 1090, and 1095
TI 05538A-15/2A	Early Warning Indications of Malfunctions
TI 05538A-35/12A	Checking Muzzle and Breech Erosion in Barrel
TI 8005-24/20	Prefire Inspection, Small Arms Weapons, Ordnance Material
TM 9-1005-237-23&P	Organizational and Direct Support Maintenance Manual (Including Repair Parts and Special Tools List) for Bayonet-Knife M6 and M7, with Bayonet-Knife Scabbard M10 and M9 Multipurpose Bayonet System
TM 9-1005-319-10	Operator's Manual for 5.56mm, M16A2 Rifle and M4/M4A1 Carbine
TM 9-1010-221-23&P	Organizational, Direct Support and General Support Maintenance Manual (Including Repair Parts and Special Tools List) for Launcher, Grenade: 40mm M203 W/E (NSN 1010-00-179-6447)
TM 9-6920-363-12&P	M261 Conversion Kit
TM 9-6920-746-12&P	M2 Practice Bolt
TM 11-5855-213-23&P	Night Vision Sight, Individual Served Weapon, AN/PVS-4
TM 750-244-7	Procedures for Destruction of Equipment in Federal Supply Classifications 1000, 1005, 1015, 1020, 1025, 1030, 1055, 1090, and 1095 to Prevent Enemy Use
TM 4700-15/1	Equipment Record Procedures
TO 00-35D-54	Materiel Deficiency Reporting and Investigating System
TO 11 A-13-10-7	Storage and Maintenance Procedures for Small Arms Ammunition
TO 11W-1-10	Historical Data Recording of Inspection, Maintenance, and Firing Data for Ground Weapons
TO 33K-1-100	(TMDE) Interval Calibration and Repair Reference Guide and Work Unit Code Manual

A-2. FIELD MANUALS.

FM 21-11 ....First Aid for Soldiers

FM 23-9 ....M16A1 and M16A2 Rifle and Rifle Marksmanship

A-3 PAMPHLETS.

DA PAM 25-30...Consolidated Index of Army Publications and Blank Forms

DA PAM 738-750......................................The Army Maintenance Management System (TAMMS)

A-4 RELATED PUBLICATIONS.

AFM 36-2227, Vol 1..................................Combat Arms Training Management and Range Operation

AMC-P 310-9...Equipment Publications Listing

AR 725-50 ....Requisitioning, Receipt, and Issue System

AR 750-1 ....Army Materiel Policies

CTA 8-100 ....Army Medical Department Expendable/Durable Items

CTA 50-970 ....Expendable/Durable Items (Except: Medical, Class V, Repair Parts and Heraldic Items)

DOD 4160.21-M-1 ....Defense Demilitarization Manual

FMFM 1-3A..Field Firing Techniques

SB 746-1..Publications for Packing Army General Supplies

SPI 00-856-6885 ....Special Packaging Instructions for M16 Rifle

A-5 FORMS.

AFTO Form 22 ....Technical Order System Publications Improvement Report

AFTO Form 105 ....Inspection, Maintenance, and Firing Data for Ground Weapons

DA Form 2028 ....Recommended Changes to Publications and Blank Forms

DA Form 2028-2.......................................Recommended Changes to Equipment Technical Manuals

DA Form 2404 ....Equipment Inspection and Maintenance Worksheet

DA Form 2408-9.......................................Equipment Control Record

DD Form 314 ....Preventive Maintenance Schedule and Record

NAVMC 10772..Recommended Changes to Technical Publications

NAVMC 11003..Ordnance Serialized Items Subsidiary Records

SF 364..Report of Discrepancy (ROD)

SF 368..Product Quality Deficiency Report

A-6 SUPPLY CATALOGS/SUPPLY LISTS.

SC 4933-95-CL-A11...............................Shop Set, Small Arms Field Maintenance

SC 5180-95-CL-A07...............................Tool Kit, Small Arms Repair

APPENDIX B

MAINTENANCE ALLOCATION CHART

Section I. INTRODUCTION

B-1. GENERAL.

a. This section provides a general explanation of all maintenance and repair functions authorized at various maintenance levels

b. The Maintenance Allocation Chart (MAC) In section II designates overall authority and responsibility for the performance of maintenance functions on the identified end Item or component. The application of the maintenance functions to the end Item or component will be consistent with the capacities and capabilities of the designated maintenance levels

c. Section III lists the tools and test equipment (both special tools and common tool sets) required for each maintenance function as referenced from section II.

d. Section IV contains supplemental instructions and explanatory notes for a particular maintenance function.

B-2. MAINTENANCE FUNCTIONS.

Maintenance functions will be limited to and defined as follows: (except for ammunition MAC')

a. Inspect. To determine the serviceability of an Item by comparing Its physical, mechanical, and or electrical characteristics with established standards through examination leg , by sight, sound, or feel).

b. Test. To verify serviceability by measuring the mechanical, pneumatic, hydraulic, or electrical characteristics of an Item and comparing those characteristics with prescribed standards

c. Service. Operations required periodically to keep an Item In proper operating condition, I e to clean (includes decontaminate, when required), to preserve, to drain, to paint, or to replenish fuel, lubricants, chemical fluids. or gases.

d. Adjust. To maintain or regulate, within prescribed limits, by bringing Into proper or exact position, or by setting the operating characteristics to specified parameters

---

'Exception is authorized for ammunition MAC to permit the redesignation/redefinition of maintenance function headings to more adequately identify ammunition maintenance functions The heading designations and definitions will be Included In the appropriate technical manual for each category of ammunition

B-2. MAINTENANCE FUNCTIONS (CONT).

e. Align. To adjust specified variable elements of an item to bring about optimum or desired performance

NOTE

Air Force Precision Measurement Equipment Laboratory (PMEL) person will calibrate small arms Inspection gages In accordance with TO 33K-1-100

f. Calibrate. To determine and cause corrections to be made or to be adjusted on Instruments or test, measuring, and diagnostic equipments used In precision measurement Consists of comparisons of two Instruments, one of which Is a certified standard of known accuracy, to detect and adjust any discrepancy in the accuracy of the Instrument being compared

g. Remove/Install. To remove and Install the same item when required to perform service or other maintenance functions. Install may be the act of emplacing, seating, or fixing Into position a spare, repair part, or module (component or assembly) In a manner to allow the proper functioning of an equipment or system.

h. Replace. To remove an unserviceable Item and Install a serviceable counterpart In Its place "Replace"' is authorized by the MAC and is shown as the 3d position code of the SMR code.

i. Repair. The application of maintenance services2, including fault location troubleshooting3, removal 'installation, and disassembly'assembly4 procedures, and maintenance actions" to Identify troubles and restore serviceability to an Item by correcting specific damage, fault, malfunction, or failure In a part, subassembly, module (component or assembly), end Item, or system.

j. Overhaul. That maintenance effort (service/action) prescribed to restore an item to a completely serviceable operational condition as required by maintenance standards In appropriate technical publications (i.e , DMWR). Overhaul is normally the highest degree of maintenance performed by the Army Overhaul does not normally return an Item to like new condition

k. Rebuild. Consists of those services/actions necessary for the restoration of unserviceable equipment to a like new condition In accordance with original manufacturing standards Rebuild is the highest degree of materiel maintenance applied to Army equipment The rebuild operation Includes the act of returning to zero those age measurements (hours miles, etc) considered In classifying Army equipment/components.

---

[2]Servlces--Inspect, test, service, adjust, align, calibrate, and or replace
[3]Fault locate troubleshoot The process of investigating and detecting the cause of equipment malfunctioning; the act of isolating a fault within a system or unit under test (UUT).
[4]Disassemble assemble encompasses the step-by-step taking apart (or breakdown) of a spare/functional group coded Item to the level of Its least competency identified as maintenance significant (.e , assigned an SMR code) for the level of maintenance under consideration '
[5]Actionswelding, grinding, riveting, straightening, facing, remachinery, and or resurfacing

B-3. EXPLANATION OF COLUMNS IN THE MAC, SECTION II.

a. **Column 1, Group Number**. Column 1 lists functional group code numbers, the purpose of which is to identify maintenance significant components, assemblies, subassemblies, and modules with the next higher a assembly. End item group number shall be "OO."

b. **Column 2, Component Assembly**. Column 2 contains the names of components. Assemblies, subassemblies, and modules forwhich maintenance is authorized.

c. **Column 3, Maintenance Function**. Column 3 lists the functions to be performed on the it em listed in Column 21. (for detailes explanation of these functions, see paragraph B-2.)

d. **Column 4, Maintenance Level**. Column 4 specifies, by the list ing of a work time figure in the appropriate subcolumn(s), the level of maintenance. If the number of complexity of the t asks within the listed maintenance function vary at different maintenance levels, appropriate work time figures will be shown for each level. The work time figure represents the average time required to restore an item (assembly, subassembly, component, module, end it4m, or system) or a serviceable condition under typicalfield operating conditions. This time includes preparation time (including any necessary disassembly assembly time), troubleshooting fault location time, and quality assurance quality control time in addition to the time required to perform the specific tasks identified for the maintenance functions authorized i the maintenance allocation chart. The symbol designations for the varies maintenance levels are as follows:

C...Operator or Crew
O...Unit Maintenance
F...Direct Support Maintenance
H...General Support Maintenance
L...Specialized Repair Activity (SRA)[6]
D...Depot Maintenance

e. **Column 5, Tools and Equipment**. Column 5 specifies, by code, those common tool sets (not individual tools) and special tools., TMDE, and support equipment required to perform the designated function.

f. **Column 6, Remarks**. This column shall, when applicable, contain a lett4er code, in alphabetic order, which shall be keyed to the remarks contained in Section IV.

[6]This maintenance level is not included in Section II, column (4) of the Maintenance Allocation Chart. To identify functions to this level of maintenance, enter a work time figure in the "H" column of section II, column (4), and use an associated reference code in the Remarks column (6). Key the c ode to Section IV . Remarks, and explain the SRA complete repair application there. The explanatory remarks)s) shall reference the specific Repair Parts and Special Tools List (RPSTL) TM which contains additional SRA criteria and the authorized spare/repair parts.

B-4. EXPLANATION OF COLUMNS IN TOOL AND TEST EQUIPMENT REQUIREMENTS, SECTION III.

a. **Column 1, Reference Code**. The tool and test equipment reference code correlates with a code used In the MAC, Section II, Column 5.

b. **Column 2, Maintenance Level**. The lowest level of maintenance authorized to use the tool or test equipment.

c. **Column 3, Nomenclature**. Name or identification of the tool or test equipment.

d. **Column 4, National Stock Number**. The National stock number of the tool or test equipment.

e. **Column 5, Tool Number**. The manufacturer's part number.

B-5. EXPLANATION OF COLUMNS IN REMARKS, SECTION IV.

a. **Column 1, Reference Code**. The code recorded In column 6, Section II.

b. **Column 2, Remarks**. This column lists information pertinent to the maintenance function being performed as indicated In the MAC, Section II.

Section II. MAINTENANCE ALLOCATION CHART FOR M16A2 RIFLE AND M4/M4A1 CARBINES

(1) GROUP NUMBER	(2) COMPONENT ASSEMBLY	(3) MAINTENANCE FUNCTION	(4) MAINTENANCE LEVEL UNIT C	UNIT O	DS F	DS H	DEPOT D	(5) TOOLS AND EQPT	(6) REMARKS CODE
00	M16A2 5.56MM RIFLE, AND M4/ M4A1 5.56MM CARBINES	Inspect	0.1	0.3	0.7			2	
		Test			0.2			2,3	
		Service	0.2	0.3	0.3				
		Replace		0.1					
		Overhaul							
00.1	CARRYING HANDLE ASSEMBLY M4/M4A1	Inspect	0.1	0.1	0.1				
		Service	0.1						
		Remove/ Install	0.1	0.1	0.1				
		Replace			0.3			1,3	
		Repair		0.1	0.2			1,3	
00.101	Rear Sight Assembly M4/ M4A1	Inspect			0.1				
		Replace			0.3			1,3	
		Repair			0.2			1,3	
01	BOLT CARRIER ASSEMBLY	Inspect	0.1	0.1	0.1				
		Service	0.1	0.1					
		Remove/ Install	0.1						
		Replace			0.1			1,2	
		Repair	01	0.1				3	
0101	Bolt Assembly	Inspect	0.1	0.1	0.1				
		Test			0.1			2	
		Service	0.1						
		Remove/ Install	0.1						
		Replace			0.1			2	
		Repair		0.1	0.1			3	

*Worktimes are included in DMWR 9-1005-249

MAINTENANCE ALLOCATION CHART (CONT)

(1) GROUP NUMBER	(2) COMPONENT ASSEMBLY	(3) MAINTENANCE FUNCTION	(4) MAINTENANCE LEVEL UNIT C	UNIT O	DS F	DS H	DEPOT D	(5) TOOLS AND EQPT	(6) REMARKS CODE
0102	Key and Bolt Carrier Assembly	Inspect	0.1		0.1				
		Service	0.1						
		Remove/ Install	0.1						
		Replace			0.1			1	
		Repair		0.1	0.1			1,3	
02	HANDLE ASSEMBLY	Inspect	0.1						
		Service	0.1						
		Remove/ Install	0.1						
		Replace			0.1		**		
		Repair			0.1			3	
03	UPPER RECEIVER AND BARREL ASSEMBLY	Inspect	0.1	0.2	0.1				
		Test			0.2			2	
		Service	0.2		0.1			1	
		Remove/ Install	0.1						
		Replace			0.5			1,2	
		Repair		0.2	0.5			1,2,3	A,B
0301	M16A2 Barrel Assembly and M4/ M4A1 Replacement Barrel	Inspect		0.1	0.1				
		Replace			0.3			1,2	
		Repair	0.1					3	
0302	Upper Receiver Assembly	Inspect			0.1				
		Replace			0 5			1.2	
		Repair			0.3			1.3	
030201	Forward Assist Assembly	Inspect			0.1				
		Replace			0.1				
		Repair			0 1			1.3	
030202	Rear Sight Assembly M16A2	Inspect			0 1				
		Replace			0.3			1,3	
		Repair			0 2			1,3	

MAINTENANCE ALLOCATION CHART (CONT)

(1) GROUP NUMBER	(2) COMPONENT ASSEMBLY	(3) MAINTENANCE FUNCTION	(4) MAINTENANCE LEVEL UNIT C	UNIT O	DS F	DS H	DEPOT D	(5) TOOLS AND EQPT	(6) REMARKS CODE
04	LOWER RECEIVER AND BUTTSTOCK ASSEMBLY	Inspect	0.1	0.1	0.1				
		Test			0.1			2	
		Service	0.2	0.2					
		Repair		0.2	0.3			1,2,3	C,D,E
0401and 0401A	Buttstock Assembly	Inspect		0.1					
		Remove/ Install		0.1				3	
		Replace		0.1				3	
		Repair		0.1				3	
0402	Hammer Assembly	Inspect			0.1				
		Remove/ Install			0.1				
		Replace			0.1				
		Repair			0.1				
0403	Trigger Assembly	Inspect			0.1				
		Remove/ Install			0.1			1,3	
		Replace			0.1			1,3	
		Repair			0.1			1,3	
040301	Trigger Subassembly M16A2 and M4	Inspect			0.1				
		Replace			0.1			1,3	
		Repair			0.1			1,3	
0404 and 0404A	Lower Receiver and Receiver Extension Assembly	Inspect			0.1				
		Test			0.1			2	
		Repair			0.2			1,3	

Section III. TOOL AND TEST EQUIPMENT REQUIREMENTS FOR M16A2 RIFLE AND M4/M4A1 CARBINES

TOOL OR TEST EQUIPMENT REF CODE	(2) MAINTENANCE CATEGORY	(3) NOMENCLATURE	(4) NATIONAL/NATO STOCK NUMBER	(5) TOOL NUMBER
1 (ARMY ONLY)	F	Shop Set, Small	4933-00-754-0664	SC 4933-95-CL-A11
2 (ARMY ONLY)	F	Tool and gage Set, DS/GS Maintenance for 5.56mm Rifle, M16 Series	4933-00-056-7106	8426685
3 (ARMY ONLY)	O	Tool Kit, Small Arms Repairman	5180-00-357-7770	SC 5180-95-CL-A07
4 (A.F ONLY)	F	Torque Wrench, ft-lb	5120-00-640-6365	A-A-411
5 (A.F ONLY)	F in.-lb	Torque Wrench,	5120-00-230-6380	T-E-12A
6 A.F ONLY)	F	Trigger Weights	4933-00-647-3696	7274758

Section IV. REMARKS

Reference Code	Remarks
A	Tool, Front Sight Post Removal and Installation
B	Depressor, Front Sight Detent
C	Tool, Pivot Pin Removal
D	Tool, Pivot Pin Installation
E	Pin, Slave

APPENDIX C

REPAIR PARTS AND SPECIAL TOOLS LIST

Section I. INTRODUCTION

C-1. SCOPE.

This RPSTL lists and authorizes spares and repair parts; special tools; special test, measurement, and diagnostic equipment (TMDE); and other special support equipment required for performance of unit and direct support maintenance of the M16A2 rifle and M4/M4A1 carbines. It authorizes the requisitioning, issue, and disposition of spares, repair parts, and special tools as indicated by the source, maintenance, and recoverability (SMR) codes.

C-2. GENERAL.

In addition to Section I, Introduction, this Repair Parts and Special Tools List is divided into the following sections:

a. **Section II Repair Parts List.** A list of spares and repair parts authorized by this RPSTL for use in the performance of maintenance. The list also includes parts which must be removed for replacement of the authorized parts. Parts lists are composed of functional groups in ascending alphanumeric sequence, with the parts in each group listed in ascending figure and item number sequence. Bulk materials are listed by item name in FIG BULK at the end of the section. Repair parts kits or sets are listed separately in their own functional group within Section II. Repair parts for repairable special tools are also listed in the section.

b. **Section III Special Tools List.** A list of special tools, special TMDE, and other special support equipment authorized by this RPSTL (as indicated by Basis of Issue (BOI) information in DESCRIPTION AND USABLE ON CODE (UOC) column) for the performance of maintenance.

c. **Section IV Cross-Reference Indexes**. A list, in National item identification number (NIIN) sequence, of all National stock numbered items appearing in the listings, followed by a list in alphanumeric sequence of all part numbers appearing in the listing. National stock numbers and part numbers are cross-referenced to each illustration figure and item number in alphanumeric sequence and cross-references NSN, CAGEC, and part number.

C-3. EXPLANATION OF COLUMNS (SECTIONS II AND III).

a. **ITEM NO. (Column (1))**. Indicates the number used to identify items called out in the illustration.

b. **SMR CODE (Column (2)).** The Source, Maintenance, and Recoverability (SMR) code is a 5-position code containing supply/requisitioning information, maintenance level authorization criteria, and disposition instruction, as shown In the following breakout:

C-3. EXPLANATION OF COLUMNS (SECTION II AND III) (CONT).

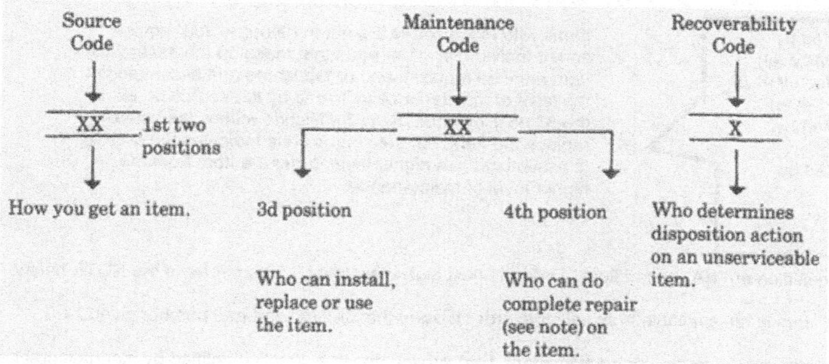

*Complete Repair: Maintenance capacity, capability, and authority to perform all corrective maintenance tasks of the "Repair" function in a use/user environment In order to restore serviceability to a failed item.

(1) Source Code. The source code tells you how to get an item needed for maintenance, repair, or overhaul of an end item/equipment. Explanations of source codes follow:

Code	Explanation
PA PB PC* * PD PE PF PG	Stocked items; use the applicable NSN to request/requisition Items with these source codes They are authorized to the level indicated by the code entered In the 3d position of the SMR code. * *NOTE: Items coded PC are subject to deterioration.
KD KF KB	Items with these codes are not to be requested/requisitioned Individually They are part of a kit which is author->zed to the maintenance level indicated In the 3d positon of the SMR code The complete kit must be requisitioned and applied.
MO-(Made at Unit/ AVUM Level) MF- (Made at DS/ AVIM Level) MH-(Made at GS Level) ML- (Made at Specialized Repair Act (SRA)) MD-(Made at Depot)	Items with these codes are not to be requested/requisitioned Individually They must be made from bulk material which Is identified by the part number In the DESCRIPTION AND USABLE ON CODE (UOC) column and listed In the Bulk Material group of the repair parts list in this RPSTL If the Item Is authorized to you by the 3d position code of the SMR code, but the source code Indicates It is made at a higher level, order the Item from the higher level of maintenance.

Code	Explanation
AO- (Assembled by unit/AVUM Level) AF- (Assembled by DS/AVIM Level) AH- (Assembled by GS Level) AL- (Assembled by SRA) AD- (Assembled by Depot)	Items with these codes are not to be requested/requisitioned Individually. The parts that make up the assembled Item must be requisitioned or fabricated and assembled at the level of maintenance indicated by the source code. If the 3d position code of the SMR code authorizes you to replace the Item, but the source code Indicates the Item Is assembled at a higher level, order the Item from the higher level of maintenance.

XA-Do not requisition an "XA"-coded item. Order Its next higher assembly. (Also, refer to the NOTE below.)

XB--If an "XB" item is not available from salvage, order It using the CAGEC and part number given.

XC-installation drawing, diagram, instruction sheet, field service drawing, that is identified by manufacturer's part number.

XD-Item is not stocked. Order an "XD"-coded Item through normal supply channels using the CAGEC and part number given, If no NSN Is available.

NOTE

Cannibalization or controlled exchange, when authorized, may be used as a source of supply for items with the above source codes, except for those source coded "XA" or those aircraft support Items restricted by requirements of AR 750-1.

(2) Maintenance Code. Maintenance codes tell you the level(s) of maintenance authorized to USE and REPAIR support items. The maintenance codes are entered In the third and fourth positions of the SMR code as follows:

(a) The maintenance code entered in the third position tells you the lowest maintenance level authorized to remove, replace, and use an Item. The maintenance code entered In the third position will indicate authorization to one of the following levels of maintenance

Code	Application/Explanation
C	-Crew or operator maintenance done within unit or aviation unit maintenance.
O	-Unit or aviation unit level can remove, replace, and use the Item.
F	-Direct support or aviation intermediate level can remove, replace, and use the Item.
H	-General support level can remove, replace, and use the Item.
L	-Specialized repair activity can remove, replace, and use the Item.
D	-Depot level can remove, replace, and use the Item

C-3. EXPLANATION OF COLUMNS (SECTIONS II AND III) (CONT).

(b) The maintenance code entered In the fourth position tells whether or not the Item is to be repaired and Identifies the lowest maintenance level with the capability to do complete repair (I e, perform all authorized repair functions). (NOTE: Some limited repair may be done on the Item at a lower level of maintenance, If authorized by the Maintenance Allocation Chart (MAC) and SMR codes I This position will contain one of the following maintenance codes

Code	Application/Explanation
O	-Unit or aviation unit is the lowest level that can do complete repair of the item
F	-Direct support or aviation Intermediate is the lowest level that can do complete repair of the Item
H	-General support is the lowest level that can do complete repair of the item .
L	-Specialized repair activity is the lowest level that can do complete repair of the item
D	-Depot is the lowest level that can do complete repair of the Item
Z	-Nonreparable. No repair Is authorized
B	-No repair is authorized. (No parts or special tools are authorized for the maintenance of a "B" coded item) However, the Item may be reconditioned by adjusting, lubricating, etc , at the user level

(3) Recoverability Code. Recoverability codes are assigned to Items to Indicate the disposition action on unserviceable items The recoverability code is entered In the fifth position of the SMR code as follows.

Recoverability Codes	Application/Explanation
Z	Nonreparable item unserviceable, condemn and dispose of the item at the level of maintenance shown In 3d position) of SMR code
O	Reparable Item. When uneconomically reparable condemn and dispose of the item at unit or aviation unit level
F	Reparable Item. When uneconomically reparable, condemn and dispose of the Item at the direct support or aviation level
H	Reparable Item. When uneconomically reparable, condemn and dispose of the Item at the .general support level

Recoverability Codes	Application/Explanation
D	-Reparable item. When beyond lower level repair capability, return to depot. Condemnation and disposal of item not authorized below depot level.
L	-Reparable item. Condemnation and disposal not authorized below Specialized Repair Activity (SRA).
A	-Item requires special handling or condemnation procedures because of specific reasons (e.g., precious metal content, high dollar value, critical material, or hazardous material). Refer to appropriate manuals/directives for specific instructions.

c. NSN (Column (3)). Indicates the National stock number (NSN) assigned to the item and which will be used for requisitioning.

d. CAGEC (Column (4)). The Contractor and Government Entity Code (CAGEC) is a 5-digit numeric code which is used to identify the manufacturer, distributor, or Government agency, etc., that supplies the Item.

e. PART NUMBER (Column (5)). Indicates the primary number used by the manufacturer (individual, company, firm, corporation, or Government activity), which controls the design and characteristics of the item by means of its engineering drawings, specifications standards, and inspection requirements to identify an item or range of items.

NOTE

When you use an NSN to requisition an item, the item you receive may have a different part number from the part ordered.

f. DESCRIPTION AND USABLE ON CODE (UOC) (Column (6)). This column includes the following Information:

(1) The Federal item name and, when required, a minimum description to identify the Item.

(2) The physical security classification of the item is indicated by the parenthetical entry which is a physical security classification abbreviation, e.g., Phy Sec C1 (C) Confidential, Phy Sec C1 (S) Secret, Phy Sec C1 (T) Top Secret.

(3) Items that are included in kits and sets are listed below the name of the kit or set.

(4) Spare/repair parts that make up an assembled item are listed immediately following the assembled item line entry.

(5) Part numbers for bulk materials are referenced in this column In the line item entry for the Item to be manufacture red/fabricated.

(6) When the item is not used with all serial numbers of the same model, the effective serial numbers are shown on the last line(s) of the description (before UOC).

(7) The usable on code, when applicable (see paragraph 5, Special Information).

C-3. EXPLANATION OF COLUMNS (SECTION II AND III) (CONT).

(8) In the Special Tools List section, the basis of issue (BOI) appears as the last line(s) in the entry for each special tool, special TMDE, and other special support equipment. When density of equipment supported exceeds density spread indicated in the basis of Issue, the total authorization is increased proportionately.

(9) The statement "END OF FIGURE" appears just below the last item description in Column 5 for a given figure In both Section II and Section III.

g. Oty (Column (7)) . The Qty (quantity per figure column) indicates the quantity of the item used in the breakout shown on the illustration figure, which is prepared for a functional group, subfunctional group, or an assembly. A "V" appearing In this column in lieu of a quantity indicates that the quantity is variable and the quantity may vary from application to application.

C-4. EXPLANATION OF COLUMNS (SECTION IV).

a. NATIONAL STOCK NUMBER (NSN) INDEX.

(1) STOCK NUMBER column. This column lists the NSN by National Item Identification Number (NIIN) sequence. The NIIN consists of the last nine digits of the NSN.

NSN
(i.e., 5305-01-674-1467). When using this column to locate an item, ignore the first 4 digits NIIN of the NSN. However, the complete NSN should be used when ordering items by stock number.

(2) FIG. column. This column lists the number of the figure where the item Is Identified/located. The figures are in numerical order in Section II and Section III.

(3) ITEM column. The item number identifies the item associated with the figure listed in the adjacent FIG. column. This item is also identified by the NSN listed on the same line.

b. PART NUMBER INDEX. Part numbers in this index are listed by part number in ascending alphanumeric sequence (i.e., vertical arrangement of letter and number combination which places the first letter or digit of each group In order A through Z, followed by the numbers 0 through 9 and each following letter or digit in like order).

(1) CAGEC column. The Contractor and Government Entity Code (CAGEC) is a 5-digit numeric code used to identify the manufacturer, distributor, or Government agency, etc., that supplies the item.

(2) PART NUMBER column. Indicates the primary number used by the manufacturer (individual, firm, corporation, or Government activity), which controls the design and characteristics of the item by means of Its engineering drawings, specifications standards, and inspection requirements to identify an Item or range of Items.

(3) STOCK NUMBER column. This column lists the NSN for the associated part number and manufacturer identified In the PART NUMBER and CAGEC columns to the left.

(4) FIG. column. This column lists the number of the figure where the item is identified/located in Sections II and III.

(5) ITEM column. The item number is that number assigned to the item as it appears in the figure referenced in the adjacent figure number column.

C-5. SPECIAL INFORMATION.

a. ASSEMBLY INSTRUCTION. Detailed assembly instructions for items source coded to be assembled from component spare/repair parts are found in Chapter 3. Items that make up the assembly are listed immediately following the assembly item entry or reference is made to an applicable figure.

b. ASSOCIATED PUBLICATIONS. The publications listed below pertain to the M16A2 rifle, M4/M4A1 carbine, and its components:

Publication	Short Title
TM 9-1005-319-10	M16A2 Rifle and M4/M4A1 Carbine

c. USABLE ON CODE. The usable on code appears in the lower left corner of the DESCRIPTION column heading, Usable on codes are shown as "UOC..." in DESCRIPTION column (justified left) on the first line applicable item description/nomenclature. Uncoded items are applicable to all models. Identification of the usable on codes used in the RPSTL are:

Code	Used On
AR8	M16A2 Rifle
AS1	M4 Carbine
AY6	M4A1 Carbine

C-6. HOW TO LOCATE REPAIR PARTS.

a. When National Stock Number or Part Number is Not Known.

(1) First. Using the table of contents, determine the assembly group or subassembly group to which the item belongs. This is necessary since figures are prepared for assembly groups and subassembly groups, and listings are divided into the same groups.

C-6. HOW TO LOCATE REPAIR PARTS (CONT).

(2) Second. Find the figure covering the assembly group or subassembly group to which the item belongs.

(3) Third. Identify the item on the figure and use the Figure and Item Number Index to find the NSN.

b. When National Stock Number or Part Number is Known.

(1) First. Using the National Stock Number or the Part Number Index, find the pertinent National Stock Number or Part Number. The NSN index is in National Item Identification Number (NIIN) sequence. (See C4.a.(1).) The pat numbers in the Part Number Index are listed in ascending alpha numeric sequence. (See C4.b.) Both indexes cross-reference you to the illustration/figure and item number of the item you are looking for.

(2) Second. Turn to the figure and item number, verify that the item is the one you're looking for, then locate the item number in the repair parts list for the figure.

C-7. ABBREVIATIONS.

Not applicable.

(1) ITEM NO	(2) SMR CODE	(3) NSN	(4) CAGEC	(5) PART NUMBER	(6) DESCRIPTION AND USABLE	(7) ON CODE (UOC)	QTY
					GROUP 00 (M16A2) RIFLE 5.56MM 9349000,(M4) CARBINE 5.56MM 9390000, AND (M4A1) CARBINE 5.56MM 12972700 FIG. C-1. (M16A2) RIFLE, (M4) AND CM4A1) CARBINES		
1	PAOOO	1005000179546	19204	8448517	HANDLE ASSEMBLY,CHA CHARGING		1
2	AFFFF		19204	8448501	BOLT CARRIER ASSEMBLY (M16A2)	UOC:AR8	1
2A	AFFFF	19200	12011849	BOLT	AND BOLT CARRI ASSEMBLY (M4 AND M4A1)	UOC:AS1,AY6	1
3	AFFFF	1005011879716	19200	9349050	UPPER RECEIVER AND BARREL ASSEMBLY (M16A2)	UOC:AR8	1
3A	AFFFF	19200	12972680	UPPER	RECEIVER AND BARREL ASSEMBLY (M4 AND M4A1)	UOC:AS1,AY6	1
4	PACZZ	1005009215004	19200	8448670	MAGAZINE,CARTRIDGE (30 ROUND)		1
5	PACZZ	1005012164510	19204	12624561	SLING,SMALL ARMS (M16A2)	UOC:AR8	1
5A	PACZZ	1005013689852	19200	12011996	SLING,SMALL ARMS (M4 AND M4A1)	UOC:AS1,AY6	1
6	XAFFA	19200	9349100	LOWER	RECEIVER AND BUTTSTOCK ASSEMBLY (M16A2)	UOC:AR8	1
6A	XAFFA		19200	9390011	LOWER RECEIVER AND BUTTSTOCK ASSEMBLY (4)	UOC:AS1	1
6B	XAFFA		19200	12972690	LOWER RECEIVER AND BUTTSTOCK ASSEMBLY (M4A1)	UOC:AY6	1
7	AFFFF		19200	12951011	CARRYING HANDLE ASS (M4 AND M4A1)	UOC:AS1,AY6	1

END OF FIGURE

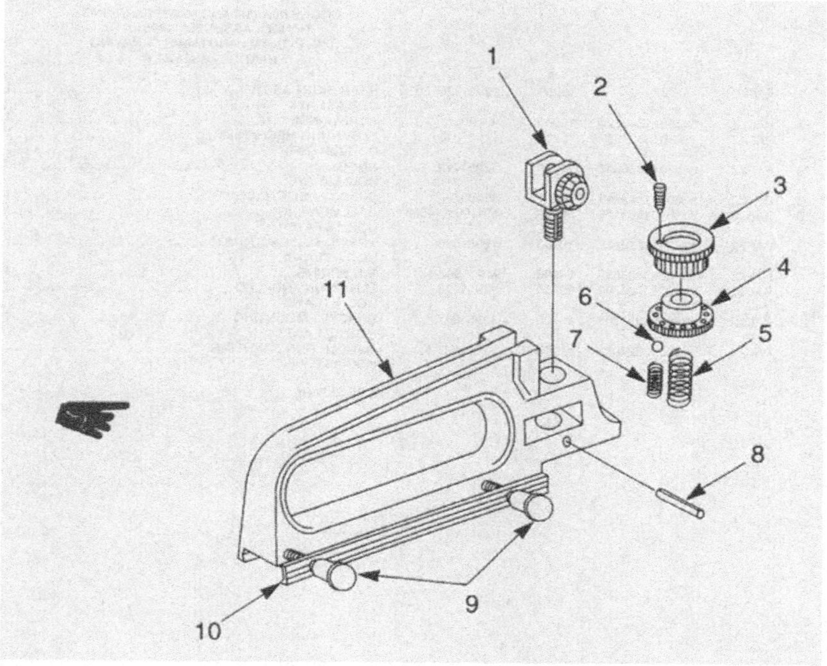

Figure C-1A. (M4 and M4A1) Carrying Handle Assembly 12951011.

(1) ITEM NO	(2) SMR CODE	(3) NSN	(4) CAGEC	(5) PART NUMBER	(6) DESCRIPTION AND USABLE ON CODE (UOC)	(7) QTY
					GROUP 00.1 (M4 AND M4A1) CARRYING HANDLE ASSEMBLY 12951011 FIG. C-1A. (M4 AND M4A1) CARRYING HANDLE ASSEMBLY	
1	AFFFF		19200	12951026	REAR SIGHT ASSY UOC:AS1,AY6	1
2	PAFZZ	5305011343622	19200	9349065	SCREW,INDEX	1
3	PAFZZ	1005013827089	19200	12951018	ELEVATING MECHANISM UOC:AS1 ,AY6	1
4	PAFZZ	5355013826801	19200	12951019	KNOB UOC:AS1,AY6	1
5	PAFZZ	5360011343710	19200	9349070	SPRING ,HELICAL,COMP	1
6	PAFZZ	3110001839175	96906	MS19060-4808	BALL,BEARING UOC:AS1,AY6	1
7	PAFZZ	5360013826802	19200	12951020	SPRING,HELICAL,COMP UOC:AS1,AY6	1
8	PAFZZ	5315008403812	96906	MS16562-121	PIN,SPRING	1
9	PAOZZ	5310013826793	19200	12951023	NUT,PLAIN,KNURLED UOC:AS1,AY6	2
10	PAOZZ	5340013823201	19200	12951017	BRACKET,MOUNTING UOC:AS1,AY6	1
11	PAFZZ	1005013827083	19200	12951021	HANDLE,GUN CARRYING UOC:AS1,AY6	1

END OF FIGURE

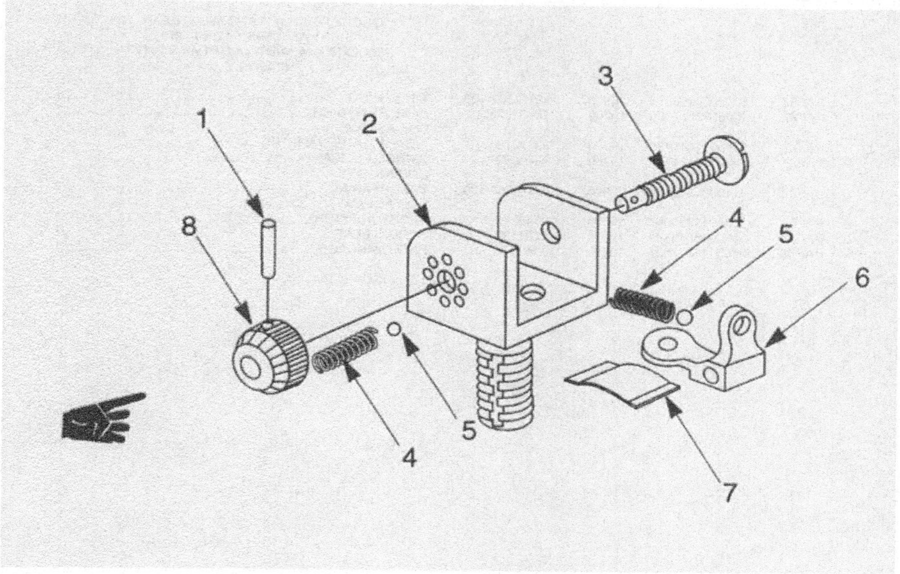

Figure C-1B. (M4 and M4A1) Rear Sight Assembly 12951026.

(1) ITEM NO	(2) SMR CODE	(3) NSN	(4) CAGEC	(5) PART NUMBER	(6) DESCRIPTION AND USABLE ON CODE (UOC)	(7) QTY
					GROUP 00.101 (M4 AND M4A1) REAR SIGHT ASSEMBLY 12951026 FIG. C-18. (M4 AND M4A1) REAR SIGHT ASSEMBLY	
1	PAFZZ	5315000586678	96906	MS16562-103	PIN,SPRING	1
2	PAFZZ	1005013827086	19200	12951028	BASE,REAR SIGHT UOC:AS1,AY6	1
3	PAFZZ	5305011441490	19200	9349076	SCREW,EXTERNALLY RE	1
4	PAFZZ	5360011481751	19200	9349069	SPRING,HELICAL,COMP UOC:AS1,AY6	2
5	PAFZZ	3110001839175	96906	MS19060-4808	BALL,BEARING UOC:AS1,AY6	2
6	PAFZZ	1005011353697	19200	9349075	APERTURE,SIGHT	1
7	PAFZZ	5360013816183	19200	12011987	SPRING,FLAT	1
8	PAFZZ	5355011343627	19200	9349077	KNOB,WINDAGE	1

END OF FIGURE

SECTION II

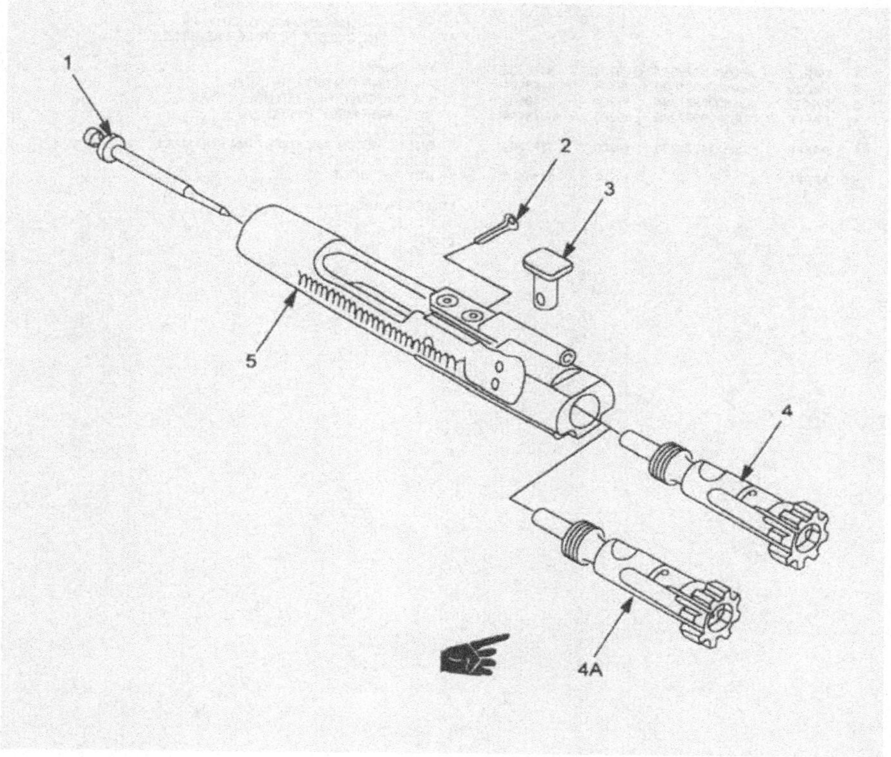

Figure C-2. Bolt Carrier Assembly (M16A2) 8448501 and (M4 and M4A1) 12011849.

(1) ITEM NO	(2) SMR CODE	(3) NSN	(4) CAGEC	(5) PART NUMBER	(6) DESCRIPTION AND USABLE ON CODE (UOC)	(7) QTY
					GROUP 01 BOLT CARRIER ASSEMBLY (M16A2) 8448501 AND (M4 AND M4A1) 12011849 FIG. C-2. BOLT CARRIER ASSEMBLY	
1	PAFZZ	1005000179547	19204	8448503	PIN,FIRING	1
2	PAOZZ	1005009991509	19204	8448504	PIN,FIRING PIN RETA RETAINING	1
3	PAOZZ	5315009927294	19204	8448502	PIN,GROOVED,HEADED BOLT CAM	1
4	PAFFF	1005009927285	19200	8448509	BOLT ASSEMBLY (M16A2) UOC:AR8	1
4A	PAFFF	1005014223770	19200	12972691	BOLT,BREECH ASSEMBLY (M4 AND M4A1) UOC:AS1,AY6	1
5	AFFFF		19204	8448505	KEY AND BOLT	1

END OF FIGURE

SECTION II

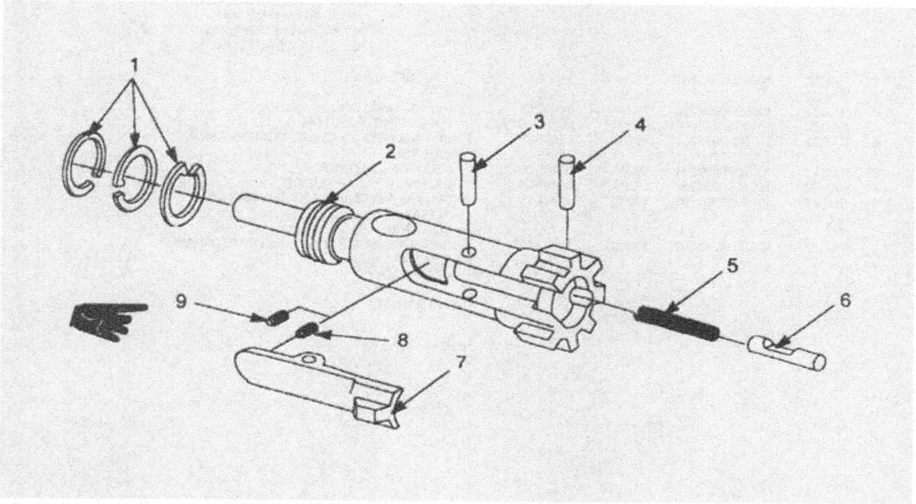

Figure C-3. Bolt Assembly (M16A2) 8448509 and (M4 and M4A1) 12972691.

(1) ITEM NO	(2) SMR CODE	(3) NSN	(4) CAGEC	(5) PART NUMBER	(6) DESCRIPTION AND USABLE ON CODE (UOC)	(7) QTY
					GROUP 0101 BOLT ASSEMBLY (M16A2) 8448509 AND (M4 AND M4A1) 12972691 FIG. C-3. BOLT ASSEMBLY	
1	PAFZZ	1005009927287	19204	8448511	RING,BOLT	3
2	XAFZZ		19204	8448510	BOLT	1
3	PAOZZ	1005009927290	19204	8448513	PIN,EXTRACTOR	1
4	PAOZZ	5315005975086	96906	MS16562-9B	PIN,SPRING EJECTOR	1
5	PAOZZ	5360009927292	19204	8448516	SPRING,HELICAL,COMP COMPRESSION, EJECTOR	1
6	PAOZZ	1005009927291	19204	8448515	EJECTOR,CARTRIDGE	1
7	PAOZZ	1005009927288	19204	8448512	EXTRACTOR,CARTRIDGE	1
8	PAOZZ	1005007603768	19200	8448755	SPRING ASSEMBLY,EXT EXTRACTOR (M16A2) UOC:AR8	1
9	PAOZZ	1005014245899	19200	12972692	SPRING ASSEMBLY,EXT EXTRACTOR (M4 AND M4A1)(BLACK) UOC:AS1 ,AY6	1

END OF FIGURE

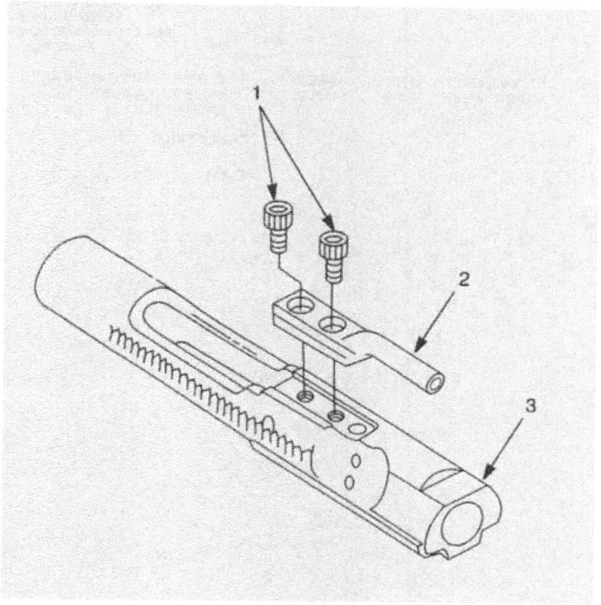

Figure C-4. Key and Bolt Carrier Assembly 8448505.

(1) ITEM NO	(2) SMR CODE	(3) NSN	(4) CAGEC	(5) PART NUMBER	(6) DESCRIPTION AND USABLE ON CODE (UOC)	(7) QTY
					GROUP 0102 KEY AND BOLT CARRIER ASSEMBLY 8448505 FIG. C-4. KEY AND BOLT CARRIER ASSEMBLY	
1	PAFZZ	5305009927284	19204	8448508	SCREW,CARRIER AND K KEY	2
2	PAFZZ	1005009927283	19200	8448506	KEY,BOLT CARRIER	1
3	PAFZZ		19200	8448507	CARRIER,BOLT	1

END OF FIGURE

SECTION II

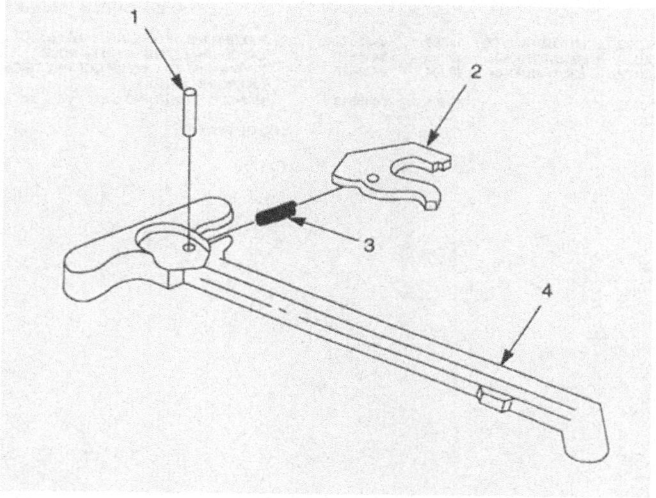

Figure C-5. Handle Assembly 8448517.

(1) ITEM NO	(2) SMR CODE	(3) NSN	(4) CAGEC	(5) PART NUMBER	(6) DESCRIPTION AND USABLE ON CODE (UOC)	(7) QTY
					GROUP 02 HANDLE ASSEMBLY 8448517 FIG. C-5. HANDLE ASSEMBLY	
1	PAOZZ	5315010489372	19204	8448521-2	PIN,SPRING CHARGING HANDLE	1
2	PAOZZ	5340009990405	19200	8448519	LATCH,CHARGING HAND HANDLE	1
3	PAOZZ	5360009990404	19204	8448520	SPRING,HELICAL,COMP COMPRESSION, CHARGING HANDLE	1
4	XAOZZ		19204	8448518	HANDLE ,CHARGING	1

END OF FIGURE

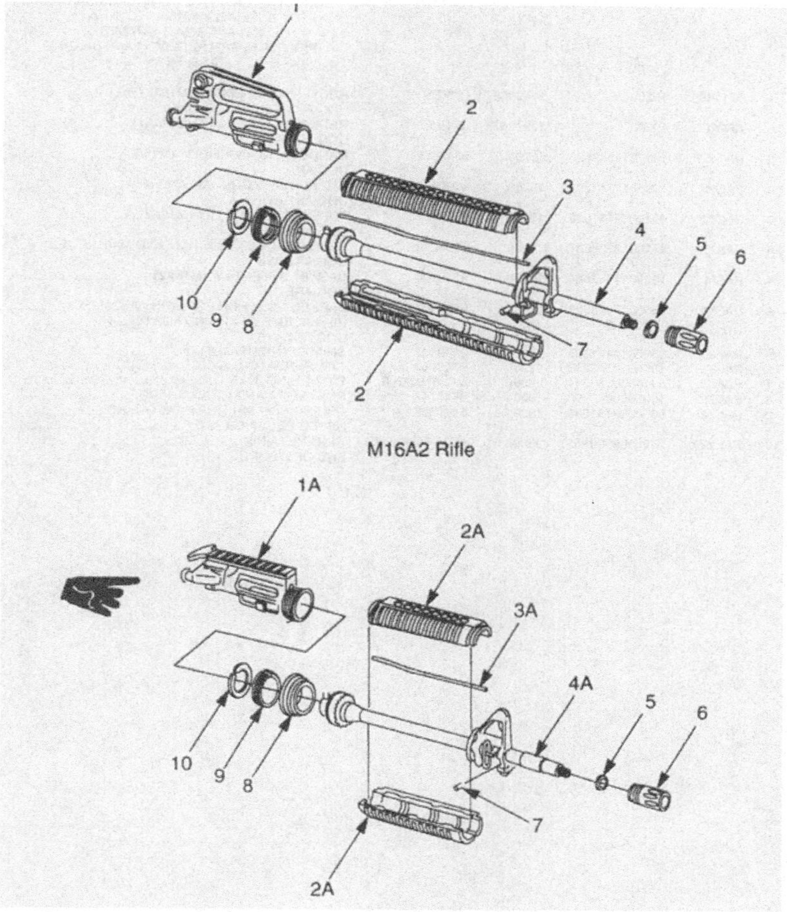

M4/M4A1 Carbine

Figure C-6. Upper Receiver and Barrel Assembly (M16A2) 9349050, and (M4 and M4A1) 12972680.

(1) ITEM NO	(2) SMR CODE	(3) NSN	(4) CAGEC	(5) PART NUMBER	(6) DESCRIPTION AND USABLE ON CODE (UOC)	(7) QTY
					GROUP 03 UPPER RECEIVER AND BARREL ASSEMBLY (M16A2) 9349050, (M4 AND M4A1] 12972680 FIG. C-6. UPPER RECEIVER AND BARREL ASSEMBLY	
1	AFFFF	19200	9349062	UPPER	RECEIVER ASSEMBLY (M16A2) UOC:AR8	1
1A	AFFFF	19200	12972675	UPPER	RECEIVER ASSY (M4 AND M4A1) UOC:AS1,AY6	1
2	PAOZZ	1005011343629	19200	9349059	HANDGUARD ASSEMBLY (M16A2) UOC:AR8	2
2A	PAOZZ	1005012342297	19200	9390003	GUARD,HAND,GUN (M4 AND M4A1) UOC:AS1,AY6	2
3	PAFZZ	4710009781038	19200	8448567	TUBE,BENT,METALLIC (M16A2) UOC:AR8	1
3A	PAFZZ	4710012338637	19200	9390016	TUBE,BENT,METALLIC (M4 AND M4A1) UOC:AS1,AY6	1
4	PAFFF	1005011467684	19200	9349124	BARREL ASSEMBLY (M16A2) UOC:AR8	1
4A	PAFFF	1005012338529	19200	9390007	BARREL AND FRONT SI SIGHT ASSEMBLY, REPLACEMENT (M4 AND M4A1) UOC:AS1,AY6	1
5	PAFZZ	5365011441496	19200	9349052	SHIM COMPENSATOR	1
6	PAFZZ	1005011343633	19200	9349051	COMPENSATOR	1
7	PAFZZ	5315000586044	96906	MS16562-106	PIN,SPRING ,GAS TUBE	1
8	PAFZZ	1005000878998	19204	8448712	RING,SLIP,HAND GUAR GUARD	1
9	PAFZZ	5360009781036	19204	8448555	SPRING,SLIP RING,HA HANDGUARD UPPER RECEIVER	1
10	PAFZZ	5325009990863	96906	M5S16626-3137	RING,RETAINING END OF FIGURE	1

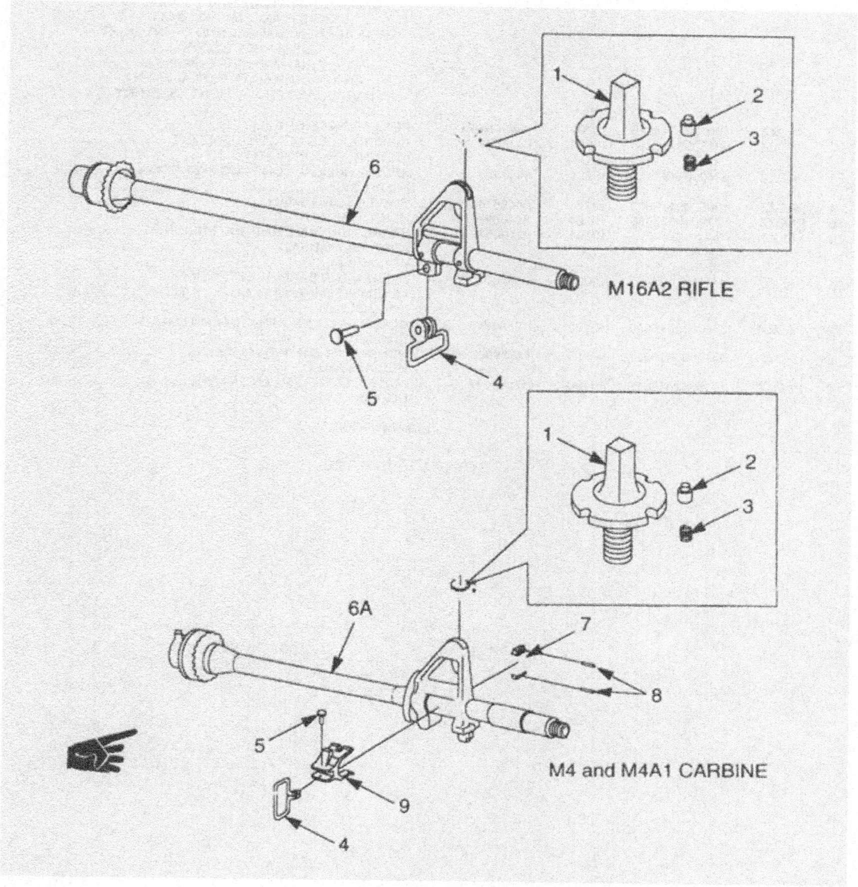

Figure C-7. (M16A2) Barrel Assembly 9349124 and (M4 and M4A1) Replacement Barrel and Front Sight Assembly 9390007.

(1) ITEM NO	(2) SMR CODE	(3) NSN	(4) CAGEC	(5) PART NUMBER	(6) DESCRIPTION AND USABLE ON CODE (UOC)	(7) QTY
					GROUP 0301 (M16A2) BARREL ASSEMBLY 9349124, AND (M4 AND M4A1) REPLACEMENT BARREL AND FRONT SIGHT ASSEMBLY 9390007 FIG. C-7. (M16A2) BARREL ASSEMBLY, AND (M4 AND M4A1) REPLACEMENT BARREL AND FRONT SIGHT ASSEMBLY	
1	PAOZZ	1005011343625	19200	9349056	POST,FRONT SIGHT	1
2	PAOZZ	5315009793930	19204	8448573	PIN,SHOULDER,HEADLE HEADLESS (DETENT, FRONT SIGHT)	1
3	PAOZZ	5360009793931	19204	8448574	SPRING,HELICAL,COMP ,COMPRESSION, FRONT SIGHT	1
4	PAOZZ	1005000179543	19204	8448571	SWIVEL,SLING,SMALL	1
5	PAOZZ	5320010637635	19204	8448697	RIVET,TUBULAR	1
6	XAFZZ		19200	9349054	BARREL AND BARREL E EXTENSION ASSEMBLY (M16A2) UOC:AR8	1
6A	XAFZZ		19200	9390009	BARREL AND BARREL EXTENSION ASSEMBLY (M4 AND M4A1) UOC:AS1,AY6	1
7	PAOZZ	5340012646530	19200	12598618	CLAMP,RIM CLENCHING (M4 AND M4A1) UOC:AS1,AY6	1
8	PAOZZ	5315006900544	96906	MS39086-93	PIN,SPRING (M4 AND M4A1) UOC:AS1,AY6	2
9	PAOZZ	1010012646517	19200	12598617	MOUNT,SWIVEL (M4 AND M4A1) UOC:AS1,AY6	1

END OF FIGURE

SECTION II

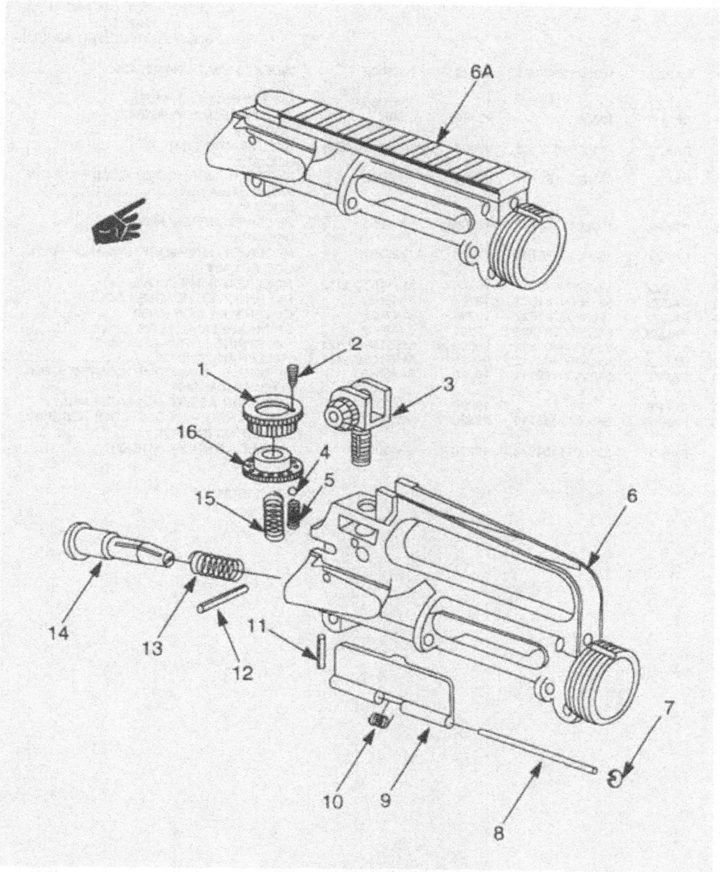

Figure C-8. Upper Receiver Assembly (M16A2) 9349062, and (M4 and M4A1) 12972675.

(1) ITEM NO	(2) SMR CODE	(3) NSN	(4) CAGEC	(5) PART NUMBER	(6) DESCRIPTION AND USABLE ON CODE (UOC)	(7) QTY
					GROUP 0302 UPPER RECEIVER ASSEMBLY (M16A2) 9349062, (M4 AND M4A1) 12972675 FIG. C-8. UPPER RECEIVER ASSEMBLY	
1	PAFZZ	1005011343621	19200	9349066	INDEX,ELEVATION (M16A2) UOC:AR8	1
2	PAFZZ	5305011343622	19200	9349065	SCREW,INDEX (M16A2)	1
3	AFFFF	19200	9349072	REAR	SIGHT ASSEMBLY (M16A2) UOC:AR8	1
4	PAFZZ	3110001839175	96906	MS19060-4808	BALL,BEARING (M16A2) UOC:AR8	1
5	PAFZZ	5360011481751	19200	9349069	SPRING,HELICAL,COMP COMPRESSION, ... INDEX (M16A2) UOC:AR8	1
6	PAFZZ	1005011343701	19200	9349063	RECEIVER,UPPER (M16A2) UOC:AR8	1
6A	PAFZZ	1005013826795	19200	12972670	RECEIVER,CARTRIDGE (M4 AND M4A1) UOC:AS1,AY6	1
7	PAOZZ	5325009990864	96906	MS16632-3012	RING,RETAINING ,COVER	1
8	PAOZZ	5315009781023	19204	8448533	PIN,GROOVED,HEADLES COVER	1
9	PAOZZ	1005009781022	19204	8448525	COVER,EJECTION PORT	1
10	PAOZZ	5360009781025	19204	8448532	SPRING,HELICAL,TORS ,COVER	1
11	PAFZZ	5315008403812	96906	M5S16562-121	PIN,SPRING FORWARD ASSIST	1
12	PAFZZ	5315008403812	96906	M5S16562-121	PIN,SPRING (M16A2)	1
13	PAFZZ	5360000179541	19200	8448540	SPRING,HELICAL,COMP COMPRESSION .... FORWARD ASSIST	1
14	AFFFF		19204	9349086	FORWARD ASSIST ASSY ASSEMBLY	1
15	PAFZZ	5360011343710	19200	9349070	SPRING,HELICAL,COMP COMPRESSION, ... ELEVATION (M16A2)	1
16	PAFZZ	5355011354972	19200	9349067	KNOB,ELEVATION (M16A2) UOC:AR8	1

END OF FIGURE

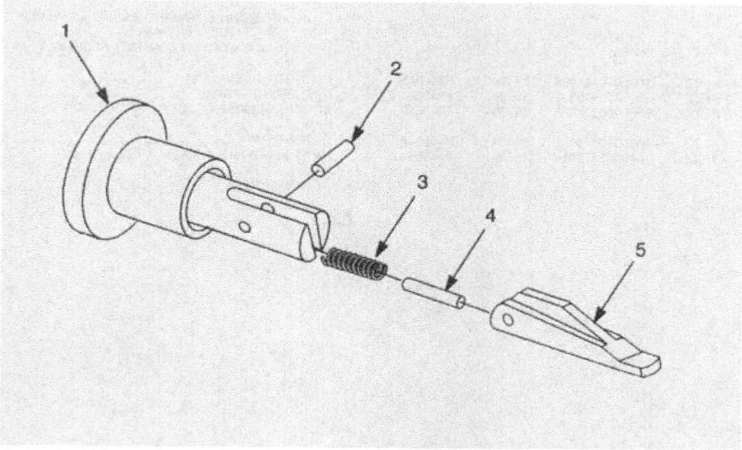

Figure C-9. Forward Assist Assembly 9349086.

(1) ITEM NO	(2) SMR CODE	(3) NSN	(4) CAGEC	(5) PART NUMBER	(6) DESCRIPTION AND USABLE ON CODE (UOC)	(7) QTY
					GROUP 030201 FORWARD ASSIST ASSEMBLY 9349086	
					FIG. C-9. FORWARD ASSIST ASSEMBLY	
1	PAFZZ	1005011441468	19200	9349085	PLUNGER ASSEMBLY	1
2	PAFZZ	5315010489372	19204	8448521-2	PIN,SPRING ,PAWL	1
3	PAFZZ	5360005238084	19200	8448542	SPRING,HELICAL,COMP COMPRESSION PAWL	1
4	PAFZZ	1005000179540	19204	8448544	DETENT,PAWL	1
5	PAFZZ	3040000179539	19204	8448543	PAWL,FORWARD ASSIST	1

END OF FIGURE

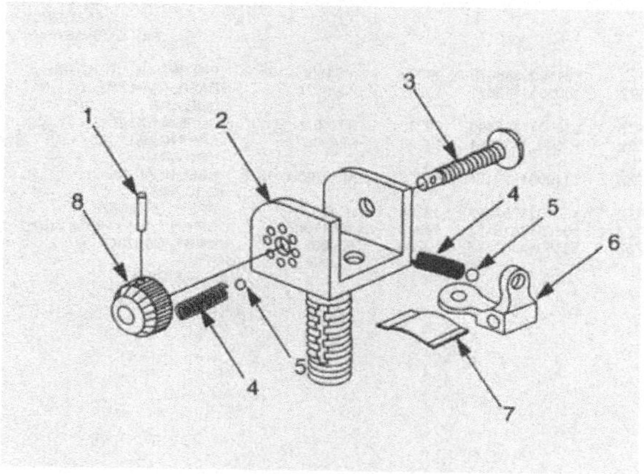

Figure C-10. (M16A2) Rear Sight Assembly 9349072.

(1) ITEM NO	(2) SMR CODE	(3) NSN	(4) CAGEC	(5) PART NUMBER	(6) DESCRIPTION AND USABLE ON CODE (UOC)	(7) QTY
					GROUP 030202 (M16A2) REAR SIGHT ASSEMBLY 9349072 FIG. C-10. REAR SIGHT ASSEMBLY	
1	PAFZZ	5315000586678	96906	MS16562-103	PIN,SPRING ,WINDAGE	1
2	PAFZZ	1005011343631	19200	9349074	BASE,REAR SIGHT UOC:ARB	1
3	PAFZZ	5305011441490	19200	9349076	SCREW,EXTERNALLY RE	1
4	PAFZZ	5360011481751	19200	9349069	SPRING,HELICAL,COMP ,REAR SIGHT UOC:AR8	2
5	PAFZZ	3110001839175	96906	M5S19060-4808	BALL,BEARING UOC:AR8	2
6	PAFZZ	1005011353697	19200	9349075	APERTURE,SIGHT	1
7	PAFZZ	5360013816183	19200	12011987	SPRING,FLAT REAR SIGHT	1
8	PAFZZ	5355011343627	19200	9349077	KNOB,WINDAGE	1

END OF FIGURE

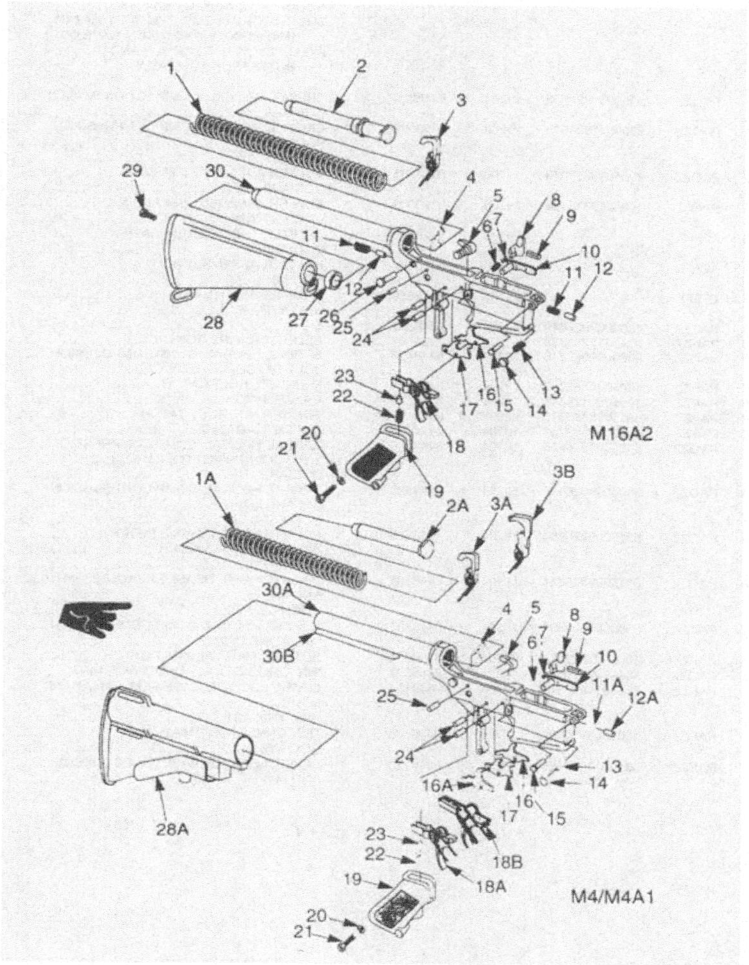

Figure C-11. Lower Receiver and Buttstock Assembly (MI 6A2) 9349100, (M4) 9390011 and (M4A1) 12972690.

(1) ITEM NO	(2) SMR CODE	(3) NSN	(4) CAGEC	(5) PART NUMBER	(6) DESCRIPTION AND USABLE ON CODE (UOC)	(7) QTY
					GROUP 04 LOWER RECEIVER AND BUTTSTOCK ASSEMBLY (M16A2) 9349100, (M4) 9390011, AND (M4A1) 12972690 FIG C-11 LOWER RECEIVER AND BUTTSTOCK ASSEMBLY	
1	PAOZZ	5360009926665	19204	8448629	SPRING, HELICAL, COMP ACTION (M16A2) UOC:AR8	1
1A	PAOZZ	5360012338617	19200	9390022	SPRING, HELICAL, COMP COMPRESSION (M4 AND M4A1) UOC:AS1, AY6	1
2	PAOZZ	1005009373078	19200	8448615	BUFFER ASSEMBLY (MI6A2) UOC:AR8	1
2A	PAOZZ	1005012313138	19200	9390023	BUFFER ASSEMBLY (M4 AND M4Al) UOC:AS1, AY6	1
3	AFFFF		19200	9349106	HAMMER ASSEMBLY (M16A2) UOC:AR8	1
3A	AFFFD		19200	9390032	HAMMER ASSEMBLY (M4) UOC:AS1	1
3B	AFFFF		19204	8448610	HAMMER ASSY (M4A1) UOC:AY6	1
4	PAFZZ	1005009926649	19200	8448595	SEAR	1
5	PAFZZ	5340012258339	19200	9381367	SELECTOR, FIRE CONTROL	1
6	PAFZZ	5360000562246	19204	8448633	SPRING, HELICAL, COMP COMPRESSION, BOLT CATCH	1
7	PAFZZ	1005000562247	19204	8448634	PLUNGER, BOLT CATCH	1
8	PAFZZ	1005000179548	19200	8448628	CATCH, BOLT	1
9	PAFZZ	5315008123312	96906119	MS16562-119	PIN, SPRING , BOLT CATCH	1
10	PAFZZ	1005000562201	19204	8448638	CATCH, MAGAZINE	1
11	PAOZZ	5360009926655	19204	8448586	SPRING, HELICAL, COMP COMPRESSION, TAKE DOWN/PIVOT PIN (M16A2) UOC:AR8	2
11A	PAOZZ	5360009926655	19204	8448586	SPRING, HELICAL, COMP COMPRESSION (M4 AND M4Al) UOC:AS1	1
12	PAOZZ	5315009926654	19204	8448585	PIN, STRAIGHT, HEADLE DETENT, TAKEDOWN PIN (M16A2) UOC:ARS	2
12A	PAOZZ	5315009926654	19204	8448585	PIN, STRAIGHT, HEADLE HEADLESS (M4 AND M4A1) UOC:AS1, AY6	1
13	PAFZZ	5360009927301	19204	8448637	SPRING, HELICAL, COMP COMPRESSION, MAGAZINE CATCH	1
14	PAFZZ	1005009927302	19204	8448636	BUTTON, MAGAZINE CAT CATCH	1
15	PAOZZ	5315000179537	19204	8448621	PIN, GROOVED, HEADED (PIVOT PIN)	1
16	PAFZZ	5340011457910	19200	9349114	LEVER, LOCK-RELEASE SEMI (M16A2 AND M4) UOC:AR8, AS1	1
16A	PAFZZ	1005009990406	19200	8448635	DISCONNECTOR (M4A1) UOC:AY6	1
17	PAFZZ	5340011441499	19200	9349113	LEVER, LOCK-RELEASE , BURST (MI6A2 AND M4)	1

(1) ITEM NO	(2) SMR CODE	(3) NSN	(4) CAGEC	(5) PART NUMBER	(6) DESCRIPTION AND USABLE ON CODE (UOC)	(7) QTY
					UOC:AR8, AS1	
18	AFFFF		19200	9349115	TRIGGER ASSEMBLY (M16A2)........	1
					UOC:AR8........	
18A	AFFFF		19200	12972697	TRIGGER ASSEMBLY (M4)........	1
					UOC:AS1........	
18B	AFFFF		19200	12972698	TRIGGER ASSEMBLY (M4A1)........	1
					UOC:AY6........	
19	PAOZZ	1005011484805	19200	9349127	GRIP, RIFLE PLASTIC, BLACK........	1
20	PAOZZ	5310005273634	96906	MS35335-61	WASHER, LOCK , RIFLE GRIP........	1
21	PAOZZ	5305012681191	88044	AN501D416-18	SCREW, MACHINE , RIFLE GRIP........	1
22	PAOZZ	5360009927292	19204	8448516	SPRING, HELICAL, COMP SAFETY........	1
23	PAOZZ	1005009926667	19204	8448631	DETENT, SAFETY........	1
24	PAFZZ	5315009927309	19204	8448609	PIN, GROOVED, HEADLES TRIGGER AND	2
					HAMMER........	
25	PAFZZ	5315009926650	19204	8448599	PIN, GROOVED, HEADLES , AUTOMATIC SEAR........	1
26	PAFZZ	5315009926653	19204	8448584	PIN, GROOVED, HEADED (M16A2)........	1
27	PAOZZ	5365012672169	19200	12597640	SPACER, STEPPED (M16A2)........	1
					UOC:AR8........	
28	PAOOO	1005011354973	19200	9349119	BUTTSTOCK ASSEMBLY, (M16A2)........	1
					UOC:AR8........	
28A	AOOOO		19200	9390012	BUTTSTOCK ASSY (M4 AND M4A1)........	1
					UOC:AS1, AY6........	
29	PAOZZ	5305011478585	19200	9349128	SCREW, MACHINE , BUTTCAP (M16A2)........	1
					UOC:AR8........	
30	XAFFA		19200	9349101	LOWER RECEIVER AND RECEIVER........	1
					EXTENSION ASSEMBLY (M16A2)........	
					UOC:AR8........	
30A	XAFFA		99999	NPN	LOWER RECEIVER ASSY ASSEMBLY (M4)........	1
					UOC:AS1........	
30B	XAFFA		99999	NPN	LOWER RECEIVER ASSY (M4Al)........	1
					UOC:AY6........	
					END OF FIGURE........	

SECTION II

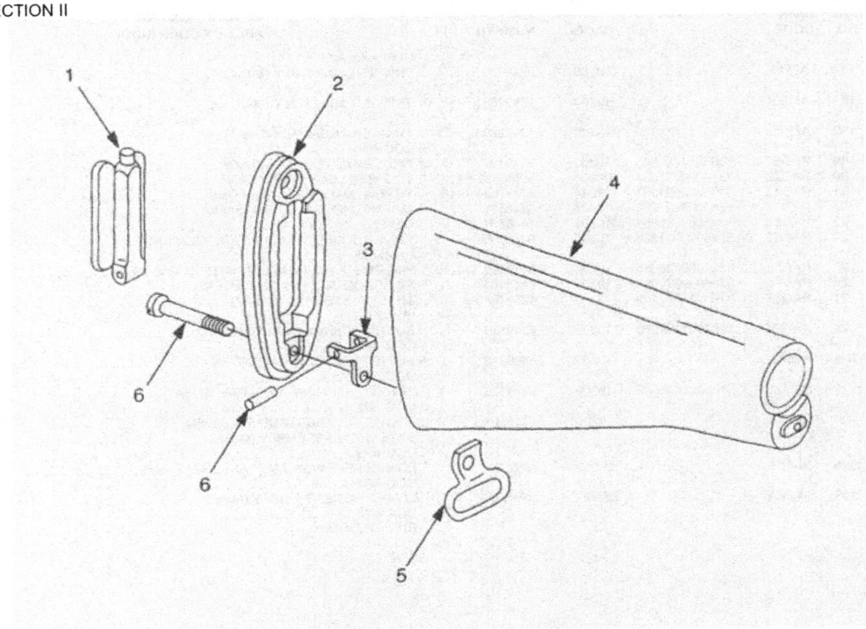

Figure C-12. (M16A2) Buttstock Assembly 9349119.

(1) ITEM NO	(2) SMR CODE	(3) NSN	(4) CAGEC	(5) PART NUMBER	(6) DESCRIPTION AND USABLE ON CODE (UOC)	(7) QTY
					GROUP 0401 (M16A2) BUTTSTOCK ASSEMBLY 9349119 FIG C-12 (M16A2) BUTTSTOCK ASSEMBLY	
1	PAOZZ	1005012288504	19200	9381380	DOOR ASSEMBLY, THUMB UOC:AR8	1
2	PAOZZ	1005011467685	19200	9349130	PLATE, BUTT, SHOULDER GUN STOCK UOC:AR8	1
3	PAOZZ	5340004633892	19200	8448653	HINGE, ACCESS DOOR BUTT PLATE UOC:AR8	1
4	XAOZZ		19200	9349121	BUTTSTOCK UOC:AR8	1
5	PAOZZ	1005004030964	19204	8448652	SWIVEL, SLING, SMALL UOC:AR8	1
6	PAOZZ	5315004633894	19204	8448655	PIN, STRAIGHT, HEADLE HEADLESS, ACCESS DOOR UOC:AR8	1
7	PAOZZ	5305011441494	19200	9349120	SCREW, MACHINE BUTT PLATE UOC:AR8	1

END OF FIGURE

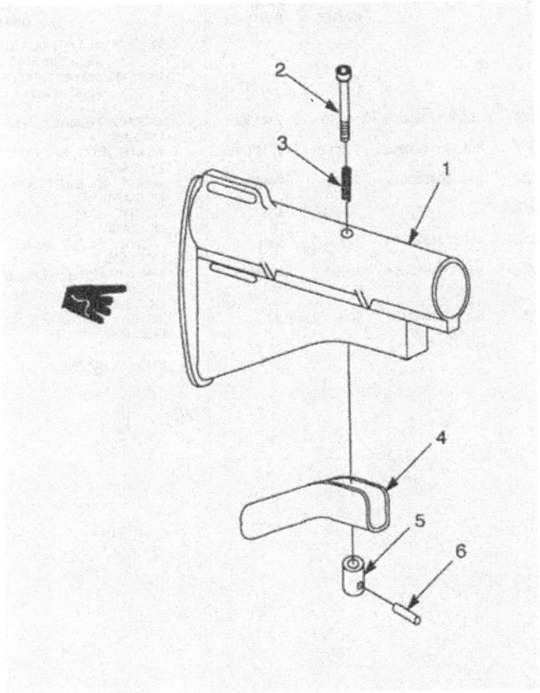

Figure C12A. (M4 and M4A1) Buttstock Assembly 9390012.

(1) ITEM NO	(2) SMR CODE	(3) NSN	(4) CAGEC	(5) PART NUMBER	(6) DESCRIPTION AND USABLE ON CODE (UOC)	(7) QTY
					GROUP 0401A (M4 AND M4A1) 9390012 BUTTSTOCK ASSEMBLY FIG C12A (M4 AND M4A1) BUTTSTOCK ASSEMBLY	
1	PAOZZ	1005012338636	19200	9390013	STOCK, GUN, SHOULDER UOC:AS1, AY6	1
2	PAOZZ	5315012338608	19200	9390025	PIN, SHOULDER, HEADLE UOC:AS1, AY6	1
3	PAOZZ	5360012338616	19200	9390027	SPRING, HELICAL, COMP COMPRESSION UOC:AS1, AY6	1
4	PAOZZ	5340012338638	19200	9390014	LEVER, LOCK-RELEASE UOC:AS1, AY6	1
5	PAOZZ	5310012338626	19200	9390026	NUT, SELF-LOCKING, RO UOC:AS1, AY6	1
6	PAOZZ	5315008439487	96906	MS16562-202	PIN, SPRING UOC:AS1, AY6	1

END OF FIGURE

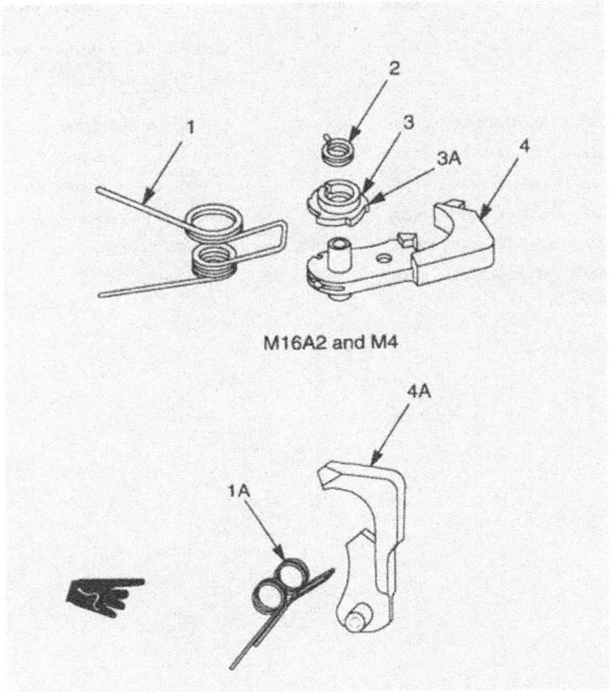

Figure C-13. Hammer Assembly(M16A2) 9349106, (M4) 9390032, and (M4A1) 8448610.

(1) ITEM NO	(2) SMR CODE	(3) NSN	(4) CAGEC	(5) PART NUMBER	(6) DESCRIPTION AND USABLE ON CODE (UOC)	(7) QTY
					GROUP 0402 HAMMER ASSEMBLY (M16A2) 9349106, (M4) 9390032, AND (M4A1) 8448610 FIG C-13 HAMMER ASSEMBLY	
1	PAFZZ	5360011441492	19200	9349107	SPRING, HELICAL, TORS HAMMER (MI6A2 AND M4) UOC:AR8, AS1	1
1A	PAFZZ	5360009926648	19204	8448611	SPRING, HELICAL, TORS (M4A1) UOC:AY6	1
2	PAFZZ	5360011365471	19200	9349109	SPRING, HELICAL, TORS BURST CAM (M16A2 AND M4) UOC:AR8, AS1	1
3	PAFZZ	1005011480172	19200	9349108	CAM, BURST (M16A2) (BLACK) UOC:AR8	1
3A	PAFZZ	3040012477969	19200	9390031	CAM, CONTROL (M4) (NICKEL/SHINY) UOC:AS1	1
4	PAFZZ	1005011343630	19200	9349110	HAMMER AND HAMMER P PIN RETAINER ASSEMBLY (MI6A2 AND M4) UOC:AR8, AS1	1
4A	PAFZZ	1005000179551	19200	8448612	HAMMER, FIRING, SMALL (M4A1) UOC:AY6	1
					END OF FIGURE	

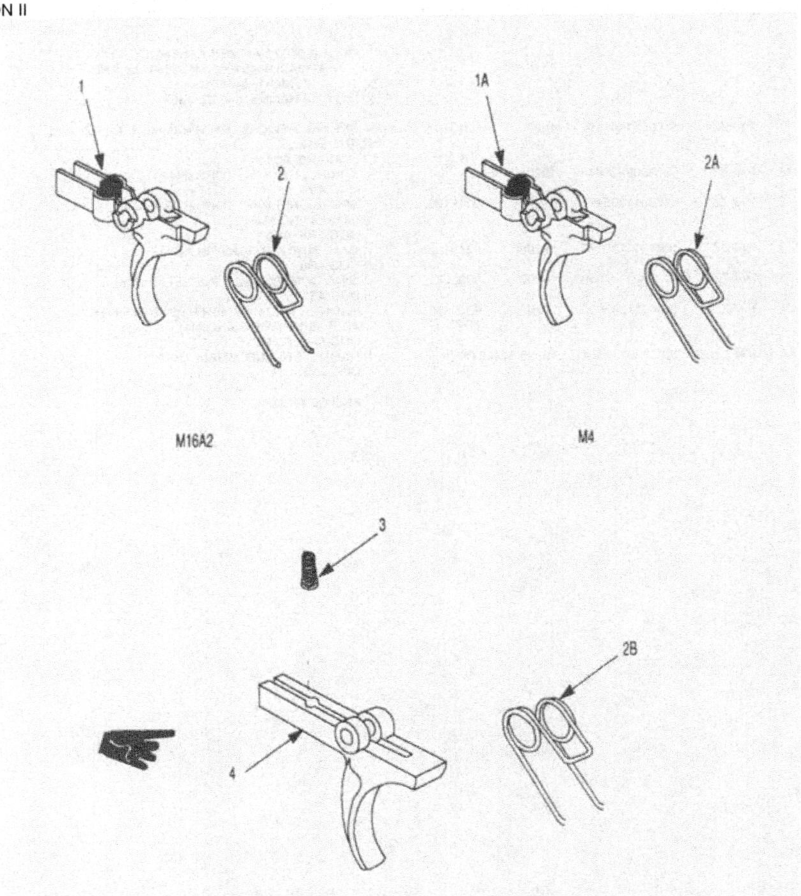

Figure C-14. Trigger Assembly(M16A2) 9349115, (M4) 12972697, and (M4A1) 12972676

(1) ITEM NO	(2) SMR CODE	(3) NSN	(4) CAGEC	(5) PART NUMBER	(6) DESCRIPTION AND USABLE ON CODE (UOC)	(7) QTY
					GROUP 0403 TRIGGER ASSEMBLY (M16A2) 9349115, (M4) 12972697, AND (M4A1) 12972676 FIG C-14 TRIGGER ASSEMBLY	
1	PAFFF	1005012192402	19200	9392518	TRIGGER SUB ASSEMBL (M16A2) UOC:AR8	1
1A	PAFFF	1005013954257	19200	12972696	TRIGGER SUBASSEMBLY (M4) UOC:AS1	1
2	PAFZZ	5360009927308	19204	8448593	SPRING, HELICAL, TORS TORSION, TRIGGER (MI6A2) UOC:AR8	1
2A	PAFZZ	5360009927308	19204	8448593	SPRING, HELICAL, TORS (M4) UOC:ASI	1
2B	PAFZZ	5360009927308	19204	8448593	SPRING, HELICAL, TORS (M4A1) UOC:AY6	1
3	PAFZZ	5360013960256	19200	12972695	SPRING, HELICAL, COMP DISCONNECT (M4A1) UOC:AY6	1
4	PAFZZ	1005009927307	19204	8448592	TRIGGER (M4A1) UOC:AY6	1

END OF FIGURE

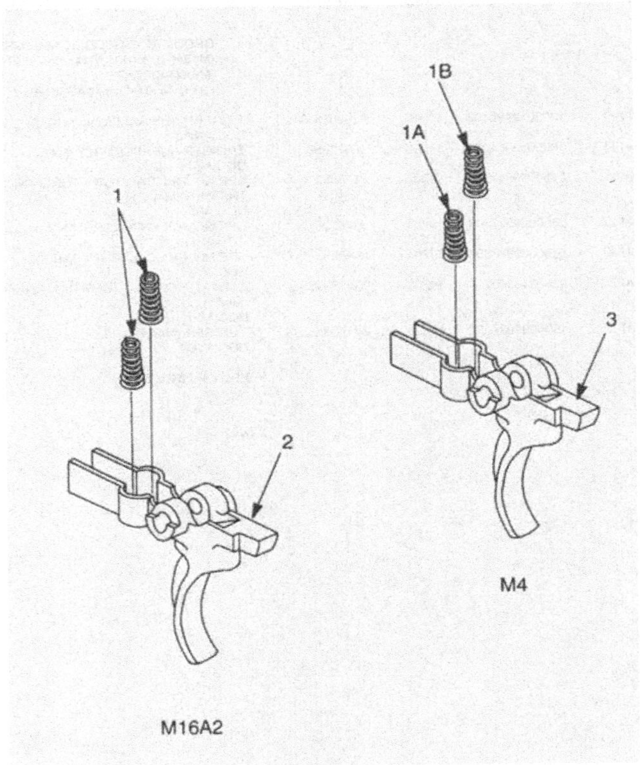

Figure C-15. Trigger Subassembly(M16A2) 9392518, (M4) 12972696.

(1) ITEM NO	(2) SMR CODE	(3) NSN	(4) CAGEC	(5) PART NUMBER	(6) DESCRIPTION AND USABLE ON CODE (UOC)	(7) QTY
					GROUP 040301 TRIGGER SUBASSEMBLY (M6A2) 9392518 AND (M4) 12972696 FIG C-15 TRIGGER SUBASSEMBLY	
1	PAFZZ	5360011350353	19200	9349116	SPRING, HELICAL, COMP DISCONNECT (M16A2) UOC:AR8	2
1A	PAFZZ	5360011350353	19200	9349116	SPRING, HELICAL, COMP DISCONNECT (M4) (NICKEL/SHINY) UOC:AS1	1
1B	PAFZZ	5360013960256	19200	12972695	SPRING, HELICAL, COMP DISCONNECT (M4) (BLACK) UOC:AS1	1
2	XAFZZ		19200	9390736	TRIGGER (M16A2) UOC:AR8	1
3	XAFZZ		19200	9390736	TRIGGER (M4) UOC:AS1	1

END OF FIGURE

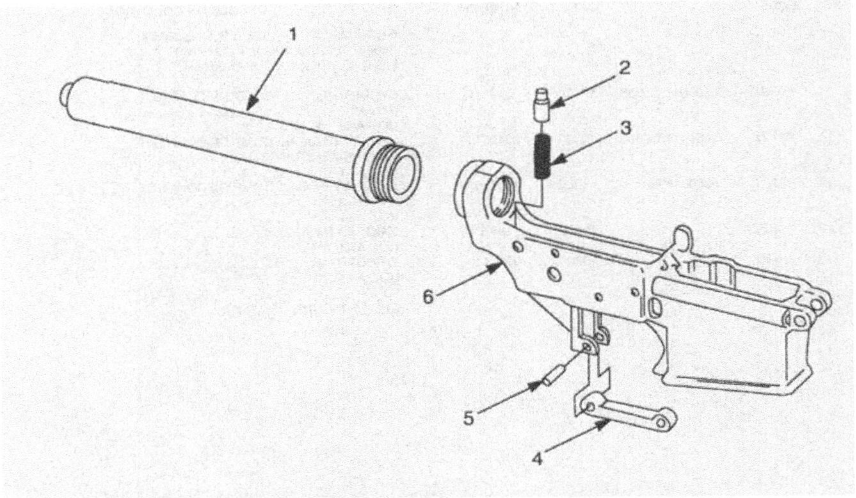

Figure C-16. (M16A2) Lower Receiver and Receiver extension Assembly 9349101.

(1) ITEM NO	(2) SMR CODE	(3) NSN	(4) CAGEC	(5) PART NUMBER	(6) DESCRIPTION AND USABLE ON CODE (UOC)	(7) QTY
					GROUP 0404 (M16A2) LOWER RECEIVER AND RECEIVER EXTENSION ASSEMBLY 9349101	
					FIG C-16 (M16A2) LOWER RECEIVER AND RECEIVER EXTENSION ASSEMBLY	
1	PAFZZ	5340009927297	19200	8448581	EXTENSION, LOWER REC RECEIVER .... (LOWER RECEIVER EXTENSION) (M1SA2)........................ UOC:AR8	1
2	PAFZZ	5315009926651	19204	8448582	PIN, SHOULDER, HEADLE BUFFER.................................... RETAINER (M16A2)... UOC:AR8	1
3	PAFZZ	5360009926652	19200	8448583	SPRING, HELICAL, COMP COMPRESSION, ... BUFFER RETAINER (NI6A2) UOC:AR8	1
4	PAFZZ	1005009927299	19204	8448587	GUARD, TRIGGER (M16A2)...	1
5	PAFZZ	5315000586081	96906	MS16562-129	PIN, SPRING TRIGGER GUARD (M16A2)...........................	1
6	XAFDA		19200	9349102	RECEIVER, (M16A2)... UOC:AR8	1

END OF FIGURE

SECTION II

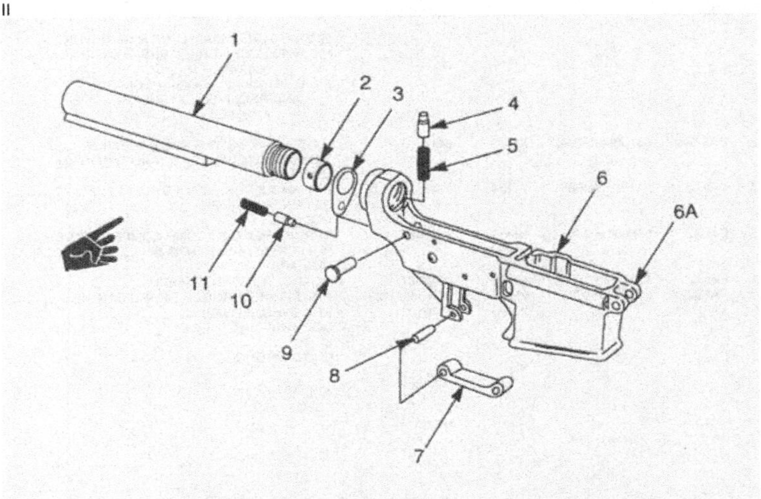

Figure C16A. (M4 and M4A1) Lower Receiver and Receiver extension Assembly, (NPN)

(1) ITEM NO	(2) SMR CODE	(3) NSN	(4) CAGEC	(5) PART NUMBER	(6) DESCRIPTION AND USABLE ON CODE (UOC)	(7) QTY
					GROUP 0404A (M4 AND M4A1) LOWER RECEIVER AND RECEIVER EXTENSION ASSEMBLY (NPN) FIG C16A (M4 AND M4A1) LOWER RECEIVER AND RECEIVER EXTENSION ASSEMBLY	
1	PAFZZ	1005012338531	19200	9390019	EXTENSION, LOWER RECEIVER (M4 AND M4A1) UOC:AS1, AY6	1
2	PAFZZ	5310012338625	19200	9390020	NUT, PLAIN, ROUND (M4 AND M4A1) UOC:AS1, AY6	1
3	PAFZZ	105012338530	19200	9390021	PLATE, RECEIVER END (M4 AND M4A1) UOC:AS1, AY6	1
4	PAFZZ	5315009926651	19204	8448582	PIN, SHOULDER, HEADLE (M4 AND M4A1) UOC:AS1, AY6	1
5	PAFZZ	5360009926652	19200	8448583	SPRING, HELICAL, COMP (M4 AND M4A1) UOC:AS1, AY6	1
6	XAFDA		19200	9390015	RECEIVER, (M4) UOC:AS1	1
6A	XAFDA		19200	12972652	LOWER RECEIVER (M4A1) UOC:AY6	1
7	PAFZZ	1005009927299	19204	8448587	GUARD, TRIGGER (M4 AND M4A1)	1
8	PAFZZ	5315000586081	96906	MS16562-129	PIN, SPRING (TRIGGER GUARD PIN) (M4 AND M4A1)	1
9	PAFZZ	5315009926653	19204	8448584	PIN, GROOVED, HEADED (TAKE DOWN) (M4 AND M4A1)	1
10	PAFZZ	5315009926654	19204	8448585	PIN, STRAIGHT, HEADLE HEADLESS (M4 AND M4A1) UOC:AS1, AY6	1
11	PAFZZ	5360009926655	19204	8448586	SPRING, HELICAL, COMP COMPRESSION (M4 AND M4A1) UOC:AS1, AY6	1

END OF FIGURE

SECTION II

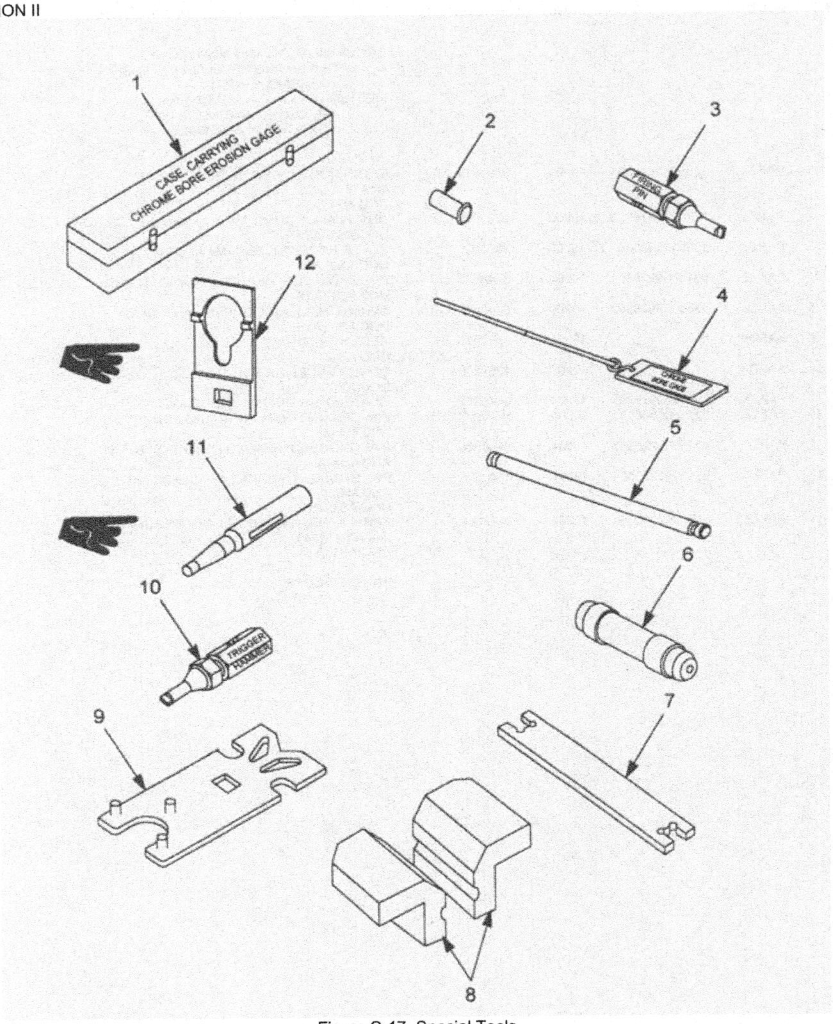

Figure C-17. Special Tools

(1) ITEM NO	(2) SMR CODE	(3) NSN	(4) CAGEC	(5) PART NUMBER	(6) DESCRIPTION AND USABLE ON CODE (UOC)	(7) QTY
					GROUP 9500 SPECIAL TOOLS FIG C-17 SPECIAL TOOLS	
	PEFZZ	1005000567106	19204	8426685	MAINTENANCE KIT, GUN DS/GS SUPPORT MAINTENANCE FOR 5.S6MM RIFLE, M16 RIFLE SERIES BOI: 2 PER SUPPORTING DSU/GSU	
1	PAFZZ	4933010355607	19204	12006359	CASE, BORE GAGE PART OF KIT P/N 8426685........	
2	PAFZZ	4933008007508	19204	8448201	..REFLECTOR TOOL, CHAMBER, PART OF KIT P/N 8426685	
3	PAFZZ	5220010755004	19200	12620101	..GAGE, PLUG, PLAIN PART OF KIT P/N 8426685........	
4	PAFZZ	5220010148183	19204	8448496	..GAGE, BARREL, EROSION BARREL EROSION PART OF KIT P/N 8426685(CHROME BARREL)........	
5	PAFZZ	5220002219391	19204	8448202	..GAGE, STRAIGHTNESS PART OF KIT P/ N 8426685	
6	PAFZZ	5220000707814	19204	7799734	..GAGE, HEADSPACE PART OF KIT P/N 8426685........	
7	PAFZZ	6695000707815	19204	7799735	..GAGE, FIRING PIN PROTRUSION PART OF KIT P/N 8426685	
8	PAFZZ	4933000709151	19204	11010032	..FIXTURE, BARREL REMOVAL PART OF KIT P/N 8426685	
9	PAFZZ	5120000709152	19204	11010033	..WRENCH, COMBINATION PART OF KIT P/ N 8426685	
10	PAFZZ	5220010439473	19204	12006472	GAGE, PLUG, TAPER CYLINDER PART OF KIT P/N 8426685	
11	PAOZZ	5315013100370	19200	12926769	KEY, MACHINE KEY, 2 PER ORGANIZATIONAL OR DIRECT SUPPORT SHOP	
12	PAFZZ	5120013246631	19200	9390035	WRENCH, SPANNER 2 PER DS/GS SUPPORT SHOP A/R FOR M4/M4A1 CARBINE UOC:AS1, AY6	
					END OF FIGURE	

CROSS-REFERENCE INDEXES

NATIONAL STOCK NUMBER INDEX

STOCK NUMBER	FIG	ITEM	STOCK NUMBER	FIG	ITEM
5315-00-017-9537	C-11	15	5315-00-979-3930	C-7	2
3040-00-017-9539	C-9	5	5360-00-979-3931	C-7	3
1005-00-017-9540	C-9	4	5360-00-992-6648	C-13	1A
5360-00-017-9541	C-8	13	1005-00-992-6649	C-11	4
1005-00-017-9543	C-7	4	5315-00-992-6650	C-11	25
1005-00-017-9546	C-1	1	5315-00-992-6651	C-16	2
1005-00-017-9547	C-2	1		C16A	4
1005-00-017-9548	C-11	8	5360-00-992-6652	C-16	3
1005-00-017-9551	C-13	4A		C16A	5
1005-00-056-2201	C-11	10	5315-00-992-6653	C-11	26
5360-00-056-2246	C-11	6		C16A	9
1005-00-056-2247	C-11	7	5315-00-992-6654	C-11	12
1005-00-056-7106	C-17			C-11	12A
5315-00-058-6044	C-6	7		C16A	10
5315-00-058-6081	C-16	5	5360-00-992-6655	C-11	11
	C16A	8		C-11	11A
5315-00-058-6678	C-1B	1		C16A	11
	C-10	1	5360-00-992-6665	C-11	1
5220-00-070-7814	C-17	6	1005-00-992-6667	C-11	23
6695-00-070-7815	C-17	7	1005-00-992-7283	C-4	2
4933-00-070-9151	C-17	8	5305-00-992-7284	C-4	1
5120-00-070-9152	C-17	9	1005-00-992-7285	C-2	4
1005-00-087-8998	C-6	8	1005-00-992-7287	C-3	1
3110-00-183-9175	C-8	4	1005-00-992-7288	C-3	7
	C-1A	6	1005-00-992-7290	C-3	3
	C-1B	5	1005-00-992-7291	C-3	6
	C-10	5	5360-00-992-7292	C-3	5
5220-00-221-9391	C-17	5		C-11	22
1005-00-403-0964	C-12	5	5315-00-992-7294	C-2	3
5340-00-463-3892	C-12	3	5340-00-992-7297	C-16	1
5315-00-463-3894	C-12	6	1005-00-992-7299	C-16	4
5360-00-523-8084	C-9	3		C16A	7
5310-00-527-3634	C-11	20	5360-00-992-7301	C-11	13
5315-00-597-5086	C-3	4	1005-00-992-7302	C-11	14
5315-00-690-0544	C-7	8	1005-00-992-7307	C-14	4
1005-00-738-6213	C-4	3	5360-00-992-7308	C-14	2
1005-00-760-3768	C-3	8		C-14	2A
4933-00-800-7508	C-17	2		C-14	2B
5315-00-812-3312	C-11	9	5315-00-992-7309	C-11	24
5315-00-840-3812	C-8	11	5360-00-999-0404	C-5	3
	C-8	12	5340-00-999-0405	C-5	2
	C-1A	8	1005-00-999-0406	C-11	16A
5315-00-843-9487	C12A	6	5325-00-999-0863	C-6	10
1005-00-921-5004	C-1	4	5325-00-999-0864	C-8	7
1005-00-937-3078	C-11	2	1005-00-999-1509	'C-2	2
1005-00-978-1022	C-8	9	5220-01-014-8183	C-17	4
5315-00-978-1023	C-8	8	4933-01-035-5607	C-17	1
5360-00-978-1025	C-8	10	5220-01-043-9473	C-17	10
5360-00-978-1036	C-6	9	5315-01-048-9372	C-5	1
4710-00-978-1038	C-6	3		C-9	2

CROSS-REFERENCE INDEXES

NATIONAL STOCK NUMBER INDEX

STOCK NUMBER	FIG	ITEM
5320-01-063-7635	C-7	5
5220-01-075-5004	C-17	3
1005-01-134-3621	C-8	1
5305-01-134-3622	C-8	2
	C-1A	2
1005-01-134-3625	C-7	1
5355-01-134-3627	C-1B	8
	C-10	8
1005-01-134-3629	C-6	2
1005-01-134-3630	C-13	4
1005-01-134-3631	C-10	2
1005-01-134-3633	C-6	6
1005-01-134-3701	C-8	6
5360-01-134-3710	C-8	15
	C-1A	5
5360-01-135-0353	C-15	1
	C-15	1A
1005-01-135-3697	C-1B	6
	C-10	6
5355-01-135-4972	C-8	16
1005-01-135-4973	C-11	28
5360-01-136-5471	C-13	2
1005-01-144-1468	C-9	1
5305-01-144-1490	C-1B	3
	C-10	3
5360-01-144-1492	C-13	1
5305-01-144-1494	C-12	7
5365-01-144-1496	C-6	5
5340-01-144-1499	C-11	17
5340-01-145-7910	C-11	16
1005-01-146-7684	C-6	4
1005-01-146-7685	C-12	2
5305-01-147-8585	C-11	29
1005-01-148-0172	C-13	3
5360-01-148-1751	C-8	5
	C-1B	4
	C-10	4
1005-01-148-4805	C-11	19
1005-01-216-4510	C-1	5
1005-01-219-2402	C-14	1
5340-01-225-8339	C-11	5
1005-01-228-8504	C-12	1
1005-01-231-3138	C-11	2A
1005-01-233-8529	C-6	4A
1005-01-233-8530	C16A	3
1005-01-233-8531	C16A	1
5315-01-233-8608	C12A	2
5360-01-233-8616	C12A	3
5360-01-233-8617	C-11	1A
5310-01-233-8625	C16A	2
5310-01-233-8626	C12A	5
1005-01-233-8636	C12A	1
4710-01-233-8637	C-6	3A
5340-01-233-8638	C12A	4
1005-01-234-2297	C-6	2A
3040-01-247-7969	C-13	3A
1010-01-264-6517	C-7	9
5340-01-264-6530	C-7	7
5365-01-267-2169	C-11	27
5305-01-268-1191	C-11	21
5315-01-310-0370	C-17	11
5120-01-324-6631	C-17	12
1005-01-368-9852	C-1	5A
5360-01-381-6183	C-1B	7
	C-10	7
5340-01-382-3201	C-1A	10
5310-01-382-6793	C-1A	9
1005-01-382-6795	C-8	6A
5355-01-382-6801	C-1A	4
5360-01-382-6802	C-1A	7
1005-01-382-7083	C-1A	11
1005-01-382-7086	C-1B	2
1005-01-382-7089	C-1A	3
1005-01-395-4257	C-14	1A
5360-01-396-0256	C-14	3
	C-15	1B
1005-01-422-3770	C-2	4A
1005-01-424-5899	C-3	9

CROSS-REFERENCE INDEXES

NATIONAL STOCK NUMBER INDEX

STOCK NUMBER	FIG	ITEM	STOCK NUMBER	FIG	ITEM
88044	AN5010416-18		5305-01-268-1191	C-11	21
96906	MS16562-103		5315-00-058-6678	C-1B	1
				C-10	1
96906	MS16562-106		5315-00-058-6044	C-6	7
96906	MS16562-119		5315-00-812-3312	C-11	9
96906	MS16562-121		5315-00-840-3812	C-8	11
				C-8	12
				C-1A	8
96906	MS16562-129		5315-00-058-6081	C-16	5
				C16A	8
96906	MS16562-202		5315-00-843-9487	C12A	6
96906	MS16562-98		5315-00-597-5086	C-3	4
96906	MS16626-3137		5325-00-999-0863	C-6	10
96906	MS16632-3012		5325-00-999-0864	C-8	7
96906	MS19060-4808		3110-00-183-9175	C-8	4
				C-1A	6
				C-1B	5
				C-10	5
96906	MS35335-61		5310-00-527-3634	C-11	20
96906	MS39086-93		5315-00-690-0544	C-7	8
99999	NPN			C-11	30A
				C-11	30B
19204	11010032		4933-00-070-9151	C-17	8
19204	11010033		5120-00-070-9152	C-17	9
19204	12006359		4933-01-035-5607	C-17	1
19204	12006472		5220-01-043-9473	C-17	10
19200	12011849			C-1	2A
19200	12011987		5360-01-381-6183	C-1B	7
				C-10	7
19200	12011996		1005-01-368-9852	C-1	5A
19200	12597640		5365-01-267-2169	C-11	27
19200	12598617		1010-01-264-6517	C-7	9
19200	12598618		5340-01-264-6530	C-7	7
19200	12620101		5220-01-075-5004	C-17	3
19204	12624561		1005-01-216-4510	C-1	5
19200	12926769		5315-01-310-0370	C-17	11
19200	12951011			C-1	7
19200	12951017		5340-01-382-3201	C-1A	10
19200	12951018		1005-01-382-7089	C-1A	3
19200	12951019		5355-01-382-6801	C-1A	4
19200	12951020		5360-01-382-6802	C-1A	7
19200	12951021		1005-01-382-7083	C-1A	11
19200	12951023		5310-01-382-6793	C-1A	9
19200	12951026			C-1A	1
19200	12951028		1005-01-382-7086	C-1B	2
19200	12972652			C16A	6A
19200	12972670		1005-01-382-6795	C-8	6A
19200	12972675			C-6	1A
19200	12972680			C-1	3A
19200	12972690			C-1	6B
19200	12972691		1005-01-422-3770	C-2	4A

CROSS-REFERENCE INDEXES

PART NUMBER INDEX

CAGEC	PART NUMBER	STOCK NUMBER	FIG	ITEM
19200	12972692	1005-01-424-5899	C-3	9
19200	12972695	5360-01-396-0256	C-14	3
			C-15	1B
19200	12972696	1005-01-395-4257	C-14	1A
19200	12972697		C-11	18A
19200	12972698		C-11	18B
19204	7799734	5220-00-070-7814	C-17	6
19204	7799735	6695-00-070-7815	C-17	7
19204	8426685	1005-00-056-7106	C-17	
19204	8448201	4933-00-800-7508	C-17	2
19204	8448202	5220-00-221-9391	C-17	5
19204	8448496	5220-01-014-8183	C-17	4
19204	8448501		C-1	2
19204	8448502	5315-00-992-7294	C-2	3
19204	8448503	1005-00-017-9547	C-2	1
19204	8448504	1005-00-999-1509	C-2	2
19204	8448505		C-2	5
19200	8448506	1005-00-992-7283	C-4	2
19200	8448507	1005-00-738-6213	C-4	3
19204	8448508	5305-00-992-7284	C-4	1
19200	8448509	1005-00-992-7285	C-2	4
19204	8448510		C-3	2
19204	8448511	1005-00-992-7287	C-3	1
19204	8448512	1005-00-992-7288	C-3	7
19204	8448513	1005-00-992-7290	C-3	3
19204	8448515	1005-00-992-7291	C-3	6
19204	8448516	5360-00-992-7292	C-3	5
			C-11	22
19204	8448517	1005-00-017-9546	C-1	1
19204	8448518		C-5	4
19200	8448519	5340-00-999-0405	C-5	2
19204	8448520	5360-00-999-0404	C-5	3
19204	8448521-2	5315-01-048-9372	C-5	1
			C-9	2
19204	8448525	1005-00-978-1022	C-8	9
19204	8448532	5360-00-978-1025	C-8	10
19204	8448533	5315-00-978-1023	C-8	8
19200	8448540	5360-00-017-9541	C-8	13
19200	8448542	5360-00-523-8084	C-9	3
19204	8448543	3040-00-017-9539	C-9	5
19204	8448544	1005-00-017-9540	C-9	4
19204	8448555	5360-00-978-1036	C-6	9
19200	8448567	4710-00-978-1038	C-6	3
19204	8448571	1005-00-017-9543	C-7	4
19204	8448573	5315-00-979-3930	C-7	2
19204	8448574	5360-00-979-3931	C-7	3
19200	8448581	5340-00-992-7297	C-16	1
19204	8448582	5315-00-992-6651	C-16	2
			C16A	4
19200	8448583	5360-00-992-6652	C-16	3
			C16A	5

CROSS-REFERENCE INDEXES

PART NUMBER INDEX.

CAGEC	PART NUMBER	STOCK NUMBER	FIG	ITEM
19204	8448584	5315-00-992-6653	C-11	26
			C16A	9
19204	8448585	5315-00-992-6654	C-11	12
			C-11	12A
			C16A	10
19204	8448586	5360-00-992-6655	C-11	11
			C-11	11A
			C16A	11
19204	8448587	1005-00-992-7299	C-16	4
			C16A	7
19204	8448592	1005-00-992-7307	C-14	4
19204	8448593	5360-00-992-7308	C-14	2
			C-14	2A
			C-14	2B
19200	8448595	1005-00-992-6649	C-11	4
19204	8448599	5315-00-992-6650	C-11	25
19204	8448609	5315-00-992-7309	C-11	24
19204	8448610		C-11	3B
19204	8448611	5360-00-992-6648	C-13	1A
19200	8448612	1005-00-017-9551	C-13	4A
19200	8448615	1005-00-937-3078	C-11	2
19204	8448621	5315-00-017-9537	C-11	15
19200	8448628	1005-00-017-9548	C-11	8
19204	8448629	5360-00-992-6665	C-11	1
19204	8448631	1005-00-992-6667	C-11	23
19204	8448633	5360-00-056-2246	C-11	6
19204	8448634	1005-00-056-2247	C-11	7
19200	8448635	1005-00-999-0406	C-1l	11A
19204	8448636	1005-00-992-7302	C-11	14
19204	8448637	5360-00-992-7301	C-11	13
19204	8448638	1005-00-056-2201	C-11	10
19204	8448652	1005-00-403-0964	C-12	5
19200	8448653	5340-00-463-3892	C-12	3
19204	8448655	5315-00-463-3894	C-12	6
19200	8448670	1005-00-921-5004	C-1	4
19204	8448697	5320-01-063-7635	C-7	5
19204	8448712	1005-00-087-8998	C-6	8
19200	8448755	1005-00-760-3768	C-3	8
19200	9349050		C-1	3
19200	9349051	1005-01-134-3633	C-6	6
19200	9349052	5365-01-144-1496	C-6	5
19200	9349054		C-7	6
19200	9349056	1005-01-134-3625	C-7	1
19200	9349059	1005-01-134-3629	C-6	2
19200	9349062		C-6	1
19200	9349063	1005-01-134-3701	C-8	6
19200	9349065	5305-01-134-3622	C-8	2
			C-1A	2
19200	9349066	1005-01-134-3621	C-8	1
19200	9349067	5355-01-135-4972	C-8	16
19200	9349069	5360-01-148-1751	C-8	5

CROSS-REFERENCE INDEXES

PART NUMBER INDEX

CAGEC	PART NUMBER	STOCK NUMBER	FIG	ITEM
19200	9349069	5360-01-148-1751	C-1B	4
			C-10	4
19200	9349070	5360-01-134-3710	C-8	15
			C-1A	5
19200	9349072		C-8	3
19200	9349074	1005-01-134-3631	C-10	2
19200	9349075	1005-01-135-3697	C-1B	6
			C-10	6
19200	9349076	5305-01-144-1490	C-1B	3
			C-10	3
19200	9349077	5355-01-134-3627	C-1B	8
			C-10	8
19200	9349085	1005-01-144-1468	C-9	1
19204	9349086		C-8	14
19200	9349100		C-1	6
19200	9349101		C-11	30
19200	9349102		C-16	6
19200	9349106		C-11	3
19200	9349107	5360-01-144-1492	C-13	1
19200	9349108	1005-01-148-0172	C-13	3
19200	9349109	5360-01-136-5471	C-13	2
19200	9349110	1005-01-134-3630	C-13	4
19200	9349113	5340-01-144-1499	C-11	17
19200	9349114	5340-01-145-7910	C-11	16
19200	9349115		C-11	18
19200	9349116	5360-01-135-0353	C-15	1
			C-15	1A
19200	9349119	1005-01-135-4973	C-11	28
19200	9349120	5305-01-144-1494	C-12	7
19200	9349121		C-12	4
19200	9349124	1005-01-146-7684	C-6	4
19200	9349127	1005-01-148-4805	C-11	19
19200	9349128	5305-01-147-8585	C-11	29
19200	9349130	1005-01-146-7685	C-12	2
19200	9381367	5340-01-225-8339	C-11	5
19200	9381380	1005-01-228-8504	C-12	1
19200	9390003	1005-01-234-2297	C-6	2A
19200	9390007	1005-01-233-8529	C-6	4A
19200	9390009		C-7	6A
19200	9390011		C-1	6A
19200	9390012		C-11	28A
19200	9390013	1005-01-233-8636	C12A	1
19200	9390014	5340-01-233-8638	C12A	4
19200	9390015		C16A	6
19200	9390016	4710-01-233-8637	C-6	3A
19200	9390019	1005-01-233-8531	C16A	1
19200	9390020	5310-01-233-8625	C16A	2
19200	9390021	1005-01-233-8530	C16A	3
19200	9390022	5360-01-233-8617	C-11	1A
19200	9390023	1005-01-231-3138	C-11	2A
19200	9390025	5315-01-233-8608	C12A	2

CROSS-REFERENCE INDEXES

PART NUMBER INDEX

CAGEC	PART NUMBER	STOCK NUMBER	FIG	ITEM
19200	9390026	5310-01-233-8626	C12A	5
19200	9390027	5360-01-233-8616	C12A	3
19200	9390031	3040-01-247-7969	C-13	3A
19200	9390032		C-11	3A
19200	9390035	5120-01-324-6631	C-17	12
19200	9390736		C-15	2
			C-15	3
19200	9392518	1005-01-219-2402	C-14	1

APPENDIX D
EXPENDABLE/DURABLE SUPPLIES AND MATERIALS LIST

Section I. INTRODUCTION

D-1. SCOPE. This appendix lists expendable durable supplies and materials you will need to operate and maintain the M16A2 Rifle This listing is for informational purposes only and Is not authority to requisition the listed Items. These items are authorized to you by CTA 50-970, Expendable, Durable Items (Except Medical, Class V, Repair Parts, and Heraldic Items), or CTA 8-100, Army Medical Department Expendable/Durable Items.

D-2. EXPLANATION OF COLUMNS.

a. Column (1)-Item Number. This number is assigned to the entry in the listing and is referenced In the narrative instructions to identify the material (e.g., "Use cloth, abrasive, crocus, Item 12, app D").

b. Column (2)-Level. This column identifies the lowest level of maintenance that requires the listed Item.

C - Operator/Crew Maintenance
O - Unit Maintenance
F - Direct Support Maintenance

c. Column (3)-National Stock Number. This is the National stock number assigned to the Item, use It to request or requisition the Item.

d. Column (4)-Description. Indicates the Federal Item name and, if required, a description to identify the item. The last line for each Item indicates the Commercial and Government Entity Code (CAGEC) in parentheses followed by the part number.

e. Column (5)-Unit of Measure (U/M). Indicates the measure used in performing the actual maintenance function. This measure is expressed by a two-character alphabetical abbreviation (e g., ea, In , pr). If the unit of measure differs from the unit of issue, requisition the lowest unit of issue that will satisfy your requirements.

Section II. EXPENDABLE/DURABLE SUPPLIES AND MATERIALS LIST

(1) ITEM NUMBER	(2) LEVEL	(3) NATIONAL STOCK NUMBER	(4) DESCRIPTION PART NUMBER AND CAGE	(5) UNIT OF MEAS
1	F	8040 00-944 7292	ADHESIVE, I<IT, (81348) MMM A 1754	KT
2	0	8020 00 244 0153	BRUSH, ARTIST'S metal ferrule, flat chisel ed(i(e 7 16 w, 1 1/8, ex- posed bristle (813481 H B 241	EA
3	F	1005-00-716 2702	BRUSH, CLEANING, SMALL ARMS (192051 7162702	EA
4	C	1005-00-903 1296	BRUSH, CLEANING, SMALL ARMS bore (19204) 11686340	EA
5	C	1005-00 999-1435	BRUSH, CLEANING, SMALL ARMS chamber (19204) 8432358	EA
6	C	1005 00-444-6602	BRUSH, CLEANING, SMALL ARMS tooth 1192041 8448462	EA
7	O	7920-00 205-2401	BRUSHCLEANING, TOOLS AND PARTS (813491 MILS43871	EA
8	O	6850-00-965-2332	CARBON REMOVING COMPOUND (813481 P-C-111	GL
9			CLEANER, LUBRICANT AND PRESERV- ATIVE (27412)	
	O	9150-01-079-6124	CLP- 4 oz (118 30 ml) bottle	EA
	O	9150-01-054 6453	CLP- 1 pt (0 47 1l bottle	EA
	O	9150-01 053-6688	CLP 7 gal (26 50 11 bottle	EA
10	C	9150 01-102 1473	CLEANER, LUBRICANT AND PRESERV- ATIVE (81349i MIL L 63460 1/2 oz (14 79 ml) bottle	EA
11	C	9920-00-292 9946	CLEANER, TOBACCO PIPE cotton turf, wire core (89855) DILLSPIPE cleaner (6 per pkg)	
			CLEANING COMPOUND, RIFLE BORE small arms bore cleaner solution (RBC) 1813491 MIL C 372	
	C	6850-00-224-6656	2 oz 159.15 mil) bottle	OZ
	O	6850-00-224-6657	8 oz (236 59 ml) can	CN
	O	6850-00-224-6663	1 gal (3.79 l) can	CN

EXPENDABLE/DURABLE SUPPLIES AND MATERIALS LIST (CONT)

(1) ITEM NUMBER	(2) LEVEL	(3) NATIONAL STOCK NUMBER	(4) DESCRIPTION PART NUMBER AND CAGE	(5) UNIT OF MEAS
13	0	5350-00-221-0872	CLOTH, ABRASIVE: (58536) A-A-1206	SH
14			Deleted See Item 23.2	
15	F		DICHLOROMETHANE, TECHNICAL: (81346) ASTM D 4701	
		6810-00-244-0290	5 gal (18.93 1) pail	CN
		6810-00-616-9188	600 lb (272.16 kg) drum	DR
16	0	6850-00-281-1985	DRY CLEANING SOLVENT: (58536) A-A-711 1 gal(3.79 I1) can	GL
17	0	8010-00-297-0560	ENAMEL: olive drab no 3407 (81348) TT-E-527 1 gal (3.791) can	GL
18	0		GLOVES, CHEMICAL AND OIL PROTECTIVE: (81348) ZZ-G-381	
		8415-00-823-7455	Size 9	PR
		8415-00-823-7456	Size10	PR
		8415-00-823-7457	Size 11	PR
19	F	9150-00-754-2595	GREASE, MOLYBDENUM DISULFIDE: (81349) MIL-G-21164	LB
20	C	1005-01-113-0321	HANDLE SECTION, CLEANING ROD, SMALL ARMS: (19204)8436776	EA
21	0	9150-01-260-2534	LUBRICANT, SOLID FILM: (81349) MIL-L-23398 16 OZ (473.18 ml) spray can	OZ
22	C	9150-00-292-9689	LUBRICATING OIL, WEAPONS: (LAW) (81349) MIL-L-14107 1 qt (0.95 1) can	QT

EXPENDABLE/DURABLE SUPPLIES AND MATERIALS LIST (CONT)

(1) ITEM NUMBER	(2) LEVEL	(3) NATIONAL STOCK NUMBER	(4) DESCRIPTION PART NUMBER AND CAGE	(5) UNIT OF MEAS
23			LUBRICATING OIL, WEAPONS: (LSA), semifluid (81349) MIL-L-46000	
	C	9150-00-935-6597	2 oz (59.15 ml) plastic btl	OZ
	C	9150-00-889-3522	4 oz (118.30 ml) plastic btl	OZ
	O	9150-00-687-4241	1 qt (0.95 l1) can	CN
	O	9150-00-753-4686	1 gal(3.79 l1) can	CN
23.1	0	5340-01-230-3181	MOUNTING BRACKET (M4/M4A1 ONLY) (19200)12556036	EA
23.2	0	8010-00-087-0102	PAINT, ENAMEL, SEMIGLOSS: paint for blank firing attachment (M1 5A2) (81348) TT-E-529 1 qt can (RED - M16A2)	EA
23.3	0	8010-01-031-1274	PAINT, ENAMEL, SEMIGLOSS: paint for blank firing attachment (M23) (81348) TT-E-529 1 pt can (YELLOW - M4/M4A1)	EA
24	0	4940-00-795-3595	PAN, WASH: (94453) 1211	EA
25	F	6850-00-826-0981	PENETRANT KIT: (81349) MIL-I-25135	KT
26	F	8135-01-019-1691	POLYETHYLENE SHEET: (84744) PE88-80-2	EA
26.1	0	1005-01-394-7677	PROTECTOR, RAIL (M4/M4A1 ONLY) (19200)12972676	EA
27	C	7920-00-205-1711	RAG, WIPING: (58536) A-A-531 50 lb (22.68 kg) bdl	LB
28	C	1005-00-050-6357	ROD SECTION, CLEANING SMALL ARMS: (19204) 8436775 (3 REQUIRED)	EA
29	F	8030-00-670-8553	SEALING COMPOUND (16059) DEVCONF	KT

EXPENDABLE/DURABLE SUPPLIES AND MATERIALS LIST (CONT)

(1) ITEM NUMBER	(2) LEVEL	(3) NATIONAL STOCK NUMBER	(4) DESCRIPTION PART NUMBER AND CAGE	(5) UNIT OF MEAS
30	C	1005-00-937-2250	SWAB HOLDER SECTION, CLEANING ROD, SMALL ARMS: (19204) 11686327	EA
31	C	1005-00-912-4248	SWAB, SMALL ARMS: (19204)11686408	SH
32	C	6920-01-395-2949	TARGET, 25 METER ZEROING, M16A2: (19200)12012024	BX

APPENDIX E

ILLUSTRATED LIST OF MANUFACTURED ITEMS

E-1. INTRODUCTION.

a. This appendix includes complete instructions for making items authorized to be manufactured or fabricated at unit or direct support maintenance.

b. A part number in alphanumeric order is provided for cross-referencing the part number of the item to be manufactured to the figure which covers fabrication criteria.

c. All bulk matenals needed for manufacture of an item are listed by part number or specification number In a tabular list on the illustration.

E-2. MANUFACTURED ITEMS PART NUMBER INDEX.

INDEX

E-3. MANUFACTURED ITEMS ILLUSTRATIONS - FORMAT AND CONTENT.

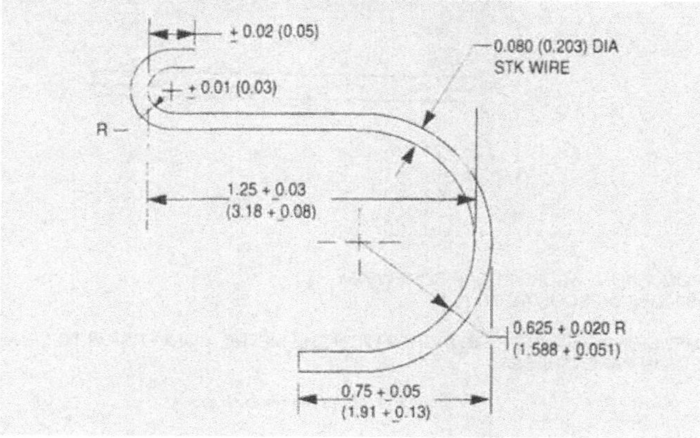

FABRICATE FROM 0.08 IN. MUSIC WIRE OR EQUIVALENT.
FINISH: NO. 5.3.1.2 OR 5.3.2.2 OF MIL-STD 17.

NOTE: ALL DIMENSIONS SHOWN ARE IN INCHES WITH THE METRIC CONVERSION TO CENTIMETERS IN PARENTHESIS.

Figure E-1. Front sight detent depressor.

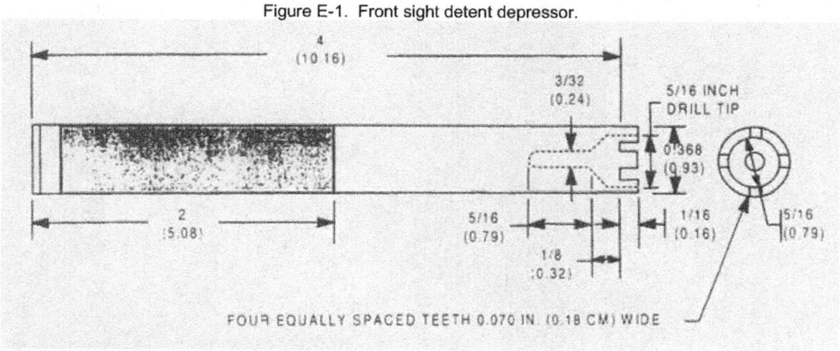

FABRICATE FROM 0.375 INCH ROUND METAL BAR,
ASTM A686. FSCM 81346, GRADE C. CLASS W2-09.
NSN 9510-00-640-4407 OR EQUIVALENT.

NOTES:
1. ALL DIMENSIONS SHOWN ARE IN INCHES WITH THE METRIC CONVERSION TO CENTIMETERS IN PARENTHESIS.
2. TEETH MUST BE HAND FILED TO FIT FRONT SIGHT POST.

Figure E-2. Front sight post removal and Installation tool.

FABRICATE FROM 1'16 IN. SOCKET HEAD SCREW KEY
NSN 5120-00-198-5398 OR EQUIVALENT.

NOTE: ALL DIMENSIONS SHOWN ARE IN INCHES WITH THE METRIC CONVERSION TO CENTIMETERS IN PARENTHESES

Figure E-3. Pivot pin removal tool.

E-3. MANUFACTURED ITEMS ILLUSTRATIONS-FORMAT AND CONTENT (CONT).

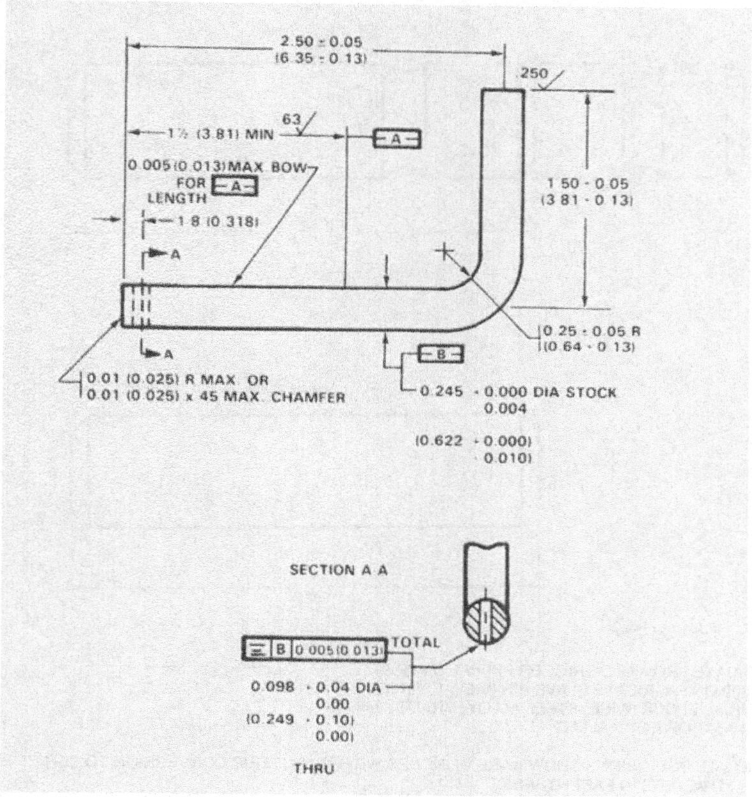

FABRICATE FROM 0 245 IN STEEL AISI 1095 OR EQUIVALENT
HARDEN AND TEMPER TO RC-57-61 FOR LENGTH -A-
FINISH 5 3 1.2 OR 5 3.2.2 OF MIL-STD-171

NOTE: ALL DIMENSIONS SHOWN ARE IN INCHES WITH THE METRIC CONVERSION TO CENTIMETERS IN PARENTHESES

Figure E-4. Pivot pin installation tool.

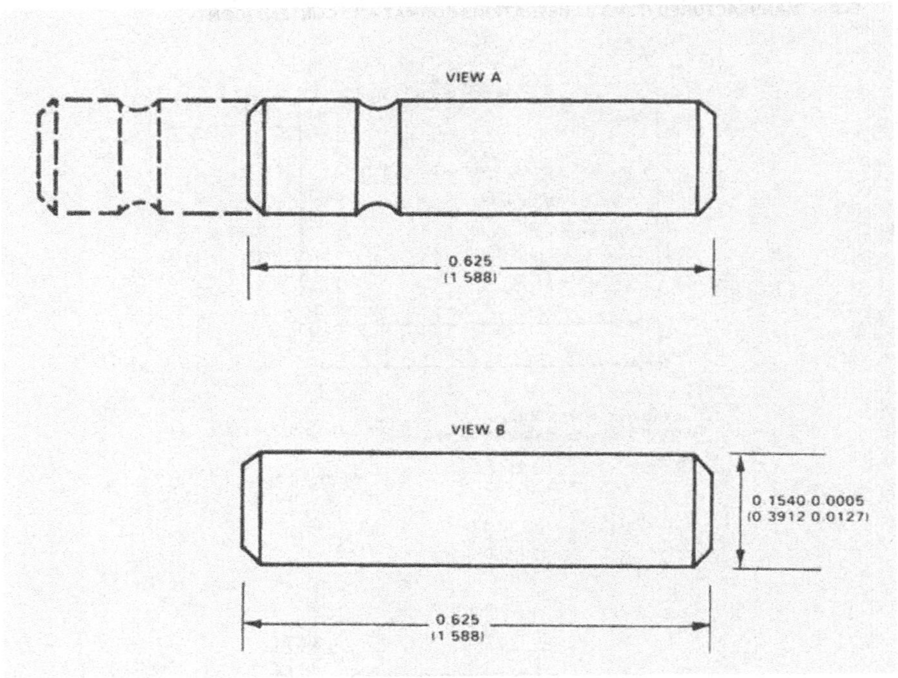

FABRICATE FROM OLD TRIGGER PIN (VIEW AI P,/N 8448609 OR FABRICATE SLAVE PIN (VIEW B) FROM MATERIAL BLOCK, WIRE, STEEL ALLOY, GRADE 4140, ASTM-A547 OR EQUIVALENT

NOTE: ALL DIMENSIONS SHOWN ARE IN INCHES WITH THE METRIC CONVERSION TO CENTIMETERS IN PARENTHESES

Figure E-5. Slave pin.

E-3. MANUFACTURED ITEMS ILLUSTRATIONS -- FORMAT AND CONTENT (CONT).

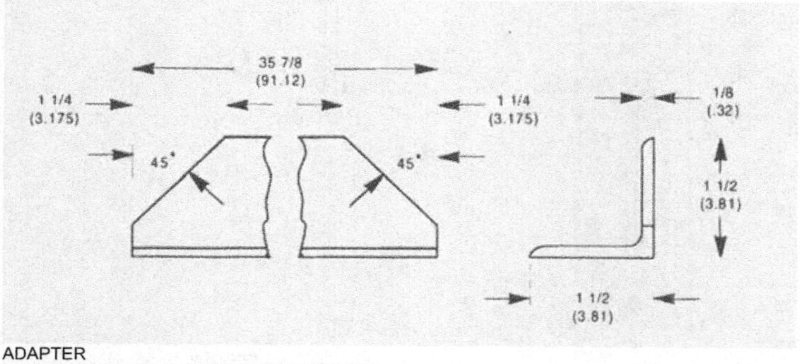

ADAPTER
FABRICATE FROM 1 1/2 INCH BY 1 1/2 INCH
BY 1/8 INCH ANGLE IRON NSN 9520-00-277-4902
OR EQUIVALENT. PAINT WITH OLIVE DRAB
ENAMEL PAINT, NSN 8010-01-350-5249, OR
EQUIVALENT.

NOTE: ALL DIMENSIONS SHOWN ARE IN INCHES WITH THE METRIC CONVERSION TO CENTIMETERS IN PARENTHESES.

Figure E-6. Adapter bar for M12 arms rack.

PIN: 060514-004

*U S Government Printing Office 1995- 648-378

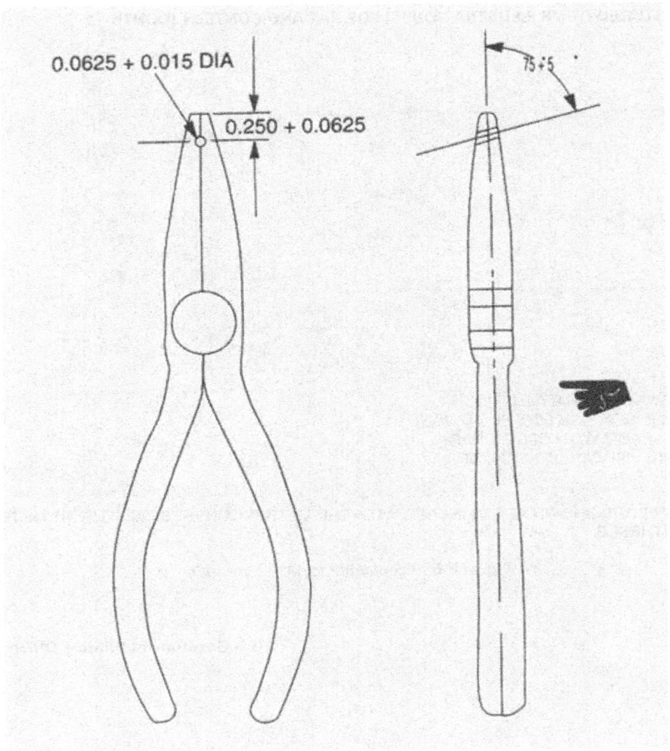

FABRICATE FROM NEEDLE NOSE PLIERS,
NSN 5120-00-268-3579 OR EQUIVALENT.

NOTE: ALL DIMENSIONS ARE IN INCHES.

Figure E-7. Modified needle nose pliers.

***U. S. GOVERNMENT PRINTING OFFICE 1997 - 545-039/4 077**

ALPHABETICAL INDEX

ALPHABETICAL INDEX (CONT)

ALPHABETICAL INDEX (CONT)

ALPHABETICAL INDEX (CONT)

Subject **Page**

Made in United States
Orlando, FL
04 November 2026

90219739R00380